# TOURS SYMPOSIUM ON NUCLEAR PHYSICS VI

## Proceedings in the Series of Tours Symposia on Nuclear Physics

| Year | Volume | Publisher | ISBN |
|------|--------|-----------|------|
| 2006 | VI | AIP Conference Proceedings Vol. 891 | 978-0-7354-0395-6 |
| 2003 | V | AIP Conference Proceedings Vol. 704 | 0-7354-0177-2 |
| 2000 | IV | AIP Conference Proceedings Vol. 561 | 1-56396-996-3 |
| 1997 | III | AIP Conference Proceedings Vol. 425 | 1-56396-749-9 |
| 1994 | II | World Scientific Publishing | 981-02-2156-8 |
| 1991 | I | World Scientific Publishing | 981-02-0892-8 |

To learn more about this title, or the AIP Conference Proceedings Series, please visit the webpage **http://proceedings.aip.org**

# TOURS SYMPOSIUM ON NUCLEAR PHYSICS VI

Tours, France  5 – 8 September 2006

*EDITORS*

M. Arnould
*ULB, Belgium*

M. Lewitowicz
*GANIL, France*

H. Emling
*GSI, Germany*

H. Akimune
M. Ohta
H. Utsunomiya
T. Wada
T. Yamagata
*Konan University, Japan*

**SPONSORING ORGANIZATIONS**
Konan University, Japan
Conseil Général d'Indre-et-Loire, France
GANIL, France
GSI, Germany
ULB, Belgium

Melville, New York, 2007
AIP CONFERENCE PROCEEDINGS ■ VOLUME 891

**Editors:**

M. Arnould
Institut d'Astronomie et d'Astrophysique
Université Libre de Bruxelles - CP 226
Boulevard du Triomphe
B-1050 Brussels
Belgium

E-mail:   marnould@astro.ulb.ac.be

M. Lewitowicz
GANIL
Blvd. H. Becquerel, BP 55027
F-14076 Caen Cedex 5
France

E-mail:   lewitowicz@ganil.fr

H. Emling
GSI
Planckstrasse 1
D-64291 Darmstadt
Germany

E-mail:   h.emling@gsi.de

H. Akimune
M. Ohta
H. Utsunomiya
T. Wada
T. Yamagata

Department of Physics
Konan University
8-9-1 Okamoto Higashinada-ku
Kobe 658-8501
Japan

E-mail:   akimune@konan-u.ac.jp
masaota@konan-u.ac.jp
hiro@konan-u.ac.jp
wada@konan-u.ac.jp
yamagata@center.konan-u.ac.jp

Authorization to photocopy items for internal or personal use, beyond the free copying permitted under the 1978 U.S. Copyright Law (see statement below), is granted by the American Institute of Physics for users registered with the Copyright Clearance Center (CCC) Transactional Reporting Service, provided that the base fee of $23.00 per copy is paid directly to CCC, 222 Rosewood Drive, Danvers, MA 01923, USA. For those organizations that have been granted a photocopy license by CCC, a separate system of payment has been arranged. The fee code for users of the Transactional Reporting Services is: 978-0-7354-0395-6/07/$23.00.

© 2007 American Institute of Physics

Permission is granted to quote from the AIP Conference Proceedings with the customary acknowledgment of the source. Republication of an article or portions thereof (e.g., extensive excerpts, figures, tables, etc.) in original form or in translation, as well as other types of reuse (e.g., in course packs) require formal permission from AIP and may be subject to fees. As a courtesy, the author of the original proceedings article should be informed of any request for republication/reuse. Permission may be obtained online using Rightslink. Locate the article online at http://proceedings.aip.org, then simply click on the Rightslink icon/"Permission for Reuse" link found in the article abstract. You may also address requests to: AIP Office of Rights and Permissions, Suite 1NO1, 2 Huntington Quadrangle, Melville, NY 11747-4502, USA; Fax: 516-576-2450; Tel.: 516-576-2268; E-mail: rights@aip.org.

L.C. Catalog Card No. 2007920800
ISBN 978-0-7354-0395-6
ISSN 0094-243X
Printed in the United States of America

# CONTENTS

Preface ... xi
Organizing Committee ... xiii
Schedule ... xv

## SUPERHEAVY ELEMENTS (SHE)

**Experiments on Synthesis of the Heaviest Element at RIKEN** ... 3
K. Morita, K. Morimoto, D. Kaji, T. Akiyama, S. Goto, H. Haba, E. Ideguchi, R. Kanumgo, K. Katori, H. Kikunaga, H. Koura, H. Kudo, T. Ohnishi, A. Ozawa, N. Sato, T. Suda, K. Sueki, F. Tokanai, H. Xu, T. Yamaguchi, A. Yoneda, A. Yoshida, and Y.-L. Zhao

**Experiments on Nuclear Structure and Synthesis of Superheavy Elements at SHIP** ... 10
F. P. Heßberger

**Mass Measurements at SHIPTRAP** ... 19
A. Martín, D. Ackermann, M. Block, A. Chaudhuri, Z. Di, S. Eliseev, D. Habs, F. Herfurth, F. Hessberger, S. Hofmann, H.-J .Kluge, G. Marx, M. Mazzocco, M. Mukherjee, J. B. Neumayr, W. Plass, S. Rahaman, C. Rauth, D. Rodríguez, C. Scheidenberger, L. Schweikhard, P. G. Thirolf, G. Vorobjev, and C. Weber

**Dynamics in the Production of Superheavy Elements** ... 27
G. G. Adamian, N. V. Antonenko, Z. Gagyi-Palffy, S. P. Ivanova, W. Scheid, and A. S. Zubov

**Chemical Investigations of Superheavy Elements—Current Results and New Techniques** ... 36
C. E. Düllmann

**Startup of Superheavy Element Chemistry at RIKEN** ... 45
H. Haba, D. Kaji, H. Kikunaga, N. Sato, T. Akiyama, K. Morimoto, A. Yoneda, K. Morita, T. Takabe, Y. Tashiro, Y. Kitamoto, K. Matsuo, D. Saika, K. Ooe, T. Kuribayashi, T. Yoshimura, A. Shinohara, and A. Toyoshima

**A Review on SHE Research at GANIL** ... 55
Ch. Stodel, N. Amar, R. Anne, G. Auger, B. Bouriquet, J.-M. Casandjian, A. Chatillon, R. Cee, E. Clément, R. Dayras, O. Dorvaux, A. Drouart, G. de France, F. de Oliveira Santos, R. de Tourreil, S. Grévy, F. Hannachi, F. Hannappe, K. Hauschild, F. P. Hessberger, S. Hofmann, A. Korichi, R. Lichtenhäler, K. Lojek, A. Lopez-Martens, A. Péghaire, J. Péter, M.-G. Saint-Laurent, Z. Sosin, L. Stuttge, Ch. Theisen, A. C. C. Villari, J.-P. Wieleczko, and A. Wieloch

**Search for a Long Lived Component in the Reaction U+U near the Coulomb Barrier** ... 60
A. C. C. Villari, C. Golabek, W. Mittig, S. Heinz, S. Bhattacharyya, D. Boilley, R. Dayras, G. De France, A. Drouart, L. Gaudefroy, L. Giot, A. Marchix, V. Maslov, M. Morjean, G. Mukherjee, A. Navin, Yu. Penionzkevich, F. Rejmund, M. Rejmund, P. Roussel-Chomaz, C. Stodel, and M. Winkler

## FUSION-FISSION DYNAMICS (FFD)

**Measurement of Evaporation Residue Cross-Sections of the Reaction $^{30}$Si+$^{238}$U at Subbarrier Energies** ... 71
K. Nishio, S. Hofmann, F. P. Heßberger, D. Ackermann, S. Antalic, V. F. Comas, Z. Gan, S. Heinz, J. A. Heredia, H. Ikezoe, J. Khuyagbaatar, B. Kindler, I. Kojouharov, P. Kuusiniemi, B. Lommel, M. Mazzocco, S. Mitsuoka, Y. Nagame, T. Ohtsuki, A. G. Popeko, S. Saro, H. J. Schött, B. Sulignano, A. Svirikhin, K. Tsukada, K. Tsuruta, and A. V. Yeremin

Recent Developments in Quasi-elastic Scattering around the Coulomb Barrier ............... 80
K. Hagino

## PHYSICS WITH EXOTIC NUCLEI (PEN)

The SPIRAL 2 Project ................................................................. 91
M. Lewitowicz *(On behalf of the SPIRAL 2 Project Group)*
RISING: Gamma-ray Spectroscopy with Radioactive Beams at GSI ........................ 99
P. Doornenbal, A. Bürger, D. Rudolph, H. Grawe, H. Hübel, P. H. Regan, P. Reiter, A. Banu, T.
Beck, F. Becker, P. Bednarczyk, L. Caceres, H. Geissel, J. Gerl, M. Górska, J. Grebosz, M.
Kavatsyuk, O. Kavatsyuk, A. Kelic, I. Kojouharov, N. Kurz, R. Lozeva, F. Montes, W.
Prokopowicz, N. Saito, T. Saito, H. Schaffner, S. Tashenov, H. Weick, E. Werner-Malento, M.
Winkler, H.-J. Wollersheim, A. Al-Khatib, L.-L. Andersson, L. Atanasova, D. L. Balabanski, M.
A. Bentley, G. Benzoni, A. Blazhev, A. Bracco, S. Brambilla, C. Brandau, P. Bringel, J. R.
Brown, F. Camera, S. Chmel, E. Clément, F. C. L. Crespi, C. Fahlander, A. B. Garnsworthy, A.
Görgen, G. Hammond, M. Hellström, R. Hoischen, H. Honma, E. K. Johansson, A. Jungclaus,
M. Kmieciek, W. Korten, A. Maj, S. Mandal, W. Meczynski, B. Million, A. Neußer, F. Nowacki,
T. Otsuka, M. Pfützner, S. Pietri, Zs. Podolyák, A. Richard, M. Seidlitz, S. J. Steer, T. Striepling,
T. Utsuno, J. Walker, N. Warr, C. Wheldon, and O. Wieland
(p,2p) Reactions on Carbon Isotopes: $p(^{9-16}C,2p)^{8-15}B$ at 250 AMeV ........................ 108
T. Kobayashi, K. Ozeki, K. Watanabe, Y. Matsuda, Y. Seki, T. Shinohara, T. Miki, Y. Naoi,
H. Otsu, S. Ishimoto, S. Suzuki, Y. Takahashi, and E. Takada
Measurement of Nuclear Moments at RIKEN ............................................. 113
H. Ueno, D. Kameda, D. Nagae, M. Takemura, K. Asahi, K. Takase, A. Yoshimi, T. Sugimoto,
T. Nagatomo, T. Arai, M. Uchida, K. Shimada, T. Inoue, J. Murata, H. Kawamura, and K. Narita
Study of Exotic Nuclei around the "Island of Inversion" ................................. 122
Z. Elekes, Zs. Dombrádi, A. Saito, N. Aoi, H. Haba, K. Demichi, Zs. Fülöp, J. Gibelin,
T. Gomi, H. Hasegawa, N. Imai, M. Ishihara, H. Iwasaki, S. Kanno, S. Kawai, T. Kishida,
T. Kubo, K. Kurita, Y. Matsuyama, S. Michimasa, T. Minemura, T. Kubo, K. Kurita,
Y. Matsuyama, S. Michimasa, T. Minemura, T. Motobayashi, M. Notani, T. K. Ohnishi,
H. J. Ong, S. Ota, A. Ozawa, H. K. Sakai, H. Sakurai, S. Shimoura, E. Takeshita, S. Takeuchi,
M. Tamaki, Y. Togano, K. Yamada, Y. Yanagisawa, and K. Yoneda
SHARAQ Project ....................................................................... 131
A. Saito, H. Sakai, S. Shimoura, T. Uesaka, T. Kawabata, K. Nakanishi, Y. Sasamoto,
E. Ideguchi, H. Yamaguchi, S. Kubono, G. P. Berg, T. Ichihara, and T. Kubo
Development of Polarized $^3$He Ion Source ............................................... 138
M. Tanaka, Y. Takahashi, T. Shimoda, S. Yasui, M. Yosoi, K. Takahisa, and N. Shimakura
The EURISOL Project ................................................................. 147
Y. Blumenfeld *(On behalf of the EURISOL Design Study)*
New Pathways to Bypass the $^{15}$O Waiting Point ....................................... 155
I. Stefan, F. de Oliveira Santos, M. G. Pellegriti, M. Angélique, J. C. Dalouzy, F. de Grancey,
M. Fadil, S. Grévy, M. Lenhardt, M. Lewitowicz, A. Navin, L. Perrot, M. G. Saint Laurent,
I. Ray, O. Sorlin, C. Stodel, J. C. Thomas, G. Dumitru, A. Buta, R. Borcea, F. Negoita,
D. Pantelica, J. C. Angélique, E. Berthoumieux, A. Coc, J. Kiener, A. Lefebvre-Schuhl,
V. Tatischeff, J. M. Daugas, O. Roig, T. Davinson, and M. Stanoiu
Quartetting in Fermion Matter and Alpha Particle Condensation in Nuclear Systems ......... 164
P. Schuck, Y. Funaki, H. Horiuchi, G. Röpke, A. Tohsaki, and T. Yamada
Studies on Exotic Nuclei by Proton-induced Direct Reaction at GSI and FAIR ............... 172
O. A. Kiselev

New Study of Reaction Cross Sections and the Nucleon Density Distribution.................. 181
    M. Fukuda, M. Takechi, M. Mihara, R. Matsumiya, K. Matsuta, T. Minamisono, T. Ohtsubo,
    T. Izumikawa, S. Momota, T. Suzuki, T. Yamaguchi, S. Nakajima, K. Kobayashi, K. Tanaka,
    T. Suda, S. Sato, M. Kanazawa, and A. Kitagawa

Precise Studies of Nucleon Density Distribution of $^6$He and $^8$He............................. 187
    M. Takechi, M. Fukuda, M. Mihara, R. Matsumiya, K. Matsuta, T. Minamisono, T. Ohtsubo,
    T. Izumikawa, S. Momota, T. Suzuki, T. Yamaguchi, S. Nakajima, K. Kobayashi,
    K. Tanaka, T. Suda, S. Sato, M. Kanazawa, and A. Kitagawa

Coulomb Excitation of $^{26}$Ne........................................................... 192
    D. Beaumel, J. Gibelin, T. Motobayashi, N. Aoi, H. Baba, Y. Blumenfeld, Z. Elekes, S. Fortier,
    N. Frascaria, N. Fukuda, T. Gomi, K. Ishikawa, Y. Kondo, T. Kubo, V. Lima, T. Nakamura,
    A. Saito, Y. Satou, E. Takeshita, S. Takeuchi, T. Teranishi, Y. Togano, A. M. Vinodkumar,
    Y. Yanagisawa, and K. Yoshida

New Developments for Isochronous Mass Measurements of Short-lived Nuclei................ 199
    R. Knöbel, S. A. Litvinov, B. Sun, K. Beckert, P. Beller, F. Bosch, D. Boutin, C. Brandau,
    L. Chen, I. J. Cullen, C. Dimopoulou, A. Dolinskii, B. Fabian, H. Geissel, M. Hausmann,
    C. Kozhuharov, J. Kurcewicz, Yu. A. Litvinov, Z. Liu, M. Mazzocco, F. Montes,
    G. Münzenberg, A. Musumarra, S. Nakajima, C. Nociforo, F. Nolden, T. Ohtsubo, A. Ozawa,
    Z. Patyk, W. R. Plaß, C. Scheidenberger, M. Shindo, M. Steck, T. Suzuki, P. M. Walker,
    H. Weick, N. Winckler, M. Winkler, and T. Yamaguchi

Exotic Cluster States in $^{12}$Be via $\alpha$-inelastic Scattering...................................... 205
    A. Saito, S. Shimoura, T. Minemura, Y. U. Matsuyama, H. Baba, N. Aoi, T. Gomi, H. Higurashi,
    K. Ieki, N. Imai, N. Iwasa, H. Iwasaki, S. Kanno, S. Kubono, M. Kunibu, S. Michimasa,
    T. Motobayashi, T. Nakamura, H. Ryuto, H. Sakurai, M. Serata, E. Takeshita, S. Takeuchi,
    T. Teranishi, K. Ue, K. Yamada, and Y. Yanagisawa

Excitation and Charged Particle Decay of Dipole Resonance Analogs in the $\alpha$ Clusters
of $^6$Li and $^7$Li............................................................................ 214
    T. Yamagata, H. Akimune, S. Nakayama, M. Fujiwara, K. Fushimi, M. B. Greenfield, K. Hara,
    K. Y. Hara, K. Hashimoto, K. Ichihara, H. Ikemizu, K. Kawase, M. Kinoshita, Y. Matsui, K.
    Nakanishi, M. Ohta, M. Sakama, M. Tanaka, H. Utsunomiya, N. Warashina, and M. Yosoi

t+t Clustering States in He-isotopes........................................................ 222
    S. Aoyama and K. Arai

Spectroscopy of $^9$He, Quasi-free Scattering $^6$He+$^4$He ........................................ 226
    R. Wolski, M. S. Golovkov, L. V. Grigorenko, A. S. Fomichev, A. V. Gorshkov, S. A. Krupko,
    A. M. Rodin, S. I. Sidorchuk, R. S. Slepnev, S. V. Stepantsov, G. M. Ter-Akopian,
    A. A. Korsheninnikov, E. Yu. Nikolskii, E. A. Kuzmin, B. G. Novatskii, D. N. Stepanov,
    S. Fortier, P. Roussel, and W. Mittig

## NUCLEAR ASTROPHYSICS (NAP)

Cross Sections for the Production of Residual Nuclides at Medium Energies: Status
and Recent Experimental and Theoretical Progress......................................... 237
    R. Michel

Spallation Nucleosynthesis by Accelerated Charged-particles in Stellar Envelopes of
Magnetic Stars............................................................................. 246
    S. Goriely

Accelerated Particle Properties in Solar Flares from Gamma-Ray Line Observations.......... 254
    J. Kiener, V. Tatischeff, G. Weidenspointner, M. Gros, and A. Belhout

Neutrino Probes of Galactic and Extragalactic Supernovae.................................. 263
    S. Ando

Heavy Element Nucleosynthesis in Jets from Collapsars............................................. 272
    S. Fujimoto, M. Hashimoto, K. Kotake, and S. Yamada
Low-energy Nuclear Reactions and the Alpha-nucleus Optical Potential: Where Do
We Stand?......................................................................................... 281
    P. Demetriou and M. Axiotis
Probing Universe with Fast Neutrons................................................................ 289
    Y. Nagai, T. Shima, A. Tomyo, M. Segawa, Y. Temma, T. Ohsaki, and M. Igashira
Underground Studies of pp and CNO ................................................................ 297
    H. Costantini
Bound Electron Screening Corrections to Reactions in Hydrogen Burning Processes .......... 306
    S. Kimura and A. Bonasera
The Electrostatic Screening of Nuclear Reactions in Dense Stellar Plasma .................... 315
    G. Shaviv
Microscopic Description of Fission Properties...................................................... 324
    H. Goutte, J.-P. Delaroche, M. Girod, and J. Libert
Photoneutron Cross Sections of Astrophysical Significance...................................... 329
    H. Utsunomiya
Study of Collective Dipole Excitations below the Giant Dipole Resonance at HI$\gamma$S............ 339
    A. P. Tonchev, C. Angell, M. Boswell, A. Chyzh, C. R. Howell, H. J. Karwowski, J. H. Kelley,
    W. Tornow, N. Tsoneva, and Y. K. Wu
Nuclear Level Densities from Drip Line to Drip Line ............................................ 348
    S. Hilaire and S. Goriely
A Status Report of the Data Evaluation in the NACRE Update and Extension Project ........ 355
    M. Katsuma
The Effects of Changes in Reaction Rates on Simulations of Nova Explosions................. 364
    S. Starrfield, C. Iliadis, W. R. Hix, F. X. Timmes, and W. M. Sparks
Understanding Nuclear "Pasta": Current Status and Future Prospects........................ 373
    G. Watanabe
The Crust of Neutron Stars........................................................................ 382
    N. Chamel
Burning Questions in Nuclear Astrophysics........................................................ 391
    B. S. Meyer

## YOUNG SCIENTIST SESSION

Neutron-Capture Nucleosynthesis in Extremely Metal-Poor Stars—Application to the
most Iron-deficient Stars HE0107-5240 and HE1327-2326 ...................................... 397
    T. Nishimura, N. Iwamoto, M. Aikawa, T. Suda, M. Y. Fujimoto, and I. Iben, Jr.
Screening Effects on Neutrino-Nucleus Reactions ................................................ 401
    F. Minato, K. Hagino, N. Takigawa, A. B. Balantekin, and Ph. Chomaz
The Importance and the Sensitivity of the Reaction $^{17}$O(n, $\gamma$)$^{18}$O in the s-Process
Nucleosynthesis .................................................................................... 405
    K. Yamamoto, T. Wada, M. Ohta, T. Nishimura, M. Y. Fujimoto, K. Katō, T. Suda, and
    M. Aikawa
Microscopic Description of Scission Configurations............................................... 409
    N. Dubray, H. Goutte, and J. F. Berger
Quantum Diffusion Approach to the Formation of a Heavy Compound Nucleus by
Heavy-ion Fusion Reactions....................................................................... 413
    K. Washiyama, N. Takigawa, and S. Ayik

Variation of Variance of Fission Fragment Mass Distribution: A Probe to Study the
Dynamics of Fusion-fission Reactions.................................................... 417
   T. K. Ghosh

## POSTER SESSIONS

Fission Fragment Mass Distribution for Nuclei in the r-Process Region..................... 423
   S. Tatsuda, K. Hashizume, T. Wada, M. Ohta, K. Sumiyoshi, K. Otsuki, T. Kajino, H. Koura,
   S. Chiba, and Y. Aritomo

Dipole Resonances in $^4$He............................................................. 427
   E. Matsumoto, S. Nakayama, R. Hayami, K. Fushimi, H. Kawasuso, K. Yasuda, T. Yamagata,
   H. Akimune, H. Ikemizu, M. Fujiwara, M. Yosoi, K. Nakanishi, K. Kawase, H. Hashimoto,
   T. Oota, K. Sagara, T. Kudoh, S. Asaji, T. Ishida, M. Tanaka, and M. B. Greenfield

Photodisintegration of $^{80}$Se and its Implications for s-Process Branching ............ 431
   A. Makinaga, H. Utsunomiya, S. Goko, T. Kaihori, S. Houhara, S. Goriely, H. Toyokawa,
   H. Harano, T. Matsumoto, H. Harada, F. Kitatani, K. Y. Hara, and Y.-W. Lui

Langevin Equation as a Stochastic Differential Equation in Nuclear Physics ............... 435
   T. Asano, T. Wada, M. Ohta, and N. Takigawa

Fission Barriers for Neutron-rich Nuclei by Means of Skyrme-Hartree-Fock-
Bogoliubov Calculation................................................................. 439
   K. Hashizume, T. Wada, M. Ohta, M. Samyn, and S. Goriely

A Reassessment of Surface Friction Model for Maximum Cold Fusion Reactions in
Superheavy Mass Region ................................................................ 443
   A. Fukushima, A. Nasirov, Y. Aritomo, T. Wada, and M. Ohta

Coherent Scattering of Neutrinos by "Nuclear Pasta" in Dense Matter...................... 447
   H. Sonoda

List of Participants .................................................................... 451
Author Index ........................................................................... 459

PREFACE

Tours Symposium on Nuclear Physics VI was held during September 5 through 8, 2006, in Tours, France, with 3 sessions for SHE (Super-heavy Elements), 1 session for FFD (Fusion-fission Dynamics), 7 sessions for PEN (Physics with Exotic Nuclei), and 8 sessions for NAP (Nuclear Astrophysics). The Proceedings of Tours Symposium 2006 includes 10 papers in SHE&FFD, 19 papers in PEN, and 19 papers in NAP based on the latest developments in the fields. A special evening session of NAP was organized for strategic discussions on the Konan-ULB compilation program moderated by K. TAKAHASHI and the core-to-core program moderated by H. UTSUNOMIYA. A round-table session was moderated by B. MEYER to discuss burning questions in nuclear astrophysics. Two parallel sessions in NAP and FFD were dedicated to 7 young scientists for oral presentations. In addition, seven posters were displayed for discussions during the coffee breaks.

We had an excursion in Château du Clos Lucé and a banquet at La Halle Eiffelin in the château. VIPs who attended the banquet from Conseil Général were Mr. Michel LEZEAU (Vice-président) and Mr. Serge BABARY (Vice-président), from l'Université François Rabelais de Tours was Mr. Jean-Paul MONGE (Vice-président), from Agence de Développement de la Touraine were Mr. Georges DUMAS (Direteur) and Mlle Angèle PLOQUIN (Chargée de mission), and from Lycée-Collège Konan were Mr. (Direteur) and Mrs. TANAKA and Mrs. CREOLA (Assistante). We had a memorial ceremony to celebrate the retirements of Prof. G. MÜNZENBERG (GSI) who has contributed to the synthesis of super-heavy elements and Prof. Y. NAGAI (Osaka Univ.) who has contributed to the neutron capture process in nuclear astrophysics. Mr. S. BABARY and Prof. M. OHTA (Konan University) made welcome addresses in the ceremony.

Tours Symposium 2006 was financially supported by Conseil Général d'Indre-et-Loire (Mr. Marc POMMEREAU, Président) for the excursion and banquet, GANIL & IN2P3 (Directeur, Prof. S. GALÈS) for publication of the Proceedings and GSI (Prof. H. EMLING.) for financial support for young scientists. This symposium was hosted by Konan University (President Prof. Y. SUGIMURA, Chief Director Prof. H. YOSHIZAWA) and supported by the Konan-ULB (Directeur, Prof. M. ARNOULD) convention. Finally, we appreciate the cooperation of Tourist Office of Tours (Ms. Marie Jo HAUTIN) for arranging accommodation hotels for our symposium.

Masahisa Ohta (Co-chairman)
Hiroaki Utsunomiya (Co-chairman)
*Konan University*

## ORGANIZING COMMITTEE

H. Akimune (*Konan, Japan*)
M. Arnould (*ULB, Belgium*)
A. Coc (*Orsay, France*)
H. Emling (*GSI, Germany*)
S. Galès (*GANIL, France*)
S. Harissopulos (*Athens, Greece*)
S. Hofmann (*GSI, Japan*)
H. Ikezoe (*JAEA, Japan*)
T. Kajino (*NAO, Japan*)
S. Kubono (*CNS, Japan*)
M. Lewitowicz (*GANIL, France*)
K. Morita (*RIKEN, Japan*)
T. Motobayashi (*RIKEN, Japan*)
Yu. Ts. Oganessian (*FLNR-JINR, Russia*)
M. Ohta (*Konan, Japan*)
H. Utsunomiya (*Konan, Japan*)
T. Wada (*Konan, Japan*)
T. Yamagata (*Konan, Japan*)

## HOST INSTITUTE

*Konan University*

## SUPPORTED BY

*Conseil Général d'Indre-et-Loire*
*GANIL*
*GSI*
*ULB*

# SCHEDULE

| | Sep. 5 (Tue) | Sep. 6 (Wed) | | Sep. 7 (Thu) | | Sep. 8 (Fri) |
|---|---|---|---|---|---|---|
| 08:45-09:00 | Opening | - | | - | | - |
| 09:00-10:30 | SHE1 | PEN 3 | | NAP1 | | NAP6 |
| 10:30-11:00 | Poster Session (CB) | | | | | |
| 11:00-12:30 | SHE2 | YS1 | YS2 | NAP2 | | NAP7 |
| 12:30-14:00 | Lunch | | | | | |
| 14:00-15:30 | PEN 1 | - | | PEN4 | NAP3 | NAP8 |
| 15:00-16:00 | Coffee break | Excursion Banquet | | Coffee break | | Closing |
| 16:00-17:30 | SHE 3 / PEN 2 | | | PEN5 | NAP4 | |
| 17:30-18:00 | Coffee break | | | Coffee break | | |
| 18:00-19:30 | FFD1 / - | | | - | NAP5 | |

SHE : Superheavy Elements  PEN : Physics with Exotic Nuclei
FFD : Fusion-Fission Dynamics  NAP : Nuclear Astrophysics

## Chairpersons

| | | | | |
|---|---|---|---|---|
| SHE1 | T. Nomura | | NAP1 | J. Kiener |
| SHE2 | G. Münzenberg | | NAP2 | S. Starrfield |
| SHE3 | T. Wada | | NAP3 | A. Tonchev |
| FFD1 | M. Ohta | | NAP4 | Y. Nagai |
| PEN1 | S. Gales | | NAP5 | K. Takahashi/H. Utsunomiya |
| PEN2 | M. Lewitowicz | | NAP6 | S. Harrisopulos |
| PEN3 | H. Emling | | NAP7 | A. Coc |
| PEN4 | T. Kobayashi | | NAP8 | B. Meyer |
| PEN5 | H. Akimune | | | |

# SUPERHEAVY ELEMENTS (SHE)

# Experiments on Synthesis of the Heaviest Element at RIKEN

K. Morita[1], K. Morimoto[1], D. Kaji[1], T. Akiyama[1,2], S. Goto[3], H. Haba[1], E. Ideguchi[4], R. Kanumgo[1], K. Katori[1], H. Kikunaga[1], H. Koura[5], H. Kudo[6], T. Ohnishi[1], A. Ozawa[7], N. Sato[8,1], T. Suda[1], K. Sueki[7], F. Tokanai[9], H. Xu[10], T. Yamaguchi[2], A. Yoneda[1], A. Yoshida[1], and Y.-L. Zhao[11]

[1] *Nishina Center for Accelerator Based Science, RIKEN, Wako, Saitama 351-0198, Japan*
[2] *Department of Physics, Saitama University, Sakura-ku, Saitama 338-8570, Japan*
[3] *Center for Instrumental Analysis, Niigata University, Ikarashi, Niigata 950-2181, Japan*
[4] *Center for Nuclear Study, University of Tokyo Wako Branch, Wako, Saitama 351-0198, Japan*
[5] *Advanced Science Research Center, Japan Atomic Energy Agency, Tokai, Ibaraki 319-1195, Japan*
[6] *Department of Chemistry, Niigata University, Ikarashi, Niigata 950-2181, Japan*
[7] *University of Tsukuba, Tsukuba, Ibaraki 305-8571, Japan*
[8] *Department of Physics, Tohoku University, Aoba-ku, Sendai 980-8578 Japan*
[9] *Department of Physics, Yamagata University, Shirakawa, Yamagata 990-8560, Japan*
[10] *Institute of Modern Physics, Chinese Academy of Science, Lanzhou 730000, China*
[11] *Institute of High Energy Physics, Chinese Academy of Science, Beijing 100039, China*

**Abstract.** At the Institute of Physical and Chemical Research (RIKEN) a series of experiments studying the productions and their decays of the heaviest elements have been performed by using a gas-filled recoil ion separator GARIS. Results on the isotope of the 112th element, $^{277}112$, and on that of the 113th element, $^{278}113$, are reviewed. Tow decay chains which are assigned to be ones originating from the isotope $^{277}112$ were observed in the $^{208}Pb(^{70}Zn, n)$ reaction. Both chains consisted of four consecutive alpha decays followed by a spontaneous fission. The results provide a confirmation of the production and decay of the isotope $^{277}112$ reported by a research group at Gesellschaft für Schwerionenforschung (GSI), Germany, produced via the same reaction by using a velocity filter. Tow decay chains, both consisted of four consecutive alpha decays followed by a spontaneous fission, were observed also in the reaction $^{209}Bi(^{70}Zn, n)$. Those are assigned to be the convincing candidate events of the decays of the isotope of the 113th element, $^{278}113$, and its daughter nuclei, $^{274}Rg$, $^{270}Mt$, $^{266}Bh$, and $^{262}Db$.

**Keywords:** confirmation of $^{277}112$, new isotopes $^{278}113$, $^{274}Rg$, and $^{270}Mt$, gas-filled recoil separator.
**PACS:** 27.90.+b, 23.60.+e, 21.10.Tg, 25.85.Ca, 25.60.Pj, 25.70.-z

## INTRODUCTION

It attracts nuclear physicists to study the nuclear properties of the heaviest elements produced by the complete fusion reaction with heavy ions. After the early theoretical predictions of the nest spherical shell closures to the known highest one, 82, to be 114 for proton's, and 126, to be 184 for neutron's (for example, ref. [1]), many experimental trials had been done to find out the next spherical magic numbers. Although the magic numbers have not been observed yet experimentally, many new isotopes of new elements have been discovered [2, -6, and references therein], providing the important information about the stability of the nuclei of large atomic numbers and neutron numbers, as well as about the fusion mechanism of massive nuclei.

At the Institute of Physical and Chemical Research (RIKEN), we have performed a series of experiments studying productions and decays of the isotopes of the heaviest elements produced by the reactions using $^{208}$Pb and $^{209}$Bi targets. The reactions studied were $^{208}$Pb($^{58}$Fe, n)$^{265}$Hs, $^{208}$Pb($^{64}$Ni, n)$^{271}$Ds, $^{209}$Bi($^{64}$Ni, n)$^{272}$Rg, $^{208}$Pb($^{70}$Zn, n)$^{277}$112, and $^{209}$Bi($^{70}$Zn, n)$^{278}$113. The isotopes $^{265}$Hs, $^{271}$Ds, $^{272}$Rg, and $^{277}$112 were firstly studied by a research group at the Gesellschaft für Schwerionenforschung (GSI), Germany, using the same reactions, with use of a velocity filter SHIP. The results obtained at RIKEN have provided clear confirmations of discoveries of these isotopes. A part of the results was published in ref. [7, 8]. In the $^{209}$Bi + $^{70}$Zn reaction, two decay chains, which can be assigned to subsequent decays originating from an isotope of the 113th element, $^{278}$113, were observed firstly. A part of the results was published in ref. [9] also.

In this article, we report the results of two reactions, $^{208}$Pb + $^{70}$Zn and $^{209}$Bi + $^{70}$Zn.

## EXPERIMENTAL PROCEDURE

The experiment had carried out from September 2003 to June 2006 taking some intermissions. Summary of the beamtime, periods, net irradiation time, beam doses, number of observed events, and the corresponding cross sections with 1σ statistical errors, is listed in Table 1. A net irradiation time for $^{209}$Bi target was 5776 hours and one for $^{208}$Pb target was 694 hours.

A $^{70}$Zn ion beam was extracted from RILAC. The lead targets were prepared by vacuum evaporation of metallic lead-208, enrichment of 98.4 %, onto carbon backing foils of 30 μg/cm$^2$ thickness. The bismuth targets were prepared by the same method.

TABLE 1. Summary of the beamtime.

| year | from ~ to | irradiation time (hours) | dose (×10$^{19}$) | number of events | cross section ×10$^{-39}$ cm$^2$ (fb) |
|---|---|---|---|---|---|
| | $^{209}$Bi + $^{70}$Zn reaction | | | | |
| 2003 | 5 Sep. ~ 29 Dec. | 1389 | 1.24 | 0 | <75 |
| 2004 | 8 Jul. ~ 2 Aug | 526 | 0.51 | 1 | $55^{+130}_{-45}$ |
| 2005 | 20 Jan. ~ 23 Jan. | 72 | 0.13 | 0 | $52^{+120}_{-43}$ |
| | 23 Mar. ~ 22 Apr. | 650 | 0.71 | 1 | $75^{+100}_{-49}$ |
| | 19 May ~ 21 May | 48 | 0.05 | 0 | $74^{+98}_{-48}$ |
| | 7 Aug. ~ 25 Aug. | 368 | 0.45 | 0 | $63^{+83}_{-40}$ |
| | 7 Sep. ~ 20 Oct. | 937 | 1.17 | 0 | $45^{+59}_{-29}$ |
| | 25 Nov. ~ 15 Dec. | 469 | 0.57 | 0 | $39^{+51}_{-25}$ |
| 2006 | 14 Mar. ~ 15 May | 1300 | 1.37 | 0 | $31^{+40}_{-20}$ |
| total | 305 days | 5776 | 6.20 | 2 | $31^{+40}_{-20}$ |
| | $^{208}$Pb + $^{70}$Zn reaction | | | | |
| 2004 | 2 Apr. ~ 24 May | 694 | 0.44 | 2 | $440^{+590}_{-290}$ |

The thickness of the metallic target layer was about 450 μg/cm$^2$ in both lead-208 and bismuth cases. Sixteen targets were mounted on a rotating wheel whose size was 30 cm in diameter. The wheel was rotated at 2000 rpm or 3000 rpm during irradiation. The beam energies were set to 346 MeV and 349 MeV at the half depth of the targets, for the $^{208}$Pb + $^{70}$Zn reaction and for the $^{209}$Bi + $^{70}$Zn reaction, respectively. Those energies were chosen to maximize the yields of one-neutron evaporation channel from the compound nuclei, according to the previous studies [7, 8].

The reaction products were separated in-flight from the beams by using a gas-filled recoil ion separator GARIS, and were guided into a detector chamber placed at the focal plane of GARIS. The magnetic rigidity (Bρ) of GARIS for evaporation residue measurements was set at 2.09 Tm for both reactions. The separator was filled with helium gas at a pressure of 86 Pa. The detector chamber was separated from the gas-filled region by a thin plastic foil and evacuated by a turbo- molecular pump down to 10$^{-4}$ Pa. After passing through the two timing counters consisted of thin plastic foils and Micro-channel-plate (MCP) detectors, the evaporation residues were implanted in the position sensitive silicon semiconductor detector set at the focal plane of GARIS. The size of the effective area of the detector was 60 mm × 60 mm. The radioactive evaporation residues were decaying sequentially into their daughter nuclei inside the crystal of the detector. Identifications of the nuclides were based on genetic correlations of the sequential decays detected in the small area in the detector. A schematic view of the GARIS is shown in Fig. 1.

Total beam dose for $^{208}$Pb + $^{70}$Zn reaction was $0.44 \times 10^{19}$ while the total dose for $^{209}$Bi + $^{70}$Zn reaction was $6.2 \times 10^{19}$.

Detailed descriptions of the experiment were given in ref. [7, 8].

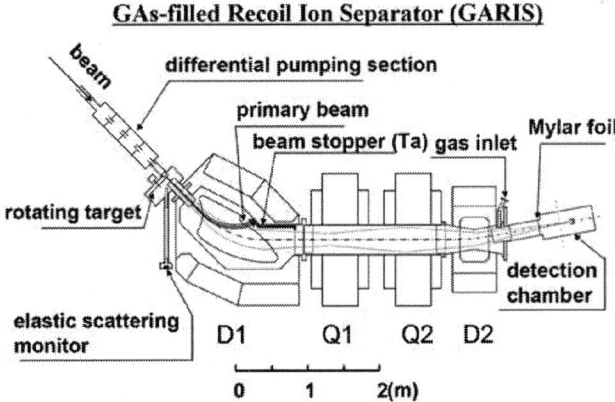

**FIGURE 1.** Schematic view of RIKEN gas-filled recoil separator GARIS. GARIS consists of four magnets in dipole- quadrupole- quadrupole- dipole configuration. The gas-filled region (separator region) and the beam transport duct are separated by a differential pumping section. The primary beam was stopped at the tantrum plate set on the wall of vacuum chamber of the first dipole magnet. Helium gas was supplied from the downstream of the second dipole magnet. Beam envelop of the evaporation

residues are indicated by dotted curve. The detection chamber was separated by 1 μm Mylar foil from the gas-filled region, and evacuated by a turbo molecular pump down to $10^{-4}$ Pa.

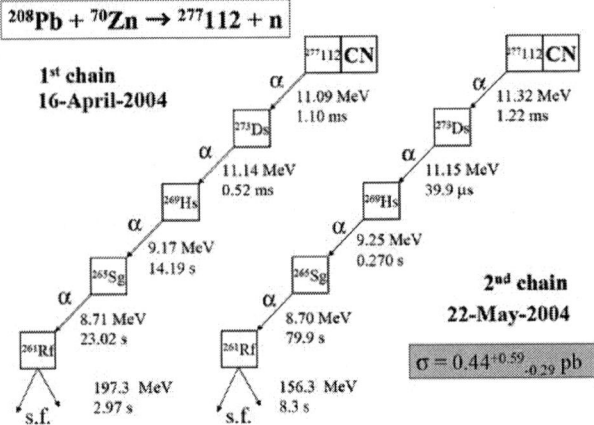

**FIGURE 2.** Decay chains observed in the $^{208}$Pb + $^{70}$Zn reaction. Decay times and decay energies are indicated. Box indicated by CN at right side of the box indicated $^{277}$112, denotes a compound nucleus. Production cross section with 1σ statistical error is also indicated.

## RESULTS AND DISCUSSION

### $^{208}$Pb + $^{70}$Zn reaction

Two decay chains were observed during the irradiation. Both chains consist of four consecutive alpha decays ($\alpha_1 - \alpha_4$) followed by spontaneous fission decays. Observed decay chains are shown in Fig. 2. Decay energies, decay times are indicated in the figure. Production cross section, deduced from the beam dose and the efficiency of the separator, was $0.44^{+0.59}_{-0.29}$ pb. The 1σ error in the cross section is only statistical one. Decay energies and the decay times are listed in Table 2 in order of the decay sequence, together with the corresponding values reported in the literatures. The assignments are also listed in the first column. Properties of the two decay chains observed in the present work, listed in third and force column of the table are quite similar to those reported in ref. [2] and [4]. Those are observed in the same reaction performed by GSI group. The decay chain reported in ref. [2] consists of six consecutive alpha decays, while the decay chain reported in ref. [4] consists of four consecutive alpha decays followed by spontaneous fission decay. Decay time distributions of four decay chains, two present ones and two observed ones by GSI group, are consistent each other, clearly indicating the decay chains have the same origin.

Decay energies of $\alpha_1$ in the present work, 11.09 ± 0.07MeV and 11.32±0.04 MeV, differ from the values obtained at GSI, 11.45MeV and 11.17 MeV, by about 0.10 MeV. Decay energies of $\alpha_2$ and $\alpha_3$ in the present work agree well with the values obtained at

GSI. Decay energies of $\alpha_4$ in the present work were 8.71±0.04MeV and 8.70±0.04 MeV, while the values obtained at GSI were only partially measured in both chains.

Based on the comparison between the decay times and the energies of the decay chain members observed in the present work and those observed in the work done by the GSI group, the present result has provided confirmations of the discovery of $^{277}$112 and its daughter nucleus $^{273}$Ds.

The $\alpha_2$ and $\alpha_4$ energies and decay times observed in the present work are also consistent to the value reported in the ref. [10]. The corresponding decay was observed in the $^{244}$Pu + $^{34}$S → $^{273}$Ds + 5n reaction performed at Flerov Laboratory of Nuclear Reaction of Joint Institute of Nuclear Research (JINR/FLNR), Russia. However, observed decay times for fifth decays, corresponding to the decay of $^{261}$Rf, of four decay chains, two present and two observed at GSI, 7.6 s in average, are differ by about one order of magnitude from the direct production of $^{261}$Rf by the $^{244}$Pu + $^{22}$Ne → $^{261}$Rf + 5n reported in ref. [11], performed also at FLNR with much higher counting statistics. The alpha energy observed in the fifth alpha in the chain in ref. [2] 8.52 MeV, differ by about 0.2MeV from the reported value in ref. [11]. Isomeric states in $^{261}$Rf can be possible.

TABLE 2. Decay energies and decay times of the isotope $^{277}$112 and its daughters. e: escape event, decay energy was not fully measured

| nuclides | | present | present | ref. [2] | ref. [4] | ref. [10] | ref. [11] |
|---|---|---|---|---|---|---|---|
| $^{277}$112 | $E_\alpha$/MeV | 11.09±.07 | 11.32±.04 | 11.45 | 11.17 | | |
| | $t_{decay}$ | 1.10 ms | 1.22 ms | 0.28 ms | 1.41 ms | | |
| $^{273}$Ds | $E_\alpha$/MeV | 11.14±.04 | 11.15±.07 | 11.08 | 11.20 | 11.35 | |
| | $t_{decay}$ | 0.52 ms | 0.04 ms | 0.11 ms | 0.31 ms | 0.39 ms | |
| $^{269}$Hs | $E_\alpha$/MeV | 9.17±.04 | 9.25±.07 | 9.23 | 9.18 | | |
| | $t_{decay}$ | 14.2 s | 0.27 s | 19.7 s | 22.0 s | | |
| $^{265}$Sg | $E_\alpha$/MeV | 8.71±.04 | 8.70±.04 | 4.60$^e$ | 0.20$^e$ | 8.63 | |
| | $t_{decay}$ | 23.0 s | 79.9 s | 74.0 s | 18.8 s | 158 s | |
| $^{261}$Rf | $E_\alpha$ or $E_{sf}$/MeV | 197 | 156 | 8.52 | 153 | | 8.30±.06 |
| | $t_{decay}$ | 2.97 s | 8.30 s | 4.70 s | 14.5 s | | 54$^{+8}_{-4}$ s |
| $^{257}$No | $E_\alpha$/MeV | - | - | 8.34 | - | 8.22 | 8.24, 8.34 |
| | $t_{decay}$ | - | - | 15.0 s | - | 384 s | 17 s |

## $^{209}$Bi + $^{70}$Zn reaction

Two decay chains were observed during the irradiation. Both chains consist of four consecutive alpha decays ($\alpha_1 - \alpha_4$) followed by spontaneous fission (SF) decays. Observed decay chains are shown in Fig. 3. The first chain was observed on 23rd of July, 2004, and the second chain was observed on 2nd of April, 2005. The decay energies and the decay times of the first chain were, ($\alpha_1$:11.68 ± 0.04 MeV, 0.344 ms), ($\alpha_2$:11.15 ± 0.07 MeV, 9.26 ms), ($\alpha_3$:10.03 ± 0.07 MeV, 7.16 ms), ($\alpha_4$:9.08 ± 0.04

MeV, 2.47 s), and (SF:204.1 MeV, 40.9 s). Those of the second chain were ($\alpha_1$:11.52 ± 0.04 MeV, 4.93 ms), ($\alpha_2$:11.31 ± 0.07 MeV, 34.3 ms), ($\alpha_3$:2.32 MeV, 1.63 s), ($\alpha_4$:9.77 ± 0.04 MeV, 1.31 s), and (SF:192.3 MeV, 0.787 s). The energy of $\alpha_3$ in the second chain (2.32 MeV) was small compare to the other decay energies. This is considered to be the event which decaying alpha was emitted near 180 degree to the incoming direction, and only part of the decay energy was measured. Considering that the coverage of solid angle of the focal plane detector is 85 % of $4\pi$, it is natural that energy of one of eight alpha decays (12.5 %) was not fully measured. The decay times of each decay sequences, ($\alpha_1$: 0.344 ms and 4.93 ms), ($\alpha_2$: 9.26 ms and 34.3 ms), ($\alpha_3$: 7.16 ms and 1.63 s), ($\alpha_4$: 2.47 s and 1.31 s), and (SF: 40.9 s and 0.787 s), are statistically consistent each other to assume that two decay chains have the same origin, though, for example, rather big difference was seen in $\alpha_3$ decays. The decay energies of each decay sequence, ($\alpha_1$: 11.68 ± 0.04 MeV and 11.52 ± 0.04 MeV), ($\alpha_2$: 11.15 ± 0.07 MeV and 11.31 ± 0.07 MeV), and ($\alpha_4$: 9.08 ± 0.04 MeV and 9.77 ± 0.04 MeV) are rather differ from each other. However, considering that the observation of the rather widely distributed decay energies of the decay chains of odd-odd nucleus $^{272}$Rg [8], those two chains are consistently assigned to the same origin.

Following the discussion described in ref. [9], the $\alpha_4$ and following SF are consistent to the sequential decay of $^{266}$Bh's and its daughter $^{262}$Db's [12]. In conclusion, the reaction product followed by the decay chain firstly observed in our experiment, was considered to be most probably due to the $^{209}$Bi($^{70}$Zn, n)$^{278}$113 reaction. The secondly observed chain was fully consistent to the firstly observed one. As a result, the members of the decay chains were consequently assigned as $^{278}$113, $^{274}$Rg, $^{270}$Mt, $^{266}$Bh, and $^{262}$Db. The decay energies and decay times of the first chain Production cross section was deduced to be $31^{+40}_{-20}$ fb ($10^{-39}$ cm$^2$). The 1$\sigma$ error in the cross section is only statistical one.

In summary, the isotope of the 113th element, $^{278}$113 and its daughters nuclei, $^{274}$Rg, and $^{270}$Mt were produced for the first time by the $^{209}$Bi + $^{70}$Zn reaction at a beam energy of 349 MeV with dose of 6.2×10$^{19}$.

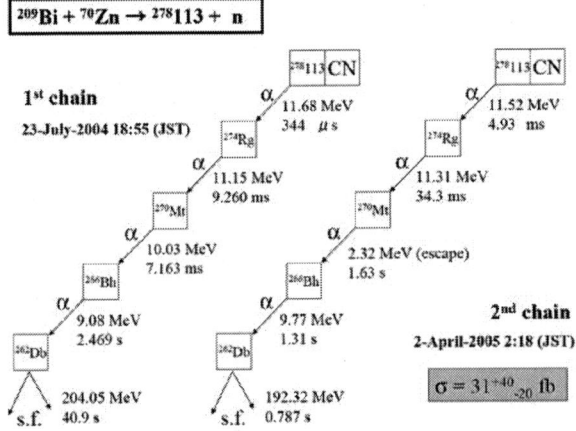

**FIGURE 3.** Decay chains observed in the $^{209}$Bi + $^{70}$Zn reaction. Decay times and decay energies are indicated. Box indicated by CN at right side of the box indicated $^{278}$113, denotes a compound nucleus. Production cross section with 1σ statistical error is also indicated.

## ACKNOWLEDGMENTS

The authors would like to thank all accelerator staff members of RIKEN Accelerator Research Facility, RARF, for their excellent operation for very long period of time.

## REFERENCES

1. A. Sobiczewski et al., Phys. Lett. 22, 500-502 (1966).
2. S. Hofmann et al., Z. Phys. **A354**, 229- (1996).
3. S. Hofmann, *Rep. Prog. Phys.* **61**, 639-689 (1998).
4. S. Hofmann and G. Münzenberg, *Rev. Mod. Phys.* **72**, 733-767 (2000).
5. Yu. Oganessian et al., Phys. Rev. **C69**, 021601(R), 029902(E) (2004), **C69**, 054607 (2004).
6. Yu. Oganessian et al., Phys. Rev. **C74**, 044602 (2006).
7. K. Morita, K. Morimoto, D. Kaji et al., *Eur. Phys. J.* **A21**, 257- (2004).
8. K. Morita, K. Morimoto, D. Kaji et al., *J. Phys. Soc. Jpn.* **73**, 1738-1744 (2004).
9. K. Morita, K. Morimoto, D. Kaji et al., *J. Phys. Soc. Jpn.* **73**, 2593-2596 (2004).
10. Yu. Lazarev et al., Phys. Rev. **C54**, 620- (1996).
11. Yu. Lazarev et al., Phys. Rev. **C62**, 064307 (2000).
12. P. A. Wilk et al., Phys. Rev. Lett. **85**, 2697- (2000).

# Experiments on Nuclear Structure and Synthesis of Superheavy Elements at SHIP

F.P.Heßberger

*Gesellschaft für Schwerionenforschung mbH, Planckstraße 1, D-64291 Darmstadt, Germany*

**Abstract.** Experiments to synthesize isotopes of heaviest elements and to investigate as well their production mechanisms as their nuclear properties have been performed during the past three decades at the velocity filter SHIP. Isotopes of six so far unknown elements with atomic numbers Z = 107-112 have been identified, excitation functions for production of transactinide isotopes in so-called 'cold' fusion reactions, using Pb- or Bi- targets and 'medium' heavy projectiles from $^{50}$Ti to $^{64}$Ni have been measured. In recent years experiments also concentrated on nuclear structure investigations by means of evaporation residue (ER)-γ- or α-γ-decay spectroscopy resulting in systematic studies of low lying Nilsson levels in odd mass nuclei and identification of high lying (E* > 1 MeV) isomeric states in neutron deficient nobelium isotopes.

**Keywords:** Superheavy Elements; Nuclear Reactions; Isomers; Nilsson Levels;
**PACS:** 25.70.Jj, 23.69+e, 23.20.Lv

## INTRODUCTION

The question for the limits of nuclear stability, specificly with respect to increasing proton (Z) and mass numbers (A) is one of the most fascinating ones in nuclear physics. It is closely connected with the structure of nuclei and the existence of closed proton and neutron shells, resulting in an enhanced stability against radioactive decay. So far the largest proton and neutron numbers for which spherical shells are verified are Z = 82 and N = 126. Early extensions of the nuclear shell model into regions beyond these numbers resulted in Z = 114 and N = 184 as proton and neutron numbers for the next spherical shells [1]. Location of the shells has been essentially confirmed by macroscopic - microscopic calculations [2,3], but more recent calculations using self-consistent nuclear models like the Skyrme-Hartree-Fock approach or relativistic mean-field model rather predict the spherical shells at proton number Z = 120 and in the neutron number range N = (172-184) [4].
From experimental side predictions on properties of superheavy nuclei can be proven by two ways. The first is the production of those nuclei and to measure their decay properties; the other way is to investigate the structure of somewhat lighter nuclei (e.g. around Z = 102, N = 152) in detail, since some single particle levels essential for the

shell gaps at Z = 114 and N = 184 come close to the Fermi level in that region, so their location here will have feedback their location at the 'superheavy shells'.

The present paper summarizes results of experiments performed at the velocity filter SHIP at GSI, Darmstadt, during the past years. They aimed as well on the synthesis of superheavy elements at Z $\geq$ 112 as at decay studies by means of $\alpha$- and $\gamma$-spectroscopy in the range Z = (101-106).

## PRODUCTION OF SUPERHEAVY ELEMENTS

Complete fusion reactions of target nuclei around doubly magic $^{208}$Pb with 'medium' heavy projectiles like e.g. $^{48}$Ca, $^{54}$Cr or $^{70}$Zn have been so far the most successful method to produce transactinide nuclei. It has been shown, however, that maximum cross sections typically decrease by a factor of roughly three, when the atomic number is incresed by one unit of Z (see e.g.[5]), reaching a value of $\sigma \approx 30$ fb for the production of $^{278}$113 in the reaction $^{209}$Bi($^{70}$Zn,n)$^{278}$113 [6]. For isotopes of elements Z > 110 reported production cross sections, however, have to be taken insofar with care, since no excitation functions have been measured and thus the values may not represent the maximum cross sections. With respect to this trend, which is explained to be essentially due to an increasing dominance of quasi fission over complete fusion at bombarding energy close to the barrier, it was suggested to use instead reactions of $^{48}$Ca projectiles with actinide targets ($^{238}$U, $^{242,244}$Pu, $^{248}$Cm etc.) to produce superheavy elements. This choice was expected to have two advantages, a) a lower probability of quasi fission, due to the more asymmetric target projectile combination, and b) the production of more neutron rich compound nuclei, thus coming closer to the predicted 'island of stability'. Of disadvantage in these reactions, however, are the higher excitation energies of the compound nuclei at the fusion barrier. While for 'Pb - induced' reactions values E* < 20 MeV are obtained for $Z_{CN}$ > 105, using actinide targets values of E* > 30 MeV are reached. With respect to this the latter reactions are commonly denoted as 'hot fusion', although this expression is somewhat misleading, since processes connected with hot nuclei, as dominance of particle emission over prompt fission due to preequilibrium particle emission and a steeper decrease of particle emission times than the decrease of fission times with increasing nuclear temperature (see e.g. [7]) are not expected to play a notable role here. Indeed, in experiments performed at the gas-filled separator (DGFRS) at FLNR-JINR, Dubna, in irradiations of targets of uranium, plutonium, americium, curium and californium isotopes, $\alpha$-decay chains, usually terminated by spontaneous fission have been observed and attributed to the decay of isotopes of elements 112 to 118 (see [8] and references therein). The cross sections appeared quite stable in the range of (1-5) pb. The latter has been explained by increased stability against prompt fission towards the predicted magic proton and neutron numbers at Z = 114 and N = 184 and was demonstrated by a comparison of cross sections for isotopes ranging from Z = 102 to Z = 116 and calculated fission barriers [8]. The arguments are, however, not straightforward. First, different target - projectile combinations are compared, thus entrance effects due to fusion dynamics are neglegted, secondly quite different maximum values for 3n and 4n deexcitation channels are reported in [8], which underlines the importance of details in the reaction mechanism. Thirdly, the number of observed decay chains attributed to isotopes of elements Z ≥ 112 is quite low, resulting in large statistical error bars. Therefore more detailed studies are necessary. It should be stressed, that the influence of predicted fission barrier hights on the evaporation residue cross sections for purely shell stabilized nuclei had been investigated at SHIP already a couple of years ago: for the reaction $^{208}$Pb($^{68}$Zn,1n)$^{265}$112 only an upper cross section limit of 1.2 pb (68% confidence level) was obtained [9], while for $^{208}$Pb($^{70}$Zn,1n)$^{267}$112 a cross section of ≈0.5 pb was observed [10], although the predicted fission barrier is considerably higher for the compound nucleus $^{276}$112 ($B_f$ = 4.8 MeV) than for $^{278}$112 ($B_f$ = 4.1 MeV) [11].

Due to the extreme importance of the results obtained at DGFRS an independent confirmation of the results is necessary also for sake of scientific clarity. So far the confirmation experiments mainly concentrated on the system $^{48}$Ca + $^{238}$U. An overview of published results is given in fig. 1. Evidently there is no agreement

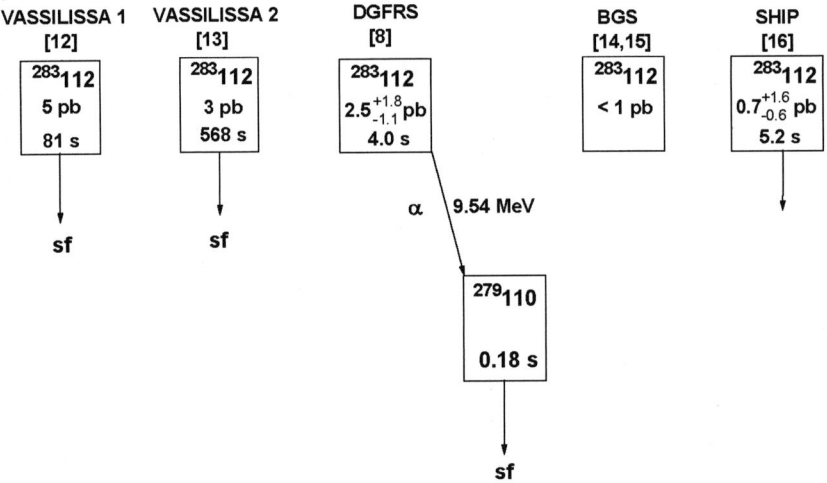

**FIGURE 1.** Comparison of the results of different experiments on the synthesis of $^{283}$112 in the reaction $^{238}$U($^{48}$Ca,3n)$^{283}$112.

between the different experiments. While at DGFRS an α-emitter of 4 s halflife and a decay energy of 9.54 MeV was attributed to $^{283}$112, at the energy filter VASSILISSA (FLNR-JINR, Dubna) [12,13] a long lived fission activity was attributed to it. At the Berkeley Gasfilled Separator (BGS) at LNBL, Berkeley, in two experiments no events that could be attributed to the decay of $^{283}$112 [14,15] were observed, resulting in production cross sections σ < 1 pb. A specific result was obtained at SHIP, GSI, Darmstadt. Here a spontaneous fission event with a mean lifetime of 5.2 s was observed. This result can be compared to that from DGFRS. Indeed in these experiments in part of the decay chains attributed to $^{283}$112 no α-decay preceeding the fission event was observed. This may have two reasons: a) the α-particle excaped the detector with an energy loss lower than the detection threshold, which was in the

DGFRS experiments about 400-500 keV [8], in the SHIP experiment <100 keV; b) $^{283}$112 has an sf branch significantly higher than the upper limit of 10% given in [8], and part of the chains with 'missing' α-decays represented indeed spontaneous fission of $^{283}$112. Under this aspect the results from DGFRS and SHIP are not in contradiction, but for an unabiguous proof the observation of at least one α - sf - correlation in an independent experiment is required. Moreover it is inevitable, since recently two of such correlations have been reported in an experiment performed at FLNR-JINR aimed on chemical separation using the production mechanism $^{242}$Pu($^{48}$Ca,3n)$^{287}$114 -- α → $^{283}$112 [17].

## NUCLEAR STRUCTURE INVESTIGATIONS OF SUPERHEAVY ELEMENTS

In recent years a second field of superheavy element research has been established successfully at SHIP: investigation of nuclear structure of transfermium nuclei by means of α - γ - or evaporation residue - γ - measurements. Different techniques have been applied so far: a) measurement of γ-radiation in prompt or delayed coincidence with α-particles. By this method information on (essentially) low lying single particle levels in 'daughter' nuclei, populated by α-decay and/or succeeding internal transitions (γ-emission or internal conversion) is obtained. These investigations were motivated by the observation that the structure at low excitation energies is essentially determined by the unpaired nucleon. Therefore similarities in the single particle level structure are observed in odd mass nuclei of even Z along the isotone lines and in odd mass nuclei of odd Z along the isotope lines. In the course of these experiments detailed decay studies have been so far performed for N = 147-153 isotones of even Z nuclei [18,19,20,21] and for A = 247-253 isotopes of mendelevium [22].

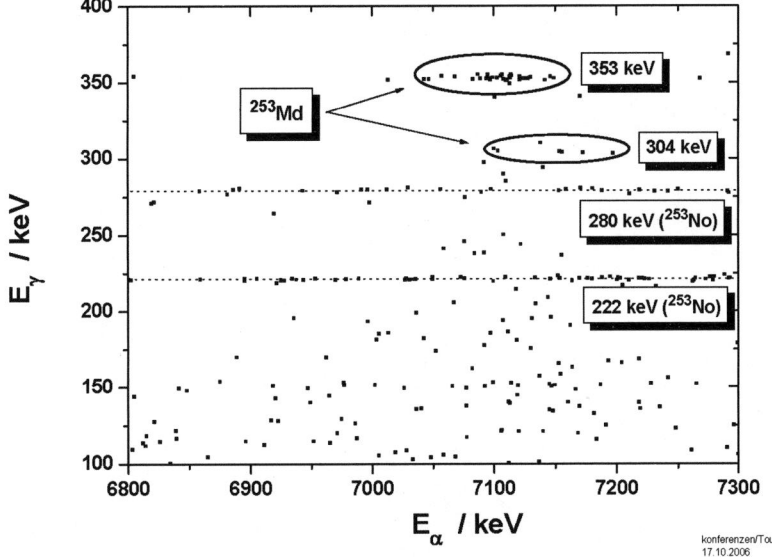

**FIGURE 2.** Alpha-gamma coincidence spectrum observed in the reaction $^{48}$Ca + $^{207}$Pb at 4.6 AMeV.

As an illustrative example α-γ-decay properties of $^{253}$Md shall be presented briefly. The isotope was first identified by Khadkhodayan et al. [23]; a half-life of ≈6 min and decay by EC was reported. First indication for a small α-branch of ≈1% was found in a previous irradiation of $^{207}$Pb with $^{48}$Ca from the observation of a so far unknown α-line of $E_\alpha$ = 7100 keV in coincidence with a γ-line of 353.2 keV [22]. The production mechanism was assumed as $^{207}$Pb($^{48}$Ca,2n)$^{253}$No -- EC → $^{253}$Md. Contrary to the other odd mass mendelevium isotopes, where two γ-transitions of ΔE ≈ 50 keV were observed in coincidence with α-particles, we here found only one α-line; so the interpretation was regarded as tentative. In a recent experiment an enhanced number of decays was observed. The α - γ - coincidence spectrum obtained in this decay study of $^{253}$Md is shown in fig. 2. Clearly, the 'second' γ-line (304 keV) is observed, thus completing the systematics presented in [22] and sketched in fig. 3.

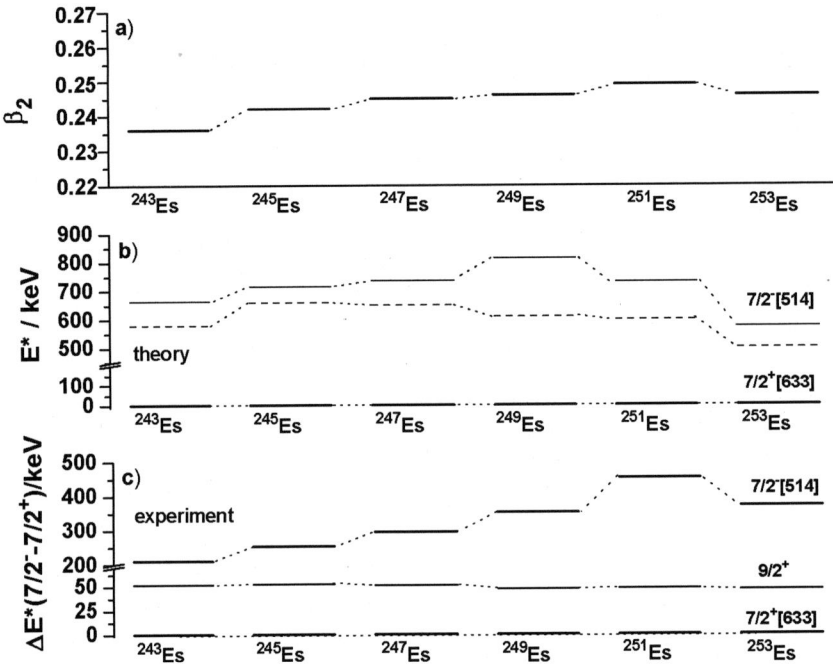

**FIGURE 3.** a) ground-state deformation $\beta_2$ as function of mass number according to calculations of Cwiok et al.[24]; b) theoretical excitation energies of the 7/2⁻[514] level according to calculations of [24] (full line) and Parkhomenko and Sobiczewski [25] (dashed line); both calculations predict 7/2⁺[633] as ground-state; c) experimental energy differences between the 7/2⁻[514] and the 7/2⁺[633] levels as well as between the 9/2⁺ member and the 7/2⁺[633] bandhead.

The ground-state deformations according to the calculations of Cwiok et al. [24] are shown in fig. 3a, the energies of the the 7/2⁻[514] proton Nilsson levels according to the calculations of Cwiok et al. [24] (full lines) and Parkhomenko and Sobiczewski

[25] (dashed lines) are presented in fig. 3b, while the experimental energy differences between the 7/2⁻[514] and the 7/2⁺[633] levels as well as the energy differences between the 9/2⁺ member and the 7/2⁺[633] bandhead are shown in fig. 3c.
Evidently the maximum deformation is expected at the neutron shell N = 152 ($^{251}$Es), which is clearly related with an energy maximum of the 7/2⁻[514] state. The difference between the 9/2⁺ state and the 7/2⁺[633] bandhead is only weakly dependent on the nuclear deformation showing a smooth trend of dercrease with increasing neutron number. The experimental trend is not fully reproduced by theories. Cwiok et al. predict a maximum in the 7/2⁻[514] excitation energy at the neutron number of N = 150 ($^{249}$Es), while the calculations of Parkhomenko et al. locate the maximum rather at N = 146,148 ($^{245,247}$Es).

## INVESTIGATION OF ISOMERIC STATES IN NEUTRON DEFICIENT NOBELIUM ISOTOPES

Specific interest has been devoted in the past years to the investigation of isomeric states. Roughly, the latter may be divided into two groups. The first is one is due to large spin differences between the ground-state and a low lying single particle level, leading to high multipolarities of the transitions and thus to long lifetimes of the excited states. In many cases even α-emission can compete with internal transitions. Mostly, the isomeric character is due to a low spin state above a high spin ground-state, as observed for e.g. $^{251m}$No, $^{247m}$Fm [19,21], where the ground-state is attributed to the Nilsson level 7/2⁺[631], the isomeric state to 1/2⁺[631].

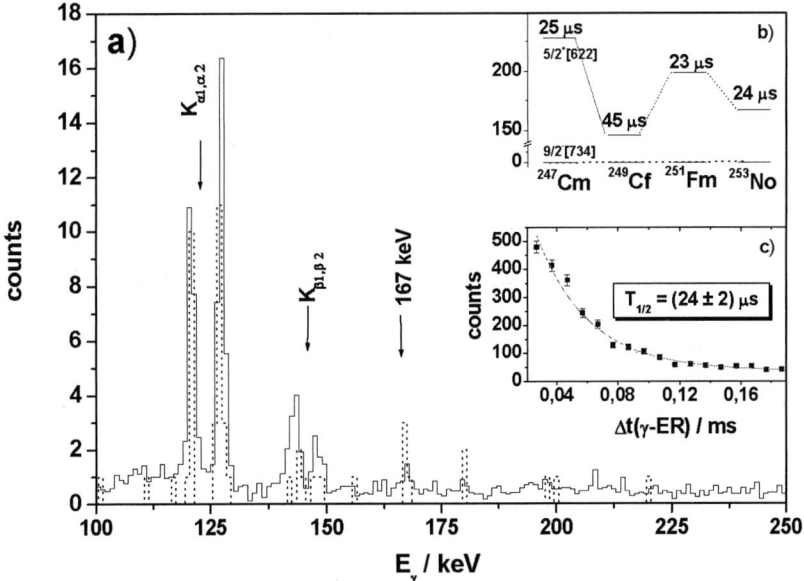

FIGURE 4. a) gamma spectrum observed in delayed coincidence with $^{253}$No evaporation residues (full line) or α-decays of $^{257}$Rf (dotted line); b) systematics of the excitation energy of the 5/2⁺[622] isomeric state in N = 151 isotones; c) decay curve of $^{253m}$No (only $K_α$ –X-rays repected)

An illustrative example for the energy systematics of low lying isomeric states is shown in fig. 4. Isomeric states based on a $5/2^+[622]$ neutron level above the $9/2^-[734]$ ground-state have been known in N = 151 isotones for a long time. While for $^{247}$Cm and $^{249}$Cf their excitation energies were known exactly, for $^{251}$Fm only an approximate value was known from the α-decay pattern of $^{255}$No [26], while for $^{253}$No only the half-life was established so far [27]. In the course our decay study of $^{255}$No [20] we observed γ-radiation from the decay of $^{251m}$Fm for the first time and determined the energy of the isomeric state more precisely as $E^* = 199$ keV. The results of our investigations of $^{253m}$No are shown in fig 4a,c. In fig. 4a the γ-spectrum observed in delayed coincidences with $^{253}$No evaporation residues, produced in the reaction $^{207}$Pb($^{48}$Ca,2n)$^{253}$No and implanted into our 'Stop-Detector' after separation from the projectile beam by SHIP is shown (full line). Besides the intense nobelium K-X-ray lines from internal conversion, a small peak is observed at 167 keV, which is attributed to the γ-decay of the isomer. Alternatively the isomeric state was populated by α-decay of $^{257}$Rf, produced by the reaction $^{208}$Pb($^{54}$Cr,n)$^{261}$Sg – α → $^{257}$Rf. The dotted line shows the γ-spectrum observed in delayed coincidence with α-decays of $^{257}$Rf. Also here a weak γ-transition at 167 keV was observed, thus proving the results from the direct population via $^{207}$Pb($^{48}$Ca,2n)$^{253}$No. The decay curve, taking into account only the $K_\alpha$ – X-rays to minimize background, is shown in fig. 4c, the systematics of level energies for the isomeric states in the N = 151 isotones is shown in fig. 4b.

The isomeric states of the second class are located at excitation energies above 1 MeV. They are understood as due to breaking nucleon pairs and excite the single nucleons into different single particle levels. They may then couple their angular momenta to high values of K. Decay into states with low K-values may be hindered strongly thus leading to long lifetimes (K-isomers). In the region around N = 152 indirect evidence had been found about thirty five years ago by Ghiorso et al. [28] for such long lived isomers in $^{254}$No ($T_{1/2}$ = 0.28 s) and $^{250}$Fm ($T_{1/2}$ = 1.8 s). Gamma decay of both isomers was recently observed in experiments performed at the RITU separator, Jyväskylä [29,30]. In the course of our experiments such isomeric states have been identified in $^{251}$No [21], $^{252}$No [5], $^{253}$No [31] and possibly also in $^{255}$No. Remarkebly are the results for the even-even isotopes; the decay patterns of $^{252m}$No and $^{250m}$Fm resemble, but are completely different to that of $^{254m}$No. While the latter was explained as two proton configuration, the similarities between the isotones $^{250m}$Fm and $^{252m}$No suggest a two neutron configuration. Another feature is the stability of $^{252m}$No against spontaneous fission. While the ground-state has already a fission branch of 25% and hence a partial fission half-life of ≈10 s, no fission from the isomeric state at E* = 1.25 MeV with a half-life of 100 ms was observed so far at a level of <5%. The enhanced stability of K-isomers was pointed out already by Xu et al. [32], who calculated the potential energy as a function of $\beta_2$ deformatiom for the ground-state of $^{256}$Fm and the 7⁻ isomeric state, which showed a significant increase of the fission barrier for the isomeric state. This behavior is similar to the known 'hindrance' of spontaneous fission in odd mass and odd-odd nuclei, i.e. their considerably longer spontaneous fission half-lives compared to neighbouring even-even nuclei. This effect can be understood as due to spin and parity conservation, which does not allow nuclei with unpaired nucleons to follow the energetically most favourable fission path and leads to an increase of the fission barrier (see e.g. [33]). As shown in [32] a similar behavior is also expected for two- (or more) quasiparticle configurations in even-even nuclei. So it seems justified to speak of an enhanced stability on a higher level.

## ACKNOWLEDGMENTS

The experiments were performed in collaboration with S. Hofmann, D. Ackermann, S. Heinz, J. Khuyagbaatar, B. Kindler, I. Kojouharov, P. Kuusiniemi, B. Lommel, R. Mann, B. Sulignano (GSI), S. Antalic, P. Cagarda, B. Streicher, S. Saro, M. Venhart (Comenius University Bratislava), R.-D. Herzberg (University of Liverpool), M. Leino, J. Uusitalo (University of Jyväskylä), A.G. Popeko, A.V. Yeremin (FLNR-JINR, Dubna). Specific thanks are devoted to H.-G. Burkhard and H.J. Schött, who are in charge for the mechanical and electrical set-up of SHIP, to J. Steiner and W. Hartmann for the production of the large area lead and bismuth targets, the UNILAC and ECR crews for the stable and high intense beams. The many fruitful discussions with A. Sobiczewski and A. Parkomenko are gratefully acknowledged.

## REFERENCES

1. H. Meldner, *Arkiv för fysik* **36**, 593- 598(1967).
2. P. Möller, J.R. Nix, W.D. Myers, W.J. Swiatecki, *Atomic Data and Nuclear Data Tables* **59**, 185-381 (1995).
3. R. Smolanczuk, A. Sobiczewski, in *Proc. EPS Conf. 'Low energy Nuclear Dynamics', St. Petersburg 1995*, edited by Yu. Ts. Oganessian et al. , World Scientific, Singapore, New Jersey, London, Hong Kong, 1995, pp 313-320.
4. M. Bender, W. Nazarewicz, P.-G- Reinhard, *Phys. Lett. B* **515**, 42-48 (2001).
5. F.P. Heßberger, *Int. Journal of Mod. Phys. E* **15**, 284-291 (2006).
6. K. Morita, K. Morimoto, D. Kaji, T. Akiyama, S. Goto, H. Haba, E. Ideguchi, R. Kanungo, K. Katori, H. Koura, H. Kudo, T. Ohnishi, A. Ozawa, T. Suda, K. Sueki, H. Xu, T. Yamaguchi, A. Yoneda, A. Yoshida, Y. Zhao, *Journ. of the Phys. Soc. of Japan* **73**, 2593-2596 (2004).
7. J. Galin, U. Jahnke, *Journ. Phys. G: Nucl. Part. Phys.* **20**, 1105-1142 (1994).
8. Yu. Ts. Oganessian, V.K. Utyonkov, Yu. V. Lobanov, F.Sh. Abdullin, A.N. Polyakov, I.V. Shirokovsky, Yu.Ts. Tsyganov, G.G. Gulbekian, S.L. Bogomolov, B.N. Gikal, A.N. Mezentsev, S. Iliev, V.G. Subbotin, A.M. Sukhov, A.A. Voinov, G.V. Buklanov, K. Subotic, V.I. Zagrebaev, M.G. Itkis, J.B. Patin, K.J. Moody, J.F. Wild, M.A. Stoyer, N.J. Stoyer, D.A. Shaughnessy, J.M. Kenneally, R.I. Il'kaev, S.P. Vesnovskii, *Phys. Rev. C* **70**, 064609 (2004).
9. F.P. Heßberger, S. Hofmann, V. Ninov, P. Armbruster, H. Folger, A. Lavrentev, M.E. Leino, G. Münzenberg, A.G. Popeko, S. Saro, Ch. Stodel, A.V. Yeremin, in *AIP Conference Proceedings 425 'Tours Symposium on Nuclear Physics II', Tours, 1995*, edited by M. Arnould, M. Lewitowicz, Yu. Ts. Oganessian, M. Ohta, H. Utsonomiya, T. Wada, American Institute of Physics, Woodbury, New York, 1998, pp. 3-15.
10. S. Hofmann, F.P. Heßberger, D. Ackermann, G. Münzenberg, S. Antalic, P. Cagarda, B. Kindler, J. Kjouharova, M. Leino, B. Lommel, R. Mann, A.G. Popeko, S. Reshitko, S. Saro, J. Uusitalo, A.V. Yeremin, *Eur. Phys. J. A* **14**, 147-157 (2002).
11. R. Smolanczuk, J. Skalski, A. Sobiczewski, *Phys. Rev. C* **52**, 1871-1880 (1995).
12. Yu.Ts. Oganessian, A.V. Yeremin, G.G. Gulbekian, S.L. Bogomolov, V.I. Chepigin, B.N. Gikal, V.A. Gorshov, M.G. Itkis, A.P. Kabachenko, V.B. Kutner, A. Yu. Lavrentev, O.N. Malyshev, A.G. Popeko, J. Rohac, R.N. Sagaidak, S. Hofmann, G, Münzenberg, M. Veselsky, S. Saro, N. Iwasa, K. Morita, *Eur. Phys. J. A* **5**, 63-68 (1999).

13. Yu. Ts. Oganessian, A.V. Yeremin, A.G. Popeko, O.N. Malyshev, A.V. Belozerov, G.V. Buklanov, M.L.Chelnokov, V.I. Chepigin, V.A. Gorshov, S. Hofmann, M.G. Itkis, A.P. Kabachenko, B. Kindler, G. Münzenberg, R.N. Sagaidak, S. Saro, H.-J. Schött, B. Streicher, A.V. Shutov, A.I. Svirikhin, G.K. Vostokin, *Eur. Phys. J. A19*, 3-6 (2004).
14. W. Loveland, K.E.Gregorich, J.B. Patin, D. Peterson, C. Rouki, P.M. Zielinski, K. Aleklett, *Phys. Rev. C* **66**, 044617 (2002).
15. K.E. Gregorich, W. Loveland, D. Peterson, P.M. Zienlinski, S.L. Nelson, Y.H. Chung, Ch. E. Düllmann, C.M. Folden III, K. Aleklett, R. Eichler, D.C. Hoffmann, J.P. Omtvedt, G.K. Pang, J.M. Schwantes, S. Soverna, P. Sprunger, R. Sudowe, R.E. Wilson, H. Nitsche, *Phys. Rev. C* **72**, 014605 (2004).
16. S. Hofmann , D.Ackermann, S. Antalic, H.G. Burkhard, R. Dressler, F.P. Heßberger, B. Kindler, I. Kojouharov, P. Kuusiniemi, M. Leino, B. Lommel, R. Mann, G. Münzenberg, K. Nishio, A.G. Popeko, S. Saro, H.J. Schött, B. Streicher, B. Sulignano, J. Uusitalo, A.V. Yeremin, *Journ. of Nucl. and Radiochem. Sciences* **7**, R25-R29 (2006).
17. R. Eichler, private communication 2006, and to be published.
18. F.P. Heßberger, S. Hofmann, D. Ackermann, V. Ninov, M. Leino, G. Münzenberg, S. Saro, A. Lavrentev, A.G. Popeko, A.V. Yeremin, Ch. Stodel. *Eur. Phys. J. A* **12**, 57-67 (2001).
19. F.P. Heßberger, S. Hofmann, D. Ackermann, P. Cagarda, R.-D. Herzberg, I. Kojouharov, P. Kuusiniemi, M. Leino, R. Mann, *Eur. Phys. J. A* **22**, 417-427 (2004).
20. F.P. Heßberger, S. Hofmann, D. Ackermann, S. Antalic, B. Kindler, I. Kjouharov, P. Kuusiniemi, M. Leino, B. Lommel, R. Mann, K. Nishio, A.G. Popeko, B. Sulignano, S. Saro, B. Streicher, M. Venhart, A.V. Yeremin, *Eur. Phys. J. A* **29**, 165-173 (2006).
21. F.P. Heßberger, S. Hofmann, D. Ackermann, S. antalic, B. Kindler, I. Kojouharov, P. Kuusiniemi, M. Leino, B. Lommel, R. Mann, K. Nishio, A.G. Popeko, B. Sulignano, S. Saro, B. Streicher, M. Venhart, A.V. Yeremin, *Eur. Phys. J. A*, (in press).
22. F.P. Heßberger, S. Antalic, B. Streicher, S. Hofmann, D. Ackermann, B. Kindler, I. Kojouharov, P. Kuusiniemi, M. Leino, B. Lommel, R. Mann, K. Nishio, S. Saro, B. Sulignano, *Eur. Phys. J. A* **26**, 233-239 (2005).
23. B. Khadkhodayan, R.A. Henderson, H.L. Hall, J.D. Leyba, K.R. Czerwinski, S.A. Kreek, N.J. Hannik, K.E. Gregorich, D.M. Lee, M.J. Nurmia, D.C. Hoffman, *Radiochimica Acta* **56**, 1-5 (1992)
24. S. Cwiok, S. Hofmann, W. Nazarewicz, *Nucl. Phys. A* **573**, 356-394 (1994).
25. A. Parkhomenko, A. Sobiczewski, *Acta Physica Polonica B* **35**, 2447-2471 (2004).
26. P.F. Dittner, C.E. Bemis Jr., D.C. Hensley, R.J. Silva, C.D. Goodman, *Phys. Rev. Lett.* **26**, 1037-1040 (1971).
27. C.E. Bemis Jr., R.J. Silva, D.C. Hensley, O.L. Keller Jr., J. R. Tarrant, L.D. Hunt, P.F. Dittner, R.L. Hahn, C.D. Goodman, *Phys. Rev. Lett.* **31**, 647-649 (1973).
28. A. Ghiorso, K. Eskola, P. Eskola, M. Nurmia, *Phys. Rev. C* **7**, 2032-2036 ( 1973).
29. R.-D. Herzberg, P.T. Greenlees, P.A. Butler, G.D. Jones, M. Venhart, I.G. Darby, S. Eeckhaudt, K. Eskola, T. Grahn, C. Gray-Jones, F.P. Heßberger, P. Jones, R. Julin, S. Juutinen, S. Ketelhut, W. Korten, M. Leino, A.-P. Leppänen, S. Moon, M. Nyman, R.D. Page, J. Pakarinen, A. Pritchard, P. Rahkila, J. Saren, A. Steer, Y. Sun, Ch. Theisen, J. Uusitalo, *Nature* **442**, 896-899 (2006).
30. J.Uusitalo, contribution to this conference.
31. F.P. Heßberger, Contrib. to *'Int. Conf. on Nuclear Structure and Related Topics'*, Dubna, 2006 (to be published).
32. F.R. Xu, E.G. Zhao, R. Wyss, P.M. Walker, *Phys. Rev. Lett.* **92**, 252501 (2004).
33. J. Randrup, C.F. Tsang, P. Möller, S.G. Nilsson, S.E. Larsson, *Nucl. Phys. A* **217**, 221-237 (1973).

# Mass Measurements at SHIPTRAP

A. Martín[1,a], D. Ackermann[1], M. Block[1,b], A. Chaudhuri[2], Z. Di[1],
S. Eliseev[1,3], D. Habs[4], F. Herfurth[1], F. Hessberger[1], S. Hofmann[1],
H.-J. Kluge[1], G. Marx[2], M. Mazzocco[1], M. Mukherjee[1,c], J.B. Neumayr[4],
W. Plass[5], S. Rahaman[1,d], C. Rauth[1], D. Rodríguez[6], C. Scheidenberger[1,5],
L. Schweikhard[2], P.G. Thirolf[4], G. Vorobjev[1,3], C. Weber[1,7,d]

[1] *Gesellschaft für Schwerionenforschung mbH, 64291 Darmstadt, Germany.*
[2] *Institut für Physik, Ernst-Moritz-Arndt-Universität, 17489 Greifswald, Germany*
[3] *St. Petersburg Nuclear Physics Institute, Gatchina 188300, Russia*
[4] *Department für Physik der Ludwig-Maximilians-Universität München, Am Coulombwall 1, 85748 Garching, Germany*
[5] *Physikalisches Institut, Justus-Liebig-Universität Gießen, Heinrich-Buff-Ring 16, 35392 Gießen, Germany*
[6] *IN2P3, LPC-ENSICAEN, 6 Bd. Marechal Juin, 14050 Caen Cedex, France*
[7] *Institut für Physik, Johannes Gutenberg-Universität, 55099 Mainz, Germany*

**Abstract.** The Penning trap facility SHIPTRAP at GSI Darmstadt was set up to perform high-precision experiments on heavy elements produced in fusion-evaporation reactions at SHIP. Up to now the masses of 52 radionuclides have been measured, 17 for the first time. In particular, the area of the rp-process endpoint, close to $^{100}$Sn, was addressed. The relative mass uncertainties were between $10^{-7}$ and $5 \cdot 10^{-8}$. Nuclides with half-lives as short as 580 ms and calculated production cross sections as low as 20 µb have been measured. The main limitation for measurements of transuranium radionuclides is their low cross section. After improving the present efficiency of SHIPTRAP, elements with production cross sections down to 1 µb will be accessible.

**Keywords:** Penning trap \ atomic mass \ binding energy.
**PACS:** 07.75 +h Mass spectrometers, 21.10.Dr Binding energies and masses

## INTRODUCTION

Over the last years, Penning traps have gained importance as tools in experimental nuclear physics [1,2]. The confinement of ions in a small volume under controlled fields in Penning traps allows to perform mass measurements of stable nuclides with a relative uncertainty of $10^{-11}$ [3]. Although the production and capture is much more difficult for short-lived nuclei, the advantages of Penning traps for mass measurements of radioactive nuclei are used at different places. Some of these Penning trap facilities

---

[a] e-mail: A.Martin@gsi.de
[b] Present address: NSCL, MSU, South Shaw Lane, East Lansing, 48824 MI, USA
[c] Present address: Universität Innsbruck, Technikerstr. 25, 6020 Innsbruck, Austria
[d] Present address: University of Jyväskylä, P.O. Box 35 (YFL), 40014 Jyväskylä, Finland

are ISOLTRAP [4] at ISOLDE/CERN, LEBIT [5] at MSU, JYFLTRAP [6] at Jyväskylä, and CPT [7] at ANL. The relative uncertainty that is reached by now for short-lived nuclei is $10^{-8}$ [5]. Via its relation with the nuclear binding energy, these high-accuracy mass values can improve our understanding of nuclear structure, the stellar nucleosynthesis of the elements and even test fundamental interactions [1,8]. The different parameters for mass formulas and the hypothesis of nuclear models can be tested comparing the predicted mass values with the measured ones. In nuclear astrophysics, the accurately measured values are of great importance because they are needed to determine the nucleosynthesis pathways and localize the borderlines of nuclear stability, i.e. fix the location of drip lines [9]. Attending to the test of the Standard Model, mass measurements have a special contribution with regard to the weak interaction, namely in the study of the superallowed transitions $0^+ \rightarrow 0^+$ testing the CVC hypothesis and the unitarity of the Cabibbo-Kobayashi-Maskawa quark mixing matrix.

The SHIPTRAP facility at GSI [10] aims for investigations of the nuclear and atomic structure by precision experiments with low-energetic ions produced in a fusion-evaporation reaction at the velocity filter SHIP [11]. In the first stage, mass measurements are performed in SHIPTRAP with a Penning trap mass spectrometer. Trap-assisted decay spectroscopy, laser spectroscopy and chemical reaction studies are foreseen for the future.

The location after SHIP offers the unique opportunity to investigate ions from the medium-heavy up to the transuranium elements. Several masses in the latter region are known only from extrapolations of systematic trends with uncertainties from 100 to 300 keV and above. The uncertainties of the estimated values grow as the distance to the experimental data increases [12], so mass measurements in this region are required. The low cross section for the production of transuranium elements makes these measurements challenging, but even the experimental determination of a few masses helps to pin down some others via α-decay chains which link the nuclides.

The first mass measurements at SHIPTRAP have focused on two different regions. Experiments carried out at the end of 2005 concentrated on neutron-deficient rare-earth nuclei near and beyond the proton drip line. Nuclides around $^{147}$Ho were measured with relative uncertainties of about $10^{-7}$. A detailed description of the results and their implications on nuclear structure and the location of the proton drip-line are presented in [13]. In February and July 2006, measurements in the end-point region of the rp-process, the Sb-Sn-Te cycle [14], were performed and the masses were determined with uncertainties on the order of 10 keV or better. These results agree with the data derived from $\beta^+$-decay studies but provide a better determination of the proton separation energies and $Q_\beta$-values, important data to improve reaction-network calculations for the rp-process.

## EXPERIMENTAL SET-UP

The ion-trap facility SHIPTRAP is coupled to the velocity filter SHIP [11] at GSI Darmstadt. A stable heavy-ion beam accelerated by the UNILAC impinges with energies between 4 and 5 MeV/u onto a foil to produce radioactive ions in fusion-eva-

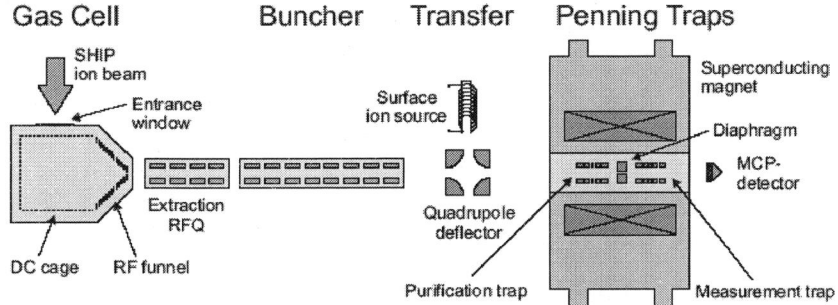

**FIGURE 1.** Schematic layout of SHIPTRAP. The fusion-evaporation reaction products from SHIP are first stopped in a gas cell, then accumulated, cooled and bunched in an RFQ buncher, and finally transferred to the double Penning trap system. After isobaric cleaning in the first trap, the cyclotron frequency is measured with the second one.

poration reactions. The low energy reaction products are separated from the primary beam by SHIP and transferred to SHIPTRAP with energies about 100 keV/u.

SHIPTRAP is composed of three different parts, as shown in the schematic overview in Fig. 1. First, the reaction products are stopped in a helium-filled buffer gas cell [15] at pressure of about 50 mbar. After entering the gas cell through a few μm thick titanium foil, they are thermalised by collisions with the buffer gas. The ions are then extracted by a combination of electrical dc and rf fields which guides them towards a nozzle, from where they are swept out of the cell by a supersonic gas jet into the extraction rf quadrupole (RFQ). This RFQ serves for differential pumping and transport of the ions to the next section: a helium gas-filled RFQ cooler and buncher. In this structure the ions are cooled within a few milliseconds, accumulated and extracted as a low-emittance bunched beam, facilitating the injection into the Penning trap system [16]. The Penning trap system is composed of two cylindrical traps placed in a superconducting magnet of 7T field strength with two homogeneous regions. The first trap is used to select one isobar with a mass-selective buffer-gas cooling technique [17]. Its resolving power is about $10^5$ for nuclides of A≈150 (Fig. 2a). The selected nuclide is transferred through a 3 mm wide and 50 mm long diaphragm into the second trap, where the mass measurement is performed via the determination of the cyclotron frequency $\nu_c$ with the time-of-flight ion-cyclotron-resonance method [18]. The ions in the trap are excited with different frequencies close to the true cyclotron frequency. When the frequency of the excitation equals the cyclotron frequency, the kinetic energy increase from the excitation is maximal and the time-of-flight to the detector is minimal. In this way, a resonance like the one shown in Fig. 2b is obtained. The sidebands are created because of the excitation profile. The mass resolving power in the measurement trap is about $10^6$, which allows to resolve isomers with excitation energies of about 100 keV at mass number A≈100. Once $\nu_c$ is known, the ion mass $m_{ion}$ can be calculated given its charge $q$, and the magnetic field strength $B$, with the relation

$$\nu_c = \frac{qB}{2\pi \cdot m_{ion}} \quad (1)$$

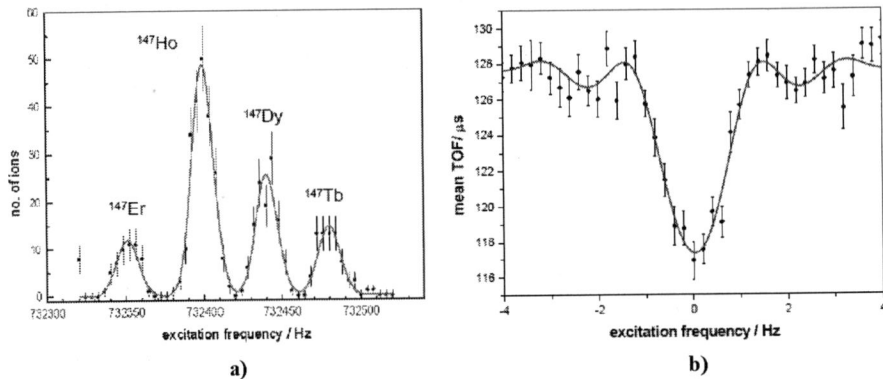

**FIGURE 2.** a) Scan of the cyclotron frequency in the first trap around A=147. The isobars are clearly separated and can be easily selected. A mass resolving power of 50000 was achieved. b) Time-of-flight Ion-Cyclotron resonance for $^{147}$Ho, excitation time 900 ms.

For the precise determination of the atomic mass, the magnetic field needs to be known with the same precision. This is achieved by a calibration measurement with a reference ion of well-known mass. As the measurement cannot be performed at the same time, the actual magnetic field is determined by a linear interpolation from two calibration measurements before and after the determination of the cyclotron frequency of the ion of interest. If $q_{ref}$, $m_{ref}$ and $\nu_{ref}$ are the charge, mass and measured cyclotron frequency of the reference ion, the atomic mass of the investigated nuclide is given by

$$m = \frac{q}{q_{ref}} \frac{\nu_{ref}}{\nu_c} \left(m_{ref} - q_{ref} m_e\right) + q m_e \qquad (2)$$

where $m_e$ is the electron mass. The atomic binding energies, of few eV, can be neglected because the typical total uncertainties for the final mass values are larger than 1 keV.

## MASS MEASUREMENTS ALONG THE RP-PROCESS PATH

The origin of the different elements and their relative abundances are one of the main fields studied in modern physics. The calculation of the path of the various stellar production mechanisms towards heavy elements strongly depends on input data from nuclear physics.

The astrophysical rp-process is important for the stellar nucleosynthesis in environments with explosive hydrogen burning at high temperatures and high densities [14]. Even if at the beginning it was not expected that the rp-process could contribute to the formation of very heavy elements, it has been proven that it contributes to the formation of heavy neutron-deficient elements up to tellurium [19]. The pathway of

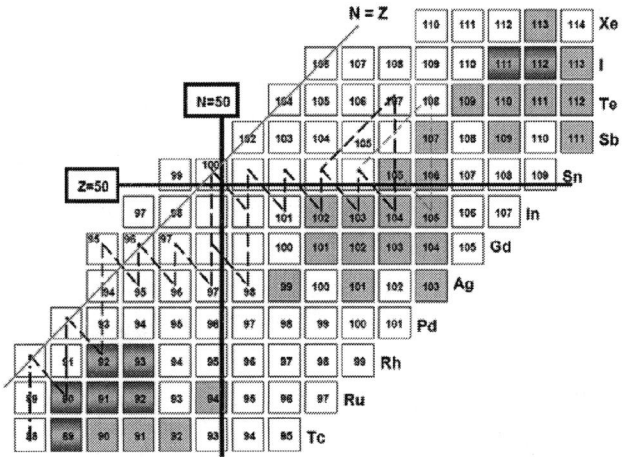

**FIGURE 3.** Predicted rp-process pathway (dashed line) [8]. The masses colored grey have been measured at SHIPTRAP. The dark grey shaded masses were experimentally determined for the first time.

the process depends strongly on the nuclear densities and the temperature in the stellar environment. The main reactions involved in the rp-process are proton capture, photo-disintegration and β-decay. Hence, the path of the rp-process depends on Q-values, branching ratios and decay constants if the stellar conditions are fixed. The network calculations generating the rp-process pathway, as discussed in [14], give the result shown in Fig. 3. The process proceeds along the Sn isotopic chain until the proton separation energy of the Sb isotopes becomes large enough to delay photo-disintegration and hence, causes the rp-process to advance towards heavier elements. This happens around $^{105}$Sn, but current mass uncertainties in the Sb isotopes, which are much higher than the necessary 10 keV, prevented a reliable prediction of the dominant SnSbTe cycle.

Many nuclei involved in the rp-process can be produced by fusion-evaporation reactions. Hence, taking advantage of the high primary beam intensities from the UNILAC accelerator and the high performance of the SHIP velocity filter, SHIPTRAP has started a program to measure the masses of nuclei close to the rp-process. During two runs in February 2006 and one in July 2006, several nuclei of interest were produced irradiating a target of $^{58}$Ni with $^{40}$Ca, $^{50}$Cr and $^{58}$Ni beams with energies from 3.6 MeV/u to 5.0 MeV/u.

The masses of 34 nuclides (grey squares in Fig. 3) were measured with uncertainties between $10^{-7}$ and $5 \cdot 10^{-8}$. One of the highlights is $^{92}$Rh, with a production cross section of only 20 μb. Preliminary results of the mass measurements for the nuclides produced with the $^{50}$Cr beam were presented in [20]. In general, the obtained mass values agree with the results of the AME 2003, but they will allow for a more accurate determination of the proton-separation energies helping to clarify the situation at the end of the rp-process. No final results are yet available because systematic errors and possible isomeric admixtures are still under investigation.

## TOWARDS THE MASS MEASUREMENT OF TRANSURANIUM ELEMENTS

The synthesis and investigation of heavy elements, with the goal to reach the island of stability predicted to be around $Z = 114\text{-}126$ has been one of the main topics in nuclear structure physics for a long time. However, due to the very low cross sections for the presently accessible production reactions, it will be very difficult to access the predicted island at the present facilities and with the available combination of beams and targets.

Reliable mass measurements for heavy and superheavy elements are important because of two reasons. First, most of the mass values for transuranium elements are determined by extrapolations [12] based on few mass values and on α-decay energies. Even if the low production cross sections make the experimental mass determination challenging, it is important to address some nuclides. Especially, new mass values at the endpoints of α-chains help to reduce the uncertainties in the entire chain. On the other hand, the actual identification of the particles via the mentioned α-decay chains is long and complicated to carry out in some cases. Since the mass of a nuclide is a unique property, it can also be used to identify it. Many of the transuranium elements have long half-lives, so, if the cross sections allow for a measurement, a Penning trap could be used as a diagnostic tool. With a production cross section of 2.2 μb, $^{254}$No is a good candidate to start with because it has been already well investigated by alpha and gamma spectroscopy and because it is connected by α-chains to many other nuclides in the region.

The nuclides currently accessible for SHIPTRAP measurements have been demonstrated by the measurement of $^{92}$Rh. The total production cross-section of $^{92}$Rh is calculated to be about 20 μb. With the implementation of several new technical developments, like the use of a high intensity source for the LINAC accelerator, a change of the detector from our actual MCP with 30% efficiency to a channeltron detector with one close to 100%, and improved conditions in the stopping cell, the measurement of $^{254}$No should be possible with the present set-up.

A future improvement of the SHIPTRAP facility especially suited for long-lived nuclides with low production rates will be the non-destructive Fourier transform ion cyclotron resonance technique [21]. With this method, the mass value can be obtained with just one ion. The technique also allows the posterior use of the sample as soon as the measurement time is shorter than the expected half-life of the ion.

## CONCLUSIONS

The SHIPTRAP facility was installed at GSI to perform precision experiments with low-energetic ions produced in fusion-evaporation reactions at the velocity filter SHIP. The masses of 52 radionuclides have been measured from October 2005 to July 2006, with relative uncertainties between $5 \cdot 10^{-8}$ and $1 \cdot 10^{-7}$. In these runs the end point region of the rp-process and the neutron-deficient rare-earth region (A ≈ 150) were investigated. The nucleus with the lowest cross section measured until now is $^{92}$Rh. The production cross section of $^{92}$Rh, calculated to be 20 μb, is only a factor of ten higher than the one of $^{254}$No. Thus, a first mass measurement of a transuranium element is in reach. Further improvements of the efficiency are possible and will allow to access more exotic nuclides.

## ACKNOWLEDGMENTS

The authors acknowledge the financial support by the EU within the contract N° RII3-CT-2004-506065 (EURONS/TRAPSPEC), as well as the help by the SHIPTRAP collaborators.

## REFERENCES

1. K. Blaum, *Phys. Rep.* **425**, 1-78 (2006)
2. L. Schweikhard, G. Bollen, (eds.) *Int. J. Mass Spectrom.* **251**(2/3) (2006)
3. R. S. Van Dyck, Jr., S. L. Zafonte, S. Van Liew, D. B. Pinegar, P. B. Schwinberg, *Phys. Rev. Lett.* **92** (2004)
4. A. Herlert, S. Baruah, K. Blaum, P. Delahaye, S. George, C. Guénaut, F. Herfurth, A. Kellerbauer, H.-J. Kluge, D. Lunney, S. Schwarz, L. Schweikhard, C. Weber, C. Yazidjian, *AIP Conf. Proc.* **831**, 152-156 (2006)
5. G. Bollen, D. Davies, M. Facina, J. Huikari, E. Kwan, P. A. Lofy, D. J. Morrissey, A. Prinke, R. Ringle, J. Savory, P. Schury, S. Schwarz, C. Sumithrarachchi, T. Sun, and L. Weissman, *Phys. Rev. Lett.* **96** (2006)
6. T. Eronen, V. Elomaa, U. Hager, J. Hakala, A. Jokinen, A. Kankainen, I. Moore, H. Penttilä, S. Rahaman, S. Rinta-Antila, A. Saastamoinen, T. Sonoda, J. Äystö, A. Bey, B. Blank, G. Canchel, C. Dossat, J. Giovinazzo, I. Matea, N. Adimi, *Phys. Lett. B* **636**, 191-196 (2006)
7. J. Clark, R. C. Barber, C. Boudreau, F. Buchinger, J. E. Crawford, S. Gulick, J. C. Hardy, A. Heinz, J. K. P. Lee, R. B. Moore, *Nucl. Instr. Meth. B* **204**, 487 (2003)
8. D. Lunney, J. M. Pearson, C. Thibault, *Rev. Mod. Phys.* **75**, 1021 (2003)
9. H. Schatz, *Int. J. Mass Spectr.* **251**, 2 (2006)
10. M. Block, D. Ackermann, D. Beck, K. Blaum, M. Breitenfeldt, A. Chauduri, A. Doemer, S. Eliseev, D. Habs, S. Heinz, F. Herfurth, F. P. Heßberger, S. Hofmann, H. Geissel, H.-J. Kluge, V. Kolhinen, G. Marx, J. B. Neumayr, M. Mukherjee, M. Petrick, W. Plaß, W. Quint, S. Rahaman, C. Rauth, D. Rodriguez, C. Scheidenberger, L. Schweikhard, M. Suhonen, P. G. Thirolf, Z. Wang, and C. Weber, *Eur. Phys. J. A* **25**, 49 (2005).
11. S. Hofmann and G.Münzenberg, *Rev. Mod. Phys.* **72**, 733 (2000).
12. G. Audi, O. Bersillon, J. Blachot, A.H. Astra, and C. Thibault, *Nucl. Phys. A* **729**(1), 3-667 (2003).
13. C. Rauth, D. Ackermann, M. Block, A. Chaudhuri, S. Eliseev, F. Herfurth, F. Heßberger, S. Hofmann, H.J. Kluge, A. Martin, M. Mukherjee, S. Rahaman, D. Rodriguez, G. Vorobjev, C. Weber and the SHIPTRAP collaboration, submitted to Eur. Phys. J. A.
14. H. Schatz, A. Aprahamian, V. Barnard, L. Bildsten, A. Cumming, M. Ouellette, T. Rauscher, F.-K. Thielemann, and M. Wiescher, *Nucl. Phys. A* **688**, 150c (2001)
15. J.B. Neumayr, L. Beck, D. Habs, S. Heinz, J. Szerypo, P.G. Thirolf, V. Varentsov, F. Voit, D. Ackermann, D. Beck, M. Block, Z. Di, S. Eliseev, H. Geissel, F. Herfurth, F. Heßberger, S. Hofmann, H.J. Kluge, M. Mukherjee, G. Münzenberg, M. Petrick, W. Quint, S. Rahaman, C. Rauth, D. Rodriguez, C. Scheidenberger, G. Sikler, Z. Wang, C. Weber, W. Plaß, M. Breitenfeld, A. Chaudhuri, G. Marx, L. Schweikhard, A.F. Dodonov, Y. Nuvikov, and M. Suhonen, *Nucl. Instr. and Meth. B* **244**, 489-500 (2006).
16. F. Herfurth, *Nucl. Instr. and Meth. B* **204**, 587-591 (2003).
17. G. Savard, S. Becker, G. Bollen, H.J. Kluge, R.B. Moore, T. Otto, L. Schweikhard, H. Stolzenberg, and U. Wiess, *Phys. Lett. A* **158**, 247-257 (1991).
18. G. Gräff, H. Kalinowsky, and J.Z. Traut, *Phys. A* **297**(1), 35-39 (1980).
19. H. Schatz, A. Aprahamian, V. Barnard, L. Bildsten, A. Cumming, M. Ouellette, T. Rauscher, F.-K. Thielemann, and M. Wiescher, *Phys. Rev. Lett.* **68**, 2 (2001).
20. G. Vorobjev, D. Ackermann, D. Beck, K. Blaum, M. Block, A. Chaudhuri, Z. Di, S. Eliseev, R. Ferrer, D. Habs, F. Herfurth, F. Heßberger, S. Hofmann, H.J. Kluge, G. Maero, A. Martin, M. Mazzocco, J. Neumayr, Y.

Novikov, W. Plaß, C. Rauth, D. Rodriguez, C. Scheidenberger, L. Schweikhard, M. Sewtz, P. Thirolf, W. Quint, and C. Weber, *Proc. Science* (2006).
21. C. Weber, K. Blaum, M. Block, R. Ferrer, F. Herfurth, H.-J. Kluge, C. Kozhuharov, G. Marx, M. Mukherjee, W. Quint, S. Rahaman, S. Stahl and the SHIPTRAP Collaboration, *Eur. Journ. Phys. A* **25**, 65-66 (2005).

# Dynamics in the production of superheavy elements

G. G. Adamian*, N. V. Antonenko*,†, Z. Gagyi-Palffy†, S. P. Ivanova*, W. Scheid† and A. S. Zubov*

*Joint Institute for Nuclear Research, 141980 Dubna (Moscow Region), Russia*
†*Institut für Theoretische Physik der Universität, 35392 Giessen, Germany*

**Abstract.** The dynamics of fusion is described by the dinuclear system concept which assumes two touching nuclei which carry out motion in the internuclear distance and exchange nucleons by transfer. The corresponding model can be applied to calculate evaporation residue cross sections for complete and incomplete fusion reactions leading to superheavy nuclei.

**Keywords:** Superheavy nuclei, fusion, quasifission, dinuclear model, mass asymmetry motion, evaporation residue cross section, survival probability, transfer reactions, master equations
**PACS:** 25.70.Jj, 24.10.-i, 24.60.-k, 27.90.+b

## INTRODUCTION

Heavy and superheavy nuclei can be produced by fusion reactions with heavy ions. We discriminate Pb or Bi based or cold fusion reactions, e. g. $^{70}$Zn + $^{208}$Pb → $^{278}$112 → $^{277}$112 + n with an evaporation residue cross section of $\sigma$ = 1 pb and an excitation energy of the $^{278}$112 compound nucleus of about 11 MeV, and actinide based or hot fusion reactions, e. g. $^{48}$Ca + $^{244}$Pu → $^{288}$114 + 4n, with the emission of more neutrons. The cross sections are small because of a strong competition between complete fusion and quasifission and small survival probabilities of the excited compound nucleus.

The dynamics of the production of superheavy nuclei can be described with the dinuclear system concept [1, 2]. This concept was introduced by V. V. Volkov [2]. A dinuclear system (DNS) is a nuclear molecule consisting of a configuration of two touching nuclei (clusters) which keep their individuality. Such a system has two main degrees of freedom which govern its dynamics: (i) the relative motion between the nuclei describing molecular resonances in the internuclear potential and the decay of the dinuclear system which is called quasifission and (ii) the transfer of nucleons between the nuclei leading to a dependence of the dynamics on the mass and charge asymmetries in fusion and fission reactions. The latter processes are described by the mass and charge asymmetry coordinates

$$\eta = \frac{A_1 - A_2}{A_1 + A_2} \quad \text{and} \quad \eta_Z = \frac{Z_1 - Z_2}{Z_1 + Z_2}. \tag{1}$$

These coordinates can be assumed as continuous or discrete quantities. For $\eta = \eta_Z = 0$ we have a symmetric clusterization with two equal nuclei, and if $\eta$ approaches the values $\pm 1$ or if $A_1$ or $A_2$ is equal to zero, a compound nucleus has been formed.

According to the dinuclear system concept, the fusion is a transfer of nucleons from the lighter nucleus to the heavier one in a touching configuration. Therefore, mainly a dynamics in the mass asymmetry degree of freedom occurs. The potential is of diabatic type with a minimum in the touching range and a repulsive part towards smaller internuclear distances prohibiting the dinuclear system to amalgamate to the compound nucleus in the relative coordinate. Such a potential can be calculated with a dynamical diabatic two-center shell model [3] and has a survival time of the order of the reaction time of $10^{-20}$ s. It can also be justified with structure calculations based on group theoretical methods [4]. A coordinate-dependence of the mass of relative motion between the nuclei leads to an energy-dependent repulsive potential after the transformation to a constant mass which screens the outer range of the potential from the inner one and, therefore, has similar properties as the diabatic potential [5].

In this article we review some aspects of the DNS model in its application to the fusion process and present recent results for evaporation residue cross sections. We like to mention that the DNS model has a large variety of applications also in nuclear structure physics [6, 7, 8, 9] and in fission [10]. For example, it is used for the description of normal-, super- and hyperdeformed bands in deformed nuclei.

## EVAPORATION RESIDUE CROSS SECTION

The cross section for the production of superheavy nuclei can be written

$$\sigma_{ER}(E_{c.m.}) = \sum_{J=0}^{J_{max}} \sigma_{cap}(E_{c.m.},J) P_{CN}(E_{c.m.},J) W_{sur}(E_{c.m.},J). \quad (2)$$

The three factors are the capture cross section, the probability for complete fusion and the survival probability. The maximal contributing angular momentum $J_{max}$ is of the order of 10 - 15. The capture cross section $\sigma_{cap}$ describes the formation of the dinuclear system at the initial stage of the reaction when the kinetic energy of the relative motion is transferred into potential and excitation energies. The DNS can decay by crossing the quasifission barrier $B_{qf}$ which is of the order of 0.5 - 5 MeV.

After its formation the DNS evolves in the mass asymmetry coordinate. The center of the mass distribution moves towards more symmetric fragmentations and its width is broadened by diffusion processes. The part of the distribution, which crosses the inner fusion barrier $B^*_{fus}$ of the driving potential $U(\eta)$, yields the probability $P_{CN}$ for complete fusion. The DNS can also decay by quasifission during its evolution. Therefore, the fusion probability $P_{CN}$ and the mass and charge distributions of the quasifission have to be treated simultaneously.

The fusion probability can be quantitatively estimated with the Kramers formula and results as

$$P_{CN} \sim \exp(-(B^*_{fus} - \min[B_{qf}, B_{sym}])/T), \quad (3)$$

where the temperature $T$ is related to the excitation energy of the DNS, and $B_{sym}$ is the barrier in $\eta$ to more symmetric configurations. $B_{sym}$ is 4-5 MeV ($> B_{qf}$) in cold fusion reactions and 0.5-1.5 MeV ( $< B_{qf}$) in hot fusion reactions. Since the inner fusion barrier increases with decreasing mass asymmetry, we find an exponential depression of the fusion probability towards symmetric projectile and target combinations in lead based reactions. In hot fusion reactions with $^{48}$Ca projectiles, $P_{CN}$ drops down with increasing mass and charge of the target nucleus. These systems run easier towards symmetric fragmentations and undergo quasifission there.

The excited compound nucleus decays by fission and emits neutrons besides negligible emissions of other particles and photons. The probability to reach the ground state of the superheavy nucleus by neutron emission is denoted as survival probability $W_{sur}$. In the case of the one-neutron emission in Pb-based reactions the survival probability is roughly the ratio $\Gamma_n/\Gamma_f$ of the widths for neutron emission and for fission because of $\Gamma_f \gg \Gamma_n$. The survival probability depends sensitively on the nuclear structure properties of the superheavy nuclei like on the level density, fission barriers and deformation [11].

With the DNS concept we reproduced the measured evaporation residue cross sections of the Pb- and actinide-based reactions with a precision of a factor of two. This concept also yields the excitation energies of the superheavy compound nuclei at the optimal bombarding energies, where the production cross sections are largest, in agreement with the experimental data.

## Isotopic dependence of production cross section

Whether the production cross section of isotopic superheavy nuclei is increasing or decreasing with the neutron number, depends on the fusion and survival probabilities. For example, the evaporation residue cross section of Ds ($Z = 110$) increases with the neutron number. The reactions $^{62}$Ni + $^{208}$Pb $\rightarrow$ $^{269}$Ds + n and $^{64}$Ni + $^{208}$Pb $\rightarrow$ $^{271}$Ds + n have cross sections of 3.5 and 15 pb, respectively. Let us discuss the isotopic dependence of the fusion and survival probabilities. When the neutron number of the projectile is increasing, the dinuclear fragmentation gets more symmetrically and the fusion probability decreases if the more symmetric DNS does not consist of more stable nuclei. Also the survival probability is of importance. For compound nuclei with closed neutron shells the survival probability is larger. Hence, the product of $P_{CN}$ and $W_{sur}$ determines whether the production cross section increases or decreases with increasing neutron number. Figure 1 shows examples for cold and hot fusion reactions [12, 13]. These calculations are very valuable and support an adequate choice of projectile and target nuclei in experiment.

## MASTER EQUATIONS FOR NUCLEON TRANSFER

The dynamics of mass and charge transfer, the fusion and the succeeding quasifission can be studied with master equations [14]. At the starting point we consider the shell model Hamiltonian of all dinuclear fragmentations of the nucleons. This Hamiltonian can be used to derive master equations for the probability $P_{Z,N}(t)$ to find the dinuclear system in a fragmentation with $Z_1 = Z$, $N_1 = N$ and $Z_2 = Z_{tot} - Z_1$, $N_2 = N_{tot} - N_1$. The

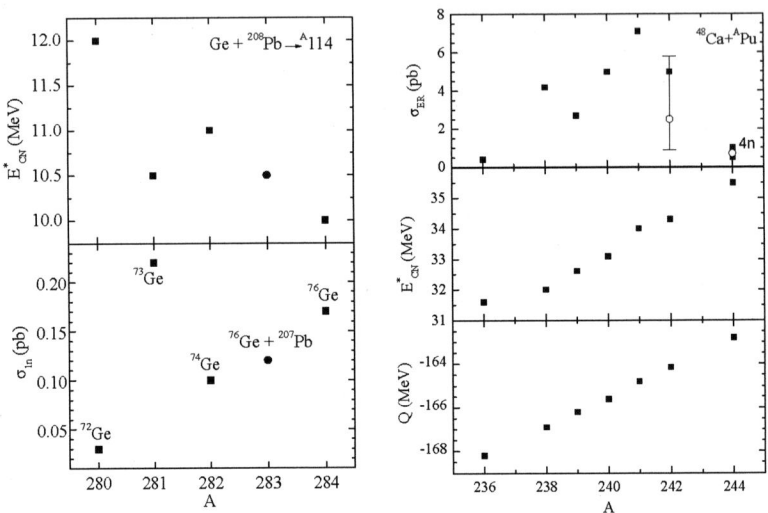

**FIGURE 1.** Excitation energy $E^*_{CN}$, evaporation residue cross section $\sigma_{1n}$, $\sigma_{3n,4n}$ and $Q$-value for Ge + $^{208}$Pb → $^A$114 (l.h.s.) and $^{48}$Ca + $^A$Pu → 114 (r.h.s.). The experimental points are from Ref. [15].

master equations are

$$\frac{d}{dt}P_{Z,N}(t) = \Delta^{(-,0)}_{Z+1,N}P_{Z+1,N}(t) + \Delta^{(+,0)}_{Z-1,N}P_{Z-1,N}(t)$$
$$+\Delta^{(0,-)}_{Z,N+1}P_{Z,N+1}(t) + \Delta^{(0,+)}_{Z,N-1}P_{Z,N-1}(t)$$
$$-\left(\Delta^{(-,0)}_{Z,N} + \Delta^{(+,0)}_{Z,N} + \Delta^{(0,-)}_{Z,N} + \Delta^{(0,+)}_{Z,N}\right)P_{Z,N}(t) - \Lambda^{qf}_{Z,N}P_{Z,N}(t). \quad (4)$$

The one-proton and one-neutron transfer rates $\Delta^{(.,.)}$ depend on the single particle energies and the temperature of the DNS where the occupation of the single particle states is taken into account by a Fermi distribution. The simultaneous transfer of more nucleons is neglected. The quantity $\Lambda^{qf}_{Z,N}$ is the rate for quasifission in the coordinate $R$ and is calculated with the Kramers formula. This rate causes a loss of the total probability $\Sigma_{Z,N}P_{Z,N}(t) \leq 1$. The DNS dynamics was also studied by Li *et al.* [16] with similar master equations.

The fusion probability is given by

$$P_{CN} = \sum_{Z<Z_{BG}, N<N_{BG}} P_{Z,N}(t_0). \quad (5)$$

It is the fraction of probability existing for $Z < Z_{BG}$ and $N < N_{BG}$ at the reaction time $t_0$, where $Z_{BG}$ and $N_{BG}$ determine the fusion barrier in the charge and neutron asymmetry coordinates. The reaction time is $t_0 \approx (3-5) \times 10^{-20}$ s and is determined by solving the balance equation for the probabilities:

$$\sum_{Z,N} \int_0^{t_0} \Lambda_{Z,N}^{qf} P_{Z,N}(t) dt = 1 - P_{CN}. \qquad (6)$$

The DNS with $Z < Z_{BG}$ and $N < N_{BG}$ evolves to the compound nucleus in a time of $10^{-21}$ s which is short compared with the decay time of the compound nucleus.

The mass and charge yields for quasifission are obtained as

$$Y(A_1) = \sum_{Z_1} \int_0^{t_0} \Lambda_{Z_1,A_1-Z_1}^{qf} P_{Z_1,A_1-Z_1}(t) dt, \qquad (7)$$

$$Y(Z_1) = \sum_{N_1} \int_0^{t_0} \Lambda_{Z_1,N_1}^{qf} P_{Z_1,N_1}(t) dt. \qquad (8)$$

## Results for quasifission

The process of quasifission which is the decay of the DNS leads to a large quantity of observable data like mass and charge distributions, distributions of total kinetic energies (TKE), variances of total kinetic energies and neutron multiplicities. Therefore, the comparison of the theoretical description with experimental data provides sensitive information about the applicability and correctness of the used model. We calculated quasifission distributions, TKEs, variances of TKE and neutron multiplicities for cold and hot fusion reactions [14] and found satisfying agreement with the experimental data of Itkis *et al.* [17]. For heavier systems, e. g. $^{48}$Ca + $^{248}$Cm, the contribution of fusion-fission products to the mass distribution can be neglected since the probability for forming a compound nucleus is very small.

## Production of asymmetric systems accompanying fusion reactions

The master equations also give probabilities for more asymmetric systems than the initial one. The cross section $\sigma(Z,N)$ for the production of a primary heavy nucleus in the asymmetric-exit-channel quasifission can be calculated as follows:

$$\sigma(Z = Z_{tot} - Z_1, N = N_{tot} - N_1) = \sigma_{cap} Y(Z,N) \qquad (9)$$

$$\text{with} \quad Y(Z,N) = \int_0^{t_0} \Lambda_{Z_1,N_1}^{qf} P_{Z_1,N_1}(t) dt.$$

The capture cross section is estimated as $\sigma_{cap} = \pi \hbar^2 J_{cap}(J_{cap}+1)T/(2\mu E_{c.m.})$ with $J_{cap} = 20$. Here, $T$ is the penetrability through the Coulomb barrier which is set about 0.5 at $E_{c.m.}$ near the barrier and 1 for larger values of $E_{c.m.}$.

In Fig. 2 we present production (transfer) cross sections for asymmetric fragmentations in the reactions $^{70,72,74,76}$Ge + $^{208}$Pb [18]. The measurement of these observable cross sections would be a proof for the fusion dynamics in the dinuclear system concept.

The incident energies correspond to the expected maxima of the excitation functions of complete fusion for the $1n$ evaporation channel. The yields of products near the initial DNS increase with decreasing neutron number because of the smaller values of $B_{qf}$.

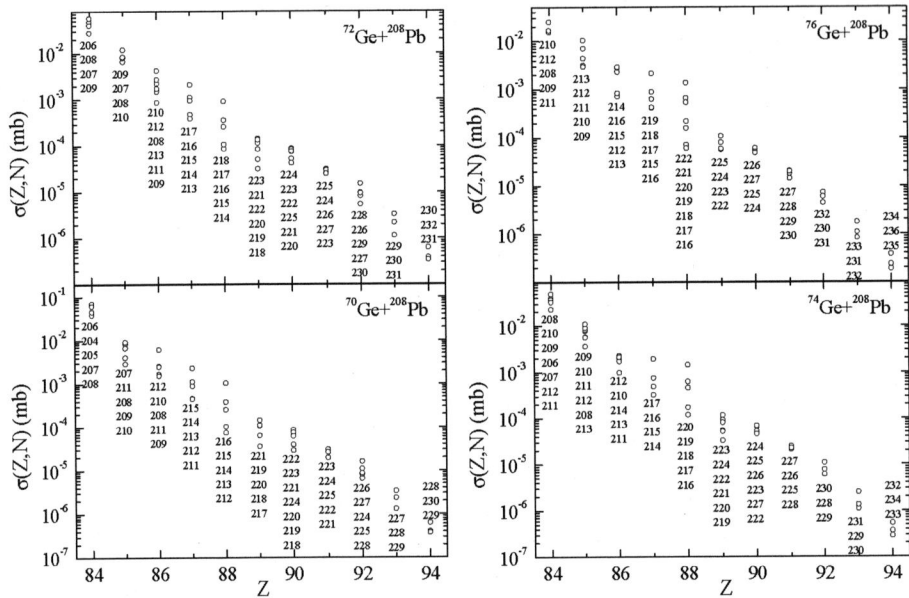

**FIGURE 2.** Calculated production cross section in the reactions $^{70,72,74,76}$Ge + $^{208}$Pb at $E_{c.m.}$ = 267.3, 270.3, 271.3 and 272.3 MeV, respectively, as a function of $Z$ and $A$ of the heavier fragment.

Figure 3 shows theoretical evaporation residue cross sections in the reactions $^{48}$Ca+$^{244,246,248}$Cm for the production of heavy nuclei in the asymmetric-exit-channel quasifission with one neutron evaporated [19]. These cross sections are calculated as

$$\sigma_{ER}(Z, N-x) = \sigma(Z,N) W_{sur}(xn). \qquad (10)$$

The probabilities $Y(Z,N)$ in Fig. 3 are estimated with master equations and a Kramers-type formula. One can produce new isotopes of superheavy nuclei with $Z = 104 - 108$ which fill the gap between the isotopes of heaviest nuclei obtained in cold and hot complete fusion reactions.

## Complete fusion reactions

The master equations are used to calculate the probability $P_{CN}$ for complete fusion. We studied the production of neutron-deficient isotopes of Pu and Cm in complete fusion reactions. In Fig. 4 we show calculated excitation functions of the evaporation residue cross sections for various $xn$ evaporation channels in the reactions $^{40,44}$Ca + $^{184,186}$W [20]. In theses reactions the isotopes $^{220-224}$Pu are produced with rather large cross sections of 2-5 nb.

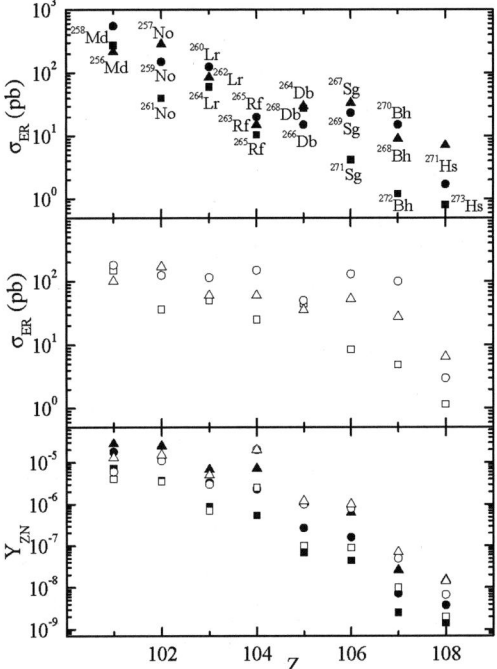

**FIGURE 3.** Calculated primary yields (lower part) and evaporation residue cross sections (middle and upper parts) are shown by triangles, circles and squares for the reactions $^{48}$Ca + $^{244,246,248}$Cm ($E_{c.m.}$=207, 205.5 and 204 MeV), respectively. The heavy fragments after $1n$ evaporation are indicated in the upper part of the figure. The closed and open symbols are calculated with the master equations and Kramers-type formula, respectively.

For comparison with experimental data we give results of calculations with master equations for the fusion reactions $^{86}$Kr+$^{134,138}$Ba in Table 1. Near the maxima of the excitation functions, the agreement of the theoretical results and experimental data is quite good. Note that in the experiment [21] not all channels are separated and the error bars are rather large.

## ACKNOWLEDGMENTS

We thank DFG (Bonn), VW-Stiftung (Hannover) and RFBR (Moscow) for supporting this work. We thank Prof. Junqing Li (Lanzhou), Prof. Enguang Zhao (Beijing), Prof. Shangui Zhou (Beijing), Prof. Wei Zuo (Lanzhou) and Dr. Ning Wang (Giessen) for valuable discussions and help.

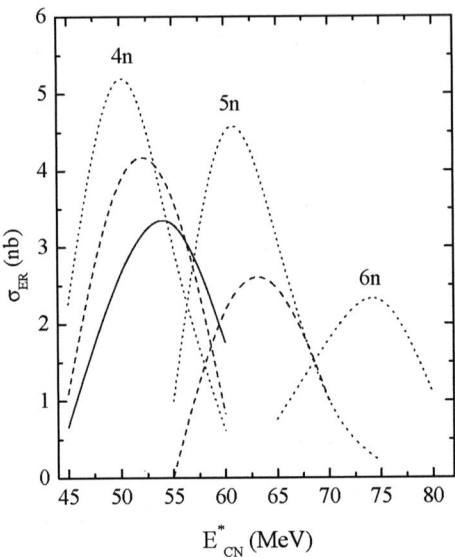

**FIGURE 4.** Calculated excitation functions of the evaporation residue cross sections for the indicated $xn$ evaporation channels in the reactions $^{40}$Ca + $^{184}$W (solid curve), $^{40}$Ca + $^{186}$W (dashed curves) and $^{44}$Ca + $^{184}$W (dotted curves).

**TABLE 1.** The calculated evaporation residue cross sections in the indicated most probable channels of the reactions $^{86}$Kr+$^{134,138}$Ba are compared with recent experimental data of Satou et al. [21].

| Reaction | $E_{c.m.}$ (MeV) | Channels | $\sigma_{ER}^{th}$ | $\sigma_{ER}^{exp}$ |
|---|---|---|---|---|
| $^{86}$Kr+$^{138}$Ba | 213.3 | $\underline{n}+\alpha n+2\alpha n$ | 14 nb | $20^{+15}_{-12}$ nb |
|  | 218.6 | $n+\underline{\alpha n}+2\alpha n$ | 18 nb | $8^{+10}_{-6}$ nb |
|  | 225.3 | $n+\alpha n+\underline{2\alpha n}$ | 94 nb | $50^{+42}_{-32}$ nb |
|  | 225.3 | $2n+\alpha 2n+2\alpha 2n$ | 59 nb | $19^{+38}_{-19}$ nb |
|  | 232.3 | $2n+\underline{\alpha 2n}+2\alpha 2n$ | 64 nb | $140^{+120}_{-90}$ nb |
|  | 237.4 | $3n+\underline{\alpha 3n}$ | 156 nb | $180^{+130}_{-100}$ nb |
| $^{86}$Kr+$^{134}$Ba | 220 | $2n+\underline{\alpha 2n}$ | 1 nb | $2^{+5}_{-2}$ nb |
|  | 220.9 | $np+\underline{\alpha np}$ | 0.7 nb | $6^{+12}_{-6}$ nb |
|  | 227 | $np+\underline{\alpha np}$ | 1.7 nb | $6^{+13}_{-6}$ nb |
|  | 229 | $np+\underline{\alpha np}$ | 3 nb |  |

# REFERENCES

1. G. G. Adamian, N. V. Antonenko, W. Scheid, and V. V. Volkov, *Nucl. Phys.* **A633**, 409 (1998).
2. V. V. Volkov, *Izv. AN SSSR ser. fiz.* **50**, 1879 (1986).
3. A. Diaz-Torres, G. G. Adamian, N. V. Antonenko, and W. Scheid, *Phys. Lett.* **B481**, 228 (2000).
4. G. G. Adamian, N. V. Antonenko, and Yu. M. Tchuvil'sky, *Phys. Lett.* **B451**, 289 (1999).
5. H. J. Fink, W. Scheid, and W. Greiner, *J. Phys. G (Nucl. Phys.)* **1**, 685 (1975).

6. T. M. Shneidman, G. G. Adamian, N. V. Antonenko, S. P. Ivanova, and W. Scheid, *Nucl. Phys.* **A671**, 119 (2000).
7. G. G. Adamian, N. V. Antonenko, R. V. Jolos, Yu. V. Palchikov, and W. Scheid, *Phys. Rev.* **C67**, 054303 (2003).
8. G. G. Adamian, N. V. Antonenko, R. V. Jolos, Yu. V. Palchikov, W. Scheid, and T. M. Shneidman, *Phys. Rev.* **C69**, 054310 (2004).
9. G. G. Adamian, N. V. Antonenko, N. Nenoff, and W. Scheid, *Phys. Rev.* **C64**, 014306 (2001).
10. A. V. Andreev, G. G. Adamian, N. V. Antonenko, and S. P. Ivanova, *Eur. Phys. J.* **A26**, 327 (2005).
11. A. S. Zubov, G. G. Adamian, N. V. Antonenko, S. P. Ivanova, and W. Scheid, *Eur. Phys. J.* **A23**, 249 (2005).
12. G. G. Adamian, N. V. Antonenko, and W. Scheid, *Phys. Rev.* **C69**, 011601 (2004).
13. G. G. Adamian, N. V. Antonenko, and W. Scheid, *Phys. Rev.* **C69**, 014607 (2004).
14. G. G. Adamian, N. V. Antonenko, and W. Scheid, *Phys. Rev.* **C68**, 034601 (2003).
15. Yu. Ts. Oganessian *et al.*, *Eur. Phys. J.* **A13**, 135 (2002); **A15**, 201 (2002).
16. W. Li, N. Wang, J. F. Li, H. Xu, W. Zuo, E. Zhao, J. Q. Li, and W. Scheid, *Europhys. Lett.* **64**, 750 (2003).
17. M. G. Itkis *et al.*, *Nucl. Phys.* **A734**, 136 (2004).
18. G. G. Adamian and N. V. Antonenko, *Phys. Rev.* **C72**, 064617 (2005).
19. G. G. Adamian, N. V. Antonenko, and A. S. Zubov, *Phys. Rev.* **C71**, 034603 (2005).
20. G. G. Adamian, N. V. Antonenko, S. P. Ivanova, W. Scheid, and A. S. Zubov, *to be published* (2006).
21. K. Satou *et al.*, *Phys. Rev.* **C73**, 034609 (2006).

# Chemical Investigations of Superheavy Elements - Current Results and New Techniques

## Christoph E. Düllmann

*Gesellschaft für Schwerionenforschung mbH, Planckstrasse 1, 64291 DARMSTADT, GERMANY*

**Abstract.** Chemical studies of the superheavy elements have progressed tremendously in recent years. This is illustrated here using the following four examples: (i) gas chemical studies of element 112, (ii) radiochemical investigations of the reaction $^{248}$Cm($^{26}$Mg,xn)$^{274-x}$Hs, (iii) complexation studies of rutherfordium, and (iv) the development of the technique of physical preseparation.

**Keywords:** Chemistry of superheavy elements, transactinides, element 112, hassium, rutherfordium, preseparation.

## INTRODUCTION

The chemical investigation of the heaviest elements has seen tremendous progress in the last few years. The chemistry of transactinides has grown to become an important subfield of nuclear chemistry, which is reflected by the fact that the first comprehensive textbook entitled "The chemistry of the superheavy elements" has recently been published [1]. Another comprehensive recent review can be found in [2]. The list of elements that are chemically investigated has been extended up to Z=108, hassium, and the chemistry of element 112 is currently being tackled. The level of detail of chemical information extracted from experiments with single atoms has increased dramatically and new techniques that promise to give access to new chemical systems are under development. Selected chemical systems allow for the investigation of nuclear reactions with cross sections as low as one picobarn, which renders these experiments as sensitive as forefront physics investigations of superheavy elements. Nuclear chemical studies can thus provide important data to the nuclear physics community for reactions where physical methods are not well suited.

Here, some of the recent highlights that are admittedly chosen somewhat arbitrarily will be presented. These are the following:

1. Gas chemical investigations of element 112
2. Radiochemical investigations of the reaction $^{248}$Cm($^{26}$Mg,xn)$^{274-x}$Hs
3. Complexation studies of Rf at JAEA
4. Physical preseparation - a new technique for transactinide chemists

# HIGHLIGHT 1: GAS CHEMICAL INVESTIGATIONS OF ELEMENT 112

Similar to the nuclear physicists who have been pursuing the experimental observation of nuclei around the long predicted "island of stability" in the vicinity of element 114 for a long time, also the chemists have paid a lot of attention to the investigation of superheavy elements. Already 30 years ago, several authors predicted element 112 (with an expected electronic configuration $s^2d^{10}$) to be very volatile and not to form metal-metal interactions with itself due to the relativistic stabilization of the spherical $7s^2$ electron pair, rendering it as inert as a noble gas [3,4]. Recent calculations of the interactions of element 112 with transition metal surfaces [5,6] suggest the formation of relatively strong metal-metal bonds. Element 112 is therefore expected to behave similarly to its supposed lighter homologue, Hg.

## Investigations of the Reaction $^{238}U(^{48}Ca,3n)^{283}112$ at Physical Recoil Separators in Dubna (RU), Darmstadt (D), and Berkeley (CA, USA)

In 1999, when a group from the Flerov Laboratory of Nuclear Reactions (FLNR) of the Joint Institute for Nuclear Research (JINR) in Dubna reported the possible observation of a long-lived isotope of element 112, this announcement prompted immediate preparatory work with respect to the chemical investigation of this element. Oganessian and co-workers [7] claimed to have observed two atoms of $^{283}112$ which were formed in the nuclear reaction $^{238}U(^{48}Ca,3n)^{283}112$ and separated in the energy filter VASSILISSA. They were reported to be formed with a cross section of $5.0^{+6.3}_{-3.2}$ pb and to decay by spontaneous fission (SF) after lifetimes of 52 s and 182 s. These reports were seemingly confirmed by studies with $^{242}Pu$ as a target that led to the observation of $^{287}114$ that decayed to $^{283}112$ which in turn again decayed by SF with a relatively long half-life [8]. In order to understand the development in the field, it is important to note that there are other, different, in fact contradictory observations concerning the production and decay properties of this isotope reported in the literature: in 2004, Oganessian *et al.* (a different group of researchers than the one from the VASSILISSA experiment, but again from the FLNR in Dubna) reported results [9] from a more extensive set of experiments using the Dubna Gas-Filled Recoil Separator (DGFRS). This study of the same nuclear reaction, $^{238}U(^{48}Ca,3n)^{283}112$, reported $^{283}112$ to be formed with a cross section of $2.5^{+1.8}_{-1.1}$ pb but to decay by emission of 9.5-MeV α-particles with a much shorter half-life of around 5 s. For a possible SF branch, an upper limit of 10% was reported [9]. Additionally, the probability of a long-lived decay mode (as observed in the experiments at VASSILISSA) was claimed to not exceed 10% based on the data obtained at the DGFRS [9]. For the chemists, this difference in lifetimes is crucial as the first chemistry experiments were designed to be sensitive to isotopes with a half-life of at least around one minute and simply could not observe the decay of a 5 s isotope. In the same year, i.e., also in 2004, the VASSILISSA group published the confirmation of their results after they had repeated the study of the $^{48}Ca+^{238}U$ reaction [10]. They observed two more SF events with lifetimes of several hundred s that were assigned to the decay of $^{283}$112.

In 2005, the same reaction, $^{238}$U($^{48}$Ca,xn)$^{286-x}$112, was studied at the velocity filter SHIP at GSI [11]. Investigations at three different beam energies yielded one candidate event which was observed at an excitation energiy $E^*$ of the compound nucleus of 34.5 MeV and corresponds to a cross section of $0.7^{+1.6}_{-0.6}$ pb. At $E^* = 32.0$ and 37.0 MeV, cross section limits (63.2% c.l., "one-event" limits) of 0.8 pb and 0.6 pb, respectively, were reported. The candidate event consisted of a correlation of an evaporation residue (EVR) followed by a SF event 7.5 s later. Since no SF branching has been reported by the DGFRS group and the half-life for the SF activity reported from VASSILISSA, these decay properties have to be considered to be in conflict with the results from Dubna. However, introduction of an SF branch based on the DGFRS data seems possible, which was raised as one possibility to reconcile the SHIP results with the ones from DGFRS [11]. At the Lawrence Berkeley National Laboratory (LBNL) in Berkeley, finally, the same reaction was studied in several experimental campaigns as well [12,13]. No event attributable to the observation of element 112 was observed and only an upper limit of roughly 1.2 pb (84% c.l.) was published [13] along with a statistical analysis that evaluated the probability that this limit was still statistically compatible with the DGFRS results. The conclusion was that this probability was less than 10 % over a generous range of beam energies and magnetic rigidities.

## First Chemical Experiments with Element 112 at Dubna (2000; 2001)

As early as 2001, Yakushev et al. [14] from the FLNR reported on an attempt to chemically characterize element 112. Based on then available data from the physics experiments, the setup was optimized for the detection of a ~3-min isotope decaying by SF. In the first experiment, the reaction products recoiling from the target were thermalized in a He filled recoil chamber and transported with the gas flow to a detection system optimized for the detection of Hg-like element 112 using Au or Pd covered PIPS detectors which were place in an array of $^3$He neutron counters to observe neutrons associated with SF. In preparatory experiments, Hg was shown to adsorb strongly on pure Au covered surfaces [14]. No SF events were observed in this study but since the collected beam dose was relatively small, no conclusive results concerning the chemical behavior of element 112 could be drawn. In a second, improved experiment [15], the detection system was extended to also cover potentially Rn-like element 112. The gas flow exiting from the Au or Pd covered detector station was introduced into an ionization chamber which was large enough for the residence time to be long enough for a several-minute isotope to decay with high probability in this volume. Still, the experiment was not sensitive to isotopes with half-lives shorter than one minute as no such decay properties for $^{283}$112 had been reported at the time of these studies. A much larger beam dose (compared to the first experiment) was accumulated and eight high-energy events were observed in the ionization chamber with none originating from the PIPS detectors where Hg-like element 112 should have adsorbed and decayed. These observations were interpreted as pointing to a Rn like behavior of element 112 [15].

## Chemical Experiments with Element 112 at GSI (2003; 2004)

A collaboration led by the group from the Paul Scherrer Institute (PSI) from Switzerland developed an improved setup to chemically investigate element 112 and performed two experimental campaigns at GSI [16]. A detection system based on the thermochromatographic method [17] was developed. The chromatographic column had a rectangular cross section. One side was formed by an array of 32 PIN diodes while the opposite side consisted of a surface covered with Au. The spacing between the Au and the PIN diodes was only 1.6 mm. Along this chromatographic channel, a negative temperature gradient was applied ranging from +35 to -185 °C. Hg adsorbed at the high-temperature entrance of the channel in a diffusion controlled way while Rn deposited all the way at the cold end of the channel. This system should therefore have allowed for an assessment of the similarity of element 112 to either Hg or Rn upon contact of an Au surface. This experiment, performed in 2003, was again designed to observe isotopes with half-lives of at least one minute. A total of 20 events with energies ≥40 MeV were registered. Seven of these formed a peak at the cold end of the column, i.e., where Rn was deposited. This peak was interpreted as resulting from a chromatographic process. The results were published as again the observation of Rn-like behavior of element 112 [18]. However, only $2\pi$ counting geometry was used, which did not allow for the observation of coincident fission fragments. The observed energies were much lower than those expected from the SF of an isotope with Z=112 based on experimental data and well known trends, e.g., according to Unik *et al.* [19] or Viola [20]. As a possible explanation for this observation, the presence of an ice layer originating from the deposition of water that was still present in the dried He gas was introduced [16,21]. Follow-up studies with an improved set-up were deemed necessary in order to confirm and corroborate these findings. In the meantime, the new results from the DGFRS studies [9] were reported and it was immediately clear that none of the chemistry experiments performed so far has been sensitive to such a short-lived isotope. Therefore, several improvements were performed in the setup [16], the most important ones being (i) the upgrade of the detection system from $2\pi$ to $4\pi$ counting geometry by employing a chromatography column formed by two opposite arrays of PIN diodes with the ones on one side being covered with a 50-nm thick layer of Au, (ii) the use of a smaller recoil chamber and an increased gas-flow rate resulting in a transport time short enough for the detection of an isotope with a few seconds half-life with relatively high probability, and (iii) employing a closed gas-loop that resulted in much dryer conditions and thus negligible coverage of the PIN-diodes with ice. The result of this experiment was that no events in accordance with any of the published decay properties for $^{283}$112 were observed, disproving the first of the GSI experiments and leaving the nuclear chemists with no knowledge about the chemical properties of element 112 and, hence, also the nuclear physicist with no independent confirmation of the production of superheavy elements in $^{48}$Ca induced reactions [16].

## Recent Experiments with Element 112 at Dubna (2006)

In the spring of 2006, a collaboration led by the group from the PSI performed another set of chemical experiment using a setup similar to the one applied in the experiments at GSI. Preliminary information indicates that two correlated decay chains that are in agreement with the results obtained at the DGFRS have been observed [22].

## Current Status of Element 112 Chemistry Experiments

To summarize the current status, it can be said that

(i) the experiments by Yakushev et al. which were only sensitive to a long-lived, i.e., T1/2 > 1 min, isotope of element 112 were interpreted as yielding indications for a Rn like behavior of element 112 [15]. If no such long-lived isotope exists, the observed signals will likely have to be discarded as evidence for the observation of element 112.

(ii) The experiments conducted at GSI did not yield any indication for the observation of element 112 and the temporary findings reported in [18] could not be confirmed in a follow-up experiment [16].

(iii) The most recent experiments were again performed at FLNR. Preliminary information indicates that two correlated decay chains that are in agreement with the results obtained at the DGFRS have been observed [22].

## HIGHLIGHT 2: RADIOCHEMICAL STUDIES OF THE REACTION $^{248}$Cm($^{26}$Mg,xn)$^{274-x}$Hs

The first chemical investigation of element 108, hassium (Hs) has been performed in 2001 [23]. It showed Hs to form a highly volatile tetroxide, $HsO_4$, and thus to behave in accordance with the trends known from the lighter homologs in group 8: tetroxides are well known for the lighter members of that group, i.e., $RuO_4$ and $OsO_4$. The formation of such highly volatile oxide compounds is an outstanding property of the group 8 elements. Non of the elements known to pose problems in the unambiguous identification of single transactinide atoms form oxides that are as volatile, which leads to a very high separation factor of this chemical system with respect to other elements, e.g., Po of at least $2 \cdot 10^4$ [24]. The small reached cross sections that were of the order of a few picobarns [25] together with the exceptional separation factor make this system suitable for nuclear reaction studies that lead to Hs isotopes with half-lives of at least a few seconds. Nuclei in the vicinity of the predicted double shell closure at Z=108 and N=162 for deformed nuclei are expected to be long-lived enough for such studies. A collaboration led by the heavy element group of the TU Munich has started an experimental campaign to measure Hs isotopes produced in the nuclear reaction $^{248}$Cm($^{26}$Mg,xn)$^{274-x}$Hs. The main emphasis was on the confirmation of the results of [23,25] by measuring an excitation function and compare the chemical behavior of Hs to the one observed in the first experiments. In 2004, the reaction $^{248}$Cm($^{26}$Mg,xn)$^{274-x}$Hs was studied at two different beam energies [26] leading to excitation energies of the compound nuclei of 40 MeV and 49 MeV at which the cross sections for the 4n and 5n channel, respectively, are predicted to be highest by HIVAP calculations. A number of decay chains with measured decay properties that are in agreement with published data for $^{269}$Hs and its daughters

[25,27,28] were observed. More recently, these studies were extended and three additional beam energies were covered, thus providing a complete excitation function for the $^{248}$Cm($^{26}$Mg,xn)$^{274-x}$Hs reaction. The results are currently being prepared for publication [26,29].

Employing a chemical separation technique proved to be a very good choice for the study of the relatively asymmetric reaction $^{248}$Cm($^{26}$Mg,xn)$^{274-x}$Hs. Physical recoil separators are known to have relatively low transmission for slow EVRs resulting from such reactions. On the other hand, the chemical technique employed here has an overall efficiency of at least 50% for isotopes with half-lives of > ~1 s. Additionally, targets with areal densities of 1 mg/cm$^2$ or even more can be used in these studies, in contrast to experiments at physical recoil separators where the optimum target thickness is known to be at most 500 µg/cm$^2$. Thus, such experiments can reach small cross sections within a relatively short time. Another important factor that should not be underestimated is that chemical experiments provide the atomic number of the investigated species, which is a very powerful feature when dealing with decay chains of unknown origin.

## HIGHLIGHT 3: RUTHERFORDIUM COMPLEXATION STUDIES AT JAEA (J)

Recently, an international collaboration reported on complexation studies with Rf that were conducted at the Japanese Atomic Energy Agency (JAEA) in Tokai, Ibaraki, Japan [30]. In the context of this contribution, I would like to put the emphasis on the technical progress rather than to give a detailed account of the (admittedly, highly interesting) chemical properties of the group 4 elements including Rf deduced from these studies. Complexation studies with Rf, the lightest transactinide, have been performed for a long time. In 1970, Silva et al. [31] reported on the extraction of Rf with ammonium α-hydroxyisobutyrate from cation exchange columns and deduced the behavior of Rf to be different from that of actinides and to be in line with what would be expected for a group 4 element. Their findings were based on the observation of 17 α-particles attributed to $^{261}$Rf and/or its daughter, $^{257}$No, which have overlapping α-particle energies. About ten years later, Hulet et al. [32] reported on the chloride complexation of Rf. Their experiments should be regarded as a big improvement as far as chemical aspects are concerned compared to the studies performed by Silva et al. But still, their findings were based on the observation of only few events, namely six α-particles that they attributed to the decay of either $^{261}$Rf or $^{257}$No. As another example, studies reported by Schumann et al. in 1998 [33] shall be mentioned. It is important to note that these few works mentioned here represent merely a - not necessarily very objectively chosen - selection of complexation studies with Rf. Schumann et al. presented detailed results for Hf, the lighter homologue of Rf, as well as Th, a so-called pseudohomologue that often exhibits chemical properties similar to the group 4 elements. Their conclusion, namely that Rf behaved more similar to Th than to Hf, was based on the observation of only 5 α-particles from $^{261}$Rf/$^{257}$No. As a final example before coming to the recent JAEA studies, the work reported by Strub et al. [34] in 2000 shall be considered. The distribution ration of Rf on the cation exchange resin material Aminex A6 was studied at five different HF concentrations, resulting in the determination of distribution ratios at two concentrations while at three more concentrations limits for this value could be determined, indicating the behavior of Rf to differ from the one of Hf and Zr in this chemical system. A total of 92 single α-particles and 18 α-α correlations were detected in the course of these studies.

Coming to the recent studies performed at JAEA, it is worth noting that this collaboration has measured an elution curve of Rf for the first time [35]. Extraction studies with a fully automated system have allowed to perform several thousand individual extraction experiments [30]: in a total of 4226 extractions, 266 α-particles, including 25 α-α correlations between $^{261}$Rf and its daughter, $^{257}$No, have been measured. The full automatization of the system together with the very detailed chemical data obtained for Rf in the investigated system constitute such a major improvement that it deserves to be mentioned as one of the recent highlights in the chemical investigations of the transactinide elements. The distribution ratio of Rf in this chemical system was determined with high accuracy at six different HF concentrations, which simply represents a new level of quality not achieved in previous studies. By transforming the data into a figure where the slope of a curve is indicative of the charge of the eluted species, the authors of [30] were able to clearly show Rf to be present as a different species, i.e., $RfF_6^{2-}$, than Hf and Zr ($MF_7^{3-}$ where M=Zr,Hf). In the meantime, Rf was investigated in additional chemical systems, with the obtained results representing the highest quality data for all of the studied systems [36].

## HIGHLIGHT 4: PHYSICAL PRESEPARATION - A NEW TECHNIQUE FOR TRANSACTINIDE CHEMISTS

A new technique that offers to allow significant advances in the field of the chemical investigations of the heaviest elements was recently introduced. In this technique, the nuclear reaction products recoiling from the target enter a physical recoil separator which isolates them from the intense primary beam as well as the majority of the undesired byproducts of the nuclear reaction. A window that is thin enough that the element of interest can penetrate it is placed at the exit of the separator. EVRs pass through this window and are then available for the chemists in a very clean environment compared to conventional chemistry experiments. This technique is called preseparation [37-40]. The pioneering work has been performed at LBNL using the BGS as a preseparator [37,38,41,42]. Studies of Rf with the automated liquid-liquid extraction system SISAK, which were unsuccessful in the past due to background problems when no preseparation was employed, were able to unambiguously detect this element once SISAK was coupled to the BGS [37,43,44].

The main advantages of the method are the following ones: (i) in conventional experiments, the intense primary heavy ion beam forms a plasma in the recoil chamber that is attached directly to the target. The absence of the beam after the preseparator guarantees plasma-free, relatively mild conditions and it is now possible to directly feed non-thermally stable molecules, like organic ones, into this so-called Recoil

Transfer Chamber [38,42]. Thus, studies of completely new compound classes of the transactinides, such as volatile metal complexes or organometallic compounds are now possible, as was demonstrated in studies of β-diketonato complexes of Zr and Hf, the lighter homologues of Rf [45]. (ii) the preseparator suppresses unwanted reaction products by several orders of magnitude. Among these are problem nuclides whose decay properties are similar as the ones of the interesting long-lived transactinide isotopes, such as $^{212}$Po. Hence, the unambiguous identification of single atoms of the transactinides is much easier to achieve or is now possible in cases where it was not before [37,43,44]. (iii) The chemist enjoys much more freedom in the choice of the chemical system to be investigated. In conventional studies, the first and foremost property of a selected chemical system had to be its high separation factor from unwanted byproducts. With the preseparator taking care of this, chemical systems can now be chosen based on completely different characteristics. As an example, extraction studies of group 4 elements (Zr, Hf) with different crown ethers are worth mentioning [46]. Zr and Hf, which exhibit notoriously similar chemical properties are known be extracted differently by some of these agents. It is therefore interesting to investigate Zr, Hf, and Rf in such a chemical system to see whether Rf behaves more similarly to Zr, or to Hf, or follows the trend established by its lighter two homologs. As crown ethers extract many divalent ions, and thus many species that are interfering with the unambiguous identification of single atoms of the superheavy elements, these studies crucially depend on preseparation.

At GSI, a new gas-filled separator called the TransActinide Separator and Chemistry Apparatus (TASCA) [47] has recently entered the commissioning phase. All components of TASCA were from the beginning on optimized with respect to the task of serving as a preseparator. Other labs are also starting up to introduce the technique, among them RIKEN with the GARIS separator [48].

## CONCLUSION

The mentioned examples from four different domains show the field of transactinide chemistry to be very lively. Tremendous advances were achieved in the last few years in many different areas. New elements become accessible to the chemists. Chemical techniques are among the most suited ones to obtain information on the atomic number of an investigated species. Cross section levels of similar magnitude as achieved in physical studies have been reached for selected chemical systems and chemical techniques were shown to nicely complement physical investigations of, e.g., asymmetric reactions. The development of fully automated systems and the introduction of new techniques guarantee the coming years to be full of exciting discoveries in the field of the heaviest elements.

## ACKNOWLEDGMENTS

I would like to thank Robert Eichler, Hiromitsu Haba, Matthias Schädel, Andreas Türler, and Alexander Yakushev for communicating recent results and Valeria Pershina for interesting discussions.

# REFERENCES

1. M. Schädel (Ed.), "The Chemistry of Superheavy Elements", Dordrecht: Kluwer Academic Publishers, 2003.
2. M. Schädel, *Angew. Chem. Int. Ed.* **45**, 368-401 (2006).
3. K.S. Pitzer, J. Chem. Phys. **63**, 1032-1033 (1975).
4. B. Eichler, Kernenergie **19**, 307-311 (1976).
5. V. Pershina et al., Chem. Phys. **311**, 139-150 (2005).
6. B. Eichler, PSI Bericht 03-01, Paul Scherrer Institut, Villigen, Switzerland, 2003, ISBN 1019-0643.
7. Yu. Ts. Oganessian et al., Eur. Phys. J. A **5**, 63-68 (1999).
8. Yu. Ts. Oganessian et al., Nature **400**, 242-245 (1999).
9. Yu. Ts. Oganessian et al., Phys. Rev. C **70**, 064609 (2004).
10. Yu. Ts. Oganessian et al., Eur. Phys. J. A **19**, 3-6 (2004).
11. S. Hofmann et al., J. Nucl. Radiochem. Sci. **7**, R25-R29 (2006).
12. W. Loveland et al., Phys. Rev. C **66**, 044617 (2002).
13. K. E. Gregorich et al., Phys. Rev. C **72**, 014605 (2005).
14. A. B. Yakushev et al., Radiochim. Acta **89**, 743-745 (2001).
15. A. B. Yakushev et al., Radiochim. Acta **91**, 433-439 (2003).
16. R. Eichler et al., Radiochim. Acta **94**, 181-191 (2006).
17. I. Zvara, Isotopenpraxis 26 (1990) 251-258.
18. H. W. Gäggeler et al., Nucl. Phys. **A734**, 208-212 (2004).
19. J.P. Unik et al., "Fragment mass and kinetic energy distributions for fissioning systems ranging from mass 230 to 256" in Proc. *Symp. Physics and Chemistry of Fission*, Rochester, NY, USA, August 13-17, 1973 (IAEA, Vienna, 1974) Vol. II, p. 19-45.
20. V.E. Viola, Nucl. Data Sheets A **1**, 391-410 (1965).
21. S. Soverna, Doctoral Thesis, University of Berne, Berne, Switzerland, 2004.
22. R. Eichler, priv. comm.
23. Ch. E. Düllmann et al., Nature **418**, 859-862 (2002).
24. Ch. E. Düllmann et al., Nucl. Instrum. Meth. **A479**, 631-639 (2002).
25. A. Türler et al., Eur. Phys. J. A **17**, 505-508 (2003).
26. J. Dvorak et al., Phys. Rev. Lett (submitted).
27. S. Hofmann et al., Eur. Phys. J. A **14**, 147-157 (2002).
28. K. Morita et al., RIKEN Accel. Prog. Rep. **38**, 69 (2005).
29. J. Dvorak et al., to be published.
30. H. Haba et al., J. Am. Chem. Soc. **126**, 5219-5224 (2004).
31. R. Silva et al., Inorg. Nucl. Chem. Lett. **6**, 871-877 (1970).
32. E. K. Hulet et al., J. Inorg. Nucl. Chem. **42**, 79-82 (1980).
33. D. Schumann et al., J. Alloys. Comp. **271-273**, 307-311 (1998).
34. E. Strub et al., Radiochim. Acta **88**, 265-271 (2000).
35. A. Toyoshima et al., J. Nucl. Radiochem. Sci. 5, 45-48 (2004).
36. M. Schädel, priv. comm.
37. J. P. Omtvedt et al., J. Nucl. Radiochem. Sci. **3**, 121-124 (2002).
38. Ch. E. Düllmann et al., Nucl. Instrum. Meth. **A551**, 528-539 (2005).
39. Ch. E. Düllmann, Czech. J. Phys. (in press).
40. Ch. E. Düllmann, Eur. Phys. J. D (submitted).
41. K.E. Gregorich (Ed), Proc. Workshop on the Physics Using Compound Nucleus Separators, E.O. Lawrence Berkeley Laboratory, Berkeley, CA, USA, April 10-12, 1997, LBNL-40483 (1997).
42. U.W. Kirbach et al., Nucl. Instrum. Meth. A **484**, 587-594 (2002).
43. L. Stavsetra et al., Nucl. Instrum. Meth. **A543**, 509-516 (2005).
44. J.P. Omtvedt et al., Eur. Phys. J. D (submitted).
45. Ch.E. Düllmann et al., to be published.
46. R. Sudowe et al., Radiochim. Acta **94**, 123-129 (2006).
47. M. Schädel et al., "The TASCA Project", GSI Scientific Report 2005, Gesellschaft für Schwerionenforschung mbH, Darmstadt, Germany, Report 2006-1, 2006, p. 262; see also http://www.gsi.de/TASCA.
48. H. Haba, priv. comm.

# Startup of Superheavy Element Chemistry at RIKEN

H. Haba,[1] D. Kaji,[1] H. Kikunaga,[1] N. Sato,[1] T. Akiyama,[1] K. Morimoto,[1] A. Yoneda,[1] K. Morita,[1] T. Takabe,[2] Y. Tashiro,[2] Y. Kitamoto,[2] K. Matsuo,[2] D. Saika,[2] K. Ooe,[2] T. Kuribayashi,[2] T. Yoshimura,[2] A. Shinohara,[2] and A. Toyoshima[3]

[1]*Nishina Center for Accelerator Based Science, RIKEN, Wako, Saitama 351-0198, Japan*
[2]*Graduate School of Science, Osaka University, Toyonaka, Osaka 560-0043, Japan*
[3]*Advanced Science Research Center, Japan Atomic Energy Agency, Tokai, Ibaraki 319-1195, Japan*

**Abstract.** Present status and perspectives of the superheavy element (SHE) chemistry at RIKEN are reviewed. A gas-jet transport system for the SHE chemistry has been installed in the focal plane of the gas-filled recoil ion separator GARIS at the RIKEN Linear Accelerator. The performance of the system was appraised using $^{206}$Fr and $^{245}$Fm produced in the $^{40}$Ar-induced reactions on $^{169}$Tm and $^{208}$Pb, respectively. The α particles of $^{206}$Fr and $^{245}$Fm separated with GARIS and transported by the gas-jet were clearly identified under desired low background condition with a rotating wheel system for α spectrometry. The high gas-jet efficiencies over 90% are independent of the beam intensity up to 2 particle μA. A gas-jet coupled SHE production system and a safety system for the usage of radioactive targets were also developed on the beam line of the RIKEN K70 AVF Cyclotron. The gas-jet transport of $^{255}$No and $^{261}$Rf produced in the $^{238}$U($^{22}$Ne,5$n$)$^{255}$No and $^{248}$Cm($^{18}$O,5$n$)$^{261}$Rf reactions, respectively, was conducted for the future Sg chemistry by $^{248}$Cm($^{22}$Ne,5$n$)$^{265}$Sg.

**Keywords:** Superheavy element chemistry, RIKEN gas-filled recoil ion separator (GARIS), Gas-jet transport system, $^{206}$Fr, $^{245}$Fm, $^{255}$No, $^{261}$Rf
**PACS:** 23.60.+e; 25.70.Gh; 25.70.Jj; 27.90.+b; 29.25.Rm

## 1. Introduction

The uppermost end of the Periodic Table has been architected with the discovery of new elements using nuclear fusion reactions at heavy-ion accelerators. With the advent of the velocity filter SHIP (Separator for Heavy-Ion Reaction Products), the discoveries of elements 107 to 112 were successful at GSI (Gesellschaft für Schwerionenforschung) using cold fusion reactions based on lead and bismuth targets. Thereafter, a new isotope of element 113 was synthesized at RIKEN in the $^{209}$Bi($^{70}$Zn,$n$)$^{278}$113 reaction using the gas-filled recoil ion separator GARIS. On the other hand, elements 112 to 116, and 118 have been produced in the $^{48}$Ca-induced hot fusion reactions on actinide targets at the Flerov Laboratory of Nuclear Reactions in Dubna. Thus, we presently know 117 elements as regularly arranged in the Periodic

Table. Recent comprehensive reviews on the syntheses of the heaviest elements are seen in Refs. [1, 2].

The elements with atomic numbers $Z \geq 104$ are called transactinide elements or recently "superheavy elements (SHEs)" [1, 3]. The chemical properties of the newly discovered and unknown elements, of course, attract a great deal of our interest. Furthermore, influences of the strong relativistic effect on valence electrons of the SHE atoms are often predicted to induce deviations in chemical properties from periodicity based on their lighter homologues in the Periodic Table [4, 5]. Thus, the investigation of the chemical properties of SHEs has become one of the most exciting and challenging research subjects in nuclear and radiochemistry [1, 3].

The production rates of the SHE nuclides, however, are extremely low, i.e. atoms per minute for elements 104 (Rf) and 105 (Db), down to atoms per hour or day for elements 106 (Sg) to 108 (Hs), and their half-lives are less than 1 min. This forces us to perform rapid and effective chemical experiments with "single atoms". For single-atom chemistry, the law of mass action is no longer valid, because the atom cannot exist in the different chemical forms taking part in the chemical equilibrium at the same time. Despite these difficulties, the experimental studies on the chemical properties of SHEs have been conducted for elements 104–108 and very recently element 112 (E112) [1, 3]. The main objectives of these studies are the placement of new elements in the Periodic Table. The results show that elements 104–108 are placed onto the expected groups 4–8 of the Periodic Table, respectively. The recent experiments of E112 as elemental state show notable indications that E112 behaves similarly to Rn not to the expected lighter homologue Hg. It is also of great importance to study detailed chemical properties of SHEs with high statistics and to compare them with properties deduced from extrapolations of periodicity and from relativistic molecular orbital calculations: the second-generation SHE chemistry. The series of the second-generation Rf experiments by an anion-exchange method in $HNO_3$, HCl, HF, and $HF/HNO_3$ was very successful at the Japan Atomic Energy Agency (JAEA) Tandem Facility [6].

Recently, we have started the SHE chemistry in RIKEN using the RIKEN Linear Accelerator (RILAC) and the K70 AVF Cyclotron. A gas-jet transport system coupled to GARIS as a preseparator was developed for the SHE chemistry. This system is expected to provide new methodologies for the SHE chemistry: identification of SHE nuclides under extremely low background condition, stable and high efficiencies of the gas-jet transport, and direct chemical reactions with a large variety of compounds. A gas-jet coupled SHE production system was also installed on the beam line of the RIKEN K70 AVF Cyclotron for the second-generation chemistry of heavier elements with $Z \geq 106$. In the following, present status and perspectives of the SHE chemistry at RIKEN are reviewed.

# 2. SHE Chemistry using GARIS as Preseparator

## 2.1. Development of Gas-jet Transport System Coupled to GARIS

The SHE atoms are produced at extremely low production rates among much larger amounts of background activities which hinder the detection of decays of the SHE nuclides of interest. Recently available high-intensity beams of more than 1 particle µA (pµA) also give rise to a serious problem in that the plasma formed by the beams significantly decreases gas-jet transport efficiencies. To overcome these situations, it has been proposed that a recoil separator for nuclear physics research on SHEs should be coupled to the chemistry system with the aid of the gas-jet transport technique [7]. With this method, background activities that cannot be effectively separated from SHEs in the usual way, such as those from Po or Rn isotopes, are largely removed. The high and stable gas-jet efficiencies are obtained without the plasma condition caused by the beam. Furthermore, chemical reactions of various compounds can be studied by directly feeding complexing reagents into the gas-jet chamber without aerosol materials. The first experiment with the recoil transfer chamber (RTC) coupled to the Berkeley Gas-filled Separator (BGS) was very successful [8, 9]. The isotope of $^{257}$Rf physically separated from the large background caused by β-particles was identified with a liquid scintillator after a liquid-liquid solvent extraction into 0.25 M dibutyl-phosphoric acid in toluene from 6 M $HNO_3$ with SISAK. Thereafter, the BGS/RTC system has been used in the model experiments of Rf [10–12] and Hs [13]. At GSI, the components of the former HElium Charge-exchange Kaleidoscope (HECK) separator are being used to set up a dedicated separator for chemistry experiments [7].

In RIKEN, the gas-filled recoil ion separator GARIS at RILAC is now in operation to search for the heaviest SHE nuclides. The isotopes of $^{271}$Ds, $^{272}$Rg, and $^{277}$112 found at GSI were confirmed with better statistics and with new spectroscopic information [14–16], and a new isotope of element 113, $^{278}$113, was successfully synthesized [17]. GARIS gives us extremely low background condition and high transport efficiencies for SHE nuclides. Recently, we have installed a gas-jet transport system in the focal plane of GARIS to start up the SHE chemistry in RIKEN. A schematic of the system is shown in Figure 1. The gas-jet chamber at an inner pressure of ~100 kPa is isolated from GARIS at ~100 Pa with a very thin Mylar-vacuum window supported with a stainless-steel honeycomb grid with 92.5% transparency and of 60 mm diameter. This grid can support the Mylar foils down to 2.4 µm thickness. As shown in Figure 1, the recoiling SHE nuclides separated with GARIS are passed through the Mylar window and are stopped in a volume of helium gas. The helium gas, often seeded with aerosol particles, is fed into the chamber through the four inlets (4 mm i.d.) and is swept out through a Teflon capillary (1.59 mm i.d.) to chemistry apparatuses. The volume of the chamber is variable for ranges of product nuclei of interest with spacer flanges: 70 mm i.d. × 30, 60, and 90 mm long. In the present work, the performance of the present system was appraised using $^{206}$Fr and $^{245}$Fm produced in the $^{40}$Ar-induced reactions on $^{169}$Tm and $^{208}$Pb, respectively.

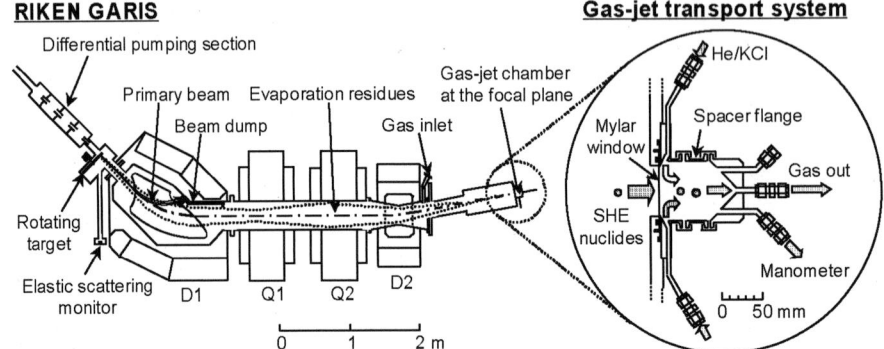

**FIGURE 1.** Schematic of the gas-jet transport system coupled to the RIKEN gas-filled recoil ion separator GARIS.

## 2.2. Model Experiments of the GARIS/Gas-jet System with $^{206}$Fr and $^{245}$Fm

The isotopes of $^{206}$Fr and $^{245}$Fm were produced in the $^{169}$Tm($^{40}$Ar,3n)$^{206}$Fr and $^{208}$Pb($^{40}$Ar,3n)$^{245}$Fm reactions, respectively, using RILAC. The metallic $^{169}$Tm and $^{208}$Pb targets of 120 and 420 μg cm$^{-2}$ thicknesses, respectively, were prepared by vacuum evaporation on a 30 μg cm$^{-2}$ carbon backing foil. Sixteen targets were mounted on a rotating wheel of 30 cm in diameter. The wheel was rotated during irradiation at 2000 or 3000 rpm. The beam energies were 170 MeV for $^{169}$Tm and 199 MeV for $^{208}$Pb at the middle of the target. The beam intensity was monitored by measuring elastically scattered projectiles with a Si PIN photodiode mounted at 45° with respect to the beam axis. The typical beam intensity was 2 pμA. The reaction products of interest were separated in-flight from the beam and the majority of the nuclear transfer products by GARIS, and were guided into the gas-jet chamber of 60 mm long through the Mylar vacuum window of 3.5 μm thickness. The separator was filled with helium gas at a pressure of 88 Pa. The magnetic rigidities of GARIS were set at 1.64 and 2.01 Tm for $^{206}$Fr and $^{245}$Fm, respectively.

In the gas-jet chamber, the reaction products separated with GARIS were stopped in helium gas, attached to KCl aerosols generated by sublimation of the KCl powder at 620 °C, and continuously transported through a Teflon capillary (1.59 mm i.d., 4 m long) to a rotating wheel system for α spectrometry, which was the compact one of the Measurement system for the Alpha-particle and spontaneous fissioN events ON-line (MANON) developed at JAEA [18] (see also Figure 5). The helium flow rate was 5 L min$^{-1}$, and the inner pressure of the chamber was 90 kPa. In MANON, the reaction products were deposited on the Mylar foils of 0.68 μm thickness and 20 mm diameter placed at the periphery of a 40-position stainless steel wheel of 420 mm diameter. After the aerosol collection, the wheel was stepped at 30- and 2-s intervals for $^{206}$Fr and $^{245}$Fm, respectively, to position the foils between seven pairs of Si PIN photodiodes (Hamamatsu S3204-09). Each detector had an active area of 18 × 18 mm$^2$ and a 38% counting efficiency for α particles. The α-particle energy resolution was 50 keV FWHM.

To evaluate the number of $^{206}$Fr and $^{245}$Fm atoms that passed through the Mylar window, the gas-jet chamber was replaced with a detector chamber equipped with a 12-strip Si detector of 60 × 60 mm$^2$ (Hamamatsu 12CH PSD). The α-particle energy resolution of PSD was 50 keV FWHM.

## 2.3. Performance of the GARIS/Gas-jet System

In Figure 2a, the α-particle spectrum measured with the 6th strip of PSD under beam-on condition is shown. The α peaks of $^{206}$Fr ($T_{1/2}$ = 15.9 s, $E_\alpha$ = 6.790 MeV [19]) and $^{205}$Fr (3.85 s, 6.915 MeV [19]) and of their daughter nuclides $^{202}$At (182 s, 6.135 MeV; 184 s, 6.228 MeV [19]) and $^{201}$At (89 s, 6.344 MeV [19]) are identified. On the other hand, the α-particle spectrum obtained with MANON after the gas-jet transport is shown in Figure 2b. The MANON spectrum was measured for 30 s with the first top detector of MANON after the 30-s aerosol collection. The broad component above 7 MeV in the PSD spectrum, which corresponds to the implantation of the evaporation residues (ERs), disappears after the gas-jet transport to MANON as expected (Figure 2b). The transport time of $^{206}$Fr was determined to 0.4 s as a difference in the rising time between the counts of the elastic scattering beam monitor and the α counts of $^{206}$Fr in the bottom detector placed at the aerosol collection site of MANON.

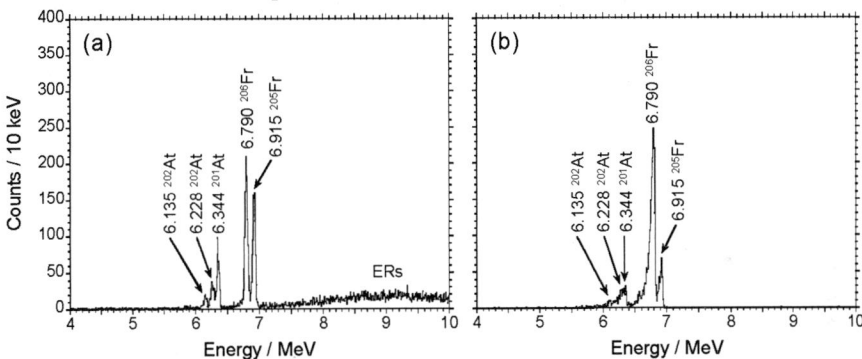

**FIGURE 2.** Alpha-particle spectra measured in the $^{169}$Tm($^{40}$Ar,3n)$^{206}$Fr experiment with (a) PSD and (b) MANON.

In Figure 3, the gas-jet efficiencies of $^{206}$Fr are shown by closed circles as a function of the $^{40}$Ar beam intensity. The high gas-jet efficiencies over 90% are obtained. Compared with open squares in Figure 3 are the data of $^{173}$W produced in the $^{nat}$Gd($^{22}$Ne,xn) reaction at the RIKEN AVF Cyclotron and transported by the gas-jet without the GARIS separation. Due to the plasma condition induced by the beam in the chamber, the gas-jet efficiencies of $^{173}$W decrease from 40% at 6.6 pnA to 25% at 0.5 pμA with an increase of the $^{22}$Ne beam intensity. Since the primary beam is separated with GARIS, such a decrease is not seen for $^{206}$Fr up to 2 pμA studied in the present work.

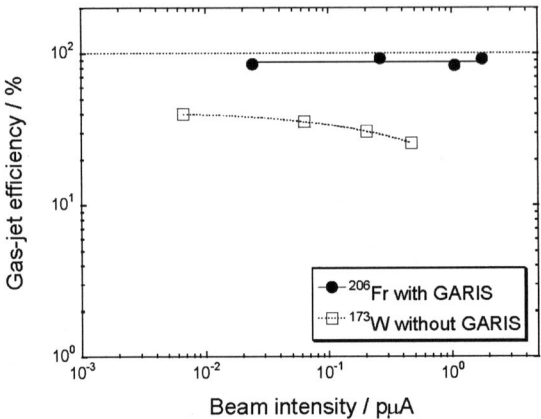

**FIGURE 3.** Variation of the gas-jet efficiency of $^{206}$Fr (closed circles) and $^{173}$W (open squares) as a function of the beam intensity.

The α particle spectra of $^{245}$Fm measured with PSD and MANON under beam-on condition are shown in Figures 4a and 4b, respectively. The beam doses of $6.55\times10^{16}$ and $9.76\times10^{16}$ were accumulated in the PSD and MANON experiments, respectively. Although large amounts of background events are measured in PSD (Figure 4a), the 8.15 MeV peak of $^{245}$Fm is clearly identified in MANON (Figure 4b). The background activities such as Po isotopes, which are produced in the transfer reactions on $^{208}$Pb, are completely removed by the GARIS/gas-jet system. The gas-jet efficiency of $^{245}$Fm is determined to be 83±9%. The transport efficiency of GARIS is 43±4% by assuming the cross section for the $^{208}$Pb($^{40}$Ar,3$n$)$^{245}$Fm reaction to be 15 nb [20].

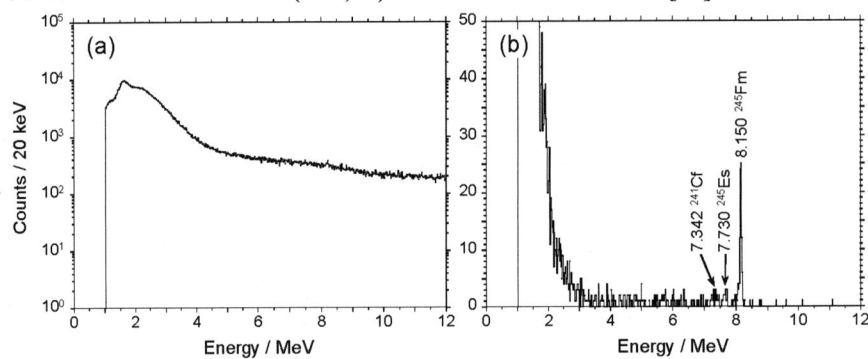

**FIGURE 4.** Alpha-particle spectra of $^{245}$Fm measured with (a) PSD and (b) MANON.

Despite of the GARIS/gas-jet transport, one can see some background events in Figure 4b, especially bellow 4 MeV. Since MANON was placed in the target room, those background events were mainly due to neutrons and/or γ rays during the irradiation. We are now constructing a chemistry laboratory isolated with a 50-cm concrete shield from the target room, where the background level is expected to be two orders of magnitude lower than that in the target room.

## 3. SHE Chemistry using the RIKEN K70 AVF Cyclotron

### 3.1. Development of the Gas-jet coupled SHE Production System

The second-generation aqueous chemistry of Rf by the ion-exchange method systematically in $HNO_3$, HCl, HF, and $HF/HNO_3$ was very successful at JAEA [6]. We plan to extend the series of the experiments to heavier elements than Rf. The ion-exchange experiments with 34-s $^{262}$Db produced in the $^{248}$Cm($^{19}$F,5n) reaction are scheduled at JAEA by developing a new ion-exchange system: Automated Ion-exchange separation apparatus coupled with the Detection system for Alpha spectroscopy II (AIDA II). For heavier elements such as Sg and Hs, we plan to use the RIKEN K70 AVF Cyclotron. Recently, we installed a gas-jet coupled target system and a safety system for the usage of radioactive targets on the beam line of the AVF Cyclotron. As shown in Figure 5, the target chamber was designed for the future chemical experiments of Sg by considering recoil ranges of $^{265}$Sg and its homologue $^{173}$W produced in the $^{22}$Ne-induced reactions on the mixed target of $^{248}$Cm and $^{nat}$Gd, respectively. In the present work, the performance of the system was investigated using the $^{238}$U($^{22}$Ne,5n)$^{255}$No and $^{248}$Cm($^{18}$O,5n)$^{261}$Rf reactions.

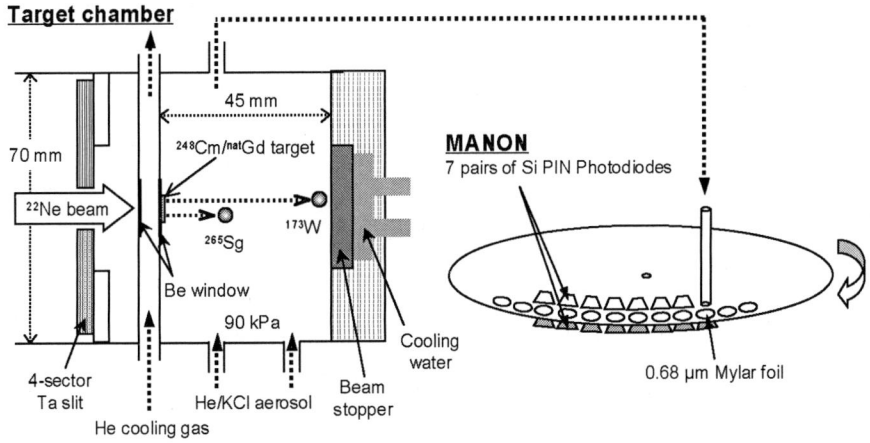

**FIGURE 5.** Schematic of the gas-jet coupled SHE production system and of the rotating wheel system MANON for α spectrometry.

## 3.2. Production and Gas-jet Transport of $^{255}$No and $^{261}$Rf

A $^{238}$U target of 630 μg cm$^{-2}$ thickness and a $^{248}$Cm target of 200 μg cm$^{-2}$ thickness were prepared by electrodeposition onto a beryllium backing foil of 2.0 mg cm$^{-2}$ thickness. The beams from the RIKEN K70 AVF Cyclotron passed through a beryllium vacuum window (2.0 mg cm$^{-2}$), the helium cooling gas (0.09 mg cm$^{-2}$), the beryllium target backing, and finally entered the target material. The beam energies on target were 105.9, 107.3, 109.0, 113.4, 116.6, and 120.9 MeV for $^{22}$Ne and 94.4 MeV for $^{18}$O. The beam intensity was approximately 350 pnA. Reaction products recoiling out of the target were stopped in helium gas (~130 kPa), attached to KCl aerosols generated by sublimation of KCl at 640 °C, and transported through a Teflon capillary (2.0 mm i.d., 45 m long) to MANON in the chemistry laboratory.

The α particles of $^{255}$No ($T_{1/2}$ = 3.1 min, $E_α$ = 7.620–8.312 MeV) [19] were identified in the α spectrum, and its radioactivity was evaluated by the two-component decay curve analysis of $^{255}$No and $^{254}$No (55 s, 8.093 MeV) [19]. The gas-jet efficiency was determined to be 50% from radioactivities of the daughter nuclide $^{255}$Fm by the catcher foil method with the aid of chemical separation. It was found that the maximum cross section of the $^{238}$U($^{22}$Ne,5n)$^{255}$No reaction is 90 nb at 113 MeV, though the excitation function measured by Donets et al. [22] shows the peak of 200 nb at 118 MeV. It is noted here that the measurements by Donets et al. [22] were made based on the α decay of the daughter nuclide $^{251}$Fm only in ~1%. The peak cross section corrected for the latest α-decay branch of 1.80% [19], about 100 nb, is good agreement with the present result. On the other hand, the beam dose of $1.36 \times 10^{17}$ was accumulated in the $^{261}$Rf experiment. A total of 61 time-correlated α pairs of $^{261}$Rf (68 s, 8.28 MeV) [19] and its daughter $^{257}$No (25 s, 8.222 and 8.323 MeV) [21] were registered in the α-energy range of interest (8.10–8.40 MeV). The cross section was evaluated to be 5 nb at 94 MeV, by assuming the gas-jet transport efficiency of $^{261}$Rf is the same as that of $^{255}$No. This cross section is comparable with the reported value of 5 nb at 97 MeV [23], but is smaller by factor of 2 than those measured at JAEA [24]. This discrepancy may result from the gas-jet transport efficiency of $^{261}$Rf assumed in each experiment.

## 4. Summary and Perspectives

We have started the SHE chemistry at RIKEN using GARIS at RILAC and the K70 AVF Cyclotron. The gas-jet transport system coupled to GARIS as a preseparator was installed for the SHE chemistry. The performance of the system was investigated using $^{206}$Fr and $^{245}$Fm produced in the $^{169}$Tm($^{40}$Ar,3n)$^{206}$Fr and $^{208}$Pb($^{40}$Ar,3n)$^{245}$Fm reactions, respectively. The α particles of $^{206}$Fr and $^{245}$Fm separated with GARIS and transported by the gas-jet were clearly identified with a rotating wheel system for α spectrometry under desired low background condition. The high gas-jet efficiencies over 90% are independent of the beam intensity up to 2 pμA as expected. These results suggest that the GARIS/gas-jet system is promising tool for the future SHE chemistry. Very recently, the gas-jet transport of $^{255}$No produced in the $^{238}$U($^{22}$Ne,5n) reaction was also successful with the gas-jet efficiency of 80%. In the future, productions of SHE nuclides with long half-lives for chemical experiments such as

$^{261}$Rf, $^{262}$Db, $^{265}$Sg, $^{269}$Hs, and $^{283}$112 will be investigated with the GARIS/gas-jet system based on the $^{238}$U and $^{248}$Cm targets. A gas chromatograph column directly coupled to GARIS, which enables isothermal-chromatographic analyses of a large variety of compounds by direct injection of chemical reagents, is under development.

The gas-jet coupled SHE production system was installed on the beam line of the AVF Cyclotron. The gas-jet transport of $^{255}$No and $^{261}$Rf produced in the $^{238}$U($^{22}$Ne,5n)$^{255}$No and $^{248}$Cm($^{18}$O,5n)$^{261}$Rf reactions, respectively, was conducted, and their cross sections were evaluated. The experiments to measure the excitation function of $^{248}$Cm($^{22}$Ne,5n)$^{265}$Sg are scheduled for the future chemical experiments of Sg. New chemistry devices such as a microchip for solvent extraction and a flow electrolytic column for electrochemistry are under development for the second-generation SHE chemistry in collaboration with Osaka University and JAEA.

## Acknowledgments

The authors express their gratitude to the crew of the RIKEN Linear Accelerator and the RIKEN K70 AVF Cyclotron for their invaluable assistance in the course of these experiments. We also thank Prof. T. Mitsugashira of Tohoku University and Prof. H. Kudo of Niigata University for preparing the $^{248}$Cm material from an old $^{252}$Cf neutron source. These researches were partially supported by the Ministry of Education, Science, Sports and Culture, Grant-in-Aid for Young Scientists (B), 16750055, 2004–2006, and by the REIMEI Research Resources of Japan Atomic Energy Research Institute, 2003.

## References

1. M. Schädel (Ed.), *The Chemistry of Superheavy Elements*, Dordrecht: Kluwer Academic Publishers, 2003.
2. S. Hofmann, G. Münzenberg, and M. Schädel, *Nucl. Phys. News* **14**, 5 (2004).
3. M. Schädel, *Angew. Chem. Int. Ed.* **45**, 368 (2006).
4. P. Pyykkö, *Chem. Rev.* **88**, 563 (1988).
5. V. G. Pershina, *Chem. Rev.* **96**, 1977 (1996).
6. Y. Nagame, K. Tsukada, M. Asai, A. Toyoshima, Y. Ishii, T. Kaneko-Sato, M. Hirata, I. Nishinaka, S. Ichikawa, H. Haba, S. Enomoto, K. Matsuo, D. Saika, Y. Kitamoto, H. Hasegawa, Y. Tani, W. Sato, A. Shinohara, M. Ito, J. Saito, S. Goto, H. Kudo, H. Kikunaga, N. Kinoshita, A. Yokoyama, K. Sueki, Y. Oura, H. Nakahara, M. Sakama, M. Schädel, W. Brüchle, and J. V. Kratz, *Radiochim. Acta* **93**, 519 (2005).
7. 5th Workshop on Recoil Separator for SHE Chemistry, September 29, 2006, Garching, Germany (http://www-w2k.gsi.de/tasca06/).
8. J. P. Omtvedt, J. Alstad, H. Breivik, J. E. Dyve, K. Eberhardt, C. M. Folden III, T. Ginter, K. E. Gregorich, E. A. Hult, M. Johansson, U. W. Kirbach, D. M. Lee, M. Mendel, A. Nähler, V. Ninov, L. A. Omtvedt, J. B. Partin, G. Skarnemark, L. Stavsetra, R. Sudowe, N. Wiehl, B. Wierczinski, P. A. Wilk, P. M. Zielinski, J. V. Kratz, N. Trautmann, H. Nitsche, and D. C. Hoffman, *J. Nucl. Radiochem. Sci.* **3**, 121 (2002).
9. L. Stavsetra, K. E. Gregorich, J. Alstad, H. Breivik, K. Eberhardt, C. M. Folden III, T. N. Ginter, M. Johansson, U. W. Kirbach, D. M. Lee, M. Mendel, L. A. Omtvedt, J. B. Patin, G. Skarnemark, R. Sudowe, P. A. Wilk, P. M. Zielinski, H. Nitsche, D. C. Hoffman, and J. P. Omtvedt, *Nucl. Instr. and Meth. A* **543**, 509 (2005).
10. Ch. E. Düllmann, G. K. Pang, C. M. Folden III, K. E. Gregorich, D. C. Hoffman, H. Nitsche, R. Sudowe, and P. M. Zielinski, "Toward Volatile Metal Complexes of Rutherfordium - Results of Test Experiments with Zr and Hf", in *Advances in Nuclear and Radiochemistry, General and Interdisciplinary* edited by S. M. Qaim and H. H. Coenen, Jülich: Forschungszentrum Jülich GmbH, 2004, Vol. 3, pp. 147-149.
11. Ch. E. Düllmann, C. M. Folden III, K. E. Gregorich, D. C. Hoffman, D. Leitner, G. K. Pang, R. Sudowe, P. M. Zielinski, and H. Nitsche, *Nucl. Instr. and Meth. A* **551**, 528 (2005).

12. R. Sudowe, M. G. Galvert, Ch. E. Düllmann, L. M. Farina, C. M. Folden III, K. E. Gregorich, S. E. H. Gallaher, D. C. Hoffman, S. L. Nelson, D. C. Phillips, J. M. Schwantes, R. E. Wilson, P. M. Zielinski, and H. Nitsche, *Radiochim. Acta* **94**, 123 (2006).
13. U. W. Kirbach, C. M. Folden III, T. N. Ginter, K. E. Gregorich, D. M. Lee, V. Ninov, J. P. Omtvedt, J. B. Patin, N. K. Seward, D. A. Strellis, R. Sudowe, A. Türler, P. A. Wilk, P. M. Zielinski, D. C. Hoffman, and H. Nitsche, *Nucl. Instr. and Meth. A* **484**, 587 (2002).
14. K. Morita, K. Morimoto, D. Kaji, H. Haba, E. Ideguchi, J. C. Peter, R. Kanungo, K. Katori, H. Koura, H. Kudo, T. Ohnishi, A. Ozawa, T. Suda, K. Sueki, I. Tanihata, H. Xu, A. V. Yeremin, A. Yoneda, A. Yoshida, Y.-L. Zhao, T. Zheng, S. Goto, and F. Tokanai, *J. Phys. Soc. Jpn.* **73**, 1738 (2004).
15. K. Morita, K. Morimoto, D. Kaji, H. Haba, E. Ideguchi, R. Kanungo, K. Katori, H. Koura, H. Kudo, T. Ohnishi, A. Ozawa, T. Suda, K. Sueki, I. Tanihata, H. Xu, A. V. Yeremin, A. Yoneda, A. Yoshida, Y.-L. Zhao, and T. Zhen, *Eur. Phys. J. A* **21**, 257 (2004).
16. K. Morita, K. Morimoto, D. Kaji, T. Akiyama, S. Goto, H. Haba, E. Ideguchi, H. Koura, H. Kudo, T. Ohnishi, A. Ozawa, T. Suda, K. Sueki, H. Xu, T. Yamaguchi, A. Yoneda, A. Yoshida, and Y.-L. Zhao, *RIKEN Accel. Prog. Rep.* **38**, 69 (2005).
17. K. Morita, K. Morimoto, D. Kaji, T. Akiyama, S. Goto, H. Haba, E. Ideguchi, R. Kanungo, K. Katori, H. Koura, H. Kudo, T. Ohnishi, A. Ozawa, T. Suda, K. Sueki, H. Xu, T. Yamaguchi, A. Yoneda, A. Yoshida, and Y.-L. Zhao, *J. Phys. Soc. Jpn.* **73**, 2593 (2004).
18. Y. Nagame, M. Asai, H. Haba, S. Goto, K. Tsukada, I. Nishinaka, K. Nishio, S. Ichikawa, A. Toyoshima, K. Akiyama, H. Nakahara, M. Sakama, M. Schädel, J. V. Kratz, H. W. Gäggeler, and A. Türler, *J. Nucl. Radiochem. Sci.* **3**, 85 (2002).
19. R. B. Firestone and V. S. Shirley, *Table of Isotopes, 8th ed.*, New York: John Wiley and Sons, 1996.
20. J. M. Nitschke, R. E. Leber, M. J. Nurmia, and A. Ghiorso, *Nucl. Phys.* **A313**, 236 (1979).
21. M. Asai, K. Tsukada, M. Sakama, S. Ichikawa, T. Ishii, Y. Nagame, I. Nishinaka, K. Akiyama, A. Osa, Y. Oura, K. Sueki, and M. Shibata, *Phys. Rev. Lett.* **95**, 102502 (2005).
22. E. D. Donets, V. A. Shchegolev, and V. A. Ermakov, *Sov. J. Nucl. Phys.* **2**, 723 (1966).
23. A. Ghiorso, M. Nurmia, K. Eskola, and P. Eskora, *Phys. Lett.* **32B**, 95 (1970).
24. Y. Nagame, M. Asai, H. Haba, S. Goto, K. Tsukada, I. Nishinaka, K. Nishio, S. Ichikawa, M. Sakama, A. Toyoshima, K. Akiyama, H. Nakahara, M. Schädel, H. W. Gäggeler, A. Türler, *J. Nucl. Radiochem. Sci.* **3**, 85 (2002).

# A review on SHE research at GANIL

Ch. Stodel[1], N. Amar[3], R. Anne[1], G. Auger[1,10], B. Bouriquet[1], J.-M. Casandjian[2], A. Chatillon[2,7], R. Cee[1], E. Clément[2], R. Dayras[2], O. Dorvaux[6], A. Drouart[2], G. de France[1], F. de Oliveira Santos[1], R. de Tourreil[1], S. Grévy[1], F. Hannachi[11], F. Hannappe[9], K. Hauschild[5], F.P. Hessberger[7], S. Hofmann[7], A. Korichi[5], R. Lichtenhäler[8], K. Lojek[4], A. Lopez-Martens[5], A. Péghaire[1], J. Péter[3], M.-G. Saint-Laurent[1], Z. Sosin[4], L. Stuttge[6], Ch. Theisen[2], A.C.C. Villari[1], J.-P. Wieleczko[1], A. Wieloch[4].

[1] *GANIL (CEA/DSM – CNRS/IN2P3), Bvd H. Becquerel, B.P. 55027, F-14076 CAEN cedex 5, France*
[2] *CEA Saclay, DSM/DAPNIA/SPhN, F-91191 Gif/Yvette cedex, France*
[3] *LPC, 6 Bd Henri Becquerel, B.P. 55027, F-14074 Caen cedex 5, France*
[4] *M. Smoluchowski Institute of Physics, Jagellonian University, Reymonta 4, Krakow 30-059, Poland*
[5] *CSNSM, Bât. 104-108, F-91405 Orsay, France*
[6] *IPHC, 23, rue du Loess B.P. 28, F-67037 Strasbourg cedex 02, France*
[7] *GSI, Plankstrasse 1, D-64291 Darmstadt, Germany*
[8] *DFN-IFUSP, Rua Do Matao, Travessa R 187, 05509-090 Cidade Universitaria, São Paulo, SP Brasil*
[9] *PNTPM, Université Libre de Bruxelles, C.P. 229, B-1050 Bruxelles, Belgium*
[10] *Collège de France, 11 place M. Berthelot, F-75231 Paris cedex 5, France*

**Abstract.** This report summarizes the experiments relative to Super-Heavy Element studies done at GANIL - CEA – CNRS since 1999. It also gives an overview of future experiments and opportunities offered by SPIRAL 2 and LINAG beams in a medium term..

**Keywords:** Fusion, Superheavy elements synthesis spectroscopy
**PACS:** 23.60.+e, 25.60.Pj, 27.90.+b

## EXPERIMENTS ON SHE AT GANIL IN 1999-2003…2006

Since 1999, several experiments studying Super-Heavy Elements (SHE) took place at GANIL either on LISE or VAMOS spectrometers or with the $4\pi$ multi-detector INDRA. These experiments covered the field of nuclear structure of SHE and reaction mechanism studies.

Concerning the synthesis of SHE, GANIL tools enabled to privilege the use of cold fusion reaction in direct and inverse kinematics using the LISE spectrometer. These experiments led to the observation of isotope of seaborgium (Z=106) and hassium (Z=108) and to the search of element 114.

Concerning the structure of SHE, in order to prove the localization of the next island of stability, the direct way is to observe properly the elements Z=114 to Z=126 and to characterize them with their decay times and other properties. But these experiments are time consuming, an other way is to get spectroscopic information for lighter elements. Such studies were performed with LISE spectrometer in order to establish the occupancy of the nuclear orbitals of $^{251}$Md. A program to study nobelium isotopes with VAMOS is also scheduled for the near future.

Concerning the reaction mechanism, which is an important issue in synthesis of SHE, it has to be considered that the latter is hindered either by quasi-fission, in the first step of the reaction, or by fission after fusion. Therefore, one way to well understand the reaction mechanism is also to study these scenarios. Such studies are mainly carried out at Dubna analyzing the fission fragments in coincidence with neutrons. To complete these studies for quantifying the fission processes, some evidence were also given by fission time measurements at GANIL with the INDRA detector.

## Production And Spectroscopy With LISE

### Small Cross-Section Measurements in Cold Fusion Reactions

Thanks to the intense primary beams of the GANIL cyclotron together with the Wien filter LISE 3 and the development of a target chamber with large rotating targets, various experiments synthesizing SHE or searching new ones were performed from 1999 to 2003. A complete set of these experiments and the experimental set-up used for such studies are described in reference [1,2] and summarized in TABLE 1.

**TABLE 1.** Summary of the experiments on synthesis of SHE

| Reaction | Energy (MeV/u) | <I> (pµA) | Events | Rejection | σ (pb) |
|---|---|---|---|---|---|
| $^{208}$Pb ($^{86}$Kr, 1n) $^{293}$118 | | 2 | 0 | $10^8$-$10^{10}$ | < 1 |
| $^{208}$Pb ($^{54}$Cr, 1-2n) $^{261,260}$Sg | 4.70, 4.76 | 0.04 | 2,8+2 | ≈ 1.7.$10^{10}$ | 590-2370 |
| $^{18}$O ($^{208}$Pb, xn) $^{226-x}$Th | 5.00 | | | $10^6$-$10^8$ | |
| $^{208}$Pb ($^{58}$Fe, 1n) $^{265}$Hs | 4.83, 4.87, 4.92 | 0.5 | 2,3,1 | > 2.$10^{10}$ | 50-70 |
| $^{208}$Pb ($^{76}$Ge, 1n) $^{283}$114 | 5.02 | 0.8 | O | >$10^{11}$ | < 1.2 |

### Spectroscopy of Transfermium Nuclei

The odd-Z isotopes $^{255}$Lr, its daughter $^{251}$Md and grand-daughter $^{247}$Es were studied in two experiments performed at GANIL and the University of Jyväskylä. The $^{255}$Lr nuclei were produced using the cold fusion reaction $^{209}$Bi ($^{48}$Ca, 2n) $^{255}$Lr at a bombarding energy of 217 MeV. The single particle structure and decay-properties were investigated using α, γ and electron spectroscopy.

Revised half-lives, α–decay branching ratio and $Q_{\alpha(gs\text{-}gs)}$ values have been deduced. The [521]1/2⁻ Nilsson orbital down-sloping from the 2f$_{5/2}$ spherical shell has been assigned to the ground state, and low-lying state of $^{255}$Lr and $^{251}$Md, respectively. The ground state of $^{251}$Md and a new isomeric state in $^{255}$Lr correspond to the [714]7/2⁻

orbital. Experimental results have been compared to new HFB calculations using the SLy4 Skyrme interaction, giving insight into the position of the $2f_{5/2}$ spherical shell.

Detailed results of these experiments can be found in references [3].

## Experiments With INDRA and VAMOS

With INDRA detector, two main experiments using $^{238}$U beams at 6.6 MeV/A were performed with Ni and Ge crystals. The aim of these experiments was to prove the formation of the compound nuclei Z=120 and 124, and Z=114 with $^{208}$Pb beams on Ge crystals. Very long fission times were measured by the blocking techniques in single crystals. A complete description of these experiments and results are given in references [4,5,6].

On the VAMOS spectrometer, an experiment was performed to search for a signature of a long living component in the collisions of $^{238}$U + $^{238}$U. Reactions with kinematics similar to fusion-fission events were observed (see A.C.C. Villari et al., this conference).

Tests to perform prompt γ-spectroscopy using the Recoil Tagging technique with EXOGAM and VAMOS were successfully achieved. The application to transfermium isotope: $^{255}$No is envisaged in the incoming scheduled GANIL experiments in order to observe the rotational bands built on low-lying states, then spins and parities could be firmly established. It will give important insight into the single-particle structure above the deformed shell closure at N=152.

## EXPERIMENTS ON SHE IN THE NEAR FUTURE AND THE USE OF SPIRAL 2 BEAMS

With SPIRAL 2, radioactive beams of I, Sb, Sn, Cd, Sr, Xe and Kr are expected to reach intensities of $10^8$ to $10^{10}$ pps. With the adequate targets, new isotopes of SHE could be produced for No (Z=102) to Hs (Z=108) with significant counting rate. The range of isotopes between these nuclei could be produced either from cold or hot fusion. Nevertheless, these beams imply to produce SHE via symmetric reactions and the effective production cross-sections for these reactions are still unknown in the SHE region [7].

In the near future, a program for studying such reactions is planned at GANIL with the present tools where a $^{136}$Xe beam will be delivered on targets of $^{96}$Zr and $^{122,124}$Sn to observe Pu and Rf isotopes. These isotopes were already produced via cold fusion reactions with $^{208}$Pb targets; the obtained cross sections and fusion excitation functions could then be compared to the previous ones and it will give some evidence on the role of mass asymmetry parameter and isospin (neutron numbers) in the reaction/synthesis mechanism.

# SYNTHESIS OF SHE WITH S³

Presently, GANIL cyclotrons deliver beams from carbon to uranium with intensities varying from 10 pµA down to 10 pnA. For the study of SHE, beams from $^{48}$Ca to $^{86}$Kr are used with intensities varying from 500 pnA to 2 pµA. With the linear accelerator LINAG and the ion source A-Phoenix (under study and development) with A/q=6 characteristic, it is expected to increase the beam intensity by a factor of 30 to 100 depending on the accelerated elements. So the opportunities to synthesize or to lead spectroscopy studies on SHE is enlarged, i.e. the sensitivity could reach the femtobarn level in a reasonable beam time (few weeks) thanks to the development of an adequate spectrometer, such as S³ (Super-Separator-Spectrometer) which is under study. The reactions envisaged are those already done the last decade at GSI, RIKEN or Dubna with the more neutron rich beams and targets. But also it would be possible to perform experiments with different isotopes, such as calcium isotopes from A= 40 to 48 to produce less neutron rich SHE (Z=112-116), which would lead reaching known isotopes already done by cold fusion. Another interesting reaction is $^{36..40}$Ar + $^{238}$U to produce isotopes of element Ds (Z=110) from A= 264 to 268 as compound nuclei, and evaporation residue A<263 for xn reaction channels. This study is interesting because it can produce new Ds isotopes allowing to study their gross properties, such as radioactive decay channels, half-lives, $Q_\alpha$ etc. It will also enable to link isotope $^{273}$Ds daughter of $^{277}$112 produced by cold fusion ($^{70}$Zn + $^{208}$Pb) to $^{279}$Ds daughter of a SHE (Z=112...116) produced by hot fusion with $^{48}$Ca beam [7]. A summary of the feasible reactions is given in TABLE 2.

**TABLE 2.** reactions with expected sensitivity for a 75 % of efficiency (transmission + detection) and 7 days of beam.

| Beam | I / A-PHOENIX pµA | Target | Compound nuclei | Sensitivity/$\sigma_{known}$ Fb |
|---|---|---|---|---|
| $^{36,40}$Ar | 80-170 | $^{238}$U | $^{274,278}$Ds* | 4 / <1600 |
| $^{86}$Kr | 70 | $^{208}$Pb | $^{294}$118* | 9 / <1000 |
| $^{70}$Zn15+ | 7 | $^{208}$Pb | $^{278}$112* | 87/ 0.5 [8] |
| $^{48}$Ca8+ | 175 | $^{238}$U, $^{244}$Pu, $^{248}$Cm | $^{286}$112*, $^{292}$114*, $^{296}$116* | 4 / 5 [9] 4/ [10] 4/ 0.7-3.7 [11] |
| $^{40}$Ca9+ | 35 | $^{238}$U, $^{244}$Pu, $^{248}$Cm | $^{278}$112*, $^{284}$114*, $^{288}$116* | 20 |

# CONCLUSIONS

In conclusion, during the last decade, various studies on SHE either on reaction mechanism, synthesis and structure were performed with GANIL facility in a large collaboration.

The experimental set-up reached a sensitivity of 0.6 pb and a beam rejection greater than $10^{10}$ with intense primary beams of the order of pµA was achieved.

LINAG and the development of the $S^3$ spectrometer offers the opportunity to reach the sub picobarn level of sensitivity, thus larger quantities of SHE isotopes will be reachable.

SPIRAL 2 beams will also enable to produce some new isotopes via symmetric reactions. A program is on the way to begin this study in order to understand the role of mass asymmetry degree of freedom in the entrance channel.

## REFERENCES

1. Ch. Stodel et al, *Proceeding World Scientifc, EXON 2004*, pp. 180-187.
2. A. Wieloch et al, *Nuc. Inst. And Meth.* **A517**, 364-371 (2004).
3. A. Chatillon et al, *Europ. Phys. Journal* **A30**, 397 (2006).
4. A. Drouart et al, *Proceeding World Scientifc, EXON 2004*, pp. 192-197.
5. D. Jacquet et al, *AIP Conf. Proc., FUSION 2006*, pp. 239.
6. M. Morjean et al, *Europ. Phys. Journal* **A**, in preparation (2006).
7. P.T. Greenlees et al, *LOI SPIRAL 2 2006*
8. S. Hofmann et al, *Europ. Phys. Journal* **A**, (2006).
9. Yu. Ts. Oganessian et al, *Eur. Phys. J.* **A5**, 63-68 (1999).
10. Yu. Ts. Oganessian et al, *Physical Review Journal* **C70**, 064609 (2004).
11. Yu. Ts. Oganessian et al, *Physical Review Journal* **C74**, 044602 (2006).

# Search for a long lived component in the reaction U+U near the Coulomb barrier

A.C.C. Villari*, C. Golabek*, W. Mittig*, S. Heinz[†], S. Bhattacharyya**,
D. Boilley*, R. Dayras[‡], G. De France*, A. Drouart[‡], L. Gaudefroy*,
L. Giot[§], A. Marchix*, V. Maslov[¶], M. Morjean*, G. Mukherjee**,
A. Navin*, Yu. Penionzkevich[¶], F. Rejmund*, M. Rejmund*,
P. Roussel-Chomaz*, C. Stodel* and M. Winkler[†]

*GANIL (IN2P3/CNRS - DSM/CEA) B.P. 55027 14076 Caen Cedex 5 France
[†]GSI Planckstr. 1 64291 Darmstadt Germany
**VEC Center I/AF Bidhan Nagar Kolkata 700064 India
[‡]SPhN - CEA/DSM/DAPNIA Gif sur Yvette Cedex France
[§]SUBATECH, 4 rue Alfred Kastler B.P. 20722 44307 Nantes CEDEX 3 France
[¶]FLNR - JINR 141980 Dubna Russian Federation

**Abstract.** We performed an experiment to search for a signature of a long living component in the collision of $^{238}U + ^{238}U$ between 6.09 and 7.35A MeV. The experiment was performed at GANIL using the spectrometer VAMOS, tuned for observing reactions with kinematics similar to fusion-fission events. Theoretical calculations indicate that if a long living component would exist for this reaction, the most probable fission channel of such a giant system would be via the emission of quasi-lead nuclei. We detected events of such a category in the focal plane of VAMOS. These events present an excitation function growing as a function of the bombarding energy.

**Keywords:** fission, giant system
**PACS:** 25.85.-w, 25.90.+k

## INTRODUCTION

The study of very heavy collision systems started in the late 70s when beams in the U region with energies above the Coulomb barrier became available at the linear accelerator of GSI. Soon, research on very heavy systems split mainly into two principal directions: the search for superheavy elements and the physics of superstrong electromagnetic fields. In both cases a di-nuclear system consisting of the two colliding nuclei is transiently created during the collision. In these earlier experiments, the emphasis has been put on the investigation of the decay channels of the di-nuclear system (production of superheavies) or on particle creation in the strong electromagnetic fields. Detailed studies of the properties of the di-nuclear system have not been performed.

For superheavy di-nuclear systems like U+U the fission barrier is no longer existing due to the strong Coulomb repulsion. Due to the absence of the fission barrier one would expect that the lifetime of the di-nuclear system is very short (about $10^{-22}$ s). However, the neutrons and protons are embedded in the nuclear potential, and hence they are moving in a potential with a strong barrier. Even protons, which are unbound by about 5 MeV in such a system, when evaluated by a liquid drop model, feel a much higher coulomb barrier of about 35 MeV. Therefore, the individual nucleons will move in this potential pocket, and will flow from one U core to the other, thus forming the giant composite system (see [1]). Thereby, in such a model, energy from the collective movement can be transformed into excitation energy via nuclear friction.

These considerations motivated us to search for a signature of such long lived giant system and, in particular, for "trigger" events to study their properties.

## DECAY CHANNELS

The above described nuclear friction effect suggests three important consequences:

1) If practically all relative collective motion has been transformed into internal excitation energy of the giant di-nuclear system, the emitted particles, in particular the fission fragments, are emitted with very low kinetic energy at the scission point. They should thus emerge out of the collision with their Coulomb barrier energy. In the language of deep inelastic scattering, this corresponds to reactions with the largest possible negative Q-values.

2) The lifetime of the di-nuclear system might be significantly longer than $10^{-22}$ s.

3) The decay of the system should be governed by statistical decay probabilities.

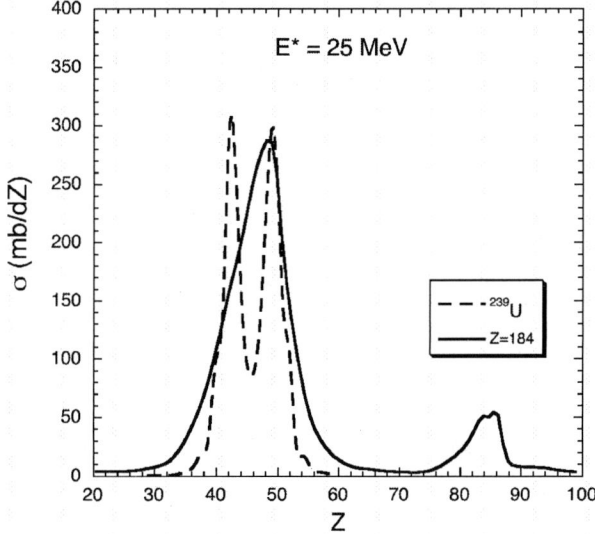

**FIGURE 1.** Calculated elemental yields after decay of $^{239}$U at 25MeV excitation anergy, and of the giant system $^{476}$184 at 25MeV excitation energy.

According to item 3) one can calculate the statistical branching ratios to different exit channels. If the excitation energy of the di-nuclear system is moderate, shell corrections should be important. In this case one expects an asymmetric fission, with preferential emission of nuclei close to the doubly magic nucleus $^{208}$Pb. This should be similar to the well known low excitation energy fission of e.g. $^{239}$U, where a pronounced double bump is observed, related to the doubly magic nucleus $^{132}$Sn. The idea that $^{208}$Pb may play an important role in fission of transuranium nuclei was raised frequently, and is partially experimentally proven [2, 3, 4, 1, 5]. In order to check this idea for this extremely heavy system, we performed a statistical calculation in a micro-canonical approach that treats fission and light particle emission in the same manner [6, 7]. The result is shown in Fig. 1. As a control of the model calculation, the fission of $^{239}$U at 25 MeV, is also shown. A population cross section of 1 barn has been assumed.

For $^{239}$U at 25 MeV excitation energy, the well known double humped fission is obtained. The same parametrisation of the model has been used to calculate the yields from the composite $^{476}$184. Besides the fission products from U and transuranium elements, a strong and isolated peak in the yield distribution is obtained around the lead region. The excitation energy of 25 MeV corresponds to an incident energy of about 7A MeV, depending on binding energy estimates. No heavy residues above the lead region are obtained, in agreement with very low cross section limits experimentally established for transuranium elements (see [8] and references therein).

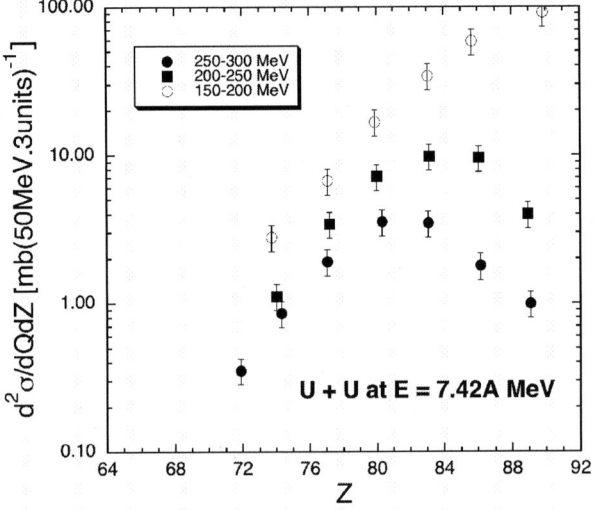

**FIGURE 2.** Double differential cross section of deep inelastic reaction residues as a function of Z for consecutive Q-bins [3].

The peak in the lead region will be well separated only in the system U+U, while in other systems such as Pb+Pb, Au+Pb, etc. it will be very difficult to isolate it from quasielastic or deep inelastic processes. Experimental results supporting the above described hypotheses can be found in [3]. In this experiment [3], inclusive one particle measurements were performed for the charge, the kinetic energy and the angular distributions of the reaction products from $^{238}$U+ $^{238}$U at 7.42A MeV. Relatively large mass transfers have been found for deep inelastic reactions with the largest Q-values. Indications for this behaviour are visible on Fig. 2 (from [3]). In this figure, one can see for the most negative Q-values a maximum in the elemental yield around Z = 82 which is in agreement with the considerations above. As a consequence, the deep inelastic reaction products usually consist of one partner with a nuclear charge number significantly lower than 92 and the corresponding partner in the transuranium region. While the lighter partner is surviving with cross sections up to the range of several mb, the produced transuranium isotopes, which in most cases undergo spontaneous fission, are predicted to survive with cross sections of nb or less. In experiments of Schaedel et al., nuclei up to Z=100 have been observed directly in U+U collisions [9]. The same is obtained in calculations performed by Zagrebaev et al. [10]

## Lifetimes of giant di-nuclear systems

There are only scarce experimental data investigating the lifetime of superheavy di-nuclear systems. J. Stroth et al. have investigated the collision system U+Au at 8.65A MeV [11]. So called "ternary events" where the Au-nucleus survives and the U-nucleus is fissioning after the collision, have been discussed. Contact times have been deduced from delta-electron spectra as well as from deflection functions (from rotation angle of the di-nuclear system) and the longest contact times have been found for reactions with the most negative Q-values. Within statistics and the discussion of a possible systematic error, contact times of up to $10^{-20}$ s have been deduced from the data. Rehm et al. investigated the reaction $^{206}$Pb + $^{110}$Pd and found nuclear contact times somewhat above $10^{-20}$ s for reactions with a large mass flow between both nuclei. Very recently, reaction times have been measured by the blocking technique in single crystals for the system $^{238}$U + Ni at 6.6A MeV and a surprisingly high cross-section of the composite systems with Z = 120 living more than $10^{-18}$ s has been found [13]. Further publications deal "only" with long lifetimes of excited nuclei in the Uranium region. In the reaction $^{238}$U + Si at 24A MeV long lifetimes for excited U-nuclei before scission have been found by Goldenbaum et al., [14] using the crystal blocking technique. Especially for excitation energies below 250 MeV, lifetimes larger than $3 \cdot 10^{-19}$ s were obtained. Similar results for lower excitation energies investigating the K-vacancy production probability are reported by Molitoris [15] as well as by Wilschut [16]. Molitoris et al. deduced lower limits for the lifetime before scission of $4 \cdot 10^{-18}$ s and more recently Wilschut found lifetimes of excited U of the order of $10^{-18}$ s. Concerning superheavy di-nuclear systems, several theoretical works have been carried through, some of them very recently, which suggest expected lifetimes in the $10^{-20}$ s range. With an overdamped Langevin equation with standard parameters, Abe [17] obtained a lifetime on the order of even $10^{-19}$ s for the composite system U+U. In a recent publication, applying a constrained molecular dynamics model [18] similarly long lifetimes have been obtained for the system Au+Au. Calculations have been performed for incident energies from 5A to 35A MeV and especially for energies around 10A MeV lifetimes reaching up to $10^{-19}$ s have been found. In a recent publication by Zagrebaev et al., the collision systems U+U, Th+Cf and U+Cm have been investigated and lifetimes longer than $10^{-20}$ s are predicted [10]. In an earlier work of Yamaji et al. with a time dependant Schroedinger equation a lifetime of the order of $10^{-20}$ s was obtained for U+U at $E_{cm}$ around 850 MeV [19].

In summary, a series of experimental and theoretical results show that nuclear systems without fission barrier (di-nuclear systems or U-like nuclei excited above the barrier) reveal lifetimes orders of magnitude longer than expected from the liquid drop model, which can most likely be explained by nuclear friction.

## EXPERIMENTAL SET-UP

As discussed above, reference [3] contains promising indications for the formation and the decay of the giant system $^{476}184$. In this reference, measurements at only one incident energy have been performed, not enough to be conclusive, and the detection device was a simple ionisation chamber. We proposed to improve these data using up-to-date high resolution devices for the detection of the heavy residues in the region around Pb. At GANIL, the VAMOS spectrometer, with a detection system especially designed for the identification of heavy nuclei was used in this new experiment. In this contribution we present preliminary mass identification results for 5 bombarding energies, ranging from 6.09 to 7.35A MeV.

The $^{238}$U beam was produced by the ECR-4M of GANIL with charge states varying from $30^+$ to $33^+$, depending on the final energy of the beam. The beam was accelerated by the cyclotrons C0 and CSS1 of GANIL up to the energies 6.09, 6.49, 6.91 and 7.35A MeV. The fifth energy (7.1A MeV) was obtained by degrading the beam energy just before the alpha-shaped spectrometer. Intensities of about 0.5 pnA impinged UF$_4$ or metallic U targets of about 400 $\mu$g/cm$^2$ placed in front of the VAMOS spectrometer. VAMOS was placed at 35 degrees covering an angular aperture of +/- 7 degrees in both the horizontal and vertical planes. The spectrometer was used in a "pure-quadrupole" mode. Just behind the two VAMOS quadrupoles, two secondary electron detectors (SED) were used for time of flight measurements. Two drift chambers and one ionization chamber allowed for trajectory reconstruction as well as $\Delta$E measurements, for Z identification. Finally, the scattered particles were stopped by a Si detection wall (300$\mu$m thick). Germanium detectors as well as another Si detector in kinematic coincidence with VAMOS was also present in the experiment. Experimental data corresponding to the latter detectors is still under analysis.

## PRELIMINARY RESULTS

We present the mass distributions obtained in the focal plane of VAMOS as a function of bombarding energy. The masses were obtained using time of flight between the first SED detector and the radio-frequency signal of the cyclotrons, corresponding to a flight path of about 4 m. In order to avoid superposition between two different beam bursts the second SED detector at about 1m from the first one helped on the construction of the time-of-flight. Time-of-flight and the energy obtained in the final Si detector were used to determine the fragment mass for each event.

The mass calibration obtained so far was checked using targets of $^{208}$Pb and natural Sn. The mass resolution in the present stage of the analysis is of the order of 3% FWHM for the U scattered particles. Note that we did not use the dispersion of the dipole, which

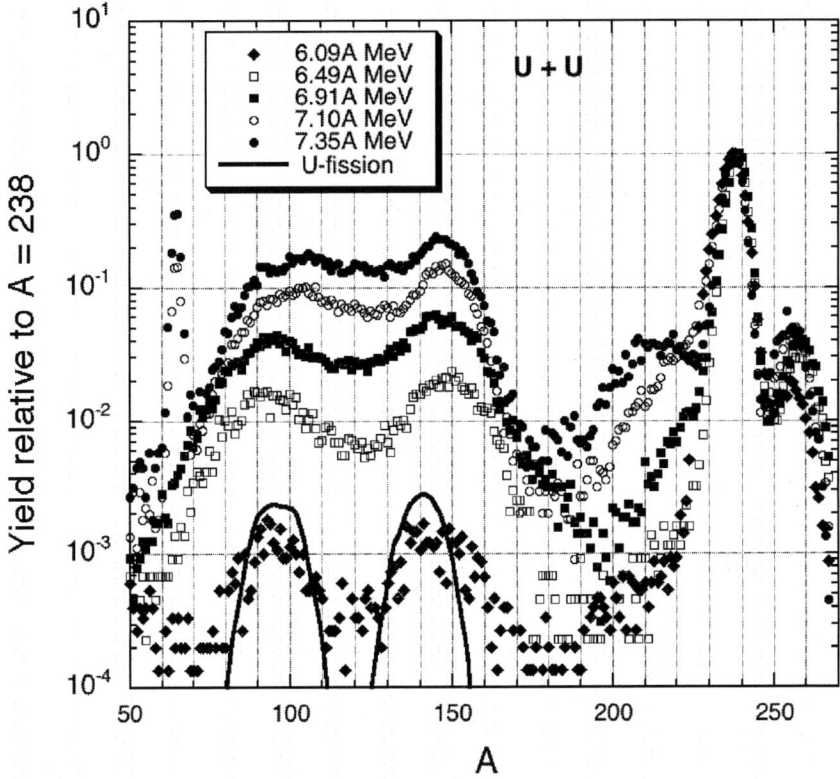

**FIGURE 3.** Mass identification spectra for different energies compared with re-normalized $^{238}$U spontaneous fission. See text for interpretation.

could in principle improve the mass resolution, but would need perfect separation of charge states. The resolution can be improved in the future by applying corrections for the energy losses in the drift and ionization chambers. No correction or condition were done up to now. The mass spectrum obtained so far is shown in Fig. 3 for 5 different bombarding energies. From left to right, one can see peaks corresponding to:

- A = 65, corresponding to a $^{65}$Cu contamination of the uranium beam for the energies 7.1 and 7.32A MeV
- A double bump for 70 < A < 170 corresponding to fission fragments from uranium. This can be compared with relative spontaneous fission yields from $^{238}$U (solid line) [20]
- At around A = 208, corresponding to quasi-lead fragments with intensity varying as a function of energy
- A = 238, corresponding to uranium ions from elastic and inelastic scattering

- At around A = 256, corresponding to a channeling effect in the silicon detector. This kind of effect is well described in [21]. Note that this is *not* a real mass, but a detector defect which simulates a heavier mass.

From figure 3 one can see clearly a *bump* in the region of quasi-lead products, which enhances significantly as a function of energy. The excitation function of this mass region is shown in Figure 4. The energy dependence of the *lead-like* production cross section in this reaction is one of the expected clues for the formation of a long-living giant system . However, it is not a proof for the existence of such long-living giant system. The completion of the excitation function - to higher energies - as well as the analysis of the angular distributions of these fragments should provide more ingredients to establish the existence or not of such a long living giant system.

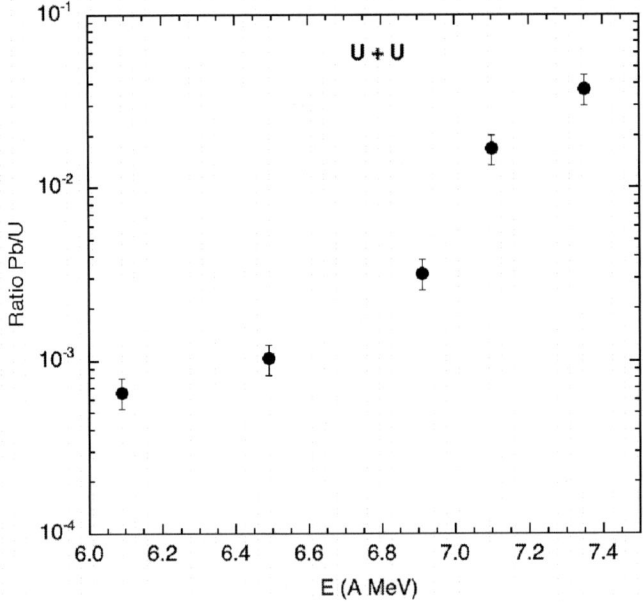

**FIGURE 4.** Excitation function of *lead-like* fragments normalized to quasi-elastic uranium yield.

## SUMMARY

We performed an experiment to search for a long living component in the collision of $^{238}$U + $^{238}$U between 6.09 and 7.35A MeV. The experiment was performed at GANIL using the spectrometer VAMOS, tuned for observing reactions with kinematics similar to fusion-fission events. Theoretical calculations indicate that if a long living component would exist for this reaction, the most probable fission channel of such a giant system would be via the emission of quasi-lead nuclei. Preliminary results show a relative increase of the yield for such events in the focal plane of VAMOS. These events present an excitation function growing as a function of the bombarding energy, as expected. The evolution of the excitation function towards higher energies as well as the fragment angular distribution should provide for additional constraints to establish the existence of such a giant system. If confirmed, we can use these events as trigger to study the physics of such a system.

## REFERENCES

1. V.I. Zagrebaev, Phys.Rev. C 64 (2001) 034606.
2. R. Kalpakchieva et al., Nukleonica 24 (1978) 417.
3. H. Freiesleben et al., Z.Phys. A 292 (1979) 171.
4. V.I. Zagrebaev ıtet al., Phys.Rev. C 65 (2001) 014607.
5. M.G. Itkis, private communication 2005.
6. W. Mittig et al., Phys. Lett. 154B (1985) 259.
7. F. Auger et al., Phys. Rev. C 35 (1987) 190.
8. M.Schaedel et al., Phys. Rev. Lett. 48 (1982) 852.
9. M. Schaedel et al., Phys.Rev.Lett 41 (1978) 469.
10. V.I. Zagrebaev et al., Phys. Rev. C 73 (2006) 031602.
11. J. Stroth et al., Z. Phys. A 357 (1997) 441.
12. K.E. Rehm et al., Nucl. Phys. A366 (1981) 477.
13. D. Jacquet et al., *Proceedings of the Int. Conf. on Reaction Mechanisms and Nuclear Structure at the Coulomb Barrier, March 19-23 2006, Venezia, Italy* AIP 853 (2006) 239.
14. F. Goldenbaum et al., Phys. Rev.Lett. 82 (1999) 5012.
15. J.D. Molitoris et al., Phys.Rev.Lett 70 (1993) 537.
16. H.W.Wilschut and V.L. Kravchuk, Nucl.Phys. A735 (2004) 156.
17. Y.Abe, private communication, Ganil 2005.
18. Toshiki Maruyama et al., Eur. Phys. J. A14 (2002)191.
19. S.Yamaji et al., J Phys. G2 (1976) L189.
20. T.R. England and B.F. Rider, Los Alamos National Laboratory, LA-UR-94-3106; ENDF-349 (1993).
21. W. Pilz et al., Nucl. Instr. Meth. Phys. Res. A419 (1998) 137.

# FUSION-FISSION DYNAMICS (FFD)

# Measurement of evaporation residue cross-sections of the reaction $^{30}$Si + $^{238}$U at subbarrier energies

K. Nishio*, S. Hofmann [1,†,**], F.P. Heßberger[†], D. Ackermann[†],
S. Antalic[‡], V.F. Comas[§], Z. Gan[¶], S. Heinz[†], J.A. Heredia[§], H. Ikezoe*,
J. Khuyagbaatar[†], B. Kindler[†], I. Kojouharov[†], P. Kuusiniemi[∥],
B. Lommel[†], M. Mazzocco[†], S. Mitsuoka*, Y. Nagame*, T. Ohtsuki[††],
A.G. Popeko[‡‡], S. Saro[‡], H.J. Schött[†], B. Sulignano[†,§§], A. Svirikhin[‡‡],
K. Tsukada*, K. Tsuruta* and A.V. Yeremin[‡‡]

*Japan Atomic Energy Agency, Tokai, Ibaraki 319-1192, Japan
[†]Gesellschaft für Schwerionenforschung mbH, D-64220 Darmstadt, Germany
**Institut für Kernphysik, Johann Wolfgang Goethe-Universität, D-60486 Frankfurt am Main, Germany
[‡]Department of Nuclear Physics, Comenius University, SK-84215 Bratislava, Slovakia
[§]Higher Instiute of Technologies and Applied Sciences, Habana 10400, Cuba
[¶]Institute of Modern Physics, Chinese Academy of Sciences, Lanzhou 730000, China
[∥]CUPP, University of Oulu, FIN-86801 Pyhäjärvi, Finland
[††]Laboratory of Nuclear Science, Tohoku University, Sendai 982-0826, Japan
[‡‡]Flerov Laboratory of Nuclear Reactions, JINR, 141 980 Dubna, Russia
[§§]Institut für Kernchemie, Johannes Gutenberg-Universität Mainz, D-55099 Mainz, Germany

**Abstract.** The reaction $^{30}$Si + $^{238}$U → $^{268}$Sg* was studied at beam energies close to the Coulomb barrier. At the above barrier energy of $E_{c.m.}$ = 144.0 MeV (center-of-mass energy at half thickness of the target), we measured three decay chains of $^{263}$Sg produced by evaporation of five neutrons. The cross-section was $(67^{+67}_{-37})$ pb. At subbarrier energy of $E_{c.m.}$ = 133.0 MeV we measured three spontaneously fissioning nuclei which we assigned to the isotope $^{264}$Sg. The half-life of $(120^{+126}_{-44})$ ms was determined. The production cross-section was $(10^{+10}_{-6})$ pb, which is about four orders magnitude larger than the calculation based on the one-dimensional barrier penetration model in the fusion process, showing fusion enhancement caused by the prolate deformation of $^{238}$U. At $E_{c.m.}$ = 128.0 MeV an upper cross-section limit of 15 pb was measured. Compared to excitation functions measured for the lighter system $^{16}$O + $^{238}$U → $^{254}$Fm*, a reduction of the fusion probability was observed at low beam energies indicating increasing competition from quasifission processes.

**Keywords:** Fusion reaction, α decay, spontaneous fission
**PACS:** 25.60.Pj, 23.60.+e, 25.85.Ca

## INTRODUCTION

Two types of fusion reactions which significantly differ by the excitation energy in the compound nucleus were successfully used in synthesis of heavy and superheavy

---
[1] Josef Buchmann-Professor Laureatus

elements (SHE). These are the cold fusion reactions based on lead or bismuth targets and the hot fusion reactions based on actinide targets. Using cold fusion reactions isotopes of elements up to proton number 112 [1] were produced at GSI in Darmstadt, Germany, up to proton number 113 [2] at RIKEN in Wako, Japan. Hot fusion reactions were used to produce isotopes of elements up to 116 and one isotope of element 118 [3, 4] were produced at FLNR in Dubna, Russia.

Another difference between the two reactions is associated with the static deformation of the target nuclei. Experimental data from measurement of excitation functions reveal that in the case of cold fusion highest cross-sections are obtained at beam energies where a contact configuration between projectile and spherical target nucleus is just reached [1]. In the case of hot fusion the cross-section maxima were measured at beam energies which are high enough so that projectile ($^{48}$Ca) and prolate target nuclei can come into contact at minimal distance (equatorial collisions) and thus form a most compact starting configuration on the way to the compound nucleus. The cross-sections drop rapidly when the energy is decreased to values, where the interaction is limited to polar collisions. In this case the probability for re-separation of the reaction partners is high.

However, the results are different in the case of light projectile $^{16}$O. In the reaction $^{16}$O + $^{238}$U, the experimental data show a large enhancement of evaporation residue (ER) cross-sections at sub-barrier energies [5], compared to the calculation based on the one-dimensional model, and the system captured inside the Coulomb barrier always comes to complete fusion even for subbarrier energies. We expect that fusion is sensitive to projectile mass (charge) at the polar collisions, whereas no sensitivity on projectile is expected for the equatorial collisions.

In order to study the anticipated phenomena, it is needed to make a systematic measurement with changing projectiles. Here, we report on the results for the reaction $^{30}$Si + $^{238}$U [6]. Our cross-section data are compared with results from fusion evaporation models taking into account the effects of the static prolate deformation of the target nucleus on the fusion process. Similar arguments were discussed and presented in earlier work on reactions using prolately deformed target nuclei of rare-earth elements [7, 8, 9, 10].

## EXPERIMENTAL METHODS

The experiment was performed at the linear accelerator UNILAC and the velocity filter SHIP of GSI in Darmstadt. The detailed information for the $^{30}$Si+$^{238}$U experiment is published in [6]. The beam of $^{30}$Si$^{6+}$ was prepared from isotopically enriched material, $^{30}$SiO, enrichment 99.5 %, at the High Charge State Injector (HLI) consisting of a 14 GHz ECR ion source, RFQ and IH structure accelerators. Average beam intensities were typically (0.7−1.0) $p\mu$A at a pulse structure of 5.0 ms wide pulses and 50 Hz repetition frequency.

The SHIP set-up is essentially the same as in previous experiments. It consists of a rotating target wheel, the velocity filter SHIP (Separator for Heavy Ion reaction Products), and the detector system in the focal plane. Detailed descriptions can be found in [1, 11, 12, 13, 14].

The uranium targets were prepared by evaporation of isotopically depleted $^{238}$UF$_4$ (<0.4 % $^{235}$U) and condensation on 45 $\mu$g/cm$^2$ carbon backing. The $^{238}$UF$_4$ target layers had thicknesses of 375–404 $\mu$g/cm$^2$, which were coated by a 15 $\mu$g/cm$^2$ thick carbon layer to prevent losses of material due to sputtering.

The transmission efficiency of SHIP was determined by using a Monte Carlo calculation, and we obtained 10 % for the probability of recoils being transmitted and implanted in the focal detector.

In the focal plane of SHIP, ERs and their subsequent $\alpha$ decay and/or spontaneous fission (sf) are detected by a position sensitive 16-strip Si PIPS detector. Escaping $\alpha$ particles or fission fragments are detected by a 'box detector' which covers an area of 85 % of the backward hemisphere. The energy resolution for fully stopped $\alpha$'s was typically 25 keV (FWHM) for the 8415 keV $\alpha$ particles from $^{252}$No. For the energy of escaping $\alpha$ particles detected in coincidence with signals from the box detector we determined a resolution of 70 keV.

A timing detector is located in front of the silicon detector array. Besides a time-of-flight measurement obtained together with signals from the silicon detector it is used for distinguishing signals from implanted ERs or background particles from radioactive decay events. The timing detector was moved out during some of the irradiations at the beginning of the experiment. In this case the radioactive decays were detected during the 15.0 ms beam off periods.

Higher energy sf fragments were measured in a second branch of the electronic circuit including low gain amplifiers. The calibration was performed with $\alpha$ particles from external $\alpha$ sources. The total kinetic energy (TKE) of sf events was obtained by summing the energies from the stop and the box detectors. Energy calibration of the TKE was performed with the known TKE of 195 MeV of $^{252}$No [15] produced in the reaction $^{48}$Ca + $^{206}$Pb. The pulse height deficit and the energy loss in the entrance windows of stop and box detectors were determined from the difference between the literature value and the measured energy based on the $\alpha$-lines calibration. In this calibration run, we found a strong dependence of the measured energies as a function of implantation depth of ERs which was regulated by the thickness of Mylar degrader foils in front of the Si detectors. The results are shown in Fig. 1. The implantation depth was calculated by the SRIM code [16]. Taking into account the calculated implantation depth 1.6 $\mu$m of ERs produced in $^{30}$Si+$^{238}$U, we found a correction of 55 MeV to get the correct value of TKE.

Behind the focal detector we mounted a clover detector consisting of four Ge crystals ($\phi$50 mm $\times$ 70 mm length). It was used to measure coincident $\gamma$- or X-rays accompanied by spontaneous fissions, whose signals were served as a strong evidence for the occurrence of sf.

The detection of correlated events is primarily based on agreement of the measured positions between implanted ERs and subsequent $\alpha$ decays or sf.

## EXPERIMENTAL RESULTS AND DISCUSSION

### Produced isotopes

We used three different beam energies of $E_{lab}$ = 163.5, 151.2 and 145.5 MeV to irradiate the uranium target. They correspond to the center-of-mass energies $E_{c.m.}$ (excitation

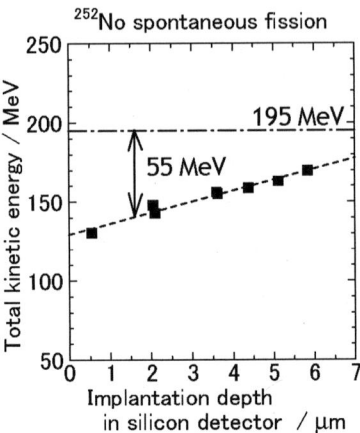

**FIGURE 1.** Sum of the energy of two fragments recorded by the stop and box detectors for the spontaneous fission of $^{252}$No, which shows the dependence on the implanted depth in the stop detector. The literature value TKE=195 MeV for $^{252}$No is indicated. Implantation depth of ERs for $^{30}$Si+$^{238}$U is also shown, and the correction energy of 55 MeV is obtained.

energies $E^*$) of 144.4(50.6), 133.0 (39.6) and 128.0 (34.5) in MeV at the half thickness of the target layer. The accumulated beam doses were $1.8 \times 10^{18}$, $4.0 \times 10^{18}$ and $1.7 \times 10^{18}$ particles, respectively.

At the highest energy of $E_{lab} = 163.5$ MeV we expect the highest cross-section for the well known isotope $^{263}$Sg [17, 18, 19, 20, 21, 22] produced as 5$n$ channel. In this energy we observed decay events which are shown in the upper row on Fig. 2, chronologically ordered from 1 to 4.

Event number 1, 2, and 4 were obtained during measurement without timing detector, resulting in the higher implantation energy, whereas number 3 was obtained during a period when the timing detector was positioned in front of the Si detector array. Subsequent to the implantation we measured within the position resolution of the detector two $\alpha$ decays in the case of event number 1 and 2, three $\alpha$ decays in the case of event number 3. Energies and half-lives agree well with the literature data for $^{263}$Sg, see references given before.

No definite conclusions can be drawn concerning the different intensity ratios of the 9.06 and 9.25 MeV $\alpha$ lines from the decay of $^{263}$Sg in the case of direct production in a fusion reaction, [17] and this work, or by population as granddaughter in the $\alpha$ decay of $^{271}$Ds [19, 20, 21, 22]. Definitely we can conclude that the two different lines originate from two different levels. The reason is that in the case of direct production of $^{263}$Sg different line intensities were measured than in the case of population by $\alpha$ decay within the chain from $^{271}$Ds. If the different $\alpha$ energies would originate from transitions of one level in $^{263}$Sg to different final states in the daughter nucleus $^{259}$Rf, then the intensity ratio would be independent from the production mechanism.

In the decay chains of $^{271}$Ds the 9.25 MeV $\alpha$ particles are observed almost exclusively [19, 20, 21, 22]. From the 22 events, whose energies are clearly assigned to the different

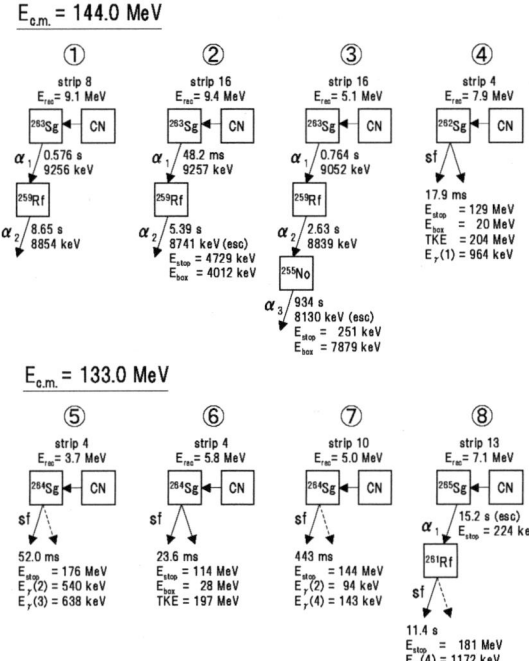

**FIGURE 2.** Decay events obtained from the reaction $^{30}$Si + $^{238}$U → $^{268}$Sg* at $E_{c.m.}$ = 144.0 MeV (upper row) and at $E_{c.m.}$ = 133.0 MeV (lower row). In the case of α decay or spontaneous fission (sf) the energies measured in the stop and box detector are given. For sf the total kinetic energy (TKE) is given, when both fission fragments were detected. Also given are the energies of γ rays coincident with sf measured in one or more of the four crystals of the Ge clover detector. The number in brackets gives the crystal number of the four fold Ge clover detector.

two levels, the majority of 20 decays belongs to the higher energy of 9.25 MeV and only 2 have the energy of 9.06 MeV. The mean half-life for each groups are determined to be $(0.56^{+0.16}_{-0.10})$ s and $(0.56^{+1.02}_{-0.22})$ s, respectively.

In the present experiment we observed two α decays at 9.25 MeV and one at 9.05 MeV. The measured lifetimes given in Fig. 2 are in good agreement with the literature data. The intensity ratio of two to one for the energetically different α rays indicates losses of the shorter lived higher energy α particles in the He-jet experiment [17]. The cross-section for the production of the three $^{263}$Sg nuclei in our experiment at the beam energy of 163.5 MeV is $(67^{+67}_{-37})$ pb.

At this highest beam energy we observed also one sf event at a lifetime of 17.9 ms $(T_{1/2} = (12.4^{+56}_{-6})$ ms), number 4 in Fig. 2. The sf event was in coincidence with a γ ray at 964 keV in one of the crystals of the clover detector. Two fragments were measured in coincidence in the stop and box detector, and we obtained 204 MeV as TKE after correction for the pulse height deficit, 55 MeV. The cross-section for the sf event is $(22^{+51}_{-18})$ pb. We tentatively assign this sf event to the isotope $^{262}$Sg produced by

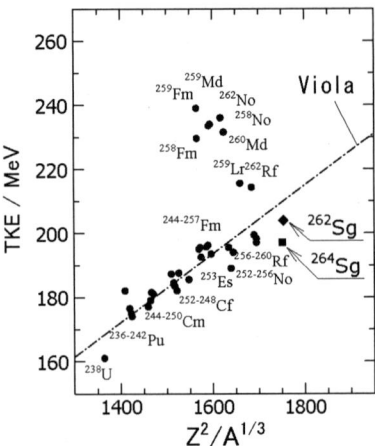

**FIGURE 3.** Total kinetic energies of the spontaneously fissionig nuclei are shown as function $z^2/A^{1/3}$. The present data for $^{262}$Sg and $^{264}$Sg fissions are shown. Dash-dotted line is the Viola formula [24].

evaporation of six neutrons. The measured TKE of 204 MeV agrees reasonably with the value of 210 MeV obtained from the empirical Viola formula [24] for $^{262}$Sg as shown in Fig. 3.

At the lower beam energy of 151.2 MeV, we observed four sf events as numbered from 5 to 8 in Fig. 2. These events were measured in the set up using the timing detector. In the case of event numbers 5, 7, and 8, $\gamma$ signals from the clover detector were coincided. In the case of event number 6, no coincident $\gamma$ ray was measured, however, both fission fragments were detected in the stop and box detector. In addition, this event occurred during the beam-off period. From the sum of the two fragment energies, we obtained a TKE of 197 MeV after correcting for the pulse height deficit.

The lifetime of the sf events number 5 to 7 is similar and relatively short. We determined a half-life of $(120^{+126}_{-44})$ ms. The events were produced at an excitation energy of 40 MeV, 11 MeV less than the value for production of $^{263}$Sg. We assign these three sf events to the sf isotope $^{264}$Sg. The TKE of 197 MeV for event number 6 has reasonable agreement with 210 MeV calculated from the Viola formula for $^{264}$Sg [24] as shown in Fig. 3.

A comparison with calculated sf half-lives reveals a shorter experimental value by about a factor of twenty. The calculated half-life given in [25] is 2.3 s. The deviation continues a trend already observed in the sf half-life of $^{262}$Sg where the experimental value is about a factor of ten shorter than the calculated one, whereas agreement between experiment and theory within a factor of two exists for the isotopes $^{260}$Sg and $^{258}$Sg [23].

The production cross-section for the three sf events assigned to $^{264}$Sg is $(10^{+10}_{-6})$ pb.

Before the 11.4 s of the sf event of number 8, we observed an escaped $\alpha$ particle with 224 keV signal. Then the first ER candidate before the sf event and escaped $\alpha$ particle was measured at 15.2 s before the $\alpha$ event. However, due to the appearance of one ER like event every 10 s on the average, the value of 15.2 s represents only a lower limit for the lifetime. We tentative assign this chain originating from the escaped $\alpha$ decay of $^{265}$Sg and spontaneous fission of $^{261}$Rf as is discussed in [6]. The production cross-section at $E_{c.m.}$ = 133.0 MeV is $(3.5^{+8.1}_{-2.9})$ pb.

At the lowest beam energy of 145.5 MeV no decay events were measured. We determined an upper cross-section limit of 15 pb at 68 % confidence level (one event would have had a cross-section of 8.2 pb).

## Discussion of the production cross-sections

Experimentally determined cross-sections of the reaction $^{30}$Si + $^{238}$U → $^{268}$Sg* are compared with two different calculations in Fig. 4. The upper part shows the fission cross-section measured at JAEA [26] by detecting single fragments. The fission fragments can originate from quasifission and compound nucleus fission, so that the cross section represents the sum of both origins. The data are compared with the results for the capture cross-section using the coupled channel code CCDEGEN [27] which is a modification of the code CCFULL [28]. In the calculation the static deformation of the target nucleus $^{238}$U with the deformation parameters $\beta_2$ = 0.275 and $\beta_4$ = 0.05 [5, 29] is taken into account. The nuclear potential was approximated by using the same parameters as in the case of the earlier studied reaction $^{16}$O + $^{238}$U → $^{254}$Fm* [5]. Also considered was the channel coupling to the $3^-$ state at 0.73 MeV in $^{238}$U and to the $2^+$ state at 2.235 MeV in $^{30}$Si. We also show in Fig. 4 the results from the CCDEGEN code using the assumption of a spherical target nucleus and no coupling to excited states in projectile and target nuclei (one dimensional model). The corresponding Coulomb barrier height is 139.7 MeV.

The coupled channel calculation (solid curve in the upper part of Fig. 4) reproduces well the experimental data points for the reaction $^{30}$Si + $^{238}$U. The one dimensional model, on the other hand, is not able to describe the data below the energy $E_{c.m.}$ = 140 MeV.

In the lower part of Fig. 4 we compare the cross-sections of the fusion evaporation channels (3n–6n) measured in this work with model calculations. Using the results of the two different models for the capture cross-section (entrance channel) as discussed above, the evaporation process of compound nucleus and surviving probability against fission was approximated by the statistical model code HIVAP [30].

The measured ER cross-section for the 5n channel, $^{263}$Sg, at $E_{c.m.}$ = 144.0 MeV agrees well with both calculations, the coupled channel calculation and the one dimensional calculation.

At the lower energy $E_{c.m.}$ = 133.0 MeV the large enhancement of the 4n ER cross-section compared to the one dimensional model confirms the validity of using the properties of colliding nucleus. Especially the enhancement is caused by the lowering of the Coulomb barrier height for collisions with the polar regions of the deformed $^{238}$U. However, the experimental 4n cross-section (10 pb) is about factor of 4 smaller than the calculated cross-section (40 pb). This indicates that there is a competition between fusion and quasifission after system is captured inside the Coulomb barrier. Also at the lowest energy $E_{c.m.}$ = 128.0 MeV the measured upper limit of the cross-section of 15 pb

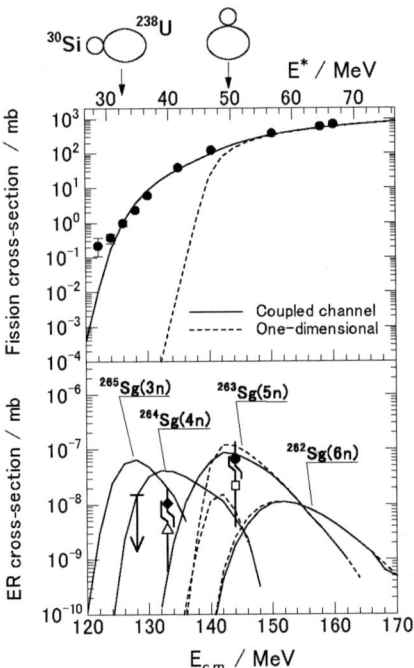

**FIGURE 4.** Fission (upper part) and evaporation residue (ER) cross-sections (lower part) of the reaction $^{30}$Si + $^{238}$U → $^{268}$Sg* as function of the center-of-mass energy $E_{c.m.}$ and excitation energy $E^*$. In the upper part, the experimental data points, taken from [26], include quasifission and compound nucleus fission. The data are compared with results for the fusion cross-section of a coupled channels calculation (full line) and of a one dimensional fusion model (dashed line). The calculated ER cross-sections in the lower part were determined from the calculated fusion cross-sections using the statistical evaporation code HIVAP [30]. At $E_{c.m.}$ = 133.0 and 144.0 MeV three events were measured in each case for the 4n (diamond) and 5n (circle) evaporation channel, respectively, and one event in each case was tentatively assigned to the 3n (open triangle) and 6n (open square) channel, respectively. The two extreme touching configurations, polar and equatorial collisions are shown. The arrows indicate the corresponding Coulomb barrier heights.

(3n + 5n channel) is about a factor 5 less than the calculated one, indicating a quasifission contribution.

Finally we compare the two extreme reactions mentioned before, namely the very asymmetric system $^{16}$O + $^{238}$U [5] which has an orientation independent fusion probability and the heaviest systems studied so far using $^{48}$Ca projectiles [3, 4], with the reaction $^{30}$Si + $^{238}$U studied here. The reaction $^{30}$Si + $^{238}$U shows an intermediate behavior. We observe a reduction, but still a measurable fusion cross-section at low energies, where a contact configuration can be reached only in polar collisions. No reduction of the ER cross-section was measured at energies which were higher than the Coulomb barrier for equatorial collisions, indicating the high probability of complete fusion when the interaction starts at this compact configuration.

## ACKNOWLEDGMENTS

We thank the UNILAC staff for preparation of the stable and intense $^{30}$Si beam. We are also grateful to W. Hartmann and J. Steiner of the GSI target laboratory for manufacturing the target wheels and H.G. Burkhard for taking care of the mechanical devices at SHIP. This work was partly supported by a Grant-in-Aid for Scientific Research of the Japan Society for the Promotion of Science.

## REFERENCES

1. S. Hofmann and G. Münzenberg, Rev. Mod. Phys. **72**, 733 (2000).
2. K. Morita et al., J. Phys. Soc. Jpn. **73**, 1738 (2004).
3. Yu.Ts. Oganessian et al., Phys. Rev. C **69**, 054607 (2004).
4. Yu.Ts. Oganessian et al., Phys. Rev. C **70**, 064609 (2004).
5. K. Nishio et al., Phys. Rev. Lett. **93**, 162701 (2004).
6. K. Nishio *et al.*, Eur. Phys. J. A **29**, 281 (2006).
7. K. Nishio *et al.*, Phys. Rev. C **62**, 014602 (2000).
8. S. Mitsuoka *et al.*, Phys. Rev. C **62**, 054603 (2000).
9. K. Nishio *et al.*, Phys. Rev. C **63**, 044610 (2001).
10. S. Mitsuoka *et al.*, Phys. Rev. C **65**, 054608 (2002).
11. G. Münzenberg et al., Nucl. Instr. and Meth. **161**, 165 (1979).
12. S. Hofmann, et al., Z. Phys. A **291**, 53 (1979).
13. H. Folger et al., Nucl. Instr. and Meth. A **362**, 64 (1995).
14. B. Lommel et al., Nucl. Instr. and Meth. A **480**, 16 (2002).
15. E.K. Hulet, Physics of Atomic Nuclei **57**, 1099 (1994).
16. http://www.srim.org/
17. A. Ghiorso et al., Phys. Rev. Lett. **33**, 1490 (1974).
18. S. Hofmann, Rep. Prog. Phys. **61**, 639 (1998).
19. S. Hofmann, J. Nucl. Rad. Sci **4**, R1 (2003).
20. T.N. Ginter et al., Phys. Rev. C **67**, 064609 (2003).
21. K. Morita et al., Eur. Phys. J. A **21**, 257 (2004).
22. C.M. Folden III et al., Phys. Rev. Lett. **93** 212702 (2004).
23. S. Hofmann et al., Eur. Phys. J. A **10**, 5 (2001).
24. V.E. Viola, Jr., Nucl. Data, Sect. A **1**, 391 (1966).
25. R. Smolanczuk et al., Phys. Rev. C **52**, 1871 (1995).
26. K. Nishio et al., unpublished.
27. K. Hagino, unpublished.
28. K. Hagino, N. Rowley, A.T. Kruppa, Computer Phys. Comm. **123**, 143 (1999).
29. D.J. Hinde et al., Phys. Rev. Lett. **74**, 1295 (1995).
30. W. Reisdorf and M. Schädel, Z. Phys. A **343**, 47 (1992).

# Recent developments in quasi-elastic scattering around the Coulomb barrier

K. Hagino

*Department of Physics, Tohoku University, Sendai 980-8578, Japan*

**Abstract.** We discuss two recent topics on heavy-ion quasi-elastic scattering at energies around the Coulomb barrier. The first topic is an application of quasi-elastic scattering at deep-subbarrier energies to extracting the surface diffuseness parameter of the nucleus-nucleus potential. The second topic is a coupled-channels analysis for the quasi-elastic barrier distribution for the $^{70}$Zn + $^{208}$Pb reaction. We show that the coupled-channels calculations which include the multi-phonon excitations in the colliding nuclei reproduce reasonably well the experimental excitation function for quasi-elastic scattering at backward angles and the barrier distribution for this reaction.

**Keywords:** Quasi-elastic scattering, barrier distribution, quantum reflection, coupled-channels method, cold fusion
**PACS:** 25.70.Bc,25.70.Jj,24.10.Eq,27.70.+w

## INTRODUCTION

The internal structure of colliding nuclei strongly influences heavy-ion collisions at energies around the Coulomb barrier. A well known example is a reaction of a deformed nucleus. In this case, the nucleus-nucleus potential depends on the orientation angle of the deformed nucleus with respect to the beam direction. Assuming that the orientation angle does not change during the collision, the cross section can then be obtained by averaging the contribution from all possible angles [1, 2, 3]. In this picture, the relative motion between the colliding nuclei experiences many distributed potential barriers depending on the orientation angle of the target nucleus, instead of a single barrier. To a good approximation, the concept of barrier distribution can be extended also to systems with a non-deformed target [4, 5, 6, 7], where the distribution originates from the coupling between the relative motion and several intrinsic degrees of freedom such as collective inelastic excitations of the colliding nuclei and/or transfer processes.

In Ref. [8], Rowley, Satchler, and Stelson argued that a barrier distribution can be directly extracted from a measured fusion excitation function $\sigma_{\text{fus}}(E)$, by taking the second derivative of the product $E\sigma_{\text{fus}}(E)$ with respect to the center-of-mass energy $E$, that is, $d^2(E\sigma_{\text{fus}})/dE^2$. This has stimulated many high precision measurements of fusion cross section, so that the second derivative is meaningful [4, 9]. The extracted barrier distributions have revealed that the concept indeed holds and the barrier distribution provides a nice tool to investigate the fusion dynamics of the entrance channel. It was also shown recently that the concept of barrier distribution still retains even in massive systems, such as $^{100}$Mo + $^{100}$Mo [10].

A similar barrier distribution can be extracted also using the quasi-elastic scattering [11, 12]. The quasi-elastic scattering is a sum of elastic, inelastic, transfer, and breakup processes, and is a good counterpart of heavy-ion fusion reaction [2]. A major difference is that the quasi-elastic scattering is related to the reflection probability of the Coulomb barrier, while the fusion is related to the transmission. Since the penetration and reflection probabilities are related to each other due to the flux conservation, similar information can be obtained both from fusion and quasi-elastic scattering.

In this contribution, we discuss two recent theoretical activities on heavy-ion quasi-elastic scattering at sub-barrier energies. We first present our recent systematic analyses on heavy-ion quasi-elastic scattering at deep-subbarrier energies, in aiming at extracting the surface diffuseness parameter of inter-nuclear potential [13, 14]. We then discuss coupled-channels calculations for the $^{70}$Zn + $^{208}$Pb reaction, for which the quasi-elastic barrier distribution has recently been obtained experimentally [15].

## QUASI-ELASTIC BARRIER DISTRIBUTIONS

Before we proceed, let us first summarize the theoretical formulas for quasi-elastic barrier distribution. In the eigenchannel representation of the coupled-channels method, the fusion and quasi-elastic cross sections are given as a weighted sum of the cross sections for uncoupled eigenchannels [2, 4, 5, 6, 7]. That is,

$$\sigma_{\text{fus}}(E) = \sum_\alpha w_\alpha \sigma_{\text{fus}}^{(\alpha)}(E), \qquad (1)$$

$$\sigma_{\text{qel}}(E,\theta) = \sum_\alpha w_\alpha \sigma_{\text{el}}^{(\alpha)}(E,\theta), \qquad (2)$$

where $\sigma_{\text{fus}}^{(\alpha)}(E)$ and $\sigma_{\text{el}}^{(\alpha)}(E,\theta)$ are the fusion and the elastic cross sections for a potential in the eigenchannel $\alpha$. Notice that the same weight factors $w_\alpha$ appear both in Eqs. (1) and (2). This is a generalization of well-known orientation average formula for a system with deformed target,

$$\sigma(E) = \int_0^1 d(\cos\theta_T)\sigma(E;\theta_T), \qquad (3)$$

where $\theta_T$ is the orientation of the deformed target and represents a continuous variable for $\alpha$ in Eqs. (1) and (2).

The idea of barrier distribution is led by the fact that the classical cross sections for fusion and quasi-elastic scattering for a single potential barrier are given by

$$\sigma_{\text{fus}}^{cl(0)}(E) = \pi R_b^2 \left(1 - \frac{V_b}{E}\right) \theta(E - V_b), \qquad (4)$$

and

$$\sigma_{\text{el}}^{cl(0)}(E,\pi) = \sigma_R(E,\pi)\,\theta(V_b - E), \qquad (5)$$

respectively[12]. Here, $R_b$ and $V_b$ are the barrier position and the barrier height for the $s$-wave scattering (thus the scattering angle is set to be $\pi$ in Eq. (5)), respectively, and $\sigma_R(E,\pi)$ is the Rutherford cross section. These yield [8, 11],

$$D_{\text{fus}}(E) \equiv \frac{d^2}{dE^2}[E\sigma_{\text{fus}}(E)] = \sum_\alpha w_\alpha \pi \left[R_b^{(\alpha)}\right]^2 \delta(E - V_b^{(\alpha)}), \qquad (6)$$

$$D_{\text{qel}}(E) \equiv -\frac{d}{dE}\left(\frac{\sigma_{\text{el}}(E,\pi)}{\sigma_R(E,\pi)}\right) = \sum_\alpha w_\alpha \delta(E - V_b^{(\alpha)}). \qquad (7)$$

Evidently, these functions provide information on how potential barrier heights are distributed, and are called fusion and quasi-elastic barrier distributions, respectively. In realistic situations, the quantum (tunneling) effect smears the delta function in Eqs. (6) and (7). Moreover, the effect of nuclear potential has to be taken into account in quasi-elastic cross sections in Eq. (7) [12]. Nevertheless, from the derivation, it is apparent that the fusion and quasi-elastic barrier distributions behave in a similar way. This is demonstrated in Fig. 5 in Ref. [12] for the $^{16}$O + $^{154}$Sm system.

In actual experiments, it is impossible to put a detector at a scattering angle $\pi$. One can, however, scale a cross section in energy by taking into account the centrifugal correction. Estimating the centrifugal potential at the distance of closest approach for the Rutherford scattering, $r_c$, the effective energy may be expressed as [11]

$$E_{\text{eff}} \sim E - \frac{\lambda_c^2 \hbar^2}{2\mu r_c^2} = 2E \frac{\sin(\theta/2)}{1 + \sin(\theta/2)}. \qquad (8)$$

Therefore, one expects that the function $-d/dE(\sigma_{\text{el}}/\sigma_R)$ evaluated at an angle $\theta$ will correspond to the quasi-elastic barrier distribution at the effective energy given by eq. (8).

## INTER-NUCLEUS POTENTIAL AND DEEP-SUBBARRIER QUASI-ELASTIC SCATTERING

Let us now discuss the application of deep-subbarrier quasi-elastic scattering to the problem of surface diffuseness anomaly in heavy-ion potential [16]. For calculations of elastic and inelastic scattering, which are sensitive only to the surface region of the nuclear potential, the diffuseness parameter of around 0.63 fm has been conventionally employed [17, 18]. This value of surface diffuseness parameter has been well accepted, partly because it is consistent with a double folding potential [19]. In contrast, a recent systematic study has shown that experimental data for heavy-ion fusion reactions require a much larger value of the diffuseness parameter, ranging between 0.75 and 1.5 fm, as long as the Woods-Saxon parameterization is employed [20].

Since quasi-elastic scattering and fusion are complementary to each other, it is of interest to investigate this problem using quasi-elastic scattering. In doing so, we are particularly interested in the deep sub-barrier region [13, 14]. At these energies, the cross sections of (quasi-)elastic scattering are close to the Rutherford cross sections, with small deviations caused by the effect of nuclear interaction, $V_N$. This effect can be taken into account by the semiclassical perturbation theory [12, 21], which leads to

$$\frac{d\sigma_{\text{el}}(E,\theta)}{d\sigma_R(E,\theta)} \sim 1 + \frac{V_N(r_c)}{ka} \frac{\sqrt{2a\pi k\eta}}{E}, \qquad (9)$$

where $k = \sqrt{2\mu E}/\hbar$, $\mu$ being the reduced mass. $\eta$ is the Sommerfeld parameter, and $r_c$ is the distance of closest approach. This formula shows that the deviation of the

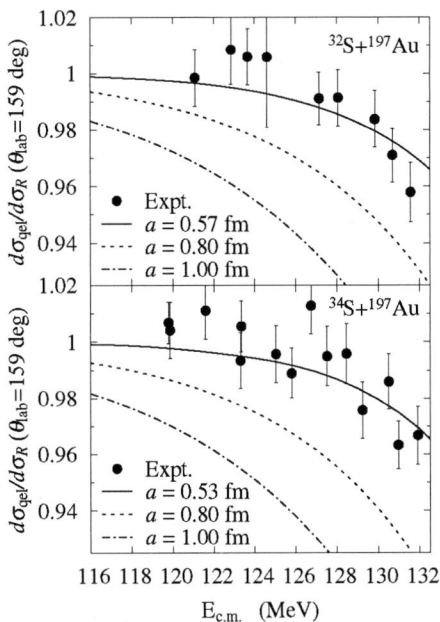

**FIGURE 1.** The ratio of the quasi-elastic to the Rutherford cross sections at $\theta_{lab} = 159°$ for the $^{32}$S + $^{197}$Au (the upper panel) reaction and for the $^{34}$S + $^{197}$Au (the lower panel) reaction.

elastic cross sections from the Rutherford ones is sensitive predominantly to the surface region of the nuclear potential, especially to the surface diffuseness parameter $a$. There is another advantage of using the deep sub-barrier data. That is, the effect of channel coupling on quasi-elastic scattering can be disregarded at these energies, since the reflection probability is almost unity irrespective of the presence of channel couplings [14]. From these considerations, it is evident that the effect of surface diffuseness parameter can be studied in a transparent and unambiguous way using the large-angle quasi-elastic scattering at deep sub-barrier energies.

Figure 1 compares the experimental data with the calculated cross sections obtained with different values of the surface diffuseness parameter for the $^{32}$S + $^{197}$Au system (the upper panel) and the $^{34}$S + $^{197}$Au system (the lower panel). In order to analyze the experimental data at deep sub-barrier energies, we use a one-dimensional optical potential with the Woods-Saxon form. Absorption following transmission through the barrier is simulated by an imaginary potential that is well localized inside the Coulomb barrier. The best fitted values for the surface diffuseness parameter are $a = 0.57 \pm 0.04$ fm and $a = 0.53 \pm 0.03$ fm for the $^{32}$S and $^{34}$S + $^{197}$Au reactions, respectively. The cross sections obtained with these surface diffuseness parameters are denoted by the solid line in the figure. The dotted and the dot-dashed lines are calculated with the diffuseness parameter of $a = 0.80$ fm and $a = 1.00$ fm, respectively. It is clear from the figure that these spherical systems favor the standard value of the surface diffuseness parameter, around $a = 0.60$ fm. The calculations with the larger diffuseness parameters underestimate the quasi-elastic cross sections and are not consistent with the energy dependence of the experimental data. We obtain a similar conclusion for the $^{32,34}$S + $^{208}$Pb and $^{16}$O + $^{208}$Pb systems[13]. This indicates that the double folding procedure is valid at least in the surface region and for spherical systems which we studied.

For deformed systems, such as $^{16}$O + $^{154}$Sm, $^{186}$W, on the other hand, we found that the surface diffuseness parameter of $a = 1.14 \pm 0.03$ fm and $0.79 \pm 0.04$ fm for the former and for the latter, respectively, is required in order to account for the experimental data [13, 14]. Although these large values of surface diffuseness parameter are consistent with that extracted from fusion, the origin of the difference between the spherical and the deformed systems is not clear. In order to clarify the difference in the diffuseness parameter, apparently further precision measurements for large-angle quasi-elastic scattering at deep sub-barrier energies are urged, especially for deformed systems.

## COUPLED-CHANNELS CALCULATIONS FOR QUASI-ELASTIC BARRIER DISTRIBUTION FOR $^{70}$ZN + $^{208}$PB REACTION

We next discuss the barrier distribution for synthesis of superheavy elements. When one discusses a fusion reaction to synthesize superheavy elements, one often refers to a single potential such as the Bass barrier [22]. On the other hand, the effect of channel coupling is in general strong for massive systems, and thus one can expect a broad distribution of potential barriers. It is thus important to study how the potential barrier is distributed for massive systems, since it is crucial to choose the right beam energy in order to effectively synthesize superheavy elements. Moreover, there is no a priori evidence why the Bass barrier is reasonable in the superheavy region. For these reasons, the quasi-elastic barrier distribution measurements have been recently performed by Mitsuoka *et al.* for systems relevant to cold fusion reactions, $^{48}$Ti, $^{54}$Cr, $^{56}$Fe, $^{64}$Ni, $^{70}$Zn + $^{208}$Pb [15]. In this section, we perform coupled-channels calculations for the $^{70}$Zn + $^{208}$Pb system.

The calculations are done with a version [23] of the coupled-channels code CCFULL [24]. This code treats the coupling to all orders in the coupling hamiltonian and employs the isocentrifugal approximation in order to reduce the dimension of the coupled-channels equations. It has been shown that the isocentrifugal approximation works well for quasi-elastic scattering at backward angles [12]. In the code, the regular boundary condition is imposed at the origin, instead of the incoming boundary condition.

Figure 2 shows the excitation function of the quasi-elastic scattering (the upper panel) and the barrier distribution (the lower panel). The deep-inelastic component has been subtracted from the experimental data using a statistical code, as is explained in Ref. [15]. The solid and dashed lines are the results of the coupled-channels and the potential model calculations, respectively. We use the Woods-Saxon potential with $V_0 = 140$ MeV, $r_0 = 1.186$ fm, and $a = 0.69$ fm for the real part and $W_0 = 50.0$ MeV, $r_w = 1.0$ fm, and $a_w = 0.1$ fm for the imaginary part. In the coupled-channels calculation, we include the double quadrupole phonon excitations in the $^{70}$Zn and the triple octupole phonon excitations in the $^{208}$Pb nucleus. In addition, we include the mutual excitation channels, [1,1], [1,2], [2,1], and [2,2], where [$n_P, n_T$] denotes the excitation channel with $n_P$ phonon state in the projectile and $n_T$ phonon state in the target nucleus. In this way, we include 10 channels

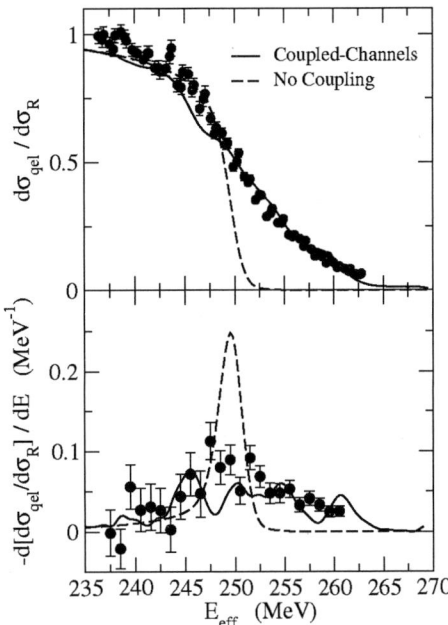

**FIGURE 2.** The ratio of the quasi-elastic to the Rutherford cross sections (the upper panel) and the quasi-elastic barrier distribution (the lower panel) for the $^{70}$Zn + $^{208}$Pb reaction. These are plotted as a function of effective energy defined by Eq. (8). The solid line is the solution of coupled-channels equations, which take into account the double quadrupole phonon excitations in the $^{70}$Zn nucleus and the triple octupole phonon excitations in the $^{208}$Pb nucleus. The dashed line shows the result without the couplings. The experimental data are taken from Ref. [15].

(including the entrance channel, [0,0]) in the calculations. The excitation energy for the single phonon state and the deformation parameter are $E_{2^+} = 0.885$ MeV and $\beta_2 = 0.228$ for the projectile nucleus $^{70}$Zn and $E_{3^-} = 2.614$ MeV and $\beta_3 = 0.11$ for the target nucleus $^{208}$Pb. We use $r_0 = 1.2$ fm for the coupling term.

In the code, the coupled-channels equations are solved by constructing $N$ linear independent solutions of the equations, where $N$ is the dimension of the coupled-channels equations. A linear superposition of these solutions is then taken to construct the physical solution, which fulfills the asymptotic boundary condition for scattering. For massive systems, it is sometimes difficult to numerically maintain the linear independence of the solutions, since the wave functions scale very differently from one channel to another. This leads to a numerical instability of the solution of the coupled-channels equations. We avoid this difficulty by taking a linear superposition of the solutions at several places, with an interval of 1 fm up to 15 fm, so that the linear independence is recovered. See Ref. [23] for details. Even though we use this prescription, we still find a small spurious oscillation in the calculated excitation function of quasi-elastic cross section due to the numerical inaccuracy, when the coupling is strong. We therefore average the calculated cross sections with a Gaussian weight with 0.5 MeV width. We have checked

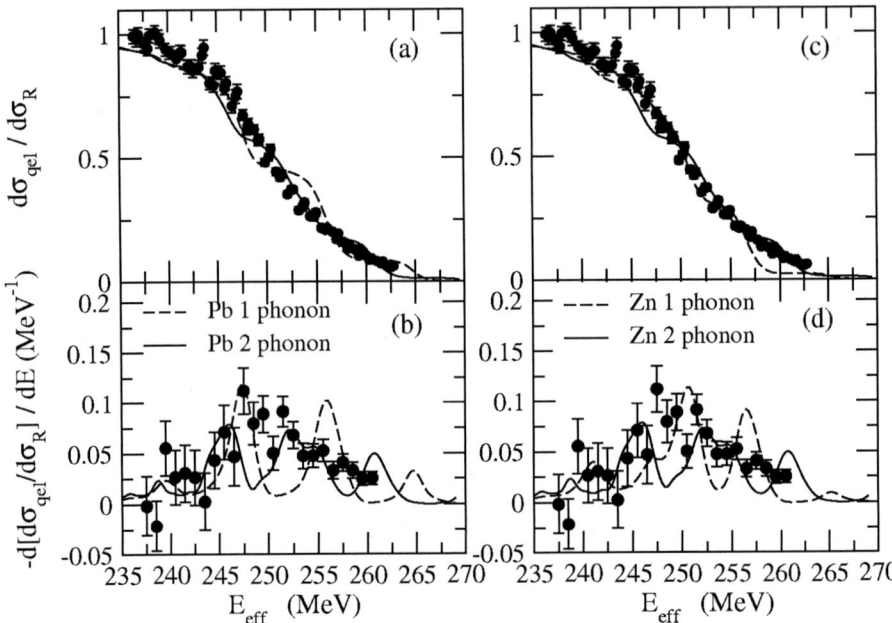

**FIGURE 3.** The ratio of the quasi-elastic to the Rutherford cross sections (3(a) and 3(c)) and the quasi-elastic barrier distribution (3(b) and 3(d)) for the $^{70}$Zn + $^{208}$Pb reaction. The figs. 3(a) and 3(b) are obtained by including the different number of octupole phonon excitations in the target nucleus as indicated in the inset, together with the double quadrupole phonon excitations in the projectile nucleus. The figs. 3(c) and 3(d) are obtained by including the different number of quadrupole phonon excitations in the projectile nucleus together with the double octupole phonon excitations in the target nucleus.

that the shape of quasi-elastic barrier distribution is insensitive to the value of the width parameter.

As we can see in the figure, the potential model calculation (the dashed line) significantly underestimate the quasi-elastic cross sections at energies above the Coulomb barrier. Also, the barrier distribution has a significantly narrow width, and is inconsistent with the experimental data. On the other hand, the coupled-channels calculation (the solid line) well reproduces the experimental data both for the excitation function and barrier distribution.

Figure 3 shows the role of multi-phonon excitations. Figures 3(a) and 3(b) are obtained by varying the number of octupole phonon excitations in the target while keeping the double phonon excitations in the projectile nucleus. On the other hand, figs. 3(c) and 3(d) are obtained by varying the number of quadrupole phonon excitation while keeping the number of octupole phonon excitations in the target nucleus to be two. These figures show that the double phonon excitations considerably alter the shape of barrier distribution as compared with the barrier distribution obtained with the single phonon excitation. For both in the projectile and in the target nuclei, the double phonon excitation leads to better agreement with the experimental data, although we find that the triple phonon excitation in the target nucleus plays a much less important role. A similar conclusion has been obtained also in Ref.[10], where the role of multi-phonon excitations was discussed for the $^{100}$Mo+$^{100}$Mo fusion reaction at energies around the Coulomb barrier.

## SUMMARY

We have discussed two recent developments in heavy-ion quasi-elastic scattering at energies around the Coulomb barrier. We first discussed the surface property of internucleus potential. We have argued that the quasi-elastic scattering at deep subbarrier energies offer a clear and almost model independent way to determine the surface diffuseness parameter, that is, the slope of asymptotic exponential tail of the potential. The value of diffuseness parameter extracted from the $^{32,34}$S + $^{197}$Au reactions is around 0.55 fm, and is consistent with the double folding potential. On the other hand, the surface diffuseness parameter extracted from systems with a deformed target, that is, $^{16}$O + $^{154}$Sm, $^{186}$W was found to be much larger (1.14 fm for the former and 0.79 fm for the latter systems). Further investigations will be required in order to clarify the system dependence of the surface diffuseness parameter. In the second part, we performed the coupled-channels analyses for a cold fusion reaction $^{70}$Zn + $^{208}$Pb, where the quasi-elastic barrier distribution was recently obtained by Mitsuoka et al.. Including the double quadrupole phonon excitations in the projectile nucleus $^{70}$Zn and the triple octupole phonon excitations in the target nucleus $^{208}$Pb in the coupled-channels calculation, we could reproduce reasonably well both the excitation function of quasi-elastic cross section and the shape of quasi-elastic barrier distribution. This indicates that the coupled-channels approach still works for the approaching phase of the reaction even in massive systems, where many degrees of freedom may be involved in the reaction [10]. It also suggests that the deep-inelastic collision can be regarded as a post-barrier phenomena, since the experimental quasi-elastic cross sections have been obtained by subtracting the deep-inelastic components. We also discussed the role of multi-phonon excitations, and showed that they play an important role in this system. The coupled-channels analyses for other cold fusion reactions, $^{48}$Ti, $^{54}$Cr, $^{56}$Fe, $^{64}$Ni + $^{208}$Pb are now in progress, and we will report on them in a separate publication.

## ACKNOWLEDGMENTS

This work is based on collaborations with N. Rowley and K. Washiyama. We thank H. Ikezoe and S. Mitsuoka for useful discussions and for sending us the experimental data before publication. This work was supported by the Grant-in-Aid for Scientific Research, Contract No. 16740139 from the Japanese Ministry of Education, Culture, Sports, Science, and Technology.

## REFERENCES

1. C.Y. Wong, Phys. Rev. Lett. **31**, 766 (1973).
2. M.V. Andres, N. Rowley, and M.A. Nagarajan, Phys. Lett. **202B**, 292 (1988).

3. T. Rumin, K. Hagino, and N. Takigawa, Phys. Rev. **C63**, 044603 (2001).
4. M. Dasgupta, D.J. Hinde, N. Rowley, and A.M. Stefanini, Annu. Rev. Nucl. Part. Sci. **48**, 401 (1998).
5. A.B. Balantekin and N. Takigawa, Rev. Mod. Phys. **70**, 77 (1998).
6. K. Hagino and A.B. Balantekin, Phys. Rev. **A70**, 032106 (2004).
7. C.H. Dasso, S. Landowne, and A. Winther, Nucl. Phys. **A405**, 381 (1983); **A407**, 221 (1983).
8. N. Rowley, G.R. Satchler, and P.H. Stelson, Phys. Lett. **B254**, 25 (1991).
9. J.R. Leigh *et al.*, Phys. Rev. **C52**, 3151 (1995).
10. N. Rowley, N. Grar, and K. Hagino, Phys. Lett. **B632**, 243 (2006).
11. H. Timmers *et al.*, Nucl. Phys. **A584**, 190 (1995).
12. K. Hagino and N. Rowley, Phys. Rev. **C69**, 054610 (2004); Brazilian J. Phys. **35**, 890 (2005).
13. K. Washiyama, K. Hagino, and M. Dasgupta, Phys. Rev. **C73**, 034607 (2006).
14. K. Hagino, T. Takehi, A. B. Balantekin, and N. Takigawa, Phys. Rev. C **71**, 044612 (2005).
15. H. Ikezoe, S. Mitsuoka, K. Nishio, K. Tsuruta, Y. Watanabe, S. Jeong, and K. Sato, in proceedings of international conference "FUSION06", AIP conf. proc. **853**, 69 (2006); S. Mitsuoka, private communications.
16. K. Hagino *et al.*, in proc. of the 4th Italy-Japan symposium on Heavy-Ion Physics, edited by S. Kubono *et al.* (World Scientific, Singapore, 2002), p.87. e-print: nucl-th/0110065.
17. R.A. Broglia and A. Winther, *Heavy Ion Reactions*, Vol. 84 in Frontiers in Physics Lecture Note Series (Addison-Wesley, Redwood City, CA, 1991).
18. P.R. Christensen and A. Winther, Phys. Lett. **65B**, 19 (1976).
19. G.R. Satchler and W.G. Love, Phys. Rep. **55**, 183 (1979).
20. J.O. Newton *et al.*, Phys. Lett. B**586**, 219 (2004); Phys. Rev. C **70**, 024605 (2004).
21. S. Landowne and H.H. Wolter, Nucl. Phys. **A351**, 171 (1981).
22. R. Bass, Phys. Rev. Lett. **39**, 265 (1977); *Nuclear Reactions with Heavy Ions*, (Springer Verlag, Berlin Heidelberg New York, 1980) p. 152.
23. K. Hagino and N. Rowley, to be published.
24. K. Hagino, N. Rowley, and A.T. Kruppa, Comp. Phys. Comm. **123**, 143 (1999).

# PHYSICS WITH EXOTIC NUCLEI (PEN)

# The SPIRAL 2 Project

## Marek Lewitowicz

*on behalf of the SPIRAL 2 project group*

*Grand Accélérateur d'Ions Lourds - GANIL
CEA/DSM-CNRS/IN2P3
BP 55027, 14076 Caen Cedex, France*

**Abstract.** The project of an important new extension of the GANIL facility – SPIRAL 2 – is shortly presented. The physics case of the facility is based on the use of high intensity stable and radioactive beams. Expected performances and main technical parameters of the facility as well as planned new experimental areas and detectors are shortly described.

**Keywords:** Radioactive nuclear beams, linear accelerators, mass separators, charged particle detectors, gamma arrays, neutron-induced reactions, fission, heavy-ion induced fusion, nuclear structure, nuclear reaction dynamics
**PACS:** 25.60.-t; 29.17.+w; 29.27.Eg; 25.85.-w

## INTRODUCTION – CURRENT GANIL FACILITY

In recent years, Rare (or Radioactive) Nuclear Beams (RNBs), i.e. beams of synthesized radioactive isotopes, have been recognized by the international scientific community as one of the most promising avenues for the development of fundamental nuclear physics and astrophysics, as well as in applications of nuclear science. Through the concept of beta-beams, RNBs might also become a way to create intense beams of pure neutrinos.

The GANIL facility[1] (Caen, France) is one of the major RNB and stable-ion facilities for nuclear physics, astrophysics and interdisciplinary research in Europe. Since the first beams delivered in 1983 the performances of the GANIL accelerator complex, shown in Fig. 1, was constantly improved with respect to the beam intensity, energy and available detection systems. Almost since the beginning of the experimental program, the facility delivered RNB produced "in-flight" at fragment separators like LISE and Alpha spectrometer.

More recently, in autumn 2001, the SPIRAL facility allowing for the production and post-acceleration of the ISOL-type RNB entered into operation. The facility, specialized in RNB of rare gases (He, Ne, Ar, Kr but also N, O and F), enlarged importantly the range of experimental possibilities dedicated to study of nuclei far from stability at GANIL. In particular, low energy an in-beam gamma-ray spectroscopy, measurements of the spectroscopic factors via transfer reactions, elastic and inelastic scattering highly benefit from pure, high intensity and high optical quality RNB post-accelerated by CIME cyclotron. The production of RNB at SPIRAL is based on a use of high intensity beams of heavy ions impinging on a universal graphite target thus using nuclear reactions in inverse kinematics.

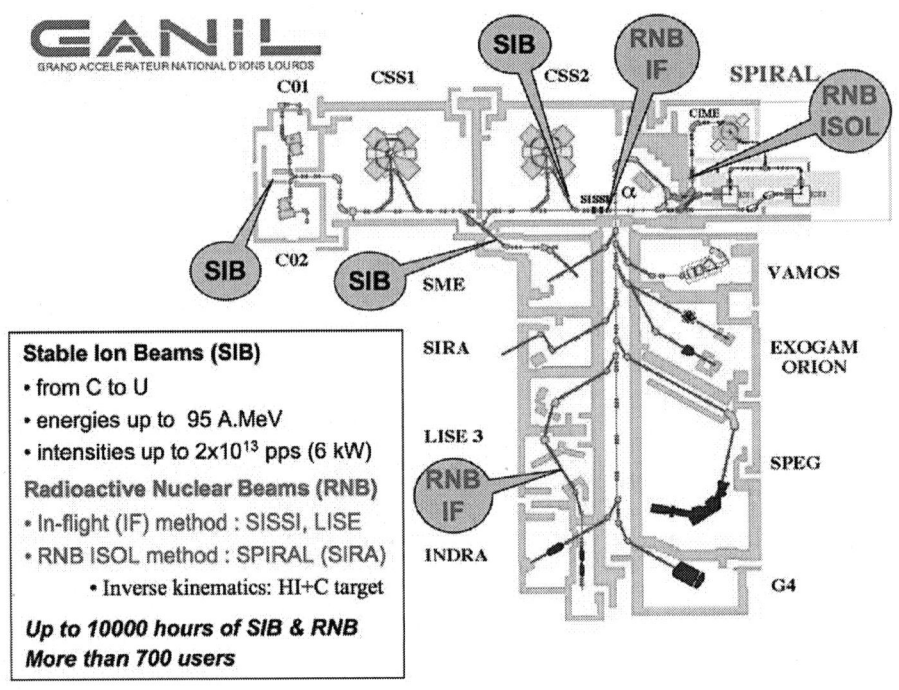

**FIGURE 1.** Layout of the current GANIL.SPIRAL facility and its main characteristics.

However, the available intensities and energies of the GANIL's stable-ion beams (see fig. 1) limit the use of high intensity RNB to relatively light nuclei (A<80).

Since the beginning of the SPIRAL project it was proposed to enlarge the range of accelerated ions by production of high intensity RNB of fission fragments. This idea, after several years of discussions and an important preliminary study phase led, in particular, in the framework of the European RTT Program was concretized in the SPIRAL 2 project[2].

The project is following the European road map for RNB facilities defined by NuPECC[3] (Nuclear Physics European Collaboration Committee – an expert committee of the European Science Foundation), which recommend the construction of two complementary next-generation RIB facilities in Europe. One is based on in-flight fragmentation (IF) as proposed for the FAIR facility at GSI (Darmstadt, Germany) and the other on the isotope-separation on-line (ISOL) method, largely developed at the CERN-ISOLDE facility over the last thirty years (the EURISOL project[6]). However, a full engineering design study and necessary R&D programme mean that the expected beginning of operation of the EURISOL facility will only be around 2020. Because of the time-line for EURISOL as well as of important unsolved yet technological issues, NuPECC recommends the construction of intermediate-generation facilities that will benefit the EURISOL project in terms of R&D and that will give the community opportunities to perform research and develop applications with RNB. Among the proposed intermediate facilities, SPIRAL2 meets the criteria of European dimension in terms of physics potential, site and size of the investment.

# LAYOUT AND PERFORMANCES OF THE SPIRAL 2 FACILITY

The SPIRAL 2 facility (fig. 2) is based on a high-power, superconducting driver LINAC, which will deliver a high-intensity, 40 MeV deuteron beam as well as a variety of heavy-ion beams with mass-to-charge ratio of 3 and energy up to 14.5 MeV/nucleon. Using a carbon converter, the 5 mA deuteron beam and a uranium carbide target, fast-neutron induced fission is expected to reach a rate of up to $10^{14}$ fissions/s. The RNB intensities in the mass range from A=60 to A=140 will be of the order of $10^6$ to $10^{11}$ particles/s (pps) surpassing by one or two order of magnitude any existing facilities in the world. For example, the intensities should reach $10^9$ pps for $^{132}$Sn and $10^{10}$ pps for $^{92}$Kr. A direct irradiation of the UC$_2$ target with beams of deuterons, $^{3,4}$He, $^{6,7}$Li, or $^{12}$C can be used if higher excitation energy leads to higher production rate for a specific nucleus of interest.

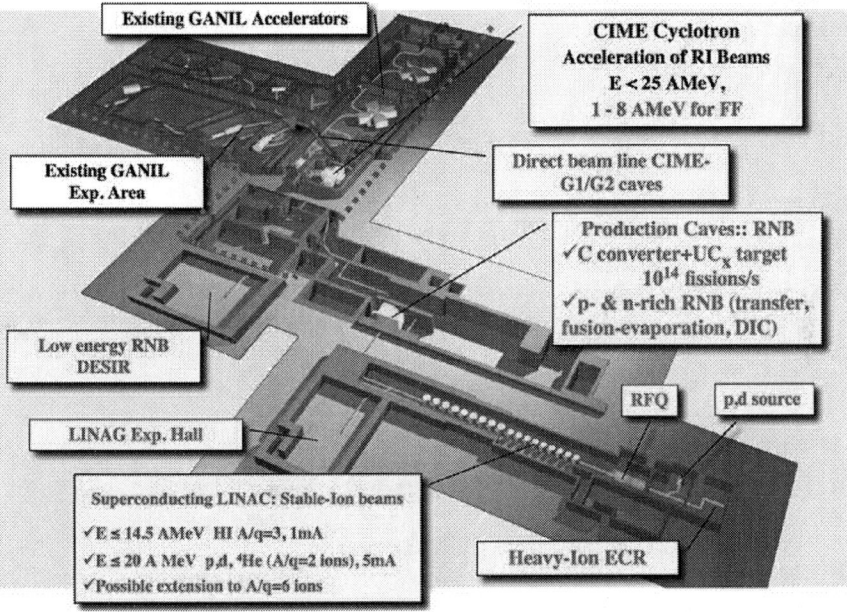

**FIGURE 2.** Layout of the SPIRAL 2 facility and its main characteristics. The dotted line corresponds to the separation between the current (upper part) and new facilities.

These neutron-rich fission RNB could be complemented by beams of nuclei near the proton drip-line, provided by fusion-evaporation or transfer reactions. For example, a production of up to $8 \times 10^4$ atoms of $^{80}$Zr per second using a 200 μA

$^{24}$Mg$^{8+}$ beam on a $^{58}$Ni target should be possible. Similarly, the heavy- and light-ion beams from LINAG can also be used directly on different production targets to produce high-intensity light RNB with the ISOL technique.

The extracted 1+ radioactive ions will be subsequently injected to the 1+/n+ charge booster (ECR ion source) and post-accelerated to energies of up to 20 MeV/nucleon (typically 6–7 MeV/nucleon for fission fragments) by the existing CIME cyclotron.

Thus, using several different production mechanisms and techniques, SPIRAL 2 would allow users to perform experiments with a wide range of neutron- and proton-rich nuclei far from the line of stability.

One of the important features of the future GANIL/SPIRAL 1/SPIRAL 2 facility will be the possibility to deliver up to five stable or radioactive beams to different users simultaneously in the energy range from keV to several tens of MeV/nucleon.

## SCIENTIFIC OPPORTUNITIES AT SPIRAL 2

In the following paragraphs, only several examples of a rich and multipurpose scientific program proposed at SPIRAL 2 will be discussed. More complete presentation of the scientific case of SPIRAL 2 can be found in the recently accomplish White Book of SPIRAL 2 (see ref. 4).

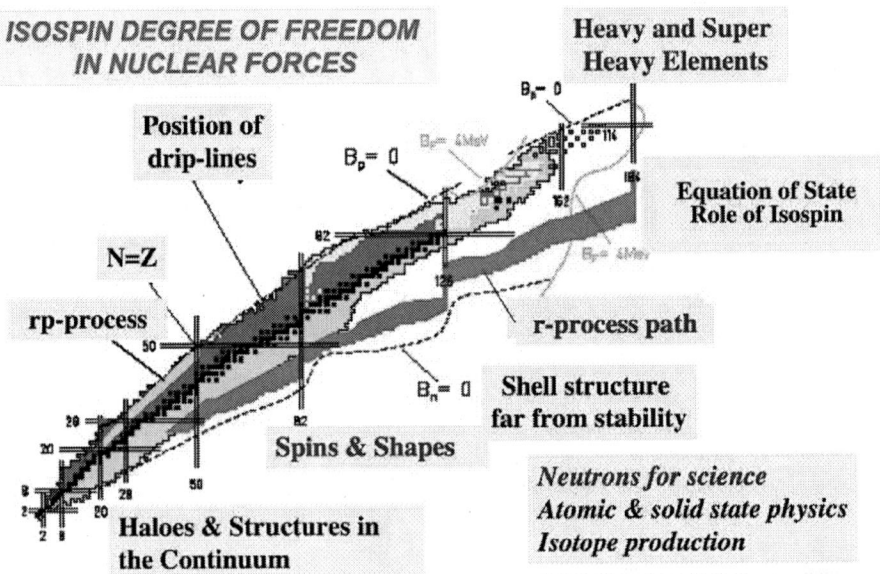

**FIGURE 3.** Main topics of the scientific case of SPIRAL 2.

## Physics Of Nuclei Far From Stability

A major part of the experimental and theoretical research program at the SPIRAL 2 facility will follow the fundamental motivation of basic nuclear science research, which is trying to establish a bridge between the nucleon-nucleon inside a nucleus and the underlying quarks and gluons. The modifications induced by the nuclear medium make the problem extremely complex. The result is that, although the reactions between nucleons in the vacuum have been studied in detail, the forces, which bind the nucleons in the nucleus, are neither well known nor really understood. However, setting up the interaction between its constituents is not enough to understand the properties of a complex object like the nucleus. One should also determine the organization of the nucleons and the phenomena at the origin of the observed structures.

This research on the nucleus and on the interactions between its constituents progresses using nuclei with unusual neutron-to-proton ratios, artificially produced in laboratories, which open up a considerable field of investigation. By allowing the study of very varied situations, the nuclei far from stability highlight the phenomena at the origin of the cohesion of the nucleus. SPIRAL 2 thanks to very high intensities of RNB will thus give access to a whole range of experiments (from elastic scattering to fusion-evaporation reactions), which are inaccessible today.

New exotic shapes and excitation modes, e.g. halo-like and molecular structures, and new modes of nuclear decay have been recently observed, while tests of fundamental symmetries, testing and refinement of the Standard Model of fundamental interactions, and exploration of the 'magic' numbers of protons and nuclei in very exotic nuclei are all enticing avenues of discovery.

## Nucleosynthesis

Nuclei were created during major events during the evolution of the Universe and are the testimony of the origin of matter: while the lightest elements were created in the first minutes following the Big Bang, heavier ones are synthesized in stars. However, these various nucleosynthesis processes involve radioactive nuclei. Due to their very short lifetimes, most of them have not survived long enough to be present on earth. To be studied these "exotic" nuclei must therefore be synthesized in laboratories. SPIRAL 2 will produce abundantly nuclei lying on or in a close proximity to the r and rp-process paths, opening up new fields of investigation. In particular, a new insight on nucleosynthesis might be achieved using transfer reactions (d,p), (p,d), etc. The production of radioactive targets in order to measure neutron capture cross-sections might help to elucidate several important problems in the understanding of the s-process.

## Neutrons For Science

One of the interesting possibilities, which will open with the SPIRAL 2 facility is related to the production of a high neutron flux in the energy range from several hundreds of keV up to about 40 MeV (fig. 4). Thanks to that, the facility will offer a unique opportunity for material irradiations and cross-section measurements, both for fission-related (notably accelerator driven systems (ADS) and Gen-IV fast reactors) and fusion-related research.

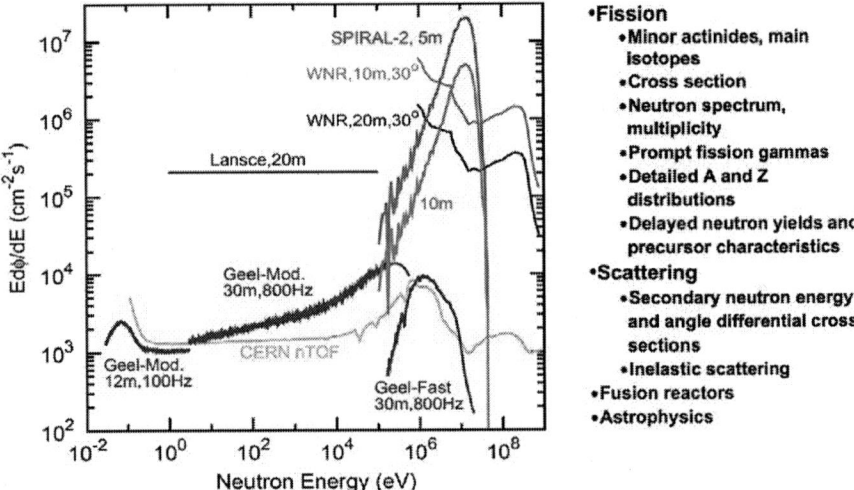

**FIGURE 4.** Neutron flux, which might be achieved at the "n-tof" facility at SPIRAL 2 compared to the existing facilities. A list of topics relevant to the scientific program at the proposed facility is shown.

The above experimental data require specific facilities like pulsed neutron beams for cross-section measurements by time-of-flight or dedicated irradiation stations for activation analysis. The high neutron fluxes with high and variable energy spectra available at SPIRAL 2 are very attractive to perform the measurements of the transmutation-incineration of nuclear waste and minor actinides in particular. The high neutron fluxes would allow the measurements of small reaction cross-sections and/or with very small targets, which might be rare, expensive, and in some cases radioactive. The energy range and conditions offered by SPIRAL 2 time-of-flight facility is complementary to other such facilities in Europe, notably GELINA of the European Commission's Joint Research Centre in Geel and the CERN based n_TOF facility. For example, the design of IFMIF facility requires reliable data up to 55 MeV neutron energy. In this case, the neutron spectrum will be determined by the activation of dosimetry foils, and for most of the suitable reactions the cross-sections have been measured once with unacceptable uncertainty or never at all and the evaluated data files are in disagreement. A lot of cross section measurements can be performed by activation technique, and the SPIRAL 2 fluxes are particularly well adapted for fast neutron measurements. Cross-sections of deuteron-induced reactions on structural materials are also needed, and a deuteron activation station in the converter cave could ensure such measurements up to 40 MeV. More details on this topic can be found in ref. 5.

# NEW EXPERIMENTAL AREAS AND NEW DETECTORS FOR SPIRAL 2

In the framework of the SPIRAL 2 project two new experimental areas will be constructed. One dedicated to the experiments with high intensity stable beams delivered by LINAG and one devoted to research program with low energy RNB proposed recently by the DESIR collaboration (fig. 5).

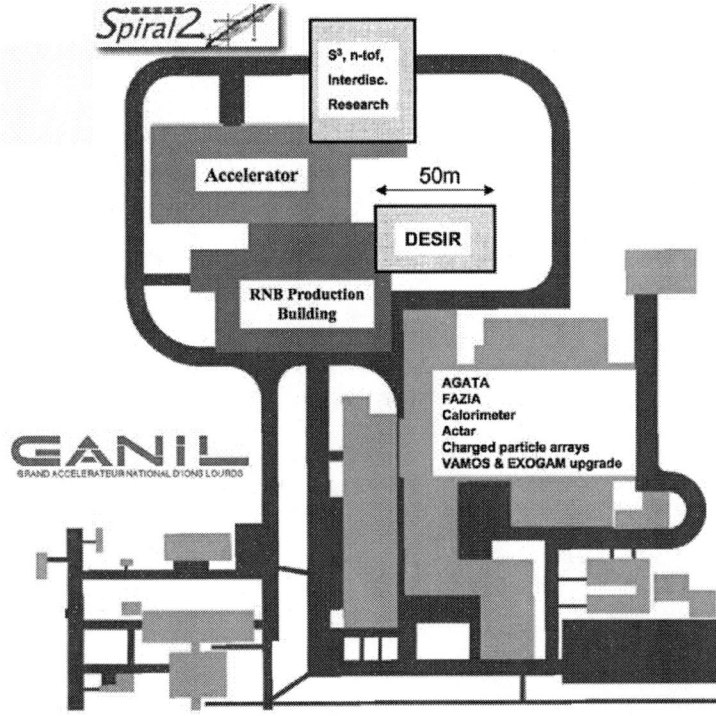

**FIGURE 5.** A preliminary layout of the GANIL/SPIRAL 1/SPIRAL 2 facility. Two new experimental areas (with contours) are shown and some of the proposed new detectors are listed.

Relatively moderate intensities and high cost of radioactive beams impose a use of the most efficient and innovative detection systems such as the magnetic spectrometer VAMOS, the 4Π gamma-array EXOGAM and AGATA as well as charged particle detectors like MAYA, MUST 2 and TIARA. Several new concepts of the detection systems (DESIR, FAZIA, SPAGA) and a new separator/spectrometer $S^3$ located in dedicated experimental halls are currently under consideration. Most of the existing detection systems and the existing experimental area should be adopted to take a full benefit of the high intensity (up to $10^{11}$ pps) RNB.

A process leading to the definition of the detectors and corresponding collaborations was initiated recently via call for letters of intent for SPIRAL 2 (see ref. 4 for details). The evaluation of the letters of intent by the SPIRAL 2 Scientific

Advisory Committee took place in the end of October 2006. It is foreseen that a call for full proposals will be launched in 2007 and signatures of Memorandums of Understanding related to the construction of new equipment will take place in 2008-2009.

## CONCLUSIONS

On May 23, 2005, the French government approved the construction of SPIRAL 2 a new 130 M€ RNB facility at GANIL. Its construction cost is shared by the French funding agencies CNRS/IN2P3 and CEA/DSM, the regional authorities of Basse-Normandie and international partners. The first beams are expected at SPIRAL 2 by 2011-2012. The project management group responsible for the construction phase was formed recently. Several agreements with international partners were signed recently. The construction of the SPIRAL 2 driver (LINAG) just started. The detailed definition of the RNB production building, of the experimental areas and of the dedicated detectors is entering in the decisive phase.

SPIRAL 2 will reinforce European leadership in the field of exotic nuclei and will serve a community of about 600 scientists.

## REFERENCES

1. www.ganil.fr
2. www.ganil.fr/research/developments/spiral2/origin.html.
3. www.nupecc.org
4. www.ganil.fr/research/developments/spiral2/index.html
5. X. Ledoux et al., Neutrons for Science, SPIRAL 2 Letter of Intent No. 17, www.ganil.fr/research/developments/spiral2/loi_texts.html
6. www.eurisol.org

# RISING: Gamma-ray Spectroscopy with Radioactive Beams at GSI

P. Doornenbal[a,b], A. Bürger[c], D. Rudolph[d], H. Grawe[b], H. Hübel[c],
P.H. Regan[e], P. Reiter[a], A. Banu[b], T. Beck[b], F. Becker[b], P. Bednarczyk[b,f],
L. Caceres[b,g], H. Geissel[b], J. Gerl[b], M. Górska[b], J. Grębosz[b,f],
M. Kavatsyuk[b], O. Kavatsyuk[b], A. Kelic[b], I. Kojouharov[b], N. Kurz[b],
R. Lozeva[b,h], F. Montes[b], W. Prokopowicz[b], N. Saito[b], T. Saito[b],
H. Schaffner[b], S. Tashenov[b], H. Weick[b], E. Werner-Malento[b,i],
M. Winkler[b], H.-J. Wollersheim[b], A. Al-Khatib[c], L.-L. Andersson[d],
L. Atanasova[h], D.L. Balabanski[j], M.A. Bentley[k], G. Benzoni[l], A. Blazhev[a],
A. Bracco[l], S. Brambilla[l], C. Brandau[b,e], P. Bringel[c], J.R. Brown[k],
F. Camera[l], S. Chmel[c], E. Clément[m], F.C.L. Crespi[l], C. Fahlander[d],
A.B. Garnsworthy[e,n], A. Görgen[m], G. Hammond[o], M. Hellström[d],
R. Hoischen[d], H. Honma[p], E.K. Johansson[d], A. Jungclaus[g], M. Kmiecik[f],
W. Korten[m], A. Maj[f], S. Mandal[q], W. Meczynski[f], B. Million[l], A. Neußer[c],
F. Nowacki[r], T. Otsuka[s,t], M. Pfützner[i], S. Pietri[e], Zs. Podolyák[e],
A. Richard[a], M. Seidlitz[a], S.J. Steer[e], T. Striepling[a], T. Utsuno[s,t],
J. Walker[g], N. Warr[a], C. Wheldon[u] and O. Wieland[l]

[a]*Institut für Kernphysik, Universität zu Köln, D-50937 Köln, Germany*
[b]*Gesellschaft für Schwerionenforschung, D-64291 Darmstadt, Germany*
[c]*Helmholtz-Institut für Strahlen- und Kernphysik, Universität Bonn, D-53115 Bonn, Germany*
[d]*Department of Physics, Lund University, S-22100 Lund, Sweden*
[e]*Department of Physics, University of Surrey, Guildford, GU2 7XH, UK*
[f]*The Henryk Niewodniczański Institute of Nuclear Physics, PL-31-342 Kraków, Poland*
[g]*Departamento de Física Teórica, Universidad Autonoma de Madrid, E-28049 Madrid, Spain*
[h]*Faculty of Physics, University of Sofia, BG-1164 Sofia, Bulgaria*
[i]*Institute of Experimental Physics, Warsaw University, PL-00-681 Warsaw, Poland*
[j]*Institute for Nuclear Research and Nuclear Energy, Bulgarian Academy of Sciences, BG-1784 Sofia, Bulgaria*
[k]*Department of Physics, University of York, York, Y01 5DD, UK*
[l]*Dipartimento di Fisica, Università di Milano, and INFN sezione di Milano, I-20133 Milano, Italy*
[m]*DAPNIA/SPhN, CEA Saclay, Gif-sur-Yvette, France*
[n]*WNSL, Yale University, New Haven, CT 06520-8124, USA*
[o]*School of Chemistry and Physics, Keele University, Staffordshire, ST5 5BG, UK*
[p]*University of Aizu, Fukushima 965-8580, Japan*
[q]*Department of Physics and Astrophysics, University of Delhi, Delhi - 110 007, India*
[r]*IReS, Université Louis Pasteur, Strasbourg, France*
[s]*Department of Physics and Center for Nuclear Study, University of Tokyo, Hongo, Tokyo 113-0033, Japan*
[t]*RIKEN, Hirosawa, Wako-shi, Saitama 351-0198, Japan*
[u]*Department SF7, Hahn-Meitner-Institut, D-14109 Berlin, Germany*

**Abstract.** The Rare Isotope Spectroscopic INvestigation at GSI (RISING) project is a major pan-European collaboration. Its physics aims are the studies of exotic nuclear matter with abnormal proton-to-neutron ratios compared with naturally occurring isotopes. RISING combines the FRagment Separator (FRS) which allows relativistic energies and projectile fragmentation reactions with EUROBALL Ge Cluster detectors for $\gamma$ spectroscopic research. The RISING setup can be used in two different configurations. Either the nuclei of interest are investigated after being stopped or the heavy ions hit a secondary target at relativistic energies and the thereby occurring excitations are studied. For the latter case, MINIBALL Ge detectors and the HECTOR array are used in addition. Example achievements of the Fast Beam setup are presented and compared to various shell model calculations, while for the Stopped Beam setup initial results are shown.

**Keywords:** Gamma-ray spectroscopy; nuclear structure; excitation probabilities; nuclear isomers
**PACS:** 21.60.Cs; 23.20.Lv; 25.70.Mn

# INTRODUCTION

One of the key issues in current nuclear structure physics research is the exploration of nuclei far away from the line of $\beta$-stability. This has led to the development of different techniques that permit the study of specific radioactive nuclei. In the case of projectile fragmentation or relativistic fission heavy ion beams are accelerated to an energy of up to $1\,A$ GeV and then strike a thick target. Since this produces a vast number of different nuclei, the fragments of interest must be selected using their magnetic rigidity after the target, which is done at the FRS at GSI [1]. Here, two pairs of dipoles are used in the so called $B\rho$-$\Delta E$-$B\rho$ mode by placing a wedge-shaped aluminum degrader at the central focal plane of the FRS and setting the second pair of dipoles according to the energy loss of the sought-after fragments. The ions passing from the intermediate to the final focus of the FRS are identified on an event-by-event basis using their magnetic rigidity $B\rho$, their time of flight between two scintillation detectors, and their energy loss in a multi sampling ionization chamber.

# RISING SETUP

When the heavy ions reach the final focus of the FRS, they can hit a secondary target, which enables the study of Coulomb excitation at relativistic energies or fragmentation processes towards even more exotic nuclei. Alternatively, they are implanted into a passive stopper followed by $\gamma$ ray measurements of decays of isomeric states produced by the fragmentation or fission process in the primary target. This paper reports on initial and selected results of these two major RISING setups, the former being the Fast Beam setup [2], the latter being the Stopped Beam setup [3–5].

## The Fast Beam Setup

In the Fast Beam setup relativistic Coulomb excitation and two-step fragmentation experiments were performed with energies in the range of 100 to 600 $A$ MeV. A $^{197}$Au reaction target with thicknesses from 0.4 to 2.0 g/cm$^2$ was used in the case of Coulomb excitation, while two-step fragmentation experiments were carried out with a 0.7 g/cm$^2$ $^9$Be target. The resulting reaction products were identified with respect to their charge and mass with the calorimeter telescope array CATE [6], consisting of $3 \times 3$ Si-CsI(Tl) modular $\Delta$E-E telescopes mounted 1400 mm downstream of the target. For the proper Doppler correction, the position sensitive CATE Si detectors and an identical Si detector placed directly after the target served as tracking detectors. In the case of Coulomb excitation, unwanted nuclear contributions could be excluded by selecting sufficiently large impact parameters. This impact parameter could be obtained by tracking the heavy ions with position sensitive multiwire detectors upstream the target and the afore mentioned Si detectors.

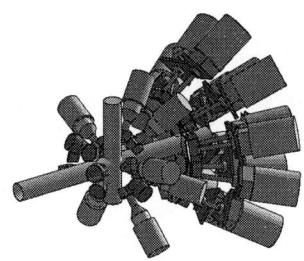

**FIGURE 1.** Drawing of the $\gamma$ ray detector setup during the Fast Beam campaign. See text for details.

In order to measure $\gamma$ rays emitted by excited states, the target area was surrounded by numerous detectors, as shown in Fig. 1: (i) 15 Cluster Ge detectors [7], positioned in three rings at extreme forward angles of 16°, 33°, and 36° at distances of 700 to 1400 mm, (ii) eight six-fold segmented MINIBALL triple Ge detectors [8] at distances of 200 to 400 mm, arranged in two rings with central angles of 51° and 85°, (iii) the HECTOR array [9, 10] at a distance of 300 mm, consisting of eight large volume BaF$_2$ detectors, situated at angles of 85° and 142°. In its least distance configuration, the efficiency for a $\gamma$ ray of 1332 keV emitted from heavy ions at 100 $A$ MeV was simulated to be 1.7% for the Cluster detectors, 3.8% for MINIBALL and 1.7% for HECTOR, not including add-back events.

## The Stopped Beam Setup

The Stopped Beam setup can be used in two configurations to measure $\gamma$ rays: Isomeric states produced in the fragmentation process are implanted into a passive stopper or heavy ions are implanted into an active $\beta$-sensitive stopper, thus enabling the search for excited states of exotic nuclei following $\beta$-decay. In contrast to the Fast Beam setup, a second degrader of variable thickness was put at the final focus of the FRS. This allowed the energy loss of the heavy ions to be tuned in such a way that the stopper could be kept at a moderate thickness.

The fifteen Ge Cluster detectors surrounding the stopper in the Stopped Beam setup are shown in Fig. 2. The Cluster detectors were placed in three rings of five detectors at angles of 51°, 90°, and 129° relative to the beam axis at a distance of approximately

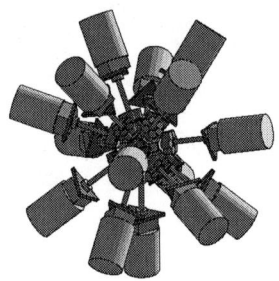

**FIGURE 2.** Drawing of the γ ray detector setup during the Stopped Beam campaign. See text for details.

22 cm from the center of the final focal plane of the FRS. The photopeak γ ray efficiency of the Stopped Beam Ge detector setup was measured to be 9(1)% at 1332 keV, not including add-back events [5]. Due to the high granularity of the Ge detector array of 105 crystals in total, a higher loss of efficiency was avoided after the prompt γ flash that was produced in the stopping process of the heavy ions. In order to identify the metastable states after the implantation, each γ ray was time stamped using a 40 MHz clock, which was part of the DGF4 timing and energy signal processing [11]. For both short-lived isomers and redundancy a conventional timing branch was installed in parallel and digitized with a short-range ($t \leq 1.0$ μs) and a long-range $t \leq 0.8$ ms VME TDC. This enabled the measurement of decays from isomeric states with half-lives in the region between several tens of ns up to 1 ms.

## SELECTED EXPERIMENTAL RESULTS

### Monopole Driven Shell Structure

During the Fast Beam campaign one of the key interests was the investigation of the monopole driven shell structure. This monopole part of the residual interaction controls the propagation of single particle energies with increasing occupation of a major shell. It causes a change of oscillator shell closures with magic numbers for very neutron rich nuclei of $N = 8, 20$ towards $N - 2 \times N_{HO} = 6, 16(14)$ [12, 13]. $N_{HO}$ is the harmonic oscillator main quantum number. The weak $N = 40$ harmonic oscillator case should shift to a $N = 32, 34$ subshell closure, where the ambiguity for $N_{HO} > 1$ stems from the presence of $j = 1/2$ orbits which strongly mix with the neighboring higher-spin orbitals [13]. The monopole residual interaction is also expected to be of isospin symmetric nature, hence its effects can be studied by comparing the nuclear structure of the $N = 20$ isotones below $^{40}$Ca with their $Z = 20$ mirror nuclei.

The RISING Fast Beam setup gives access to excitation energies $E_{2_1^+}$ of $I^\pi = 2_1^+$ and $B(E2; 2_1^+ \to 0^+)$ values that can both be used as signatures for shell structure. Two type of experiments were performed: A Coulomb excitation experiment of the neutron rich $^{54,56,58}$Cr, which are located in between the $N = 40$ subshell closure across a deformed region to spherical nuclei at $N = 28$, second, and a two-step fragmentation experiment to investigate the mirror energy difference, defined as $\Delta E_M = E_x(I, T_z = -T) - E_x(I, T_z = +T)$, between $^{36}$Ca and $^{36}$S.

## The Subshell Closure at $N = 32, 34$ — Coulomb Excitation of $^{54,56,58}$Cr

Experimentally, possible subshell closures may develop at $N = 32, 34$ in neutron rich Ca ($Z = 20$) isotopes as indicated by a rise in the $2_1^+$ energy of $^{52}$Ca [14]. The Ti and Cr ($Z = 22, 24$) isotopes exhibit a maximum of those energies at $N = 32$ [15–17].

Besides the $2_1^+$ energies, $B(E2; 2_1^+ \to 0^+)$ values provide crucial information to test the evolution of subshell structures. Therefore, three experiments were performed to measure the Coulomb excitation of $^{54}$Cr, $^{56}$Cr and $^{58}$Cr, where the known $B(E2; 2_1^+ \to 0^+)$ value in $^{54}$Cr served as normalization and reference for possible systematic errors in the analysis.

A primary beam of $^{86}$Kr with an energy of 480 $A$ MeV was incident on a 2.5 g/cm$^2$ $^9$Be target. Out of the fragmentation products, the Cr isotopes were selected and incident on a 1.0 g/cm$^2$ $^{197}$Au target at energies of around 136 $A$ MeV. More details of the experiment are given in Ref. [18]. The obtained Doppler corrected Cr spectra yield values of 8.7(30) W.u. for $^{56}$Cr and 14.8(42) W.u. for $^{58}$Cr. These results are shown in Fig. 3 together with the experimental $B(E2; 2_1^+ \to 0^+)$ and $2_1^+$ systematics of the Ti and Cr isotopes and are compared to shell model calculations (KB3G, GXPF1, GXPF1A) [19–21].

The local peak in the $N = 32$ $2_1^+$ energies is confirmed by a minimum of the $B(E2; 2_1^+ \to 0^+)$ values in the present experiment and a recent result for Ti isotopes [19]. For $N = 34$, however, the gap is not developed in Cr and Ti which leaves $^{54,56}$Ca as

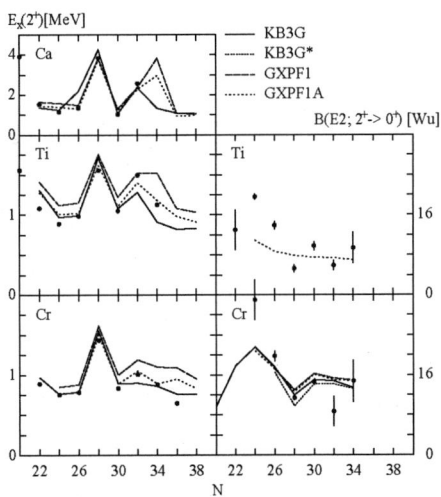

**FIGURE 3.** Experimental $2_1^+$ energies and $B(E2; 2_1^+ \to 0^+)$ values of neutron rich Ca, Ti and Cr isotopes in comparison to different shell model calculations.

the crucial experimental probes. The shell model calculations reproduce the variation in the $2_1^+$ energies but fail to reproduce the $B(E2;2_1^+ \to 0^+)$ values which stay almost unchanged in the different approaches from $N = 30$ to 34.

## Mirror Symmetry in $A = 36$, $T = 2$ Nuclei

Going along the $N = 20$ isotones south of $^{40}$Ca, the shell stabilization of $^{36}$S, $^{34}$Si and the shell quenching in $^{32}$Mg are expected to be caused by the monopole part of the two-body interaction. This scenario is anticipated to be symmetric in isospin and may not or little affected by neutron binding energy differences [12, 13]. It can be verified in the $N = 20$ mirror region along the light Ca ($Z = 20$) isotopes. From the Ca isotopes, detailed spectroscopy exists only for $^{38}$Ca [22], while no excited states are known for the $N = 16$ isotope, thus the mirror nucleus of $^{36}$S is of high interest.

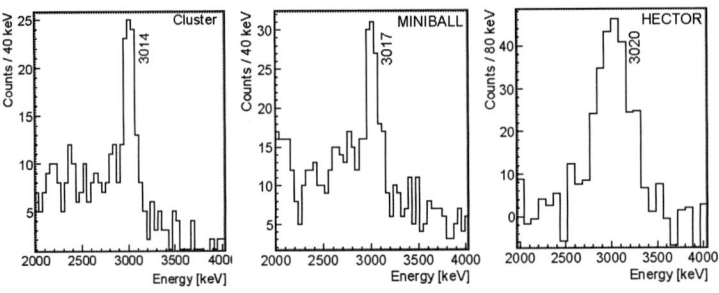

**FIGURE 4.** Doppler corrected $^{36}$Ca gated $\gamma$ ray spectra measured with the Cluster, MINIBALL and HECTOR detectors. For the HECTOR array the background was subtracted.

A beam of $^{40}$Ca at an energy of 420 $A$ MeV was bombarded on a 4.0 g/cm$^2$ $^9$Be target. As a secondary beam $^{37}$Ca was selected and incident on a 0.7 g/cm$^2$ $^9$Be secondary target to populate excited states in $^{36}$Ca. In Fig. 4 Doppler corrected $\gamma$ ray spectra of $^{36}$Ca are shown for the MINIBALL, Cluster and HECTOR detectors. The energy of the $2_1^+ \to 0^+$ transition was determined to 3015(16) keV, yielding a value of $\Delta E_M = -276(16)$ keV for the $A = 36$, $T = 2$ mirror pair. This value is significantly larger than mirror energy differences observed for $T = 1$ states in the $sd$ and $pf$ shell [15]. Other known $T = 2$ mirrors in the $sd$ shell, $A = 24$ and 32, also exhibit much smaller mirror energy differences of $-102(11)$ keV and $-117(12)$ keV, respectively [23, 24].

In our approach to understand the large $\Delta E_M$, we have used the experimental single-particle energies from the $A = 17$, $T = 1/2$ mirrors and applied these onto a modified isospin symmetric USD interaction [25, 26] in a shell model calculation. Monopole corrections were applied to reproduce the $Z, N = 14, 16$ shell gaps, the $I^\pi = 2_1^+$ excitation energies and the $^{40}$Ca single hole energies. The results of this calculation are shown in Table 1 for $^{36}$Ca and $^{36}$S, yielding a value of $\Delta E_M = -268$ keV. This is in close agreement to the experimental result and shows that the experimental single-particle energies may account empirically for the one-body part of Thomas-Ehrman and/or Coulomb effects [27, 28], since the isospin symmetry is preserved in the interactions' two-body matrix elements but not in the single-particle energies used.

**TABLE 1.** Comparison of shell model calculation with experimental values for $^{36}$Ca and $^{36}$S.

|  | $E_{2_1^+}$ [keV] | | $\pi$-gap [MeV] | | $\nu$-gap [MeV] | |
|---|---|---|---|---|---|---|
|  | Exp.* | SM† | Exp. | SM | Exp. | SM |
| $^{36}$Ca | 3015(16)** | 3290 | 4.55(30) |  | 4.16(9) | 3.999 |
| $^{36}$S | 3290.9(3) | 3558 | 4.524(2)‡ | 4.244 | 5.585 |  |

\* From Ref. [15].
† Shell model calculation, see text for details.
\*\* This work.
‡ Coulomb Corrected.

## First Results from the Stopped Beam Setup

The Stopped Beam campaign started with the investigation of heavy odd-odd $N = Z$ nuclei along the proton drip-line [3]. It was followed by isomeric decay studies in the region of the doubly magic $^{56}$Ni, $^{132}$Sn and $^{208}$Pb [29, 30]. We will focus on the results of $^{54}$Ni, produced after the fragmentation of a $^{58}$Ni primary beam, which demonstrate nicely the excellent possibilities of combining the FRS with the RISING $\gamma$ ray spectrometer in its Stopped Beam configuration.

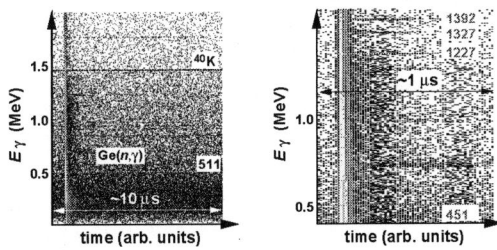

**FIGURE 5.** $^{54}$Ni gated two-dimensional matrices of $\gamma$ energy versus time after implantation.

After the identification of $^{54}$Ni reaching the final focal plane with the FRS detectors, a correlation matrix between the $\gamma$ energy and the time after the implantation can be generated. Two examples are shown in Fig. 5. On the left hand, a more general view provides an indication of the so-called prompt flash (vertical line), which marks the implantation time $t = 0$, as well as horizontal lines arising from, for example, room background or $(n,\gamma)$ reactions in the Ge detectors. But more than that, also distinct and as a function of time fading horizontal lines are visible, indicating decays from isomeric states. Some of these are highlighted on the right hand side of Fig. 5, which zooms into the energy-time region of interest for $^{54}$Ni. By setting cuts on distinct times after the implantation and comparing the intensities, information on the half-lives of the isomers can be obtained. Moreover, provided that the spectra are rich enough in statistics, $\gamma\gamma$ coincidence measurements can help to examine the level structure. This is done in Fig. 6 where in the upper background subtracted spectrum a simple projection of the two-dimensional panel is shown for the time range $0.05 \ \mu s \leq t \leq 1.0 \ \mu s$. In this spectrum six discrete $\gamma$ transitions at 146, 451, 1227, 1327, 1392, and 3241 keV are visible and all have a lifetime of $\tau \sim 220$ ns. In the lower panel, however, gates on the already established $6^+ \rightarrow 4^+ \rightarrow 2^+ \rightarrow 0^+$ cascade [31–33] are set at energies of 451, 1227, and 1392 keV. Since the lines observed at 146 and 3241 keV are clearly in coincidence with the cascade, they are suggested to be the $10^+ \rightarrow 8^+$ and $8^+ \rightarrow 6^+$ transitions [34], what also follows from the known mirror isomer $^{54}$Fe [15]. Even a weak line at 3386 keV is visible, which can be associated to the small $10^+ \rightarrow 6^+$ $E4$ branch. The most surprising result is, however, that none of the observed $\gamma$ rays comes in coincidence with the 1327 keV line, which is seen only in the singles spectrum. Since it has the same energy as the $9/2^- \rightarrow 7/2^-$ ground state transition in $^{53}$Co, it is suggested that this $9/2^-$ state in $^{53}$Co can be populated via a direct proton decay ($Q_p \sim 1.3$ MeV) from the isomeric state in $^{54}$Ni [34].

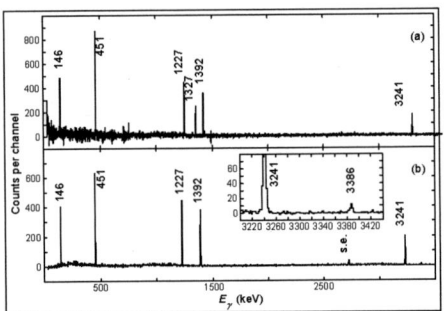

**FIGURE 6.** The upper panel (a) shows the energy projection of the two-dimensional matrix of $^{54}$Ni for the range $0.05~\mu s \leq t \leq 1.0~\mu s$. In the lower panel (b), the $\gamma\gamma$ coincidence with one of the transitions at 451, 1227, and 1392 keV demonstrates the correlation to other observed lines.

## SUMMARY

The shown exemplary results demonstrate the possibilities of high resolution $\gamma$ ray spectroscopy at relativistic energies utilizing the two-step fragmentation or Coulomb excitation technique with RISING, as has been shown for the results of $^{54,56,58}$Cr and $^{36}$Ca. A wealth of other interesting results have been obtained for example in relativistic Coulomb excitation of $^{108,112}$Sn, $^{134}$Ce, and $^{136}$Nd [35, 36]. For the Stopped Beam campaign, a wide range of nuclei have been populated in isomeric states following fragmentation and fission: For example the long sought-after $^{82}$Nb, $^{86}$Tc, $^{130}$Cd, and $^{204}$Pt [29, 30, 37, 38]. Here, $^{54}$Ni was chosen to illustrate the large capabilities of a highly efficient $\gamma$ ray spectrometer used in combination with the FRS. In the future, a series of active stopper experiments is foreseen to perform $\beta$-delayed $\gamma$ ray spectroscopy.

## ACKNOWLEDGMENTS

The collaboration would like to thank the EUROBALL Owners Committee, the MINI-BALL collaboration and the HECTOR collaboration for providing their $\gamma$ detectors to the RISING project. We also acknowledge the high beam intensities provided by the accelerator department at GSI. This permits us to study very exotic nuclei. This work is supported by the European Commission contract No. 506065 (EURONS), the German BMBF under grant Nos. 06BN-109, 06K-167, the Swedish Research Council, the Polish State Committee for Scientific Research (KBN grant No. 620/E-77/SPB/GSI/P-03/DWM105/2004-2007), the Bulgarian Science Fund under grant No. VUF06/05, and the EPRSC(UK).

## REFERENCES

1. H. Geissel et al., Nucl. Instr. Meth. B **70** (1992) 286.
2. H.J. Wollersheim et al., Nucl. Instr. Meth. A **537** (2005) 637.
3. P.H. Regan, et al., Proc. of the NN06 conference, Nucl. Phys. **A**, in press.
4. S. Pietri et al., Proc. of the CAARI'06 conference, to be published in Nucl. Inst. Meth. **B**.
5. S. Pietri et al., Proc. of the 41$^{st}$ Zakopane School of Physics, to be published in Act. Phys. Pol. **B**.
6. R. Lozeva et al., Nucl. Instr. Meth. A **562** (2006) 298.
7. J. Eberth et al., Nucl. Inst. Meth. A **369** (1996) 135.
8. J. Eberth et al., Prog. Part. Nucl. Phys. **46** (2001) 389.
9. A. Maj et al., Nucl. Phys. A **571** (1994) 185.
10. F. Camera, Ph. D. Thesis, University of Milano, Italy, 1992.
11. M. Pfützner et al., Nucl. Inst. Meth. A **493** (2002) 155.
12. T. Otsuka et al., Phys. Rev. Lett. **87** (2001) 082502.
13. H. Grawe, Act. Phys. Pol. **B 34** (2003) 2267.
14. A. Huck et al., Phys. Rev. C **31** (1985) 2226.
15. ENSDF database, http://www.nndc.bnl.gov/ensdf/.
16. S.N. Liddick et al., Phys. Rev. Lett. **92** (2004) 072502.
17. P.F. Mantica et al., Phys. Rev. C **67** (2003) 014311.
18. A. Bürger et al., Phys. Lett. **B 622** (2005) 29.
19. D.C. Dinca et al., Phys. Rev. C **71** (2005) 041302.
20. E. Caurier et al., Eur. Phys. J. **A 15** (2002) 145.
21. M. Honma et al., Phys. Rev. C **69** (2004) 034335.
22. P.D. Cottle et al., Phys. Rev. C **60** (1999) 031301.
23. S. Kanno et al., Prog. Theor. Phys. (Kyoto), Suppl. **146** (2002) 575.
24. P.D. Cottle et al., Phys. Rev. Lett. **88** (2002) 172502.
25. B.A. Brown, B.H. Wildenthal, Ann. Rev. of Nucl. Part. Sci. **38** (1988) 29.
26. Y. Utsuno et al., Phys. Rev. C **60** (1999) 054315.
27. R.G. Thomas, Phys. Rev. **88** (1952) 1109.
28. J.B. Ehrman, Phys. Rev. **81** (1951) 412.
29. M. Górska, A. Jungclaus, M. Pfützner et al., to be published.
30. Zs. Podolyák et al., Proc. of the RNB7 conference, to be published in Eur. Phys. J. **A**.
31. K.L. Yurkewicz et al., Phys. Rev. C **70** (2004) 054319.
32. K. Yamada et al., Eur. Phys. J. **A25** S1 (2005) 409.
33. A. Gadea et al., Phys. Rev. Lett. **97** (2006) 152501.
34. D. Rudolph et al., to be published.
35. A. Banu et al., Phys. Rev. C **72** (2005) 061305(R).
36. T. Saito et al., submitted to Phys. Rev. Lett.
37. L. Caceres et al., Proc. of the 41$^{st}$ Zakopane School of Physics, to be published in Act. Phys. Pol. **B**.
38. A.B. Garnsworthy et al., Proc. of the 41$^{st}$ Zakopane School of Physics, to be published in Act. Phys. Pol. **B**.

> # (p,2p) Reactions on Carbon Isotopes: p($^{9-16}$C,2p) $^{8-15}$B at 250AMeV

Toshio Kobayashi, Kazutaka Ozeki[1], Kiwamu Watanabe, Yohei Matsuda, Yoko Seki, Tokukazu Shinohara, Toshiya Miki, Yuki Naoi, Hideki Otsu[2], Shigeru Ishimoto[3], Shoji Suzuki[3], Yutaka Takahashi[4], and E. Takada[5]

*Department of Physics, Tohoku University, 2-1 Aoba, Aramaki, Aoba, Sendai 980-8578, Japan*
[1] *CYRIC, Tohoku University, 2-1 Aoba, Aramaki, Aoba, Sendai 980-8578, Japan*
[2] *RIKEN, 2-1 Hirosawa, Wako, Saitama 351-0198, Japan*
[3] *KEK, 1-1 Oho, Tsukuba, Ibaraki 305-0801, Japan*
[4] *RCNP, Osaka University, 10-1 Mihogaoka, Ibaraki, Osaka 567-0047, Japan*
[5] *NIRS, 4-9-1 Anagawa, Inage, Chiba, Chiba 263-8555, Japan*

**Abstract.** Proton knockout reactions on carbon isotopes, p($^{9-16}$C,2p)$^{8-15}$B, at 250MeV/A were performed for systematic information on weakly to strongly bound 1p valence protons and deeply bound 1s protons. Various information such as energy gap between 1s and 1p orbits, widths of momentum distributions for 1s and 1p orbits, relative (p,2p) yields are obtained.

**Keywords:** knockout reaction, single-particle property, momentum distribution.
**PACS:** 25.60.-t

## INTRODUCTION

Proton knockout reaction, (p,2p), in the inverse kinematics is one of the experimental methods to study the single-particle properties of bound protons [1] in unstable nuclei. Bound proton in the beam is knocked out by quasifree proton-proton scattering, and the hole state is produced. By measuring four-momenta of two protons in the final state, momentum distribution, angular momentum, and separation energy are deduced. Measurement in the inverse kinematics provides additional information. Since hole state is produced in the beam velocity, decay mode of the hole states are measured with high efficiency by detecting particles in the forward direction.

Carbon isotopes, $^9$C to $^{16}$C, are selected as targets. By going from proton-rich side to neutron-rich side, neutron separation energy of valence orbit comes down from 20 MeV to 1 MeV, while proton separation energy ($S_p$) of the valence orbit (p$_{3/2}$) goes up from 1.3 MeV for $^9$C to 22.6 MeV for $^{16}$C. Protons in core orbit (s$_{1/2}$) are very strongly bound from about 20 MeV to 50 MeV. Therefore (p,2p) reactions on carbon isotopes provide a chance to study the variation of proton single-particle orbits, from weakly bound orbit to strongly bound orbit for 1p$_{3/2}$, and very strongly bound orbits for 1s$_{1/2}$.

# EXPERIMENT

Measurements were performed at HIMAC (Heavy Ion Medical Accelerator in Chiba) accelerator facility in NIRS (National Institute of Radiological Sciences). $^{12}$C and $^{18}$O beams accelerated up to 350-400 MeV/A by the synchrotron were used to produce secondary carbon isotope beams at 250MeV/A. Using the secondary beam line (SB2), carbon isotopes are separated and momentum tagged. Beam intensity was $10^4$ to $10^5$ particles/spill, depending on isotopes. Experimental setup at F3 is shown in Fig. 1. Detector system consists of beam detectors, solid hydrogen target of 5mm thickness, two-arm proton telescopes, and forward magnetic spectrometer. Solid hydrogen target was essential to avoid background and to reduce multiple Coulomb scattering. Proton telescopes are set at 39 degrees to detect protons up to 210 MeV. Forward particles are measured and identified by a combination of magnetic analysis, charge measurement, and TOF measurement.

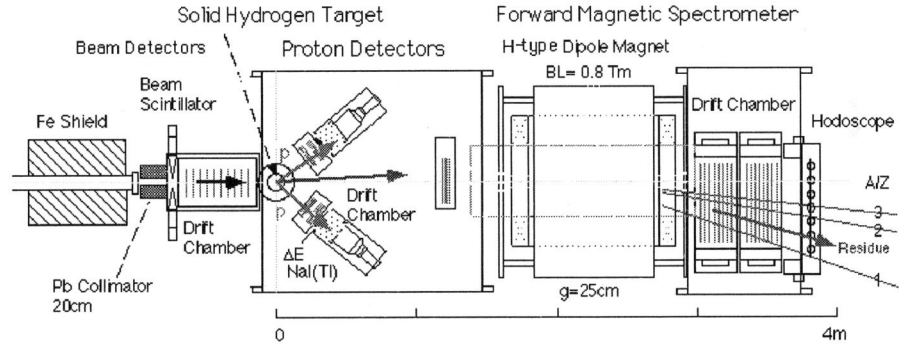

**FIGURE 1.** Experimental Setup at F3.

# Results

Proton separation energy ($S_p$) distributions are obtained from four momenta of two protons. The spectra consist of sharp peak corresponding to p-hole states and broad bump corresponding to s-hole states. Transitions to Boron ground state are selected by requiring mass-identified Boron isotopes, $^{A-1}$B, detected in the forward direction and setting appropriate gate on $S_p$ spectra. The separation-energy resolution is about 1.3 MeV rms in the present measurement, mainly limited by the angular resolution of the detector system. Transitions to s-hole states are selected by tagging no Boron isotopes in the forward direction and set gate above proton emission threshold on $S_p$ spectra. The latter selection requires that hole states decay by charged-particle emission. Results are shown in Fig. 2. Separation energy distribution for $^{12}$C is consistent with that obtained by (p,2p) reactions on $^{12}$C using 400MeV proton beams [2].

From present measurement, s-hole states in carbon isotopes are systematically observed. After correcting for the acceptance, $S_p$ distributions are fitted by the functional form, and the peak value and the width of the s-hole states are obtained.

Energy gap between 1p and 1s orbits is shown in Fig. 3. Gap energy has a minimum at around A=12 and becomes wider on both proton-rich and neutron-rich sides. Attractive force between $\nu 1p_{1/2}$ and $\pi 1p_{3/2}$ may explain the increase from $^{12}$C to neutron-rich side. Since the identification of s-hole states in $^{9}$C has some ambiguities at the moment, and the data for $^{9}$C is tentative.

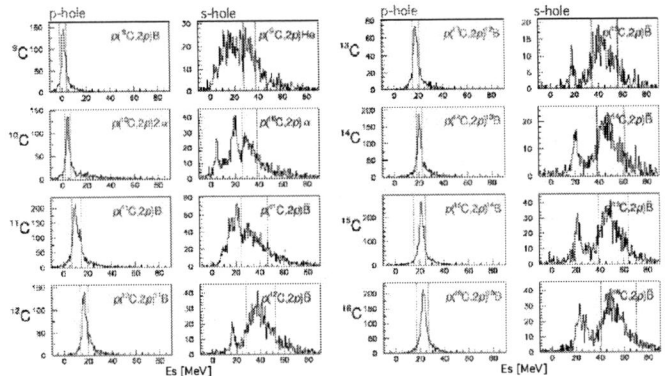

**FIGURE 2.** Separation-energy distributions for p-hole states (left) and s-hole states (right).

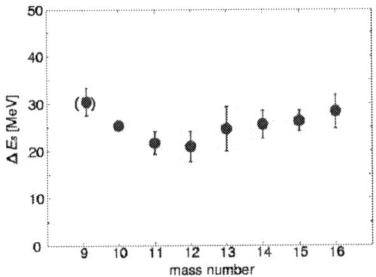

**FIGURE 3.** Energy gap between 1s and 1p orbits in carbon isotopes.

Momentum distributions of bound protons in 1s and 1p orbit are obtained by setting appropriate gates on $S_p$ spectra as shown in Fig. 2. Effects of finite geometrical acceptance are corrected for by selecting regions in the momentum space. Two kinds of momentum distributions, $d\sigma/dq$ and $d\sigma/dq_z$, are obtained after correcting for the acceptance, and fitted by using the functional form, $d^3\sigma/d\vec{q}^{\,3} \propto q^{2l} \exp(-q^2/\sigma_l^2)$ for $l$=0 and 1, assuming Harmonic-oscillator-type wave functions, where $z$ and $l$ are beam direction and angular momentum, respectively. Angular momentum $l$ can be uniquely identified, except few cases, by comparing reduced $\chi^2$ of the fitting. Although momentum widths of s-hole states for $^{13}$C to $^{15}$C can not be deduced from $d\sigma/dq$ distribution due to statistics, momentum widths for other cases are consistent with each other from two methods. The widths $\sigma_l$ are summarized in Fig. 4. The momentum widths for $^{12}$C are consistent with those obtained by $^{12}$C(e,e'p) reactions [3].

Momentum width of p-orbit increases monotonically from proton-rich to neutron-rich side, reflecting weak binding nature of valence protons in the proton-rich side. Momentum width of s-orbit is rather constant below A=12, and starts to increase above A=12. Simple potential model after adjusting the proton separation energy qualitatively reproduces the tendency of momentum widths of p-orbits. On the other hand, the same model gives constant momentum width, independent of the separation energy, for s-orbits, contradicting with the experimental widths.

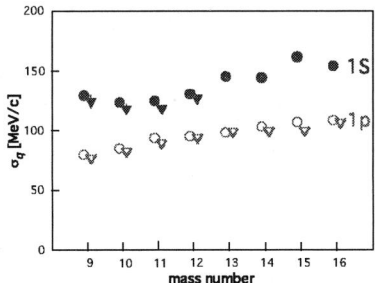

**FIGURE 4.** Momentum width $\sigma_1$ (Harmonic oscillator-type) of p-orbits (open) and s-orbits (closed). Triangles are obtained from $d\sigma/dq$ distributions, and circles from $d\sigma/dq_z$ distributions.

From the momentum widths of s-orbit, root mean square charge radius of 1s protons in the carbon isotopes can be deduced, by assuming Harmonic oscillator-type wave function. The charge radius decreases from 2.0 fm in the proton-rich side to 1.5 fm in the neutron-rich side. The present observation indicates that the proton radius of the core is shrinking towards neutron-rich side.

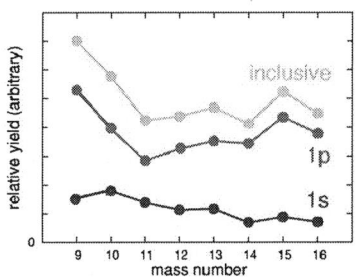

**FIGURE 5.** Relative (p,2p) yield.

Relative (p,2p) yields are obtained after the acceptance correction. The results are summarized in Fig. 5. Total yield is rather constant between $^{11}$C and $^{16}$C, and the yield is about 50% larger for $^{9}$C. When yields are divided into p orbit and s orbit contributions, it is the p orbit contribution, which is increasing at $^{9}$C. Since the yield is expected to be proportional to the spectroscopic factor, present observation also indicates that the spectroscopic factor for the weakly bound valence proton is larger, as in the case of knockout reactions [4]. There is a tendency that the yield for s-orbits are decreasing towards neutron-rich side. Since very-strongly bound s orbit is involved, it is not clear that the reason for the reduction is due to the variation of the spectroscopic factor of deeply bound protons or due to the reaction mechanism.

## Summary

Proton knockout (p,2p) reactions from $^{9-16}$C isotopes at 250 MeV/ were performed for systematic information on weakly to strongly bound 1p protons and deeply bound 1s protons. Separation-energy distributions are measured with 1.3 MeV (rms) resolutions. S and p hole states are identified and separated by tagging the decay mode of the hole states. Energy gap between 1s and 1p orbits are rather constant at about 20 MeV, and tends to become wider on both proton and neutron-rich sides. Momentum distribution widths for 1p valence orbit are roughly consistent with simple calculations. On the other hand, momentum distribution widths of 1s protons show increase toward neutron-rich side, indicating that the charge radii of 1s protons are shrinking towards the neutron-rich side. Relative (p,2p) yields indicate weakly-bund valence proton in $^9$C have larger spectroscopic factors compared with the valence protons in other carbon isotopes.

## ACKNOWLEDGMENTS

The experiment was supported by the Research Project with Heavy Ions at NIRS-HIMAC.

## REFERENCES

1. Gerhard Jacob and Th.A.J. Maris, Rev. Mod. Phys. **38**, 121 (1966).
   Gerhard Jacob and TH.A.J. Maris, Rev. Mod. Phys. **45**, 6 (1973).
2. M. Yosoi et al., Phys. Lett. **B551**, 255 (2003).
3. J. Mougey et al., Nucl. Phys. **A262**, 462 (1976).
4. P.G. Hansen and J.A. Tostevin, "Direct Ractions with Exotic Nuclei", Annu. Rev. Nucl. Sci. **53**, 219-261 (2003).

# Measurement of nuclear moments at RIKEN

H. Ueno*, D. Kameda*, D. Nagae†, M. Takemura†, K. Asahi†*, K. Takase†,
A. Yoshimi*, T. Sugimoto*, T. Nagatomo*, T. Arai†, M. Uchida†,
K. Shimada†, T. Inoue†, J. Murata**, H. Kawamura** and K. Narita**

*RIKEN, 2-1 Hirosawa, Wako, Saitama 351-0198, Japan
†Department of Physics, Tokyo Institute of Technology, Meguro-ku, Tokyo 152-8551, Japan
**Department of Physics, Rikkyo University, Nishi-Ikebukuro 3-34-1, Toshima-ku, Tokyo 171-8501, Japan

**Abstract.** Based on the technique of fragment-induced spin polarization combined with the β-NMR method, we have recently carried out experiments at RIKEN to measure nuclear moments of neutron-rich aluminum isotopes. In the measurements, the nuclear magnetic-dipole moments of $^{30,32}$Al and the electric-quadrupole moments of $^{31,32}$Al have been determined. The obtained magnetic moments, as well as the other known magnetic moments of aluminum isotopes, agree well with shell model calculations with the USD interaction. The obtained quadrupole moments are smaller than those of $^{27,28}$Al, suggesting spherical shapes of $^{31,32}$Al. These results seem to suggest that $^{31}$Al and $^{32}$Al are located outside the *island of inversion*. In near future, studies on nuclear moments will be extended further in the RIKEN RIBF project, in which medium- and heavy-mass regions come within our scope by using superconducting fragment separator BigRIPS. Also, an upgrade program for the existing fragment separator RIPS is in progress. In the RIBF project, nuclear moments will be measured for isomer states as well as ground states.

## INTRODUCTION

Nuclear-electromagnetic moments are one of the basic probes to obtain information on the nuclear structure. It has been revealed that spin-oriented radioactive-isotope beams (RIBs) can be produced in the projectile-fragmentation reaction [1, 2, 3, 4], which offers us the opportunity of studying the structures of nuclei far from the stability line, through the measurement of electromagnetic nuclear moments. So far, measurements have been carried out in the *p*-shell neutron-rich nuclear region. The obtained experimental nuclear moments have been shown quite effective in discussing the effect of neutron excess on their nuclear structure, e.g., in terms of configuration mixing [5, 6] and the isospin dependence of the effective charges [7]. In order to extend the observation to the neutron-rich *sd*-shell region, we have recently measured ground-state nuclear moments of neutron-rich aluminum isotopes.

In the region of neutron-rich *sd*-shell nuclei, intriguing phenomena have been reported [8, 9, 10, 11, 12], which have been discussed in association with the ground-state nuclear deformation induced by the inversion of amplitudes between *sd*-normal and *pf*-intruder configurations [13]. Theoretically, two different large-scale shell models have recently been proposed to describe the nuclear structure around the *island of inversion*. The gap energies between 0p-0h and 2p-2h configurations were calculated for aluminium isotopes as well as for the nuclei inside the island of inversion [14]. The predicted gap energy takes a minimum at $^{33}$Al among the aluminium isotopes, although the gap energies of all the aluminium isotopes are slightly positive for this interaction. On the other hand, the probabilities of 0p-0h, 2p-2h, and 4p-4h configurations are calculated for the $N = 20$ isotones based on the Monte-Carlo shell model [15]. In these calculation, the amplitudes of normal configurations are taken over by those of the intruders at $^{33}$Al, indicating that $^{33}$Al represents a turning point along the atomic number Z. Thus, aluminium isotopes are considered as good examples to investigate change of nuclear structure. Microscopic studies of nuclei close to the *island of inversion*, as well as of those inside it, are needed for understanding the mechanism of its occurrence.

In our recent work, the ground-state $\mu$ moments of $^{30\text{-}32}$Al and the electric-quadrupole moments ($Q$) of $^{31, 32}$Al have been measured by the $\beta$-NMR method combined with the fragment-induced spin-polarization technique [3, 4]. The $\mu$ measurement for $^{31}$Al is a re-measurement with an accuracy significantly improved from the previous measurement [16], while the $\mu$ for $^{30, 32}$Al and $Q$ for $^{31, 32}$Al are new.

At RIKEN, the RI Beam Factory (RIBF) project [17] is in progress, where intense beams are provided at the energy $E \sim 350\,A$ MeV. It is expected that measurements of nuclear moments can be carried out productively at RIBF. Since spin-aligned RIBs [1, 2] can be produced efficiently using the superconducting fragment separator BigRIPS [18], the systematic $\mu$-moment measurement of isomeric states for unstable nuclei are under preparation. For the production of spin-polarized RIBs in the measurement of ground-state $\mu$ moments, it is essential that the beam energy is as low as $E \sim 100\,A$ MeV [3, 4]. Thus, an upgrade program for the existing fragment separator RIPS [19] is in progress in order to deliver $E \sim 115\,A$ MeV beams to RIPS in the phase-II program of RIBF. Details of the upgrade program are given later.

## EXPERIMENTAL PROCEDURE

Experiments were performed using RIPS [19]. Beams of $^{30\text{-}32}$Al were obtained from the fragmentation of $^{40}$Ar projectiles at an energy of $E = 95\,A$ MeV on a $^{93}$Nb target. In order to have spin-polarized aluminum-isotope beams, the emission angle and the outgoing momentum were selected. Then the spin-polarized fragments were introduced into an NMR apparatus located at the final focus of RIPS. They were implanted into a single crystal of $\alpha$-Al$_2$O$_3$, which was kept in a vacuum chamber and cooled to a temperature of $T < 100$ K to assure the preservation of spin polarization during the $\beta$-ray counting period as long as $\tau[^{30}\text{Al}] = 5.2$ s. The $\alpha$-Al$_2$O$_3$ single crystal was also used in the nuclear quadrupole resonance (NQR) measurements to determine the $Q$ moments of $^{31}$Al and $^{32}$Al, where frequency shifts from the Larmor frequency due to the $eqQ$ interaction were measured. Since the frequency shifts were expected to be very small, their $\mu$ moments needed to be known with higher accuracy compared with the reported values [16, 20]. Thus, the precision measurements of $\mu[^{31, 32}\text{Al}]$ were carried out prior to the $Q$-moment measurements using a Si stopper at room temperature. It provides relaxation times longer than their lifetimes $\tau[^{31}\text{Al}] = 930$ ms and $\tau[^{32}\text{Al}] = 48$ ms [21]. In the measurements, the $\beta$-NMR method [22] was adopted, in which signals of NMR and NQR are observed through a change in the $\beta$-ray asymmetry.

# RESULT AND DISCUSSION

## Magnetic moments of $^{30\text{-}32}$Al

Figures 1 (b) and (c) show the obtained NMR spectra in the $\mu$-moment measurements for $^{31}$Al and $^{32}$Al, respectively. Measured $\beta$-decay up/down ratios are plotted together with those obtained without the oscillating magnetic field, as a function of frequency. The previously reported NMR spectrum for $^{30}$Al is also shown in Fig. 1(a). Based on the curve-fitting analyses, we have obtained experimental values of $|\mu_{\exp}[^{31}\text{Al}]| = 3.824(8)$ $\mu_N$ and $|\mu_{\exp}[^{32}\text{Al}]| = 1.951(5)$ $\mu_N$. They agree with reported $\mu$ moments [16, 20], and, moreover, their accuracies have been improved to a level that the measured $\mu$ values give a sufficient basis for precision $Q$-moment measurements.

All experimental $\mu$ moments of neutron-rich aluminum isotopes, shown in Fig. 2 (a), agree with the shell-model calculations carried out using the USD interaction [23, 24]. This result indicates that their dominant component of nuclear configurations can be described within a $sd$ model space for $^{30\text{-}32}$Al. This observation agree with the measurements carried out at GANIL, where the $\mu$ moments of neutron-rich aluminum isotopes $^{31\text{-}34}$Al have been determined [16, 25]. The obtained $\mu$ moments of $^{33, 34}$Al deviate from the shell-model predictions carried out within a $0\hbar\omega$ model space, indicating the effect from intruder configurations [25], while agreement with conventional shell-model predictions [23, 24] is observed for $\mu_{\exp}[^{31, 32}\text{Al}]$ [16, 25].

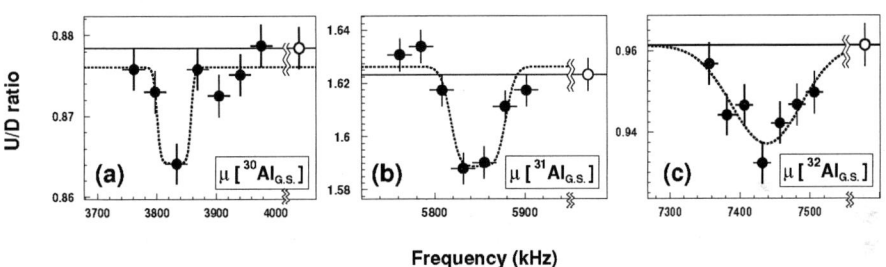

**FIGURE 1.** Obtained NMR spectra in the precision measurements for (a) $^{30}$Al, (b) $^{31}$Al, and (c) $^{32}$Al. Measured $\beta$-ray up/down ratios are plotted by solid circles and those obtained without the oscillating magnetic field are shown by open circles.

## Electric-quadrupole moments of $^{31, 32}$Al

The structure change in neutron-rich aluminum isotopes can be further investigated through the measurement of their $Q$ moments because of its high sensitivity to nuclear deformation. Figures 3 (a) and (b) show the obtained NQR spectra for $^{31}$Al and $^{32}$Al, respectively. In a manner similar to the above $\mu$ measurements, we have determined experimental $Q$ moments $|Q_{\exp}[^{31}\text{Al}]| = 104(9)$ $e\cdot$mb [26] and $|Q_{\exp}[^{32}\text{Al}]| = 24(2)$

$e$·mb [27]. Details of the measurements and analyses will be given in Refs. [26, 27].

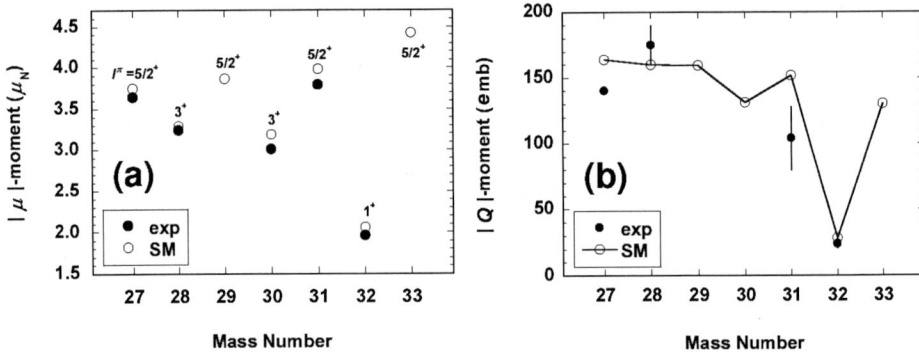

**FIGURE 2.** Comparison of experimental (a) magnetic-dipole moments and (b) electric-quadrupole moments of neutron-rich aluminum isotopes with shell-model predictions based on the USD interaction.

The obtained values of $|Q_{\exp}[^{31}\text{Al}]|$ and $|Q_{\exp}[^{32}\text{Al}]|$ are plotted in Fig. 2(b) together with known $Q$ moments for other aluminum isotopes as a function of mass number $A$. As seen in Fig. 2(b), the value of $|Q_{\exp}|$ stays almost constant at $|Q| \sim 150$ $e$·mb [28, 29] for $A = 27$ and 28, but becomes smaller at $A = 31$. At $A = 32$, $|Q_{\exp}|$ takes a very small value of 24(2) $e$·mb. These observations suggest that the shape of the neutron-rich aluminium isotopes becomes spherical when approaching $N = 19$, and thus $^{32}$Al is a normal nucleus, although anomalous level ordering has been reported for the isomeric state at $E_x = 956$ keV [30]. It is interesting to note that the observed *normal* properties of $^{32}$Al are in contrast to those of $^{31}$Mg, for which anomalous nuclear structure has been reported [31], despite that these two nuclei differ only in the proton number.

Then, the $|Q_{\exp}|$ value are compared with shell-model calculations. As shown in Fig. 2(b), the above observed mass-number dependence of the $|Q_{\exp}|$ values is well reproduced except for the present $|Q_{\exp}[^{31}\text{Al}]|$ value: the $|Q_{\exp}[^{31}\text{Al}]|$ value deviates from the shell-model prediction by 30 %. The reason for this discrepancy has not been understood. The nuclear configuration predicted for $^{31}$Al seems reasonable, since the reported experimental $\mu$ moment as well as the level structure [16] is well reproduced. Also, the deviation would not be associated with the *island of inversion*, since the direction of the deviation is opposite to what is expected from a deformed shape. This discrepancy can not be explained even if the large-scale shell model [15] is adopted. Similarly to the above USD configuration, the predicted $^{31}$Al configuration in this model shows *normal-sd* properties, where the total amplitude of *normal-sd* component is predicted to be as large as $\sim 85$ % [32]. For the understanding of this discrepancy, further consideration is needed.

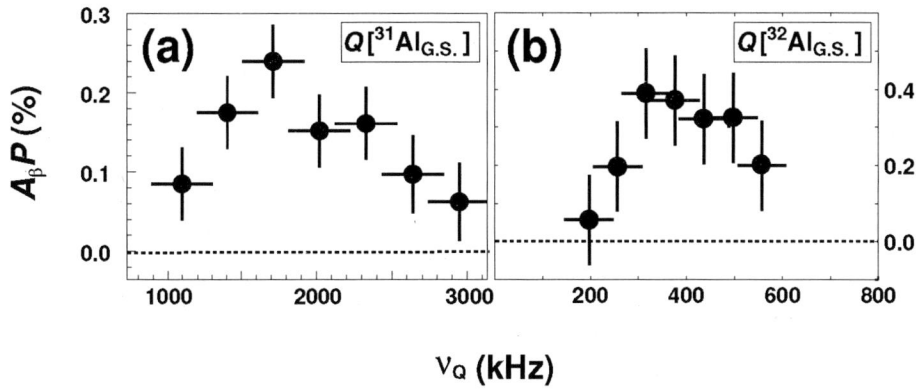

**FIGURE 3.** Obtained NQR spectra in the precision measurements for (a) $^{31}$Al and (b) $^{32}$Al.

## NUCLEAR-MOMENT MEASUREMENT AT RIBF

In the RIBF project [17] intense primary beams can be provided over the whole range in the atomic number at the energy $E = 400$ and $350\,A$ MeV for lighter- ($A < 40$) and heavier-mass ions, respectively, after the final acceleration by Superconducting Ring Cyclotron (SRC) [33] in the cyclotron cascade. By using BigRIPS, a large variety of RIBs can be produced. Measurements of $\mu$ moments for the isomeric state of unstable nuclei will be effectively carried out with BigRIPS, since higher energies are preferred in view of the production yield of RIBs, the population of isomeric states, and production of spin-aligned RIBs [1, 2]. Measurements are under preparation.

In addition to the phase-I program, in which BigRIPS and the ZeroDegree spectrometer are equipped, several experimental devices have been proposed as illustrated in Fig. 4. The upgrade program of RIPS has been also proposed among them, as denoted by *RI-Spin Laboratory* in Fig. 4. In the RIBF configuration, beams are accelerated up to the energy of $E = 115\,A$ MeV with Intermediate-Stage Ring Cyclotron (IRC) [34]. This energy is high enough to produce RI beams via projectile-fragmentation reaction. Therefore, after a beam-transport line from IRC to RIPS shown in Fig. 4 is constructed [35], intense beams will also be provided to RIPS. The maximum magnetic rigidity of the beam-transport line from IRC to RIPS is designed to be $B\rho = 4.2$ Tm [35]. This value is sufficient for RIPS to accept beams from IRC, except for the very heavy ions near uranium.

In this configuration, RIPS further enhances research opportunities on spin-related subjects. Compared with presently obtained RIB intensities with the AVF-RRC acceleration, their intensities are drastically increased as shown in Fig. 5. We note that beams at $E = 115\,A$ MeV allow for a scheme to produce spin-oriented RIBs [1, 2, 3, 4] and to implant them into sample materials with limited thickness. Thus, in addition to studies of nuclear structure through electromagnetic moments, $\beta$ decay, $\beta$-$\gamma$ spectroscopy, and material science studies would be conducted based on spin-related research techniques such as $\beta$-NMR and $\gamma$-PAD/PAC.

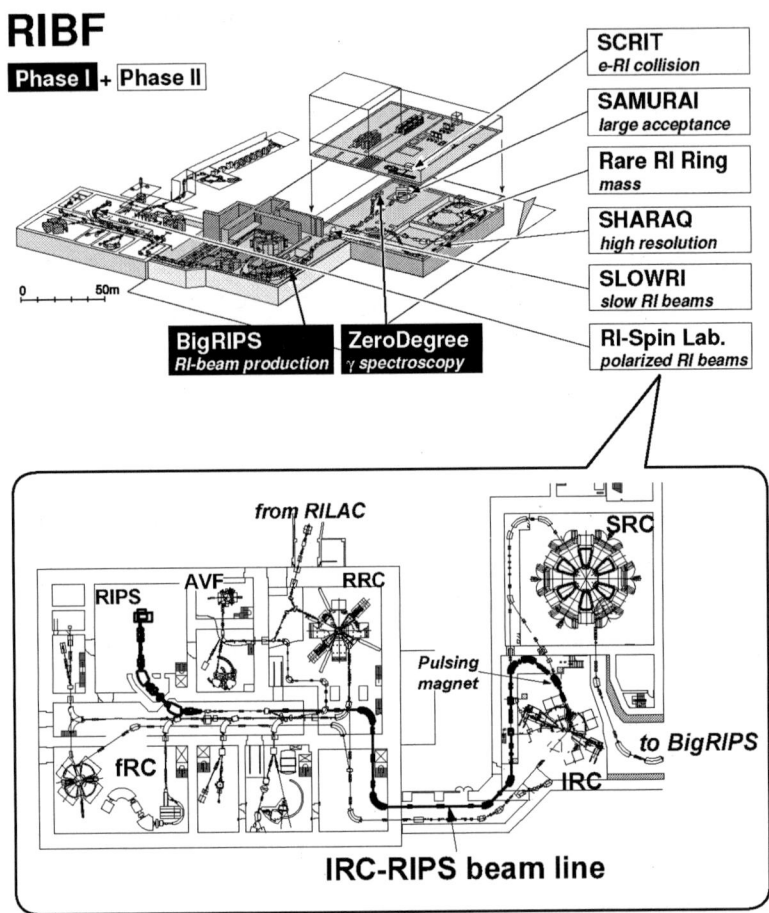

**FIGURE 4.** A schematic plan view of the RIBF project. A layout of the beam transport line from IRC to RIPS, which is included in the spin-polarized RI beam project, is shown in the lower inlet.

In order to enhance and fully capitalize the unique and valuable experimental opportunities provided by this IRC-RIPS configuration, a time-sharing beam delivery to BigRIPS and RIPS has been proposed. A pulsing magnet, which has already been installed downstream of IRC as shown in Fig. 4, can be used for this purpose. This magnet is able to be operated not only in a DC mode but also in a pulse mode. The designed switching time of the pulsing magnet is 10 ms. Within this period, the magnet current is stabilized to better than $1 \times 10^{-4}$ [36, 37]. One example of the operation scheme for the beam-delivery sequence is 10 ms (switching) - 100 ms (flat top) - 10 ms (switching) - 900 ms (flat base). In a flat top interval, the pulsing magnet is excited and the beam is transported to the IRC-RIPS beam line. In a flat base interval, the magnet is switched off, thus the beam being transported to SRC just through the magnet.

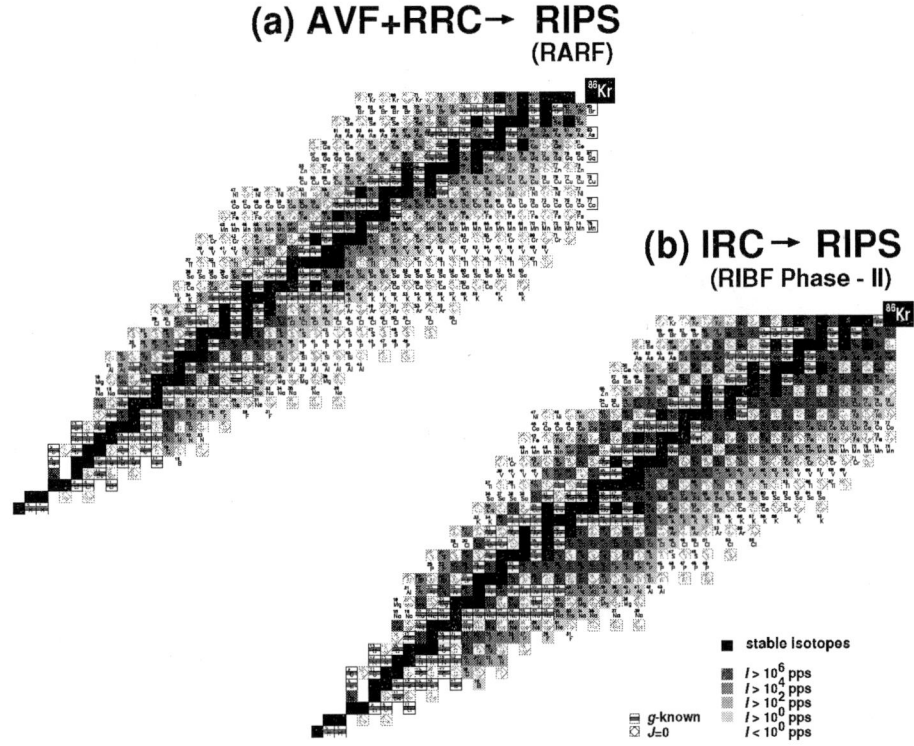

**FIGURE 5.** Comparison of RI-beam production rates in the projectile-fragmentation reaction. Upper partial nuclear chart (a) shows the case of present AVF-RRC acceleration. The panel (b) shows intensities estimated assuming $I = 1$ p$\mu$A and $E = 115$ $A$ MeV projectiles of all stable elements. In both figures the isotopes whose ground-state nuclear moment have been measured are indicated.

## SUMMARY

The structure of nuclei located near the *island of inversion* have been studied through the measurement of static electromagnetic moments of aluminum isotopes, since aluminium isotopes are located near/on the border of the *island of inversion* so that they can be considered as a good subject of investigation of the change of nuclear structure. Based on the technique of fragment-induced spin polarization combined with the $\beta$-NMR method, we have carried out the measurement of $\mu$ moments of $^{30, 32}$Al and $Q$ moments of $^{31, 32}$Al. The obtained $\mu$ moments agree well with shell model calculations with the USD interaction as well as other known $\mu$ moments of aluminum isotopes. The obtained $Q$ moments are smaller than those of $^{27, 28}$Al, suggesting spherical shapes of $^{31, 32}$Al. These observations suggest that these aluminum isotopes up to $^{32}$Al are located outside the *island of inversion*.

Recently, $\mu$ moments of $^{33,\ 34}$Al have been determined [25] at GANIL, which show deviation from the conventional shell-model prediction within a $0\hbar\omega$ model space. Since $Q$ moments give us knowledge on the nuclear deformation, which is essential for the understanding of the occurrence of *island of inversion*, their $Q$-moment measurements are desirable, and in fact they are under preparation.

In the RIBF project, high intensity RIBs will be produced in medium- and heavy-mass regions, and studies on nuclear moments should be extended dramatically. Measurements of isomer-state nuclear moments will be carried out with BigRIPS. Also, an upgrade of RIPS has been proposed in the RIBF phase-II program, in which high intensity beams at energy $E = 115\ A$ MeV are delivered to RIPS by installing a beam-transport line from IRC to RIPS. Compared with presently obtained RIB intensities, their intensities are drastically increased. This upgrade should enhance research opportunities for spin-related subjects, since it enables the implantation of spin-oriented RIBs in sample materials with rather limited thickness.

## REFERENCES

1. K. Asahi *et al.*, Phys. Rev. C **43**, 456 (1991)
2. W.-D. Schmidt-Ott *et al.*, Z. Phys. A **350**, 215 (1994)
3. K. Asahi *et al.*, Phys. Lett. B **251**, 488 (1990).
4. H. Okuno *et al.*, Phys. Lett. B **335**, 29 (1994).
5. H. Ueno *et al.*, Phys. Rev. C **53**, 2142 (1996)
6. K. Asahi *et al.*, Nucl. Phys. A **704**, 88c (2002) and references theirin.
7. H. Ogawa *et al.*, Phys. Rev. C **67** 064308 (2003)
8. C. Thibault *et al.*, Phys. Rev. C **12**, 644 (1975).
9. C. Détraz *et al.*, Nucl. Phys. A **394**, 378 (1983).
10. C. Détraz *et al.*, Phys. Rev. C **19**, 164 (1979).
11. D. Guillemaud *et al.*, Nucl. Phys. A **426**, 37 (1984).
12. T. Motobayashi *et al.*, Phys. Lett. B **346**, 9 (1995).
13. E.K. Warburton, J.A. Becker, B.A. Brown, Phys. Rev. C **41**, 1147 (1990).
14. E. Caurier, F. Nowacki, A. Poves, and J. Retamosa, Phys. Rev. C **58**, 2033 (1998).
15. Y. Utsuno, T. Otsuka, T. Mizusaki, and M. Honma Phys. Rev. C **64**, 011301(R) (2001).
16. D. Borremans *et al.*, Phys. Lett. B **537**, 45 (2002).
17. Y. Yano, *Proc. 17th Int. Conf. on Cyclotrons and their Applications*, Tokyo Japan, October 18-22, 2004, ed. A. Goto and Y. Yano, Particle Accelerator Society of Japan, 169 (2005).
18. T. Kubo, Nucl. Instr. Meth. B **204**, 97 (2003).
19. T. Kubo *et al.*, Nucl. Instr. Meth. B **70**, 309 (1992).
20. H. Ueno *et al.*, Phys. Lett. B **615**, 186 (2005).
21. T. Minamisono *et al.*, Phys. Rev. C **14**, 376 (1976).
22. K. Sugimoto, A. Mizouchi, K. Nakai and K. Matsuta, J. Phys. Soc. Japan **21**, 213 (1966).
23. B.A. Brown A. Etchegoyen and W.D.M. Rae, OXBASH, MSU Cyclotron Laboratory Report No. **524**, (1986).
24. B.H. Wildenthal, Prog. Part. Nucl. Phys. **11**, 5 (1984).
25. G. Neyens *et al.*, to be apper in Proc. of The Seventh International Conference on Radioactive Nuclear Beams, Cortina d'Ampezzo, Italy, July 3-7, 2006.
26. D. Nagae *et al.*, *to be published*.
27. D. Kameda *et al.*, *to be published*.
28. D. Sundholm and J. Olsen, Phys. Rev. Lett. **68**, 927 (1992).
29. H.-J. Stöckmann *et al.*, Hyp. Int. **4**, 170 (1978).
30. M. Robinson *et al.* Phys. Rev. C **53**, R1465 (1996). Hyp. Int. **4**, 170 (1978).
31. G. Neyens *et al.*, Phys. Rev. Lett. **94**, 022501 (2005).

32. Y. Utsuno, *private communication* (2006)
33. H. Okuno it et al., *Proc. 17th Int. Conf. on Cyclotrons and their Applications*, Tokyo Japan, October 18-22, 2004, ed. A. Goto and Y. Yano, Particle Accelerator Society of Japan, 373 (2005).
34. J. Ohnishi et al., *Proc. 17th Int. Conf. on Cyclotrons and their Applications*, Tokyo Japan, October 18-22, 2004, ed. A. Goto and Y. Yano, Particle Accelerator Society of Japan, 197 (2005).
35. N. Fukunishi *et al.*, RIKEN Accel. Prog. Rep. **35**, 283 (2002)
36. H. Kouzu *et al.*, RIKEN Accel. Prog. Rep. **34**, 355 (2001)
37. K. Kusaka *et al.*, RIKEN Accel. Prog. Rep. **35**, 297 (2002)

# Study of exotic nuclei around the "island of inversion"

Z. Elekes*, Zs. Dombrádi*, A. Saito[†], N. Aoi**, H. Baba[†], K. Demichi[‡], Zs. Fülöp*, J. Gibelin[§], T. Gomi[‡], H. Hasegawa[‡], N. Imai**, M. Ishihara**, H. Iwasaki[†], S.Kanno[‡], S. Kawai[‡], T. Kishida**, T. Kubo**, K. Kurita[‡], Y. Matsuyama[‡], S. Michimasa**, T. Minemura**, T. Motobayashi**, M. Notani[†], T.K Ohnishi[†], H.J. Ong[†], S. Ota[¶], A. Ozawa[‖], H.K. Sakai[‡], H. Sakurai[†], S. Shimoura[†], E. Takeshita[‡], S. Takeuchi**, M. Tamaki[†], Y. Togano[‡], K. Yamada[‡], Y. Yanagisawa** and K. Yoneda**

*Institute of Nuclear Research of the Hungarian Academy of Sciences, P.O. Box 51, Debrecen H-4001, Hungary
[†]University of Tokyo, Tokyo 1130033, Japan
**The Institute of Physical and Chemical Research, 2-1 Hirosawa, Wako, Saitama 351-0198, Japan
[‡]Rikkyo University, 3 Nishi-Ikebukuro, Toshima, Tokyo 171, Japan
[§]Institut de Physique Nucléare, 15 rue Georges Clemenceau, 91406 Orsay, France
[¶]Kyoto University, Kyoto 606-8501, Japan
[‖]Tsukuba University, Tennoudai 1-1-1, Tsukuba-shi, Ibaraki 305-8571, Japan

**Abstract.**
**Keywords:** γ-spectroscopy, $^{27}$F, $^{27,28}$Ne, $^{30,31}$Na, $^{33,34}$Mg, inelastic proton scattering, radioactive beam
**PACS:** 21.60.Cs; 23.20.Lv; 25.40.Ep; 27.30.+t; 29.30.Kv

## INTRODUCTION

Experimental data accumulated since the late seventies on a missing $N=20$ shell closure in the Mg–Na region launched the idea of the collapse of the usual shell model ordering of the single particle states in neutron–rich nuclei [1, 2]. According to the early calculations, the effective single particle energy of the $f_{7/2}$ orbit becomes lower than that of the $d_{3/2}$ one in $^{28}$O [2]. However, systematic investigations have revealed that the observed phenomena can also be described by considering a strong correlation energy associated with the proton–neutron $T=0$ interaction leading to a large deformation [1, 3] without assuming a significant change of the single particle energy structure. The deformed 2p–2h states may intrude below the normal spherical states and form an "island of inversion" of definite borders with only nine nuclei included. In these so-called USD calculations, the effective interaction, giving a reasonable description of nuclei close to the stability, leads to an effective $N=20$ shell gap changing from 7 MeV at $Z=20$ to about 5 MeV at $Z=8$ [3, 4]. This shell gap is large enough to conserve the spherical $N=20$ shell closure.

As an alternative approach, the Monte Carlo diagonalization method has also been introduced in the shell model for the region of light neutron–rich nuclei adding the two lower $fp$ shell orbits ($1f_{7/2}$ and $2p_{3/2}$) to the $sd$ shell model space (Monte Carlo shell

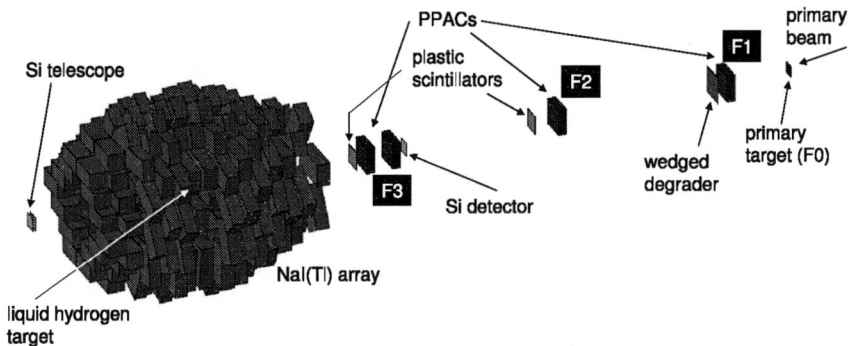

**FIGURE 1.** Layout of the experimental setup.

model – MCSM) [5]. The use of the enlarged valence space allows the mixing of the $sd$ and $fp$ configurations and gives a good description of the available experimental data close to the "island of inversion" and even beyond [5, 6, 7]. However, its effective interaction leads to a rapid decrease of the shell gap to 1.2 MeV at $Z=8$ [8] suggesting the disappearance of the $N=20$ shell gap. As a consequence of this rapidly decreasing shell gap, the MCSM predicts a much wider "island of inversion" than the models with a closed $N=20$ shell. In this model, the crossing of the intruder and normal configurations takes place at $N=18$ resulting in a deformed ground state even at this neutron number and low energy intruder states up to $N=17$.

Recently, the observation of two excited states at 1249 keV and 1588 keV in $^{29}$Na [9] provided new data that supports the MCSM prediction of having low–lying $fp$ states mixed with the normal ones at $N=18$ [9] and of a small $N=20$ shell gap. Therefore, in order to contribute to the mapping of the borders of the "island of inversion" and the clarification of the $N=20$ shell gap size, inelastic proton scattering investigations were carried out on nuclei in and around the island, i.e., $^{28}$Ne, $^{30,31}$Na and $^{33,34}$Mg nuclei. This process was also suitable to deduce the deformation of these isotopes and check the neutron decoupling phenomenon [10, 11, 12] in this region by comparing the proton and neutron distributions. The models can be further checked by investigating odd-nuclei out of the "island of inversion" according to the USD shell model since the direct observation of a single particle state from the $fp$ shell could pose a stringent test on the $N=20$ shell gap. Thus, we studied the level structure of $^{27}$F by (p,p') reaction and that of $^{27}$Ne by neutron knock-out reaction.

## EXPERIMENTAL

The experiment was carried out at the RIKEN Accelerator Research Facility. The schematic view of the experimental setup can be seen in Fig. 1. A $^{40}$Ar primary beam of 94 MeV/nucleon energy with 60 pnA intensity was transported to a $^{181}$Ta production target of 0.5 mm thickness. The RIPS [13] fragment separator analyzed the momentum and mass of the reaction products. The secondary radioactive beams were produced in two individual runs with different settings of B$\rho$ values. An aluminum wedged de-

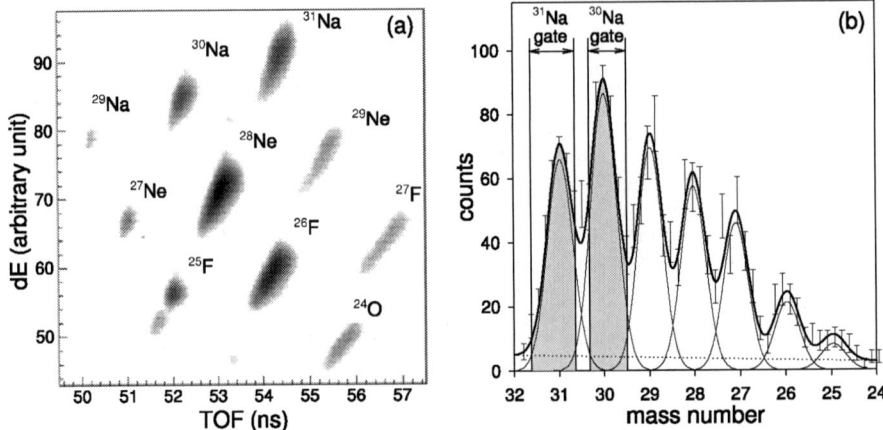

**FIGURE 2.** (a) Isotope separation of the incoming "cocktail" beam for the first run. (b) Separation of sodium isotopes using $\Delta E$–$E$ information in the silicon telescope produced in coincidence with $\gamma$-rays. The bold solid line is a sum of 7 Gaussian functions and a polynomial background. The individual Gaussians and the background function are also plotted with thin solid and dotted lines, respectively.

grader of 221 mg/cm² was put at the momentum dispersive focal plane (F1) for purifying the constituents. In the first run of $^{27}$F, $^{28}$Ne and $^{30,31}$Na, the secondary beam included neutron–rich O, F, Ne and Na nuclei with $A/Z \approx 3$ while mainly Mg and Al isotopes were mixed in the second run of $^{33,34}$Mg. The fragment separator was set to its full 6% momentum acceptance to achieve as high beam intensities as possible. The total intensity was about 100 particle/s (pps) for both runs, while the $^{27}$F, $^{28}$Ne, $^{30,31}$Na, $^{33,34}$Mg intensities reached 4, 20, 8, 6, 3 and 2 pps, respectively, on average. The identification of incident beam species was performed on an event–by–event basis by means of energy loss, time-of-flight (TOF) and magnetic rigidity ($B\rho$) [14]. Determining the $B\rho$ values by a parallel plate avalanche counter (PPAC), the position of the fragments at F1 was also measured. The sensitive area of this PPAC was 15x10 cm² which covered the total momentum range of the secondary beam. Two plastic scintillators of 1 mm thicknesses were placed at the first and second focal planes (F2 and F3) to measure the TOF. One silicon detector with thickness of 0.35 mm was inserted at F3 for energy loss determination. The separation of the isotopes was complete, which is demonstrated in Fig. 2a for the Na run.

The secondary beam hit a liquid hydrogen target [15] of 30 mm diameter the thickness of which was 24 mm and its entrance and exit windows were made of 6.6 $\mu$m Aramid foil. The average areal density of the hydrogen cooled down to 22 K was 210 mg/cm². The mean energy of the isotopes in the target was around 50 MeV/nucleon. Two PPACs at F3 upstream of the target monitored the position of the incident particles. The beam spot size was 24 mm both in horizontal and vertical directions. The reaction products and scattered particles were detected and identified by a PPAC and a silicon telescope of three layers with thicknesses of 0.5, 0.5 and 1 mm located about 80 cm downstream of the target. Each layer was made of a 2x2 matrix of detectors the active area of which was

48x48 mm$^2$. The $Z$ identification was performed by TOF–energy loss method where the TOF was taken between the PPACs upstream and downstream of the secondary target. The isotope separation was done by use of the $\Delta E$–$E$ method. The mass spectra are dominated by the beam particles, however requiring coincidence with $\gamma$ rays, we could eliminate the non–interacting or elastically scattered part of the beam making the $\Delta E$–$E$ method sensitive enough. It is demonstrated in Fig. 2b where the linearized mass spectrum of sodium isotopes is shown for the 2x2 matrix Si–telescope (events from $^{30}$Na and $^{31}$Na beams are added). The $\Delta E$–$E$ curves in each detector were linearized by second degree polynomial functions. In case of Mg isotopes, similar separation could be achieved.

The de–exciting $\gamma$ rays emitted by the inelastically scattered nuclei were detected by the DALI2 setup consisting of 146 NaI(Tl) scintillators [16] surrounding the target. The energy calibration of the detectors was made by standard $^{22}$Na, $^{60}$Co and $^{137}$Cs radioactive sources. The intrinsic energy resolution of the array was 10% (FWHM) for a 662 keV energy $\gamma$ ray.

## RESULTS AND DISCUSSION

The Doppler–corrected $\gamma$ ray spectra for $^1$H($^{27}$F,$^{27}$F) (Fig. 3), $^1$H($^{28}$Ne,$^{27}$Ne) (Fig. 4a), $^1$H($^{28}$Ne,$^{28}$Ne) (Fig. 4b), $^1$H($^{30}$Na,$^{30}$Na) (Fig. 5a), $^1$H($^{31}$Na,$^{31}$Na) (Fig. 5b) and $^1$H($^{33}$Mg,$^{33}$Mg) (Fig. 6a), $^1$H($^{34}$Mg,$^{34}$Mg) (Fig. 6b) reactions were produced by putting an additional gate on the time spectra of the NaI(Tl) detectors selecting the prompt events. By fitting the spectra with Gaussian functions and smooth exponential backgrounds, first, the positions of the peaks were determined to be 504 keV and 777 keV for $^{27}$F, 765(20) keV and 904(21) keV for $^{27}$Ne, 1319(22) keV and 1711(30) keV for $^{28}$Ne, 403(18) keV for $^{30}$Na, 370(12) keV for $^{31}$Na and 483(17) keV for $^{33}$Mg, 685(16) keV for $^{34}$Mg.

The peaks for $^{27}$F are detected for the first time in the present experiment. For $^{27}$Ne, a 772(7) keV line was observed before in a fragmentation reaction [17] while a peak at 870(16) keV was recently detected in the $^{12}$C($^{28}$Ne,$^{27}$Ne) reaction [18]. The energies determined for $^{28}$Ne are in reasonable agreement with the values 1289(9) keV and 1719(11) keV in Ref. [17] and 1320(20) keV in Ref. [19]. In Ref. [17] the 1711 keV transition is connected to the 1319 keV one establishing a state at 3030 keV. For $^{30}$Na, a single peak was detected earlier [20] at 433(16) keV. It slightly differs from our value of 403(18) keV, however they overlap within 1$\sigma$ limit. For $^{31}$Na, a 350(20) keV peak was previously observed in a reaction on $^{197}$Au target [21] which coincides with our value of 370(12) keV. For $^{33}$Mg, several peaks were detected in a $\beta$ decay study earlier [23] including 484.1(1) keV one matching our line at 483(17) keV. The energy for the first excited state of $^{34}$Mg was previously determined to be 660(10) keV [24] and 656(7) keV [25] which overlap with the present value (685(16) keV) within 1$\sigma$ limit.

In order to derive the cross sections for the production of the $\gamma$ rays in proton inelastic scattering, the above peak positions were fed into the detector simulation software GEANT4 [26] and the resultant response curves plus smooth polynomial backgrounds were used to analyze the experimental spectra. The observed cross sections are listed in Table 1.

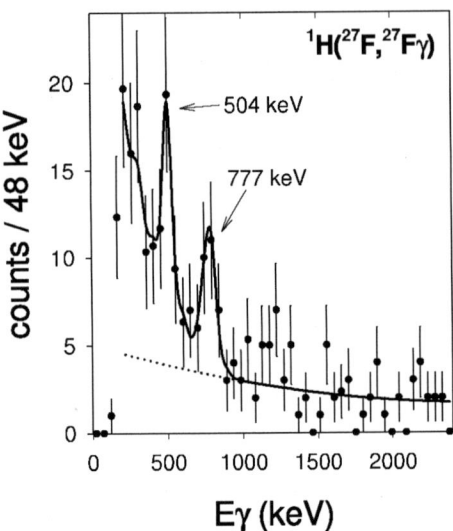

**FIGURE 3.** Doppler–corrected spectra of γ rays emerging from $^1$H($^{27}$F,$^{27}$F) reaction. The solid line is the final fit including the spectrum curves from GEANT4 simulation and additional smooth polynomial backgrounds plotted as separate dotted lines for each nucleus.

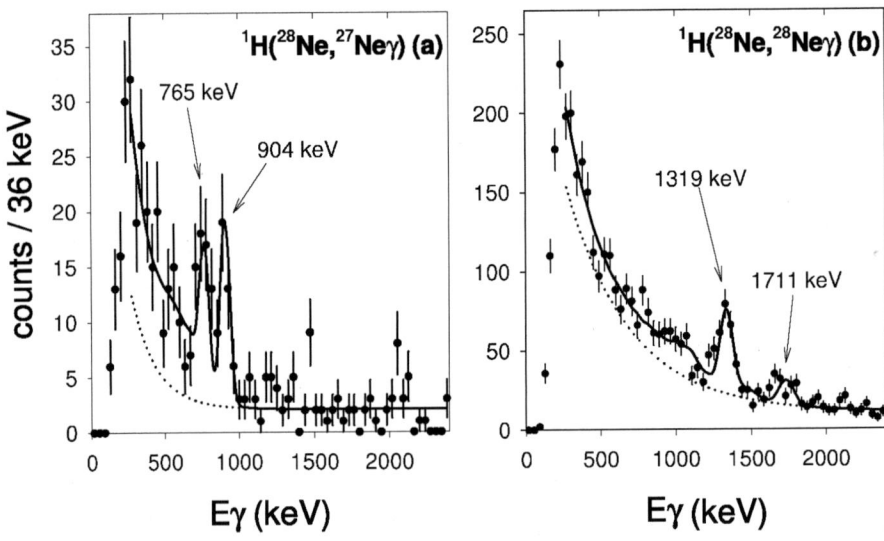

**FIGURE 4.** Doppler–corrected spectra of γ rays emerging from $^1$H($^{28}$Ne,$^{27}$Ne) (a), $^1$H($^{28}$Ne,$^{28}$Ne) (b) reactions. The solid line is the final fit including the spectrum curves from GEANT4 simulation and additional smooth polynomial backgrounds plotted as separate dotted lines for each nucleus.

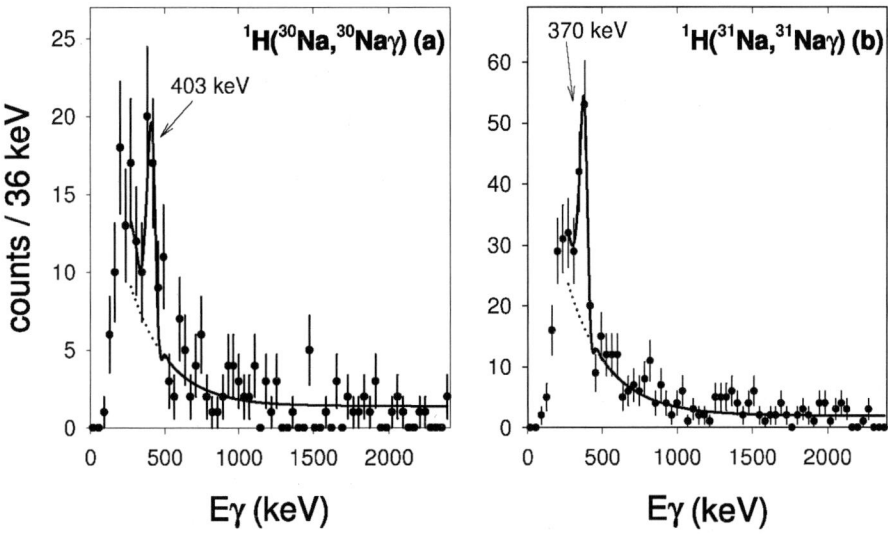

**FIGURE 5.** Doppler–corrected spectra of γ rays emerging from $^1$H($^{30}$Ne,$^{30}$Ne) (a), $^1$H($^{31}$Ne,$^{31}$Ne) (b) reactions. The solid line is the final fit including the spectrum curves from GEANT4 simulation and additional smooth polynomial backgrounds plotted as separate dotted lines for each nucleus.

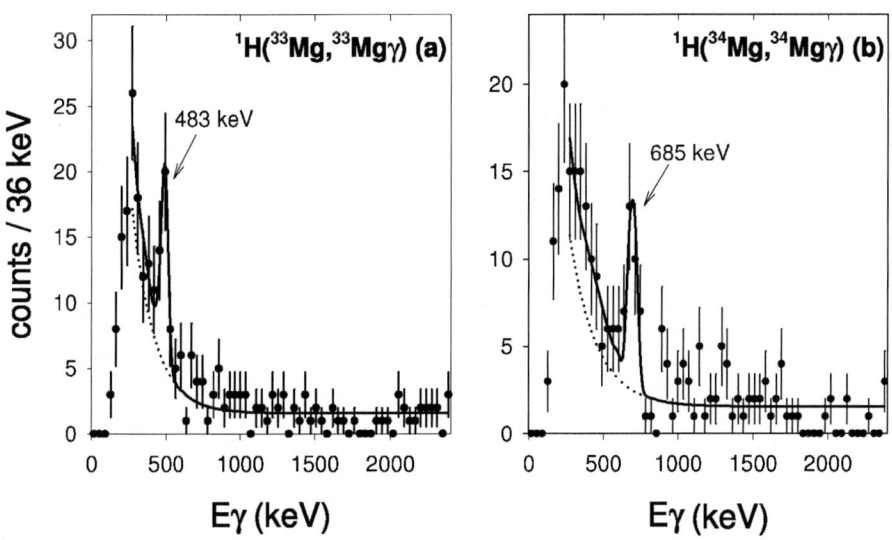

**FIGURE 6.** Doppler–corrected spectra of γ rays emerging from $^1$H($^{33}$Mg,$^{33}$Mg) (a), $^1$H($^{34}$Mg,$^{34}$Mg) (b) reactions. The solid line is the final fit including the spectrum curves from GEANT4 simulation and additional smooth polynomial backgrounds plotted as separate dotted lines for each nucleus.

**TABLE 1.** Angle integrated cross sections of the (p,p') process corrected for the detection efficiency and deformations for $^{28}$Ne, $^{30,31}$Na and $^{33,34}$Mg nuclei.

| isotope (peak) | cross section | $\beta_M$ | $\beta_C$ | $\beta_n$ |
|---|---|---|---|---|
| $^{28}$Ne (1319 keV) | 22±5 mb | 0.25±0.05 ($2_1^+ \to 0_{gs}^+$) | 0.36±0.03 [18] | 0.23±0.05 |
| $^{30}$Na (403 keV) | 18±4 mb | 0.32±0.04 ($3_1^+ \to 2_{gs}^+$) | 0.41±0.10 [20] | 0.30±0.05 |
| $^{31}$Na (370 keV) | 24±4 mb | 0.56±0.05 ($5/2_1^+ \to 3/2_{gs}^+$) | 0.66±0.16 [20] | 0.54±0.07 |
| $^{33}$Mg (483 keV) | 33±10 mb | 0.47±0.08 ($7/2_1^+ \to 5/2_{gs}^+$) | 0.52±0.12 [22] | 0.46±0.10 |
| $^{34}$Mg (685 keV) | 111±37 mb | 0.68±0.16 ($2_1^+ \to 0_{gs}^+$) | 0.58±0.06 [25] | 0.70±0.13 |

From a distorted wave analysis of the cross sections, we derived "matter" deformation parameters ($\beta_M$). In the calculations, the standard collective form factors were applied and the global phenomenological parameter set CH89 proposed in [27] was employed for the optical potential. The "matter" deformation parameters deduced in this way can be also found in Table 1. The mass deformations of these nuclei are consistent with their charge deformations determined from Coulomb excitation experiments, which can be also seen in Table 1.

Based on the "matter" and charge deformation parameters and using the formula of

$$(Nb_n + Zb_p)\, \delta_M = Nb_n \delta_n + Zb_p \delta_p, \qquad (1)$$

the neutron deformation parameters (see Table 1) could also be extracted where $\delta_{n,p} = \beta_{n,p} R$ (assuming $\delta_p = \delta_C$) and $R$, $N$ and $Z$ are the nuclear radius, the neutron and the proton numbers, respectively, while $b_n/b_p = 3$ are the sensitivity parameters for protons and neutrons of our (p,p') probe. The results show that $^{30,31}$Na and $^{33,34}$Mg nuclei are largely deformed; the deformation of the proton and neutron distributions are similar and cannot be distinguished at the present experimental uncertainties. However, both the neutron and the proton deformations of $^{28}$Ne are much smaller than is characteristic of nuclei in the "island of inversion". A strong decrease of the neutron effective charge has been observed in nuclei far from the valley of stability like $^{16}$C [10] or $^{17}$B [12]. Using the reduced effective charges of Sagawa and Asahi for $^{28}$Ne [28], both the B(E2) and the neutron transition matrix elements can be described in a correct way in the MCSM [29], but the neutron transition probability becomes strongly underestimated in the USD shell model.

The validity of the above models can also be checked by the investigation of odd nuclei out of the "island of inversion" in the USD calculations. For $^{27}$F, the USD shell model predicts no bound excited state while the MCSM expects the first excited state to be bound at 1.2 MeV (see Fig. 7a). On the basis of the decay properties, the 777 keV transition may be a reasonable candidate for the transition between the $1/2^+$ state of the MCSM and the ground state. The existence of the other bound excited state calls for further theoretical effort however at least one of the bound excited states is interpreted by the MCSM.

For $^{27}$Ne, the situation is similar. Comparing our level scheme with the one calculated in the USD shell model and the MCSM in Fig. 7b, it is seen that one of the excited states may correspond to the $1/2^+$ state predicted by both models; the other one should come from an out of the *sd* shell model space excitation. Indeed, the additional excited state may correspond to the theoretical $3/2^-$ one of the MCSM, associated with the $\nu p_{3/2}$

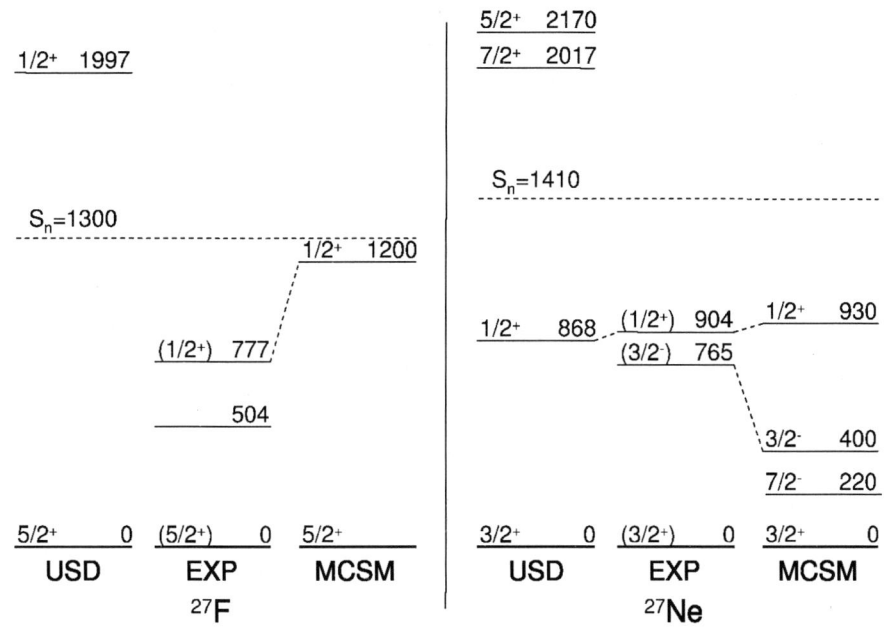

**FIGURE 7.** (a) Experimentally determined low–lying levels of $^{27}$F nucleus plotted together with the predictions of USD [30] and MCSM shell model calculations. (b) Similar plot for $^{27}$Ne.

configuration since it decays by a prompt transition to the presumed $3/2^+$ ground state. (The decay of the $7/2^-$ state predicted by the MCSM to the $3/2^+$ ground state cannot be visible in the present experiment due to its long lifetime even if it is bound.) Thus, the observation of two low–lying excited states in $^{27}$Ne is in agreement with the MCSM prediction.

## CONCLUSIONS AND SUMMARY

Summarizing our results, we have studied the structure of $^{27}$F, $^{28}$Ne, $^{30,31}$Na and $^{33,34}$Mg nuclei by proton inelastic scattering while $^{27}$Ne was excited and studied via neutron–knock–out reaction. From the cross section of the inelastic scattering process the mass deformations were deduced for $^{28}$Ne, $^{30,31}$Na and $^{33,34}$Mg isotopes, and from a comparison with the results of the Coulomb excitation experiments, the neutron deformations were also derived. $^{30,31}$Na and $^{33,34}$Mg nuclei have significant mass deformation in agreement with the conclusions of the Coulomb excitation studies, and the neutron deformations do not differ from those of the charge deformations within experimental error, showing that the proton and neutron deformations are in phase as it is expected for usual deformed nuclei. However, the neutron and proton deformations for $^{28}$Ne are found to be much smaller than is characteristic of nuclei in the "island of inversion", which could be interpreted by the MCSM calculations using reduced effective charges.

The results for $^{27}$F and $^{27}$Ne supports the validity of the MCSM calculations and of a small $N=20$ shell gap.

## ACKNOWLEDGMENTS

We would like to thank the RIKEN Ring Cyclotron staff for their assist during the experiment. One of authors (Z. E.) is grateful for the JSPS Fellowship Program in RIKEN and thanks the support from OTKA F60348. The European authors thank the kind hospitality and support from RIKEN. The present work was partly supported by the Grant-in-Aid for Scientific Research (No. 1520417) by the Ministry of Education, Culture, Sports, Science and Technology and by OTKA T38404, T42733 and T46901.

## REFERENCES

1. B. H. Wildenthal, et al., *Phys. Rev.*, **C22**, 2260 (1980).
2. M. H. Wildenthal, et al., *J. Phys.*, **G9**, L165 (1983).
3. E. K. Warburton, et al., *Phys. Rev.*, **C41**, 1147 (1990).
4. J. Retamosa, et al., *Phys. Rev.*, **C55**, 1266 (1997).
5. Y. Utsuno, et al., *Phys. Rev.*, **C60**, 054315 (1999).
6. Y. Utsuno, et al., *Phys. Rev.*, **C70**, 044307 (2004).
7. Y. Utsuno, et al., *Phys. Rev.*, **C64**, 011301 (2001).
8. E. Caurier, et al., *Nucl. Phys.*, **A693**, 374 (2001).
9. V. Tripathi, et al., *Phys. Rev. Lett.*, **94**, 162501 (2004).
10. Z. Elekes, et al., *Phys. Lett.*, **B586**, 34 (2004).
11. N. Imai, et al., *Phys. Rev. Lett.*, **92**, 062501 (2004).
12. Zs. Dombrádi, et al., *Phys. Lett.*, **B621**, 81 (2005).
13. T. Kubo, et al., *Nucl. Instrum. Meth.*, **B70**, 309 (1992).
14. H. Sakurai, et al., *Phys. Lett.*, **B448**, 180 (1999).
15. H. Ryuto, et al., *Nucl. Instrum. Meth.*, **A555**, 1 (2005).
16. S. Takeuchi, et al., *RIKEN Acc. Prog. Rep.*, **36**, 148 (2003).
17. M. Belleguic, et al., *Phys. Rev.*, **C72**, 054316 (2005).
18. H. Iwasaki, et al., *Phys. Lett.*, **B599**, 17 (2004).
19. B. V. Pritychenko, et al., *Phys. Lett.*, **B461**, 322 (1999).
20. B. V. Pritychenko, et al., *Phys. Rev.*, **C66**, 024325 (2002).
21. B. V. Pritychenko, et al., *Phys. Rev.*, **C63**, 011305 (2001).
22. B. V. Pritychenko, et al., *Phys. Rev.*, **C65**, 061304 (2002).
23. S. Nummela, et al., *Phys. Rev.*, **C64**, 054313 (2001).
24. K. Yoneda, et al., *Phys. Lett.*, **B499**, 233 (2001).
25. H. Iwasaki, et al., *Phys. Lett.*, **B522**, 227 (2001).
26. S. Agostinelli, et al., *Nucl. Instrum. Meth.*, **A506**, 230 (2003).
27. R. L. Warner, et al., *Phys. Rep.*, **201**, 57 (1991).
28. H. Sagawa, et al., *Phys. Rev.*, **C63**, 064310 (2001).
29. T. Otsuka, et al., *private communication based on PRC60(1999)054315*.
30. B. A. Brown, *http://www.nscl.msu.edu/~brown/sde.htm*

# SHARAQ project

A. Saito*, H. Sakai*, S. Shimoura[†], T. Uesaka[†], T. Kawabata[†],
K. Nakanishi[†], Y. Sasamoto[†], E. Ideguchi[†], H. Yamaguchi[†], S. Kubono[†],
G. P. Berg**, T. Ichihara[‡] and T. Kubo[‡]

*Department of Physics, University of Tokyo, 7-3-1 Hongo, Bunkyo, Tokyo 113-0033, Japan
[†]Center for Nuclear Study, University of Tokyo, RIKEN Campus, 2-1 Hirosawa, Wako, Saitama 351-0198, Japan
**Department of Physics, University of Notre-Dame, Indiana 46556, U.S.A.
[‡]RIKEN, 2-1 Hirosawa, Wako, Saitama 351-0198, Japan

**Abstract.** The SHARAQ spectrometer designed for high-resolution spectroscopies with RI beams is now under construction at the RI Beam Factory in RIKEN. Descriptions of the proposed physics programs, specifications of the spectrometer, dispersion-matched beam-line, beam-line detectors, and the construction schedule are presented.

**Keywords:** High-resolution spectrometer, dispersion matching, RI beam
**PACS:** 25.60.-t Reactions induced by unstable nuclei, 29.30.-h Spectrometers and spectroscopic techniques

## INTRODUCTION

The high-resolution SHARAQ spectrometer is now under construction at the RI Beam Factory (RIBF) in RIKEN (Fig. 1). The spectrometer will be used for high-resolution spectroscopies of unstable nuclei, in combination with a variety of radioactive isotope (RI) beams from BigRIPS [1].

Physics programs proposed with the SHARAQ spectrometer are based on a new technique of missing mass spectroscopy [2]. In the new technique, the RI beam is used as a "probe" to study new excitation modes such as double Gamow-Teller (DGT) state and the isovector spin monopole resonance (IVSMR). Since the RI beams have a large internal energy, the $Q$-value of the CX reaction can have a large positive value, namely the reaction is exothermic. The exothermic CX reactions induced by the RI beams open new possibilities to access highly excited states such as double Gamow-Teller (DGT) state and the isovector spin monopole resonance (IVSMR). The tetra neutron states [4] and the super heavy hydrogens [5, 6] can be also investigated by the CX reactions. The energy range of the RI beams at RIBF, 100–300 $A$ MeV, is well suited for the spectroscopic studies since the spin-isospin interaction is the strongest relative to the spin-isospin independent ones [3] at these energies. The required energy and angular resolutions to measure these reactions are $\sim 1$ MeV and $\sim 1$ mrad, respectively. To perform such experiments with high resolutions, light unstable nuclei with large internal energies, e.g. $^8$He, $^{12}$N, etc., are good candidates of the RI beams to minimize the effects of the multiple scattering and the energy-loss difference between the incident and outgoing particles.

In addition to these new experimental studies, the SHARAQ spectrometer will be used

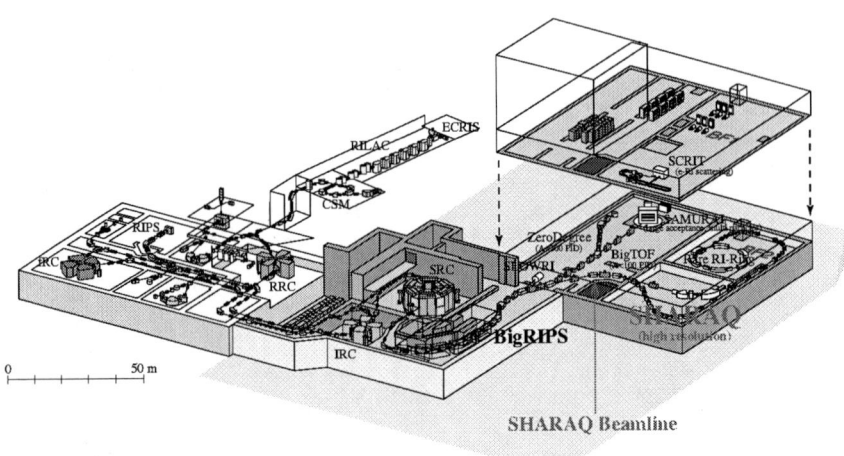

**FIGURE 1.** RI Beam Factory.

for the experiments of unstable nuclei in inverse kinematics. One of the most important role of the SHARAQ spectrometer is the high-resolution analysis of reaction products. For inelastic scattering, breakup, or knockout reactions of nuclei in heavier mass region, the high-resolution analysis is needed for good particle identifications. Other possibility is the measurements of the elastic scatterings of unstable nuclei which were hardly possible without the high-resolution spectrometer.

Since the RI beams generally have a large emittance, the dispersion matching technique must be introduced to perform high resolution measurements. If the dispersions of the SHARAQ spectrometer and its beam-line are properly matched, the missing-mass resolution can be significantly improved even in the presence of large momentum spread of the RI beam. The beam-line detectors are required to have large efficiencies for light RI beams, e.g. $^{8}$He at a few hundred MeV per nucleon, and to have good position resolution and high counting rates. In the following sections, the specifications of the SHARAQ spectrometer, dispersion-matched beam-line, beam-line detectors, and construction schedule are described.

## SHARAQ SPECTROMETER

The SHARAQ spectrometer is designed to achieve a resolving power $p/\delta p$ of $1.5 \times 10^4$ and an angular resolution better than 1 mrad for particle with magnetic rigidity of 6.8 Tm at maximum. The maximum rigidity is selected for the exothermic reaction of neutron-rich nuclei. Since the ejectile in the CX reaction induced by the neutron-rich nucleus is near to or on the stability line, the required maximum rigidity is less than that of BigRIPS (9 Tm). A view of the SHARAQ spectrometer is shown in Fig. 2. Normal conducting dipole magnets with an orbital radius of 4.4 m combined with super-conducting doublet quadrupoles and one normal-conducting quadrupole are used to meet the design criterion. The first doublet quadrupoles (SDQ: Q1 and Q2) are super-conducting magnets. Its design is, for the most part, inherited from that of super-conducting triplet quadrupoles (STQ) of BigRIPS [1]. Subsequent normal-conducting dipole (D1) and quadrupole (Q3) magnets are recycled from the decommissioned SMART spectrograph [7]. The last dipole magnet (D2) will be newly constructed. Details of the design are described in other reports [8]. The D2 magnet is a 60-deg. bending magnet with a pole gap of 200 mm. Its exit is tilted by 30 deg. in horizontal focusing direction to reduce a distance from the D2 magnet to the focal plane. The spectrometer can rotate around its target position from $-2$ deg. to 15 deg. for finite angle measurements. The specifications of the SHARAQ spectrometer are listed in Table 1. Details of the spectrometer are described in other reports [9, 10].

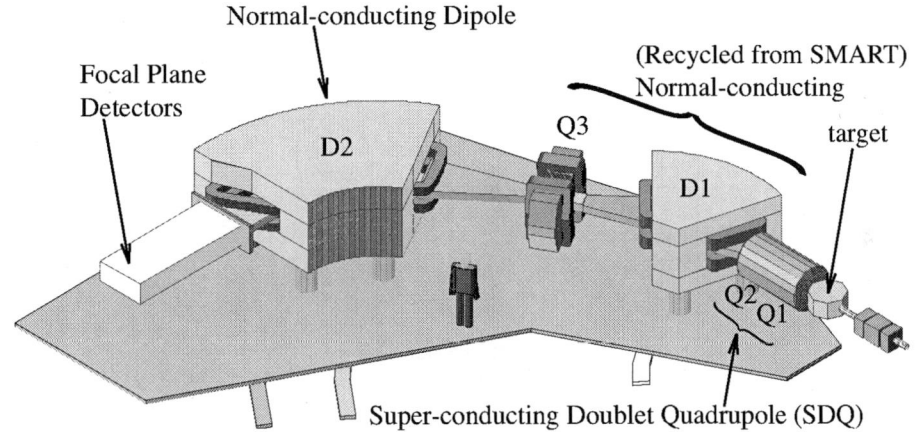

**FIGURE 2.** The SHARAQ spectrometer.

**TABLE 1.** Specifications of the SHARAQ spectrometer.

| | |
|---|---|
| Maximum rigidity | 6.8 Tm |
| | (c.f. 9 Tm in BigRIPS) |
| Momentum resolution | $p/\delta p \sim 1.5 \times 10^4$ |
| Angular resolution | $\sim 1$ mrad |
| Momentum acceptance | $\pm 1\%$ |
| Angular acceptance | > a few msr |
| Rotating angle | from $-2$ deg. to $+15$ deg. |

## DISPERSION-MATCHED BEAM-LINE

Figure 3 shows the dispersion-matched beam-line. The RI beams produced at F0 are transported to the SHARAQ target through the BigRIPS beam-line and the SHARAQ beam-line. The beam-line is designed to fulfill dispersion-matching condition when combined with the SHARAQ spectrometer. Simultaneous achievement of lateral dispersion matching and angular dispersion matching conditions are crucially important in the use of the SHARAQ spectrometer for RI beams which necessarily accompany a large momentum spread. A dispersion matching technique and/or event-by-event tagging of the beam momentum will be introduced for the purpose to compensate the energy spread of the RI beam. Figure 4 shows the calculated optics of the dispersive beam transport. The RI beam is produced by primary reactions at F0. From F0 to F3, the beam-line is used to separate the RI beam from other fragments. The starting point of the dispersive beam transport was set to F3. Details of the dispersion-matched beam-line is described in other report [11].

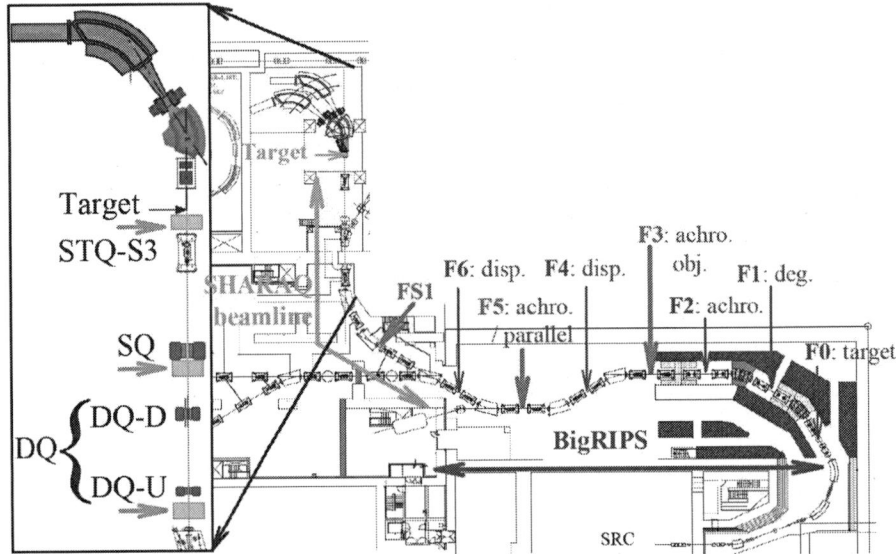

**FIGURE 3.** Dispersion-matched beam-line for the SHARAQ spectrometer. The RI beams produced at F0 are transported to the SHARAQ target through the BigRIPS beam-line and the SHARAQ beam-line. The dispersive beam transport starts from the achromatic focus at F3.

## BEAM-LINE DETECTORS

The beam-line detectors are used for two purposes. One is the tuning of the dispersion-matched beam transport, the other is the measurement of the beam particles event-by-event.

The beam-tuning needs a certain detection efficiency and a position resolution better than 1 mm. The standard BigRIPS detector system consisting of parallel plate avalanche counters (PPAC's) [12] and plastic scintillators will be installed at each focal plane from F0 to F6 (Fig. 3). Additional tracking and timing detectors will be installed at each focus from F-S1 to the SHARAQ target. The PPAC's can be used for tracking the RI beams with heavier masses since its efficiencies are generally large enough.

The event-by-event tracking measurement needs several requirements. The efficiency of the detector should be as large as possible even for light RI beams such as $^8$He at a few hundred MeV per nucleon. To perform high-resolution measurements, the thickness

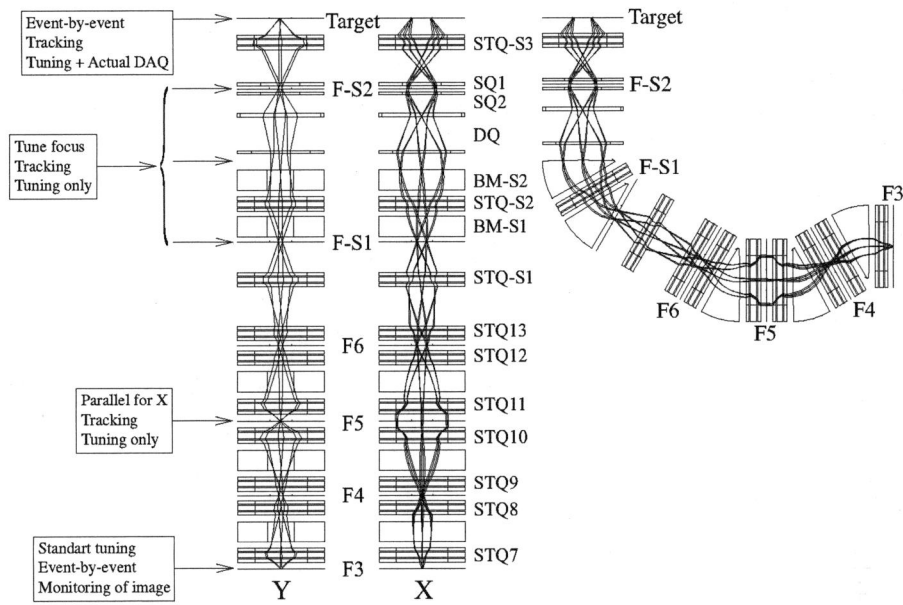

**FIGURE 4.** Calculated optics of the dispersive beam transport from F3 to the SHARAQ target.

of the detector should be as small as possible to minimize multiple scattering effects and the energy-loss difference between the incoming and the outgoing particles. The event-by-event measurement will be needed at F3 for monitoring image and at SHARAQ target to measure the momentum vector of the RI beam.

In order to realize the performance described above, low-pressure multi-wire drift chambers (MWDC's) are being developed. The multiple scattering can be reduced to ~0.1 mrad when the isobutan gas with 10–20 kPa is used. The thickness of the gas corresponds to $10^{-4}$ of the radiation length. The expected position resolution is less than 300 $\mu$m in FWHM. The maximum counting rate is 1 MHz for the dispersive beam transport. The specifications are summarized in Table 2. In addition, we are developing low-pressure MWDC with stripped cathodes (MWDC-SC). The structure of the detector is shown in Fig. 5. The cathodes are stripped with 2.5 mm pitch and the signals are read out using the delay-line (DL) developed for the PPAC [12]. The number of channels can be reduced with the DL-readout. The prototypes of them are now being tested.

## SCHEDULE

The SHARAQ spectrometer will be put into operation in 2008. The construction of the D2 magnet is planned to finish in March 2007. In the first half of FY2007, all the magnet together with a rotating stage will be installed at the experimental vault of the RIBF building. Following the completion of the installation, field map measurements

**TABLE 2.** Specifications of the beam-line detectors.

|  | Low-pressure MWDC | Low-pressure MWDC-SC |
|---|---|---|
| position resolution | <300 $\mu$m[†] |  |
| efficiency | >95%[†] |  |
| multiple scattering | ~0.1 mrad[†] | ~0.1 mrad[†] |
| counting rate | ~1 MHz[†] |  |
| anode wire | 12.5 $\mu$m$^\phi$ | 20 $\mu$m$^\phi$ |
| potential wire | 75 $\mu$m$^\phi$ | 75 $\mu$m$^\phi$ |
| cell size | 5×5 mm$^2$ | 6×6 mm$^2$ |
| anode-cathode pitch | 2.4 mm | 2.5 mm |
| cathode | 2 $\mu$m$^t$ | 1.5 $\mu$m$^t$, 2.5-mm pitch |
| gas | i-C$_4$H$_{10}$ | i-C$_4$H$_{10}$ |
| gas pressure | 10–20 kPa[†] | 10–20 kPa[†] |

[†] designed value.

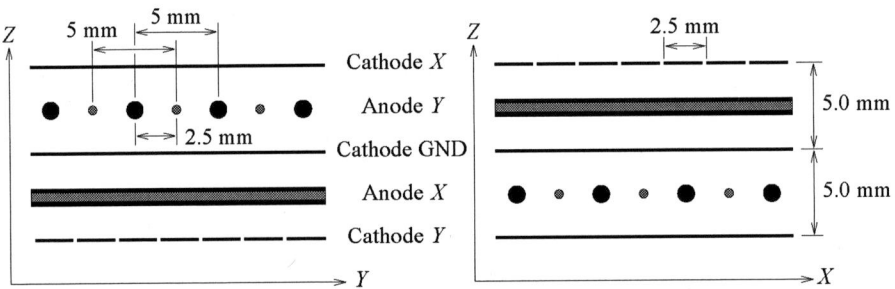

**FIGURE 5.** Structure of the low-pressure MWDC-SC.

of the magnets will be made in the second half of FY2007. Design and construction of beam-line magnets, together with development of beam-line and focal plane detectors will be carried out in parallel by the end of FY2007.

## SUMMARY

An overview of the SHARAQ project is presented. The high-resolution spectrometer and the dispersion-matched beam-line enable us to perform high-resolution spectroscopies with RI beams. Several physics programs with the SHARAQ spectrometer and other possibilities to use the SHARAQ spectrometer are presented. The basic designs of the SHARAQ spectrometer and the dispersion-matched beam-line are fixed. The developments of beam-line detectors, low-pressure MWDC and low-pressure MWDC-SC, are now in progress. The construction will be completed in FY2007 and experiments are scheduled to start in FY2008.

# ACKNOWLEDGMENTS

This work is supported by the Grant-in-Aid of Specially Promoted Research (Grant No. 17002003) of the Ministry of Education, Culture, Sports, Science, and Technology of Japan.

# REFERENCES

1. T. Kubo, *Nucl. Instrum. Meth. in Phys. Res.*, **B204**, 97, (2003).
2. S. Shimoura, Proposal for RIBF Technical Advisory Committee 2005.
3. W. G. Love and M. A. Franey, *Phys. Rev.*, **C24**, 1073 (1981).
4. F. M. Marqués, M. Labiche, N. A. Orr, J. C. Angélique, L. Axelsson, B. Benoit, U. C. Bergmann, M. J. G. Borge, W. N. Catford, S. P. G. Chappell, N. M. Clarke, G. Costa, N. Curtis, A. D'Arrigo, E. de Góes Brenna, F. de Oliveira Santos, O. Dorvaux, G. Fazio, M. Freer, B. R. Fulton, G. Giardina, S. Grévy, D. Guillemaud-Mueller, F. Hanappe, B. Heusch, B. Jonson, C. Le Brun, S. Leenhardt, M. Lewitowicz, M. J. López, K. Markenroth, A. C. Mueller, T. Nilsson, A. Ninane, G. Nyman, I. Piqueras, K. Riisager, M. G. Saint Laurent, F. Sarazin, S. M. Singer, O. Sorlin, and L. Stuttgé, *Phys. Rev.*, **C65**, 044006 (2002).
5. A. A. Korsheninnikov, M. S. Golovkov, I. Tanihata, A. M. Rodin, A. S. Fomichev, S. I. Sidorchuk, S.V. Stepantsov, M. L. Chelnokov, V. A. Gorshkov, D. D. Bogdanov, R. Wolski, G. M. Ter-Akopian, Yu. Ts. Oganessian, W. Mittig, P. Roussel-Chomaz, H. Savajols, E. A. Kuzmin, E. Yu. Nikolskii, and A. A. Ogloblin, *Phys. Rev. Lett.*, **87**, 092501 (2001).
6. A. A. Korsheninnikov, E. Yu. Nikolskii, E. A. Kuzmin, A. Ozawa, K. Morimoto, F. Tokanai, R. Kanungo, I. Tanihata, N. K. Timofeyuk, M. S. Golovkov, A. S. Fomichev, A.M. Rodin, M. L. Chelnokov, G. M. Ter-Akopian, W. Mittig, P. Roussel-Chomaz, H. Savajols, E. Pollacco, A. A. Ogloblin, and M. V. Zhukov, *Phys. Rev. Lett.*, **90**, 082501 (2003).
7. T. Ichihara, T. Niizeki, H. Okamura, H. Ohnuma, H. Sakai, Y. Fuchi, K. Hatanaka, M. Hosaka, S. Ishida, K. Kato, S. Kato, H. Kawashima, S. Kubono, S. Miyamoto, H. Orihara, N. Sakamoto, S. Takaku, Y. Tajima, M. H. Tanaka, H. Toyokawa, T. Uesaka, T. Yamamoto, T. Yamashita, M. Yosoi, and M. Ishihara, *Nucl. Phys.*, **A569**, 287c (1994).
8. T. Uesaka, G. P. Berg, S. Shimoura, T. Kawabata, and H. Sakai, *CNS Ann. Rep. 2005*, 65 (2006).
9. T. Uesaka, S. Shimoura, H. Sakai, T. Kawabata, G. P. A. Berg, A. Saito, K. Nakanishi, Y. Sasamoto, T. Kubo, S. Kubono, E. Ideguchi, and H. Yamaguchi, *Eur. Phys. J.*, **A** (Proc. of RNB7 conference), in press.
10. T. Uesaka, S. Shimoura, T. Kawabata, H. Sakai, S. Kubono, E. Ideguchi, H. Yamaguchi, T. Kubo, and G. P. Berg, *CNS Ann. Rep. 2005*, 61 (2006).
11. T. Kawabata, T. Kubo, H. Sakai, S. Shimoura, and T. Uesaka, *CNS Ann. Rep. 2005*, 63 (2006).
12. H. Kumagai, A. Ozawa, N. Fukuda, K. Sümmerer, and I. Tanihata, *Nucl. Instrum. Meth. in Phys. Res.*, **A470**, 562 (2004).

# Development of polarized $^3$He ion source

M. Tanaka*, Y. Takahashi†, T. Shimoda†, S. Yasui**, M. Yosoi‡,
K. Takahisa‡ and N. Shimakura§

*Department of Clinical Technology, Kobe Tokiwa College, Ohtani-cho 2-6-2, Nagata-ku, Kobe 653-0083, Japan
†Department of Physics, Graduate School of Science, Osaka University, 1-1, Machikaneyama, Toyonaka, Osaka 560-0043, Japan
**Department of Physics, Tokyo Institute of Technology, O-okayama 2-12-1, Tokyo 152-8551, Japan
‡Research Center for Nuclear Physics, Osaka University, Mihogaoka 10-1, Ibaraki, Osaka 567-0047, Japan
§Department of Chemistry, Niigata University, Nino-cho 8050, Niigata 950-2181, Japan

**Abstract.** A long history on the polarized $^3$He ion source developed at RCNP is presented. We started with an "OPPIS" (Optical Pumping Polarized Ion Source) and later found the fundamental difficulties in the OPPIS. To overcome them an "EPPIS" (Electron Pumping Polarized Ion Source) was proposed and its validity was experimentally proven. However, a serious technical disadvantage was also found in the EPPIS. To avoid this disadvantage we proposed a new concept, "SEPIS" (Spin Exchange Polarized Ion Source), which uses an enhanced spin-exchange cross section theoretically expected at low $^3$He$^+$ incident energies for the $^3$He$^+$ + Rb system.

Next, we describe the present status of the SEPIS development; construction of a bench test device allowing the measurements of not only the spin-exchange cross sections $\sigma_{se}$ but also the electron capture cross sections $\sigma_{ec}$ for the $^3$He$^+$ + Rb system. The latest experimental data on $\sigma_{ec}$ are presented and compared with other previous experimental data and the theoretical calculations. A design study of the SEPIS for practical use in nuclear (cyclotron) and particle physics (synchrotron) is shortly mentioned.

Finally, we mention possibility to polarize ions heavier than $^3$He as an application of SEPIS. The theoretical calculation showed that $\sigma_{se}$ comparable to that for the $^3$He$^+$ + Rb is expected for the Li$^{2+}$ + Rb system, which suggests that the SEPIS will hopefully be a general tool to polarize any heavy ions.

## INTRODUCTION

It is well known that a polarized $^3$He nucleus is approximately regarded as a polarized neutron. To use a polarized $^3$He$^{2+}$ as a new probe of nuclear physics inaccessible by other probes, we have developed a $^3$He ion source over the decade at RCNP. In spite of our enormous efforts, the development is still under way. However, the high energy physicists still think to use the polarized $^3$He beam as a counterpart of the polarized proton beam to investigate the spin structure of nucleon. Further, nuclear and astrophysicists are also requiring the polarized $^3$He beam to carry out a variety of experimental programs in nuclear- and astro-physics. One of urgent topics is the establishment of the isovector spin-dipole resonance (IVSDR) in nuclei, because a role of the neutral and charged weak currents is influential to the neutrino nuclear synthesis in supernovae as pointed out by Kolbe, Langanke et al. [1]. For this purpose, the ($^3$He ,t)

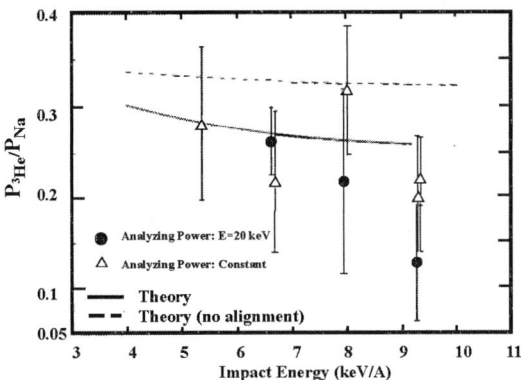

**FIGURE 1.** Energy dependence of polarization transfer coefficient for the $^3$He$^+$ + Na system. Na vapor was polarized by means of laser optical pumping with a dye laser pumped by an Ar ion laser.

reactions with polarized $^3$He$^{2+}$ beams will hopefully offer a most sensitive probe for this subject.

Under the above circumstances, we believe that it is still valuable to develop a polarized $^3$He ion source to be dedicated to particle physics as well as nuclear physics even now. In this paper, we present, at first, a long history on the development of the polarizee $^3$He ion source performed at Osaka University emphasizing on the many failures and how to overcome them. As a consequence of enormous efforts, we could finally arrive at a new concept of SEPIS, i.e., Spin Exchange Polarized Ion Source which will be hopefully one of the most powerful polarized $^3$He ion sources in the next generation.

After presenting the long history, we present a somewhat detailed description on the SEPIS. A bench test device to experimentally prove the principle of SEPIS was constructed and many fundamental data were taken on, for example, the performance of an ECR ion source, polarimeter, and electron capture cross sections. Consequently, we are almost ready to measure the energy dependence of the spin-exchange cross sections for the $^3$He$^+$ + Rb system in a wide range of incident $^3$He$^+$ ion energy.

In what follows, we touch a design of a practical SEPIS as a future plan. As discussed later, one of the important subjects for a practical SEPIS is how to efficiently ionize the polarized $^3$He$^+$ ion without depolarization. For this purpose, we propose two types of ionization methods, i.e. stripping after acceleration and use of a tubular EBIS ionizer with a strong axial magnetic field.

Finally, we address the possibility to polarize ions heavier than $^3$He by means of the SEPIS. For this purpose, we calculated $\sigma_{se}$ for the $^6$Li$^{2+}$ + Rb system and found that the large $\sigma_{se}$ comparable to the $^3$He$^+$ + Rb system.

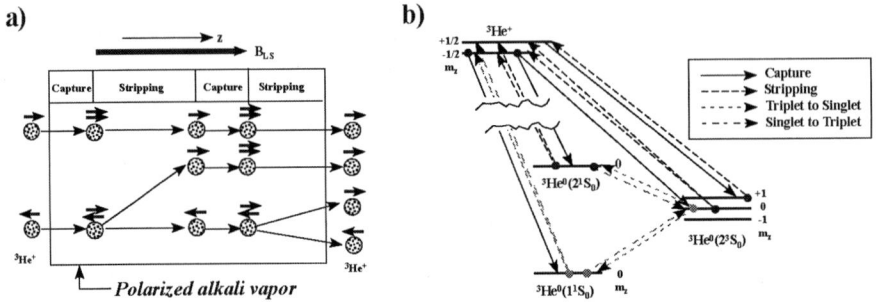

**FIGURE 2.** A schematic picture showing the principle of the EPPIS

## HISTORY OF DEVELOPMENT

### Optical Pumping Polarized Ion Source (OPPIS)

Encouraged by the pioneering work by Mori et al. [2], over the two decades ago, we decided to develop an OPPIS (Optical Pumping Polarized Ion Source) type polarized $^3$He ion source which uses a capture of polarized electron of an alkali atom (Na) by a $^3$He$^{2+}$ ion under the presence of a static magnetic field and subsequent creation of the nuclear polarization by the hyperfine interactions. To optimize the performance of the $^3$He OPPIS, we measured an energy dependence of polarization transfer coefficient $P_T$ defined by the ratio of the $^3$He nuclear polarization to the atomic polarization of Na atom. In addition to a minor energy dependence of $P_T$, a substantial reduction of the nuclear polarization ($P_T \sim \frac{1}{3}$) was observed as shown in Fig. 1 [3]. This large reduction was due to the depolarization through the spin orbit coupling of transferred electron during the cascade photon emission. It was suggested that an enormously strong decoupling field ($\sim$32 T) was required to avoid this depolarization. However, to use such a strong magnetic field was not practical under the present circumstance.

To overcome this difficulty, we proposed, then, a new polarized $^3$He ion source, in which the depolarization due to the spin orbit coupling was reduced. This type of ion source was later called an EPPIS (Electron Pumping Polarized Ion Source) [4].

### Electron Pumping Polarized Ion Source (EPPIS)

The EPPIS uses a multiple electron capture and stripping between a $^3$He$^+$ ion and $^3$He atoms in a polarized alkali vapor, where no electron capture of $^3$He$^{2+}$ ions which induced a large depolarization is present since the contributing states of $^3$He are either a $^3$He$^+$ ion or neutral $^3$He atom as illustrated in a) of Fig. 2. The technical term, "electron pumping" was named on the analogy of the principle of the optical pumping as shown in b) of Fig. 2: Let's start with a substate $m_z$=-1/2 state of a $^3$He$^+$ ion. As a result of a polarized electron capture, mainly a $^3$He$^0$($2^3S_1$) state with $m_z$=0 is populated (solid curves), and a subsequent electron stripping reaction (dashed curves) populates

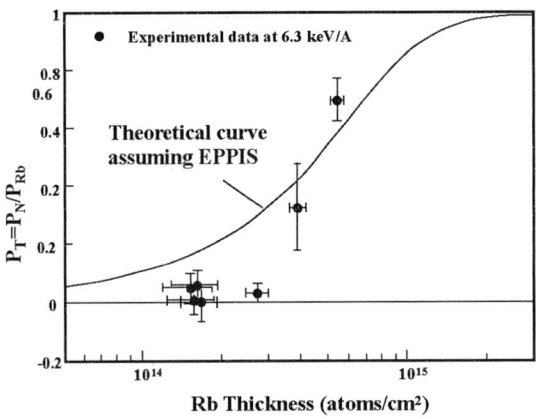

**FIGURE 3.** Experimental results on the EPPIS compared with the theory of the EPPIS

both $m_z=+1/2$ and $-1/2$ states of a $^3$He$^+$ ion with an equal probability. Repeating these processes many times, the $m_z=+1/2$ state of $^3$He$^+$ ion is predominantly populated. This process is similar to the principle of the optical pumping, in which circularly polarized photons are used instead of polarized electrons. Therefore, we named our ion source an EPPIS. To check the validity of the EPPIS principle we constructed a bench test device: We used a Rb vapor polarized by means of optical pumping with a Ti:Sapphire laser pumped by 20-W Ar ion laser instead of polarized Na vapor used in the OPPIS development. To create a large nuclear polarization for $^3$He$^+$ ions, a thick Rb vapor ($\leq 10^{15}$ atoms/cm$^2$) with a large polarization and a superconducting solenoidal magnet generating an axial magnetic field stronger than 6 T were introduced. The observed thickness dependence of the $^3$He$^+$ nuclear polarization was successfully reproduced by the EPPIS theory [5] as shown in Fig. 3. However, it was found that the EPPIS required a polarized alkali vapor thickness much higher than that used for the OPPIS. This is a serious drawback of the EPPIS method.

Meanwhile, in the course of the EPPIS development, an extremely large spin-exchange cross sections $\sigma_{se}$ ( $>10^{-14}$ cm$^2$ ) between a $^3$He$^+$ ion and Rb atom was theoretically predicted at a low $^3$He$^+$ incident energy region ($\sim$1 keV/A) by the semi-classical close-coupling method based on the molecular orbital expansion [6] as shown in Fig. 4. If this is true, large spin-exchange cross sections might be used for polarizing a $^3$He$^+$ ion and subsequently a $^3$He nucleus much easier than the EPPIS. We named this type polarized $^3$He ion source "SEPIS" (Spin Exchange Polarized Ion Source) [7]. In what follows, we will describe the SEPIS emphasizing on a historical view of the spin-exchange processes themselves, description of the SEPIS principle, a bench test device to check the validity of SEPIS, and future development.

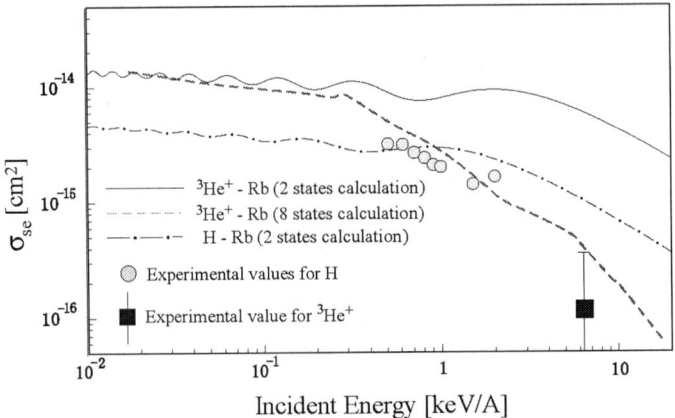

**FIGURE 4.** Energy dependence of the spin exchange cross sections for the $^3\text{He}^+$ + Rb system. The solid curve The experimental data is our previous result on the EPPIS.

## PRESENT AND FUTURE DEVELOPMENT

### SEPIS

Importance of the spin-exchange processes in atomic collisions was, firstly, pointed out in the radio astronomy more than 50 years ago by Purcel et al. [8]. They claimed that the 21-cm radiation corresponding to the hyperfine splitting of the hydrogen atom from the universe is due to the spin-exchange excitation of the hydrogen atom by an electron impact. In a similar way, the SEPIS uses the spin-exchange collisions between a polarized electron of an alkali atom and an unpolarized electron of $^3\text{He}^+$ ion. One of the candidates for this is employed for the $^3\text{He}^+$ + Rb system, and a simulation calculation was performed. The calculated results are shown in Fig. 5, where the $^3\text{He}^+$ polarization (dotted curve) increases according as an increase of the polarized Rb vapor thickness owing to the spin-exchange collisions. It is noteworthy to mention that the SEPIS needs the Rb vapor with a thickness of about $2 \times 10^{14}$ atoms/cm$^2$, which is amazingly an order of magnitude less than that needed for the EPPIS. This is the greatest advantage of the SEPIS relative to the EPPIS. Though the SEPIS seems to be a promising method, there may exist a drawback concerning the $^3\text{He}^+$ beam intensity extractable. The $^3\text{He}^+$ beam intensity extracted from the SEPIS is limited mainly by the large electron capture processes at low impact $^3\text{He}^+$ energies as given by Eq. (1),

$$^3\text{He}^+ + \text{Rb} \rightarrow\, ^3\text{He} + \text{Rb}^+. \tag{1}$$

The calculated results of the Rb vapor thickness dependence of the fractional output $^3\text{He}^+$ beam intensity are shown as a solid curve in Fig. 5, where the curve was drawn assuming the experimental data of $\sigma_{ec}$ for the Cs + $^4\text{He}^+$ system [9]. The reason why we employed the data for the Cs + $^4\text{He}^+$ system instead of Rb + $^3\text{He}^+$ system is that no data of $\sigma_{ec}$ was available for the latter system. Therefore, we must experimentally

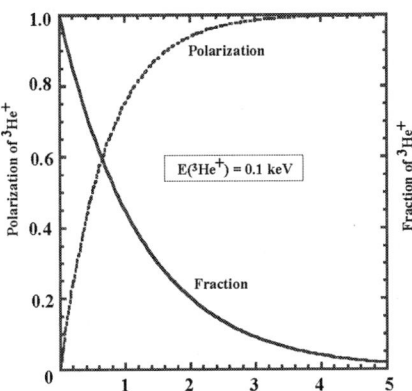

**FIGURE 5.** $^3$He$^+$ polarization (dotted curve) and fraction of $^3$He$^+$ beam current (solid curve) plotted as a function Rb vapor thickness

settle the values of $\sigma_{ec}$ for the Rb + $^3$He$^+$ system in the present work. To enable the measurements of both $\sigma_{se}$ and $\sigma_{ec}$ at low $^3$He$^+$ incident energies possibly lower than 1 keV, we have constructed a device as shown in Fig. 6. To allow the measurement of the incident energy dependence, the Rb cell was insulated and applied a high voltage.

The measurement of $\sigma_{ec}$ was performed with this bench test device. An output $^3$He$^+$ beam intensity $I$ penetrating the Rb cell was measured by changing the Rb vapor thickness, and $\sigma_{ec}$ was extracted assuming

$$I = I_0 e^{-\sigma_{ec} \times L}, \quad (2)$$

where $L$ is a Rb vapor thickness and $I_0$ is an input $^3$He$^+$ beam intensity. The Rb vapor thickness $L$ was observed by measuring a Faraday rotation angle of a probe laser penetrating the Rb vapor in the presence of the magnetic field. The observed $\sigma_{ec}$ were summarized in Fig. 7. It is interesting to note that the observed data are systematically shifted to the higher energy in comparison with the data for the Cs + $\alpha^+$ system. It was also found that the observed $\sigma_{ec}$ is qualitatively reproduced by the theory [6]. Though the measurement of $\sigma_{ec}$ is not a direct observation of the spin-exchange processes, this result indirectly supports the SEPIS principle. We are currently preparing the measurement of $\sigma_{se}$. For this purpose, we have introduced a solid state green laser (10 W) to pump a Ti:Sapphire laser. A new Rb vapor cell which keeps the liquid Rb for a long time was fabricated and will soon be installed in the bench test device.

## Future Prospect

When the SEPIS principle will be experimentally proven, we will begin designing a practical SEPIS. In designing the practical SEPIS, we must pay particular attention on ionization of the polarized $^3$He$^+$ ions, since we must inject not polarized $^3$He$^+$ beams but

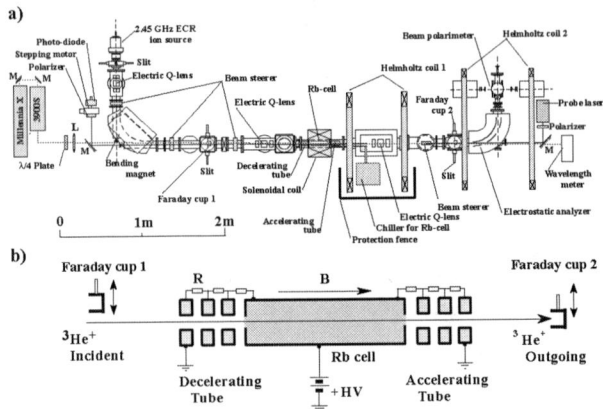

**FIGURE 6.** Experimental equipment under construction to prove the principle of the SEPIS

polarized $^3$He$^{2+}$ beams in the accelerators. Taking the above point into account we offer two types of SEPIS as schematically shown in Fig. 8, the detail of which is presented in our latest paper [7]. Here, the polarized $^3$He$^+$ ion is ionized either by stripping after pre-acceleration( a) ) or ionization by a tubular EBIS ionizer with a holding field larger than 2 T ( b) ). The former option is suitable for acceleration by cyclotrons with a CW operation, while the latter one is suitable for accelerations by synchrotrons with a pulsed operation. In case of the tubular EBIS ionizer, a strong axial magnetic field larger than 2 T should be used to avoid the depolarization during the ionization processes.

## Application of SEPIS

Lastly, we briefly touch a possible application of the SEPIS method. We propose here to polarize heavy ions. In Fig. 9, we show the calculated result on $\sigma_{se}$ for the $^6$Li$^{2+}$ + Rb system plotted as a function of the $^6$Li$^{2+}$ energy. It was demonstrated that $\sigma_{se}$ exceeding $10^{-14}$ cm$^2$ was expected at a low energy region in a similar way to the case of the $^3$He$^+$ + Rb system. This is an encouraging result. Hopefully, the heavy ions including radioisotope beams can be polarized by means of the SEPIS.

## ACKNOWLEDGMENTS

The present work is a part of our continuing project over the decade. Our project has been performed in collaboration with many people from the world. In particular, we are grateful to Profs. L. W. Anderson (Wisconsin), Yu. Plis (Dubna), R. Morgenstern (KVI, the Netherlands), T. Yamagata (Konan, Japan), M. Fujiwara (RCNP, Japan), S. Nakayama (Tokushima, Japan), K. Katori (Riken, Japan), T. Itahashi (Osaka, Japan), M. Kondo (RCNP, Osaka), and H. Ogata (RCNP, Osaka) for their kind collaborations.

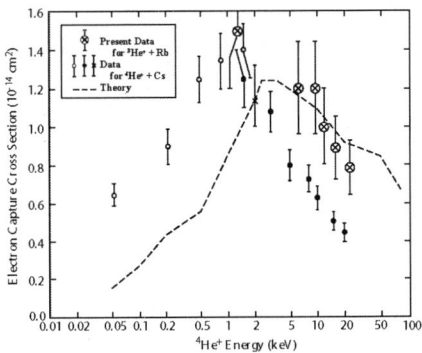

**FIGURE 7.** The electron capture cross section $\sigma_c$ for $^3$He$^+$ on Rb plotted as a function of $^4$He$^+$ energy in the energy range 0.05-23 keV, showing our measurement (symbol: ⊗), measurements by Lorents and Peterson (symbol: o), by Schlachter et al. (symbol: ●), and by Donally and Thoeming (symbol: ×). Except for our measurement, all of the experimental data are taken for the $^4$He$^+$ + Cs system. The dashed curve is the results of theoretical calculation.

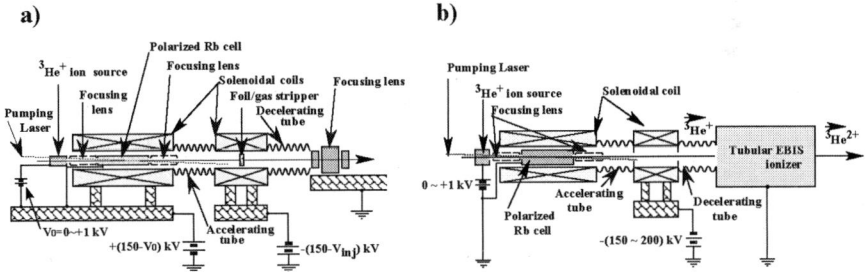

**FIGURE 8.** The layout of the proposed practical SEPIS. a) SEPIS based on the ionization by the stripping after acceleration. b) SEPIS based on the ionization by the tubular EBIS ionizer.

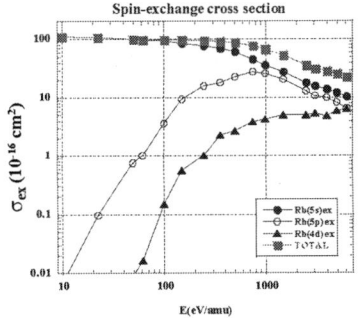

**FIGURE 9.** The theoretical calculation of $\sigma_{se}$ obtained for the $^6Li^{2+}$ Rb system.

We thanks many graduate and under graduate students from time to time during our development. The present work was financially supported by the budget from Kobe Tokiwa, Osaka University, and JSPS.

## REFERENCES

1. E. Kolbe, K. Langanke, and P. Vogel, Nucl. Phys. **A652** (1999) 91.
2. Y. Mori, A. Takagi, K. Ikegami, S. Fukumoto, and A. Ueno, J. Phys. Soc. Jpn. **55** (1986) 453.
3. M. Tanaka, N. Shimakura, T. Oshima, K. Katori, M. Fujiwara, H. Ogata, and M. Kondo, Phys. Rev. **A50** (1994) 1184.
4. M. Tanaka, M. Fujiwara, S. Nakayama, and L. W. Anderson, Phys. Rev. **A52** (1995) 392.
5. M. Tanaka, T. Yamagata, K. Yonehara, T. Takeuchi, Y. Arimoto, M. Fujiwara, Y. A. Plis, L. W. Anderson, and R. Morgenstern, Phys. Rev. **A60** (1999) R3354.
6. Y. Arimoto, N. Shimakura, T. Yamagata, K. Yonehara, and M. Tanaka Phys. Rev. **A64** (2001) 062714-1.
7. M. Tanaka, Yu. A. Plis, E. D. Donets, N. Shimakura, Y. Arimoto, T. Yamagata, and K. Yonehara, Nucl. Instr. and Meth. **A537** (2005) 501.
8. E. M. Purcel, and G. B. Field, Astrophys. J. **124** (1956) 542.
9. A.S. Schlachter, S. H. Loyd, P. J. Bjorkholm, L. W. Anderson, and W. Haeberli , Phys. Rev. **174** (1968) 201-211.
10. M. Tanaka, Y. Takahashi, T. Shimoda, T. Furukawa, M. Yosoi, K. Takahisa, N. Shimakura, and S. Yasui, Nucl. Instr. and Meth. **A** (2006) in press.

# The EURISOL Project

*Y. Blumenfeld, on behalf of the EURISOL Design Study*
*Institut de Physique Nucléaire, IN2P3-CNRS, 91406 Orsay Cedex, France*

**Abstract.** The European plan for Radioactive Beam Facilities aims for the construction of two "next generation" facilities: FAIR, a projectile fragmentation facility to be located at GSI, Darmstadt, Germany, and EURISOL, a high power ISOL facility. The basic layout of EURISOL will be described. The most challenging technical parts of such a facility are currently being designed and protoyped within a pan-European design study supported by the European Union. The organization of this study will be outlined and the first conclusions will be discussed.

**Keywords:** Radioactive Ion Beam facilities, EURISOL, European Collaboration, superconducting linear accelerators, high power fission targets.
**PACS:** 29.17+w, 29.25.Rm

## INTRODUCTION

The advent of radioactive beam facilities has given a new impetus to nuclear structure physics during the last two decades. It has led to several major unexpected discoveries such as the existence of dilute neutron matter in halo nuclei, the modification of shell structure and magic numbers far from stability, proton and two-proton radioactivity, new regions of shape coexistence … However, after this fruitful first survey, further progress is hampered by the weak beam intensities of current installations which correlate with the difficulty to reach the confines of nuclear binding where new phenomena are predicted, and where the r-process path for nuclear synthesis is expected to be located. The advancement of Radioactive Ion Beam (RIB) science calls for the development of so-called next-generation facilities, which will provide beam intensities several (2-4) orders of magnitude higher than presently available.

There are two main methods to produce RIB: projectile fragmentation and Isotope Separation On Line (ISOL). The former delivers high energy beams far from stability with generally mediocre optical qualities, while the latter produces excellent quality beams with high intensities, but its efficiency decreases for the most exotic very short lived species. The evident complementary nature of the two approaches has led NuPECC, the European Coordination Committee for Nuclear Physics, to advise building two next generation facilities in Europe, one fragmentation and one ISOL installation [1]. The fragmentation facility will be FAIR [2] in Darmstadt, Germany, while the ISOL facility is christened EURISOL. The construction of FAIR is scheduled to start in 2007. The technical challenges for EURISOL were found to be too great to envisage rapid construction, and a European roadmap towards the ultimate facility EURISOL was proposed. This roadmap will be explained in the next section, followed by a description of the main challenges along the road towards EURISOL and how the community has undertaken to tackle them in the framework of the EURISOL Design Study. Some preliminary results concerning accelerators and targets will then be discussed.

## THE EUROPEAN ISOL ROADMAP

In order to continuously update the scientific case for ISOL facilities and meet the technical challenges involved in the EURISOL concept, the European RIB community identified in 2003 three major directions to be pursued concurrently until the end of the decade:
- Exploit vigorously the current ISOL facilities: EXCYT (Catania), Louvain la Neuve, SPIRAL, REX-ISOLDE... in order to further the scientific justification of the field and train a generation of young scientists who will construct and utilize the facilities of the future.
- Construct the intermediate generation ISOL facilities which will constitute a unique testing ground for many technical solutions to be implemented in EURISOL. Four such facilities are currently planned: HIE-ISOLDE at CERN, an upgrade of REX-ISOLDE; MAFF, a reactor based installation at Munich; SPES at Legnaro; and SPIRAL2 at GANIL, Caen, possibly the most ambitious of the four.
- Produce detailed engineering designs and prototypes of the most challenging parts of the EURISOL facility in the framework of the EURISOL Design Study, a pan-European Research and Development project supported by the European Union.

This roadmap is expected to lead to a detailed design of an ultimate ISOL facility for which construction could start during the first years of the next decade.

## THE EURISOL CONCEPT AND CHALLENGES

A baseline concept for a next generation ISOL facility was devised in the European EURISOL RTD program which ran between 2000 and 2004, and for which the detailed conclusions can be found in [3]. A schematic drawing of the concept is shown on fig. 1. The driver is a 1 GeV Continuous Wave superconducting LINAC which can accelerate a proton beam with an intensity of 5mA corresponding to 5MW power. Capability for accelerating $^3$He, deuterons and A/Q=2 ions, will be included. The proton beam will impinge on a liquid Hg converter to produce a copious amount of neutrons in order to induce close to $10^{16}$ fissions per second in a $UC_X$ target with a total mass which could reach several kg. Spallation targets capable of absorbing directly 100 kW of beam power will also be available. After ionization, beams will be purified and reaccelerated by a superconducting LINAC with minimum beam losses. The final energy of the RIB can be adjusted continuously from rest to 150 AMeV for $^{132}$Sn. EURISOL is designed to provide a large energy range for a wide selection of isotopes which will allow physicists to combine a unprecedented variety of complementary probes for the study of exotic nuclei.

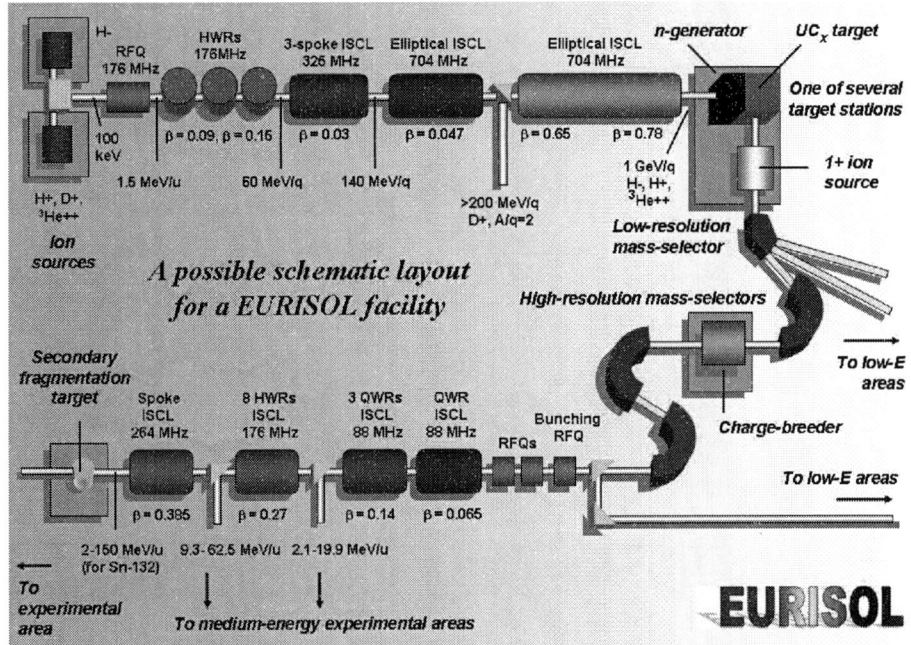

FIGURE 1. Schematic view of the EURISOL concept

It should be immediately apparent that the design of such an ambitious facility represents a huge endeavor for scientists and engineers. Some of the main challenges are summarized below:
- Design a 5MW; 1GeV proton driver with additional capability of 200 AMeV deuterons and A/Q=2 Heavy Ions; build and test prototypes of the cavities.
- Design a liquid Hg converter which will accept 5 MW of beam power.
- Design a UCx target which will make the most efficient use of the neutrons produced.
- Evaluate the safety constraints of the above set up.
- Design an efficient multi-user beam distribution system.
- Design a superconducting HI LINAC capable of accelerating $^{132}$Sn up to 150 AMeV with minimum beam losses.
- Investigate technologies for the instrumentation of the future
- Provide a conceptual study for a beta-beam neutrino facility.

Tackling successfully these problems calls for combining unique expertise from many laboratories and institutions throughout Europe and beyond, which is the goal of the EURISOL Design Study presented in the next section.

## THE EURISOL DESIGN STUDY

The EURISOL Design Study was initiated by a call to all European laboratories involved in RIB physics to contribute their expertise to the design of EURISOL and to build and test prototypes of the most challenging elements. Twenty institutions and laboratories from 14 European countries bid to become full participants in the project. In addition, another 20 laboratories from Europe, Asia and North America agreed to take part as contributors conveying their unique know-how on specific points. The European Union agreed to support this study with a contribution of about 9M€, out of a total cost estimated at 30 M€. The participants in the Design Study are listed in table 1.

**TABLE 1.** Participants in EURISOL Design Study

| Participant | Country | Participant | Country |
|---|---|---|---|
| GANIL (coordinator) | France | Inst. Physics Vilnius | Lithuania |
| CNRS/IN2P3 | France | Warsaw University | Poland |
| INFN | Italy | Inst. Phys. Bratislava | Slovakia |
| CERN | Europe | U. Liverpool | United Kingdom |
| U. C. Louvain | Belgium | GSI Darmstadt | Germany |
| CEA | France | U. Santiago | Spain |
| NIPNE | Romania | CCLRC Daresbury | United Kingdom |
| U. Jyväskylä | Finland | Paul Scherrer Institute | Switzerland |
| L.M.U. München | Germany | Inst. Phys. Latvia | Latvia |
| FZ Jülich | Germany | Stockholm. U. MSL | Sweden |

In order to cover as efficiently as possible the studies of the various parts of the facility, the Design Study consists of 12 tasks, split into 5 topical areas. Each task is led by an institution, which organizes the collaboration between the participant and contributing laboratories with expertise in the field. Table 2 lists the topical areas and the tasks.

**TABLE 2.** Topical areas and tasks of the EURISOL Design Study

| Topical Area | Tasks | Lead Institution |
|---|---|---|
| Management | Management | GANIL |
| Accelerators | Driver accelerator | INFN |
|  | Post Accelerator | GANIL |
|  | Superconducting cavity prototyping | IN2P3 |
| Targets | Liquid Hg Converter | CERN |
|  | Direct targets | CERN |
|  | High power fission target | INFN |
| Physics, beams and safety | Safety and radioprotection | CEA |
|  | Beam Preparation | U. Jyväskylä |
|  | Beam Intensities | GSI |
|  | Physics and Instrumentation | U. Liverpool |
| Beta Beams | Beta Beam Conceptual Design Study | CERN |

The Design Study started on February 1, 2005 for a period of 4 years. Due to the challenging nature of the tasks, most of the deliverables are foreseen for the last two years of the contract. Nevertheless, some important elements can already be presented, and will be discussed in the next section.

# SOME PRELIMINARY RESULTS

This section aims to give a partial view of some of the accomplishments of the Design Study as of today. More detailed results can be found on the website www.eurisol.org.

## Extended Capabilities of the Driver Accelerator

The baseline design for the driver is a 1GeV 5MW CW superconducting proton LINAC. Nevertheless, the ISOL technique can be implemented with many beams, and one of the milestones of the DS was to conclude concerning the eventual benefits of the use of heavier incident particles such as deuterons, $^3$He and even heavy ions. To this aim a collaboration between the participants of the target, accelerator and beam intensity tasks was formed. It was concluded that extended capabilities of the proton driver accelerator can bring important enhancements to the performances, versatility and scientific reach of the EURISOL facility, particularly by enhancing the variety of intense beams available. The optimization of the driver LINAC can provide the possibility of these improvements at the price of a modest cost increase. While the baseline EURISOL beam should remain a 1 GeV 5 MW proton beam, the additional capabilities of $^3$He at 2 GeV, deuterons at 250 MeV, heavy ions of A/Q=2 at 125AMeV should be fully included in the detailed engineering design. In addition the possibility of H$^-$ acceleration and stripping would increase the multi-user capability of the facility by allowing the simultaneous use of at least two target stations. However, heavy ion acceleration for A/Q>2 will not be considered, because it would entail a very different accelerator design which would be much more costly and could no longer be fully optimized for proton acceleration.

**FIGURE 2.** 2-gap spoke cavities equipped with their cold tuning system.

The chosen design of the driver includes in sequence an RFQ, Half Wave Resonators (HWR), Spoke Resonators, and finally elliptical cavities for the highest velocities. The most novel of these cavities are the HWR and the Spoke Cavities. A prototype for the latter is being constructed at IPN Orsay and a schematic drawing is shown on fig.2.

## Multi-MegaWatt Converter and Fission Target

The proton to neutron converter and fission target set-up which should accept up to several MW of beam power is doubtless the most challenging part of the EURISOL facility. The main charges are to design a liquid Hg converter capable of efficiently using the incident beam while absorbing most of its power to avoid the necessity of a sophisticated beam stop. The fission target itself, which will contain several kg of Uranium Carbide, is sufficiently challenging to be the subject of a separate task. Work is currently ongoing and only preliminary ideas can be put forth as of now.

A Hg loop with an entrance window has been chosen as the baseline design for the converter. Methods to alleviate the thermal stress on the window are currently under study. A windowless liquid film model produced by a honeycomb structure is an alternative being investigated. The UCx target would be composed of several bars of dimensions 40x6x3 cm. Each bar would produce $10^{15}$ fissions per second under normal running conditions. Release times from such large targets need to be studied and optimized, and tests are planned at the ISOLDE facility at CERN. A schematic view of the current design of the converter and target is displayed on fig.3.

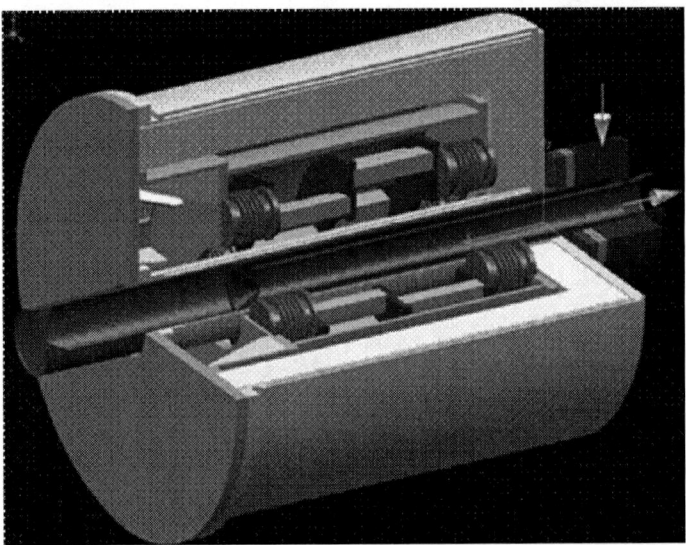

FIGURE 3. Schematic view of Hg converter and UCx target. The proton beam enters from the left.

## Conceptual Design of a Beta-Beam Facility

In order to investigate CP violation in the lepton sector and the $\Theta_{13}$ mixing angle, the neutrino community has proposed to study and build a beta-beam facility. The so-called beta-beam concept for accelerator-driven neutrino experiments envisages the production of a pure beam of electron neutrinos (or their antiparticles) through the beta-decay of radioactive ions $^{6}$He and $^{18}$Ne circulating in a high-energy storage ring. A conceptual design study of such a facility is an integral part of the EURISOL DS. It would make use of the EURISOL driver and targets in order to produce $5 \cdot 10^{13}$ particles per second of $^{6}$He and $5 \cdot 10^{12}$ of $^{18}$Ne. These radioactive ions, after acceleration in a dedicated Linear Accelerator, would then be injected into the CERN PS and SPS in order to be accelerated up to a relativistic factor $\gamma = 100$ before being stacked in the decay ring. The resulting pure electron or anti-electron flavor neutrino beam would then fly towards a large detector located for example in the Fréjus underground laboratory. A schematic view of the beta-beam layout is presented on fig.4.

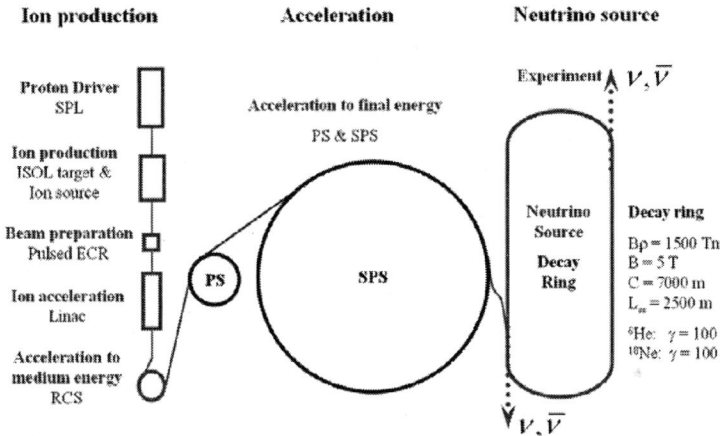

FIGURE 4. Schematic view of the beta-beam layout.

## SUMMARY AND OUTLOOK

A Pan-European Design Study, supported by the European Union, for an ambitious and challenging next generation ISOL RIB facility, EURISOL, has started since February 2005. Detailed engineering design and prototyping work is underway and will be pursued until 2009.

In parallel, different possible locations for building such a facility will be critically examined. Three types of sites have been identified:
- A national laboratory with a pre-existing radioactive beam facility. EURISOL would then constitute a major upgrade of the existing facility. GANIL in Caen, France could be an example for such a choice.
- An international research organization such as CERN. CERN would in addition be the natural location for a beta-beam facility to be built.

- A green field site with no previous nuclear physics facility.

The charge of the DS is to provide all relevant input to allow the community and funding bodies to make an informed choice.

It is estimated that the technical advances necessary should be reached in time for construction of EURISOL to start around the year 2013. European physicists could then benefit at the end of the next decade from a RIB facility boasting unequaled intensities and isotopic range; designed, built and funded through true pan-European collaboration.

## ACKNOWLEDGMENTS

We acknowledge the financial support of the European Community under the FP6 "Research Infrastructure Action - Structuring the European Research Area" EURISOL DS Project Contract no. 515768 RIDS. The EC is not liable for any use that can be made of the information contained herein.

## REFERENCES

1. http://www.nupecc.org/pub/NuPECC_Roadmap.org
2. http://www.gsi.de/fair
3. The Eurisol Report Contract HPRI-CT-1999-500001; http://www.ganil.fr/eurisol/Final_Report/EURISOL-REPORT.pdf

# New pathways to bypass the $^{15}$O waiting point

I. Stefan, F. de Oliveira Santos, M.G. Pellegriti, M. Angélique, J.C. Dalouzy, F. de Grancey, M. Fadil, S. Grévy, M. Lenhardt, M. Lewitowicz, A. Navin, L. Perrot, M.G. Saint Laurent, I. Ray, O. Sorlin, C. Stodel, J.C. Thomas[*], G. Dumitru, A. Buta, R. Borcea, F. Negoita, D. Pantelica[†], J.C. Angélique[**], E. Berthoumieux[‡], A. Coc, J. Kiener, A. Lefebvre-Schuhl, V. Tatischeff[§], J.M. Daugas, O. Roig[¶], T. Davisson[‖] and M. Stanoiu[††]

[*]*Grand Accélérateur National d'Ions Lourds UMR 6415 B.P. 5027 F-14076 Caen Cedex, France*
[†]*Horia Hulubei National Institute of Physics and Nuclear Engineering P.O. Box MG6 Bucharest-Margurele, Romania*
[**]*Laboratoire de Physique Corpusculaire CNRS-IN2P3 UMR 6534 ISMRA et Université de Caen F-14050 Caen, France*
[‡]*CEA Saclay DSM/DAPNIA/SPHN F-91191 Gif-sur-Yvette, France*
[§]*CSNSM UMR 8609 CNRS-IN2P3/Univ.Paris-Sud Bât. 104 91405 Orsay Campus, France*
[¶]*CEA/DIF/DPTA/PN BP 12 91680 Bruyères le Châtel, France*
[‖]*Department of Physics and Astronomy University of Edinburgh Edinburgh EH9 3JZ, United Kingdom*
[††]*Institut de Physique Nucléaire UMR 8608 CNRS-IN2P3/Univ.Paris-Sud F-91406 Orsay, France*

**Abstract.** Two reactions $^{15}$O$(p,\beta^+)^{16}$O and $^{15}$O$(p,\gamma)(\beta^+)^{16}$O are proposed as new pathways to bypass the $^{15}$O waiting point in astrophysical context. The later reaction is found to have a surprisingly high cross section, approximately $10^{10}$ times higher than the first reaction. These cross sections were calculated after precise measurements of energies and widths of the proton-unbound $^{16}$F low lying states, obtained using the H($^{15}$O,p)$^{15}$O reaction. The large $(p,\gamma)(\beta^+)$ cross section can be understood to arise from the more efficient feeding of the low energy wing of the ground state resonance by the gamma decay. The implications of the new reactions are discussed.

**Keywords:** Resonant elastic scattering, waiting point, unbound nucleus, $^{16}$F
**PACS:** 25.60.-t,97.10.Cv,25.70.Ef,25.40.Cm,21.10.-k,27.20.+n

## INTRODUCTION

Unbound nuclei play a major role in astrophysics. The proton-unbound $^2$He and the alpha-unbound $^8$Be nuclei illustrate this fact. The former is involved in the p$(p,\beta^+)$d reaction, first reaction of the *pp* chain of reactions governing the energy generation in the sun [1, 2]. The latter, whose lifetime is about $10^{-16}$ seconds, is involved in the triple alpha reaction [2] which is at the origin of the formation of all the heavier elements. The proton-unbound nuclei $^{15}$F and $^{16}$F probably play an important role in X-ray bursts. These astronomical events are known to happen in close binary systems, where accretion takes place from an extended companion star on the surface of a neutron star (type I X-ray bursts). The accreted matter is compressed until it reaches sufficiently high pressure conditions to trigger a thermonuclear runaway. In these explosive events, the carbon and nitrogen elements are mainly transformed into $^{14}$O and $^{15}$O by successive proton captures [3, 4]. Then, the pathway for new proton captures is hindered by the proton-unbound nuclei $^{15}$F and $^{16}$F. The reaction flux and the energy generation are then limited by the relatively slow $\beta^+$-decay of $^{14}$O ($t_{1/2}$=71 s) and $^{15}$O ($t_{1/2}$=122 s), which create waiting points. The sudden and intense release of energy observed in X-ray bursts requires to circumvent the limited energy generation in breakout reactions. The $^{15}$O($\alpha,\gamma$)$^{19}$Ne reaction is considered to be one of the key reactions in this context [3, 4]. It makes the transition into the nucleosynthetic $rp$ process (rapid proton capture) which may be responsible for an increased rate of energy generation and the synthesis of heavier elements. In such explosive environments, $^{16}$F is strongly populated in the ground state (g.s.) or in the first excited state, and leads to an equilibrium between formation and decay of this proton-unbound nucleus. From time to time before the proton is emitted, $^{16}$F can capture another proton thus producing the $^{17}$Ne particle stable isotope. This two-proton capture process was calculated to be significant for extreme densities (larger than $10^{11}$ g/cm$^3$) [5]. In this paper, $\beta^+$-decay of $^{16}$F to $^{16}$O is proposed as an alternative channel. Two reactions channels $^{15}$O($p,\beta^+$)$^{16}$O and $^{15}$O($p,\gamma$)($\beta^+$)$^{16}$O are studied. Both reactions eventually proceed through the $\beta^+$-decay of the intermediate unbound $^{16}$F g.s., which is fed directly by a proton capture or indirectly through a proton capture to the first excited state followed by a $\gamma$-emission. This is the first time that a ($p,\gamma$)($\beta^+$) reaction is proposed, which is a sequence of reactions that involves an intermediate unbound nucleus. When the $\gamma$-decay occurs to the low energy wing of the $^{16}$F g.s. resonance the subsequent proton emission is dramatically hindered due to the fact that the low energy proton has to tunnel through the Coulomb potential of the $^{15}$O nucleus. The calculation of these reaction cross sections requires the measurement of the energies, widths, spins and parities of the low lying states of $^{16}$F. These were obtained from the measurement of the H($^{15}$O,$p$)$^{15}$O resonant elastic excitation function using low energy $^{15}$O beam at the SPIRAL facility.

## SPECTROSCOPY OF $^{16}$F

The beam of radioactive $^{15}$O nuclei was produced at the SPIRAL-GANIL facility through the projectile fragmentation of a 95 A.MeV $^{16}$O primary beam on a thick carbon target. Mean intensities of $10^7$ pps at an energy of 1.2 A.MeV were obtained after post acceleration by the CIME cyclotron. A beam contamination of less than 1 % of $^{15}$N was achieved using a vertical betatron oscillation selection device [6] and a suitable degrader in the analysis line of LISE spectrometer [7] where the measurements were made. Two stable beams, $^{14}$N and $^{15}$N, were also used in similar experimental conditions for calibrations. The excitation function for the elastic scattering at these low energies can be described by the Rutherford scattering, but shows "anomalies", i.e. various resonances that are related to individual states in the compound nucleus. The principle of the measurement is described in [8, 9] and references therein. A 31(1) $\mu$m thick polyethylene (CH$_2$)$_n$ target was used, thick enough to stop the beam inside. The scattered protons were detected by a silicon detector, placed at forward angles (180° in the center of mass frame) within an angular acceptance of 2°. Protons were identified using their energy and time-of-flight. The energy resolution was 4 keV in the center of mass (c.m.) frame. Fig. 1 shows the excitation function for the H($^{15}$O,$p$)$^{15}$O reaction measured from 0.450 MeV to 1.1 MeV.

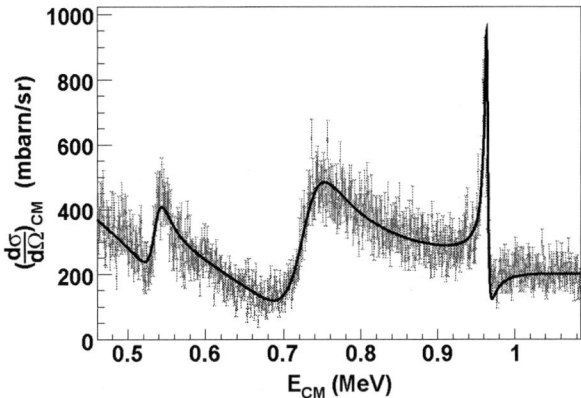

**FIGURE 1.** Excitation function for the H($^{15}$O,$p$)$^{15}$O reaction at 180° in c.m.. The line is a result of an R-matrix calculation using parameters from Table I.

The measured cross section was reproduced by an R-matrix [10] calculation using the code ANARKI [11] which is seen to be in a good agreement with the data. A value of $S_p$ = -534 ± 5 keV was obtained for the proton separation energy in agreement with the recommended value [12]. The R-matrix analysis was also used to extract the properties of the first three states in $^{16}$F, given in Table I. A significant difference was found between the present and the recommended value of the width for the first excited state [13]. This width is an important parameter used in the calculations presented in the next sections.

**TABLE 1.** Measured properties for the low-lying states in $^{16}$F.

| | Recommended values [13] | | This work | | |
|---|---|---|---|---|---|
| $J^\pi$ | $E_x$ (keV) | $\Gamma_p$ (keV) | $E_{CM}$ (keV) | $E_x$(keV) | $\Gamma_p$ (keV) |
| $0^-$ | 0 | 40 ± 20 | 534 ± 5 | 0 | 25 ± 10 |
| $1^-$ | 193 ± 6 | < 40 | 732 ± 10 | 198 ± 10 | 70 ± 5 |
| $2^-$ | 424 ± 5 | 40 ± 30 | 958 ± 2 | 425 ± 2 | 6 ± 3 |

# CALCULATION OF $^{15}$O($p,\beta^+$)$^{16}$O

The calculation of the $^{15}$O($p,\beta^+$)$^{16}$O cross section was made using the properties of the $^{16}$F g.s. resonance measured in the present work and the Breit-Wigner formula for a single-level resonance [2]:

$$\sigma(E_p) = \frac{1}{4\pi}\lambda^2 \frac{2J_r+1}{(2J_i+1)(2J_f+1)} \frac{\Gamma_{in}\Gamma_{out}}{(E_p-E_R)^2 + (\frac{\Gamma_{Tot}}{2})^2} \quad (1)$$

where $\lambda$ is the de Broglie wavelength, $J$ are the spins, and $E_R, \Gamma_{Tot}, \Gamma_{in}, \Gamma_{out}$ being the resonance energy, total width, and partial widths of the incoming and outgoing channels.

In the $(p,\beta^+)$ case, $E_R = E_{g.s.}$ the energy of the g.s. resonance, $\Gamma_{in} = \Gamma_p^{g.s.}$ the proton width, and $\Gamma_{out} = \Gamma_\beta$ corresponds to the $\beta^+$-decay partial width. The energy dependance of the proton width $\Gamma_p^{g.s.}(E_p)$ for the incoming channel was taken into account using the relation:

$$\Gamma_p^{g.s.}(E_p) = \Gamma_p^{g.s.}(E_{g.s.}) \sqrt{\frac{E_p}{E_{g.s.}} \frac{P(E_p)}{P(E_{g.s.})}} \qquad (2)$$

where $\Gamma_p^{g.s.}(E_{g.s.})$ is the proton width at the resonance energy and $P(E_p)$ is the penetrability function under the Coulomb potential barrier. A partial lifetime for $^{16}F(\beta^+)$ of 1 second and a negligible branching ratio to the $^{15}O(p,\beta^+)^{12}C+\alpha$ final decay channel were assumed. This assumption is supported by the $\beta^-$-decay properties measured in the mirror nucleus $^{16}N$ [13]. The $\beta^+$-decay partial width was taken as a constant since the energy dependence of the Fermi function is small due to large $Q_{\beta^+}$ value ($Q_{\beta^+}$=15417(8) keV [12]). The calculated $^{15}O(p,\beta^+)^{16}O$ cross section is shown in Fig. 2 as a function of the c.m. energy.

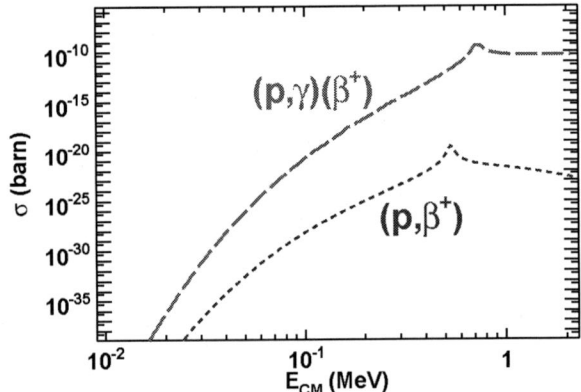

**FIGURE 2.** $^{15}O(p,\beta^+)^{16}O$ and $^{15}O(p,\gamma)(\beta^+)^{16}O$ cross sections are shown as a function of the c.m. energy.

The maximum of the cross section is observed at the energy of 534 keV corresponding to the $^{16}F$ g.s. resonance. At this energy the $(p,\beta^+)$ cross section is very small, about $10^{-20}$ barns, since $^{16}F$ mainly decays by proton emission, which is $\simeq 10^{20}$ times stronger than the $\beta^+$-decay (since $\Gamma_p^{g.s.}(E_{g.s.}) = 25$ keV and $\Gamma_\beta = 0.66 \; 10^{-18}$ keV).

## CALCULATION OF $^{15}O(p,\gamma)(\beta^+)^{16}O$

The calculation of the $^{15}O(p,\gamma)(\beta^+)^{16}O$ reaction was performed sequentially, a schematic representation of this reaction is shown in Fig. 3. Proton capture reaction to the first excited state of $^{16}F$ is considered, followed by a $\gamma$ decay to the g.s. resonance, from which a $\beta^+$-decay branching ratio is taken into account.

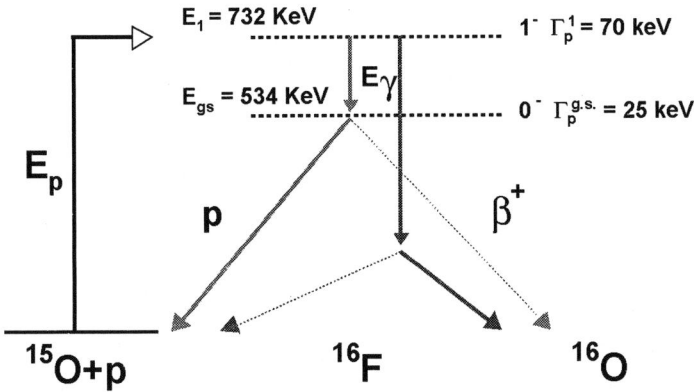

**FIGURE 3.** Schematic representation of the $^{15}O(p,\gamma)(\beta^+)^{16}O$ reaction (see text). Two cases are represented. In red, $\gamma$-transitions populate the $^{16}F$ g.s. at the resonance energy. In that case, $^{16}F$ mainly decays by proton emission. In blue, high energy $\gamma$-transitions populate the low energy wing of the g.s. resonance. In that case, $\beta^+$-decay dominates.

It is assumed that the cross section $\sigma_{p\gamma\beta}(E_p)$ for the $(p,\gamma)(\beta^+)$ reaction at the energy $E_p$ is an integration of the differential cross section over all possible energies of the $\gamma$ transition (since the g.s. has a large width):

$$\sigma_{p\gamma\beta}(E_p) = \int \sigma_{p\gamma}(E_p,E_\gamma) P_\gamma(E_\gamma) P_\beta(E_p,E_\gamma) dE_\gamma \qquad (3)$$

where $\sigma_{p\gamma}(E_p,E_\gamma)$ is the cross section to capture the proton at the energy $E_p$ and to emit a $\gamma$-ray with an energy $E_\gamma$, $P_\gamma(E_\gamma)dE_\gamma$ is the strength function, that is the probability for the $\gamma$-ray to have an energy between $E_\gamma$ and $E_\gamma+dE_\gamma$, and $P_\beta(E_p,E_\gamma)$ is the branching ratio function for the $^{16}F$ nucleus to decay by $\beta^+$-ray emission. The first term $\sigma_{p\gamma}(E_p,E_\gamma)$ is calculated using a Breit-Wigner formula with the following parameters $E_1$, $\Gamma^1_{Tot}(E_p,E_\gamma)$, $\Gamma^1_p(E_p)$, $\Gamma^1_\gamma(E_\gamma)$ being the energy, total width, proton width and $\gamma$ width for the resonance corresponding to the first excited state of $^{16}F$. The $\gamma$-ray is emitted from a $1^-$ state to the $0^-$ g.s., which corresponds to a M1 transition, whose energy dependence of the $\gamma$ width $\Gamma^1_\gamma(E_\gamma)$ is:

$$\Gamma^1_\gamma(E_\gamma) = \Gamma^1_\gamma(E_1 - E_{g.s.})\{\frac{E_\gamma}{E_1 - E_{g.s.}}\}^3 \qquad (4)$$

A $\gamma$ lifetime of 1 ps was obtained from the mirror nucleus [13], which corresponds to the partial width $\Gamma^1_\gamma(E_1 - E_{g.s.}) = 0.66 \ 10^{-3} eV$. The strength function of the $^{16}F$ g.s. resonance was calculated assuming a Breit-Wigner parametrization:

$$P_\gamma(E_\gamma)dE_\gamma = \frac{1}{N} \frac{dE_\gamma}{(\Delta E)^2 + (\frac{\Gamma^{g.s.}_{Tot}(E_p-E_\gamma)}{2})^2} \qquad (5)$$

where the normalization constant is:

$$N = \int \frac{1}{(\Delta E)^2 + (\frac{\Gamma_{Tot}^{g.s.}(E_p - E\gamma)}{2})^2} dE_\gamma \qquad (6)$$

with $\Delta E = E_p - E_\gamma - E_{g.s.}$ and $\Gamma_{Tot}^{g.s.}(E_p - E_\gamma) = \Gamma_\beta + \Gamma_p^{g.s.}(E_p - E_\gamma)$ is the total width of the g.s. resonance. The $\beta$ branching ratio is calculated using:

$$P_\beta(E_p, E_\gamma) = \frac{\Gamma_\beta}{\Gamma_\beta + \Gamma_p^{g.s.}(E_p - E_\gamma)} \qquad (7)$$

Naively, one might have expected to obtain a small cross section for the $(p,\gamma)(\beta^+)$ reaction, similar to the $(p,\beta^+)$ one, since $\gamma$- and $\beta$-widths are much smaller than proton-widths. Contrary to naive expectations, the $(p,\gamma)(\beta^+)$ cross section is about $10^{10}$ times larger than the $(p,\beta^+)$ cross section, as shown in Fig. 2. The large ratio can be explained in the following way. As it has been shown previously, there is only one $(p,\beta^+)$ reaction for $10^{20}$ $(p,p)$ reactions. In the $(p,\gamma)(\beta^+)$ case, one $\gamma$-ray is emitted for $10^8$ captured protons (from the ratio of the widths) and about one $\gamma$-transition over $10^3$ populates the low energy wing of the g.s. resonance (less than 15 keV above the proton-emission threshold) where it is almost always followed by a $\beta^+$-decay ($P_\beta \simeq 1$). This implies that one captured proton over $10^9$ induces a $(p,\gamma)(\beta^+)$ reaction, that is a factor $10^9$ times larger than in the $(p,\beta^+)$ reaction. This is the main explanation of the large factor between the $(p,\gamma)(\beta^+)$ and the $(p,\beta^+)$ reaction cross sections.

## ASTROPHYSICAL CONSEQUENCES

At a given temperature T of the gas inside the star, protons exhibit a Maxwellian distribution and the reaction rates $N_A <\sigma v>$ are calculated by integrating numerically the Maxwellian-averaged cross sections $\sigma(E_p)$ over all possible proton energies. The obtained reaction rates are shown in Fig. 4 (a) as a function of the temperature. The rate of the $(p,\beta^+)$ reaction is negligible compared to that of the reaction $(p,\gamma)(\beta^+)$ at all temperatures. To evaluate the impact of this latter reaction, it has to be compared with the competing $\beta^+$-decay of $^{15}$O and the $^{15}$O$(\alpha,\gamma)^{19}$Ne alpha capture reaction. Fig. 4 (b) shows the temperature and density conditions where the $(p,\gamma)(\beta^+)$ reaction represents 10 to 50 % of the total reaction flux initiated by the $^{15}$O nucleus. Boxes delimit conditions where novae and X-ray bursts might happen. For the lowest temperatures (< $10^8$ K), the $(p,\gamma)(\beta^+)$ reaction requires extreme densities (> $10^{10}$ g cm$^{-3}$) to compete with the $^{15}$O($\beta^+$)-decay. For the highest temperatures (> 1.1 $10^9$ K), the $(\alpha,\gamma)$ reaction always dominates. In X-ray bursts, the $(p,\gamma)(\beta^+)$ reaction might represent up to 30 % of the total flux. Within the uncertainties of the calculations, the $(p,\gamma)(\beta^+)$ reaction could be faster than the $(\alpha,\gamma)$ reaction for temperatures up to $10^9$ K. A more precise evaluation depends on the $(\alpha,\gamma)$ reaction rate (not well known) and on the relative abundances in hydrogen and helium, since one reaction consumes protons and the other alpha particles. In these extreme conditions, a new cycle of reactions is operating: $^{15}$O$(p,\gamma)(\beta^+)^{16}$O$(p,\gamma)^{17}$F$(p,\gamma)^{18}$Ne$(\beta^+)^{18}$F$(p,\alpha)^{15}$O. This new cycle could speed-up

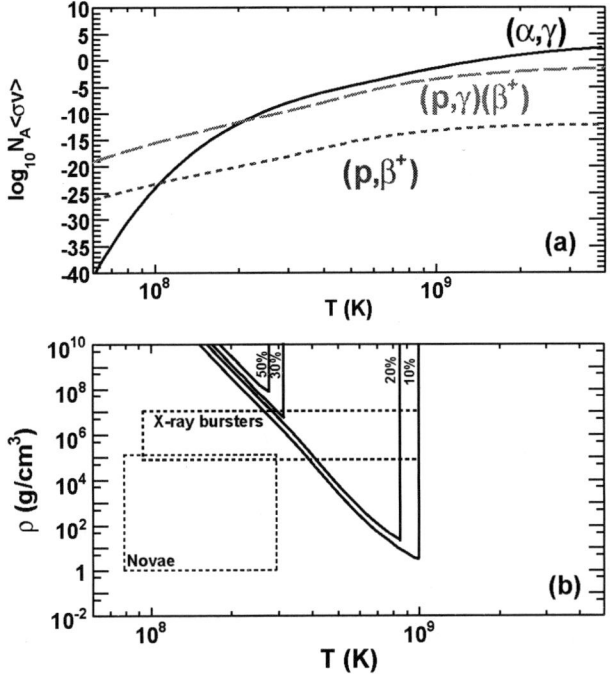

**FIGURE 4.** (a) $^{15}O(p,\beta^+)^{16}O$ and $^{15}O(p,\gamma)(\beta^+)^{16}O$ reaction rates are shown as a function of the temperature. The $^{15}O(\alpha,\gamma)^{19}Ne$ reaction rate is also shown for comparison. (b) Density versus temperature conditions where the $^{15}O(p,\gamma)(\beta^+)^{16}O$ reaction represents 10 to 50 % of the total reaction flux initiated by the $^{15}O$ nucleus.

the CNO cycle and occur complementary to breakout reactions. The role of this new proposed cycle of reactions remains to be studied more carefully under various X-ray bursts conditions.

## DISCUSSION

In the following, several aspects of this study are discussed:

1. **Uncertainties and their evaluated effects.** The position and width of the low lying $^{16}F$ states were measured with a high precision (see Table 1). The effect of the uncertainties in these measured parameters results in a change by less than a factor of 2 in the calculated cross sections. The calculated $(p,\gamma)(\beta^+)$ cross section is insensitive to the $^{16}F$ $\beta^+$-decay lifetime, as a variation by a factor of 100 causes the cross section to change by only a factor of 2. The lifetime of the $\gamma$-transition is a sensitive parameter since the $(p,\gamma)(\beta^+)$ cross section is almost directly proportional to this parameter. A value measured in the mirror nucleus was used, but this

assumption works only to within a factor of 10 [14]. The other excited states in $^{16}$F were also studied and found to be negligible.

2. **Alternative reactions.** Non-resonant direct capture contributions, quantum interferences, continuum couplings [15], intrastate transitions [16] and other reaction mechanisms were not studied and remain to be evaluated. Moreover, high temperature and density environments would correspond to conditions where the alternative $^{16}$F$(p,\gamma)^{17}$Ne reaction might be competitive with the temperature independent $^{16}$F$(\beta^+)^{16}$O decay. The $^{15}$O$(p,\gamma)(p,\gamma)^{17}$Ne cross section was estimated using the same formalism as used for the $(p,\gamma)(\beta^+)$ reaction. The first reaction was found to compete with the second in X-ray bursts conditions. However, these calculations are only a first estimate since hitherto unknown widths of the excited states in $^{17}$Ne are used.

3. **Validity of equ. (3).** The validity of equ. (3) to calculate the $(p,\gamma)(\beta^+)$ cross section is questionable. This equation is equivalent to the assertion that the $(p,\gamma)(\beta^+)$ reaction is a pure sequential process. The triple alpha reaction is a similar kind of reaction, that involves an intermediate unbound nucleus ($^8$Be). Implicitly, the calculation of this reaction is performed using the same assertion as used in our study, that is the alpha are captured sequentially [17]. The effect of the assumption made in equ. (3) could be evaluated by using a three-body formalism, as proposed for example by L. V. Grigorenko [18] to calculate the $^{15}$O$(2p,\gamma)^{17}$Ne reaction.

4. **Validity of equ. (5) and (6).** Equations (5) and (6) describe the $^{16}$F ground state resonance using a Breit-Wigner shape. Implicitly, a Breit-Wigner shape means that the state decays according to an exponential law. This is a first approximation which is certainly not correct [19]. The change of the decay law, causing the change of the strength function, would induce severe corrections on the results. The study of this peculiar aspect is strongly required. In any case, the feeding of the low energy tail of the $^{16}$F ground state resonance by $\gamma$-transitions will remain larger than its feeding by direct proton captures. Moreover, the proton emission threshold is probably not a limit for the $\gamma$-transitions, thus allowing the feeding of the $^{16}$F resonance in proton-bound energies. All these affects remain to be observed.

5. **Will $^{16}$F survive a long time ?** In quantum mechanics, probability and lifetime are often synonyms. Because time-reversal invariance holds for electromagnetic and nuclear forces, the partial width is independent of the direction of the process. The partial width for the formation of the compound nucleus $^{16}$F through an entrance channel $^{15}$O+p at an energy E is identical with the partial width for the decay of the $^{16}$F nucleus produced at the same energy. The partial width for the formation of $^{16}$F being reduced at low energy by the Coulomb barrier, the proton decay width is reduced as well. Does that mean $^{16}$F will survive a long time when populated in the low energy tail of the ground state ? A time long enough that the beta-decay can compete with the proton emission ? This remains to be observed experimentally.

## ACKNOWLEDGMENTS

We thank the GANIL crew for delivering the $^{15}$O beam, and M. Płoszajczak for stimulating discussions. This work has been supported by the IN2P3-IFIN-HH Program.

## REFERENCES

1. H. Bethe, and C. Critchfield, *Phys. Rev.* **54**, 248 (1938).
2. C. S. Rolfs, and W. S. Rodney, *Cauldrons in the Cosmos*, The University of Chicago Press, Chicago, 1988.
3. R. Wallace, and S. Woosley, *Astrophys. J. Suppl. Ser.* **45**, 389 (1981).
4. M. Wiescher, H. Schatz, and A. Champagne, *Phil. Trans. Roy. Soc. London A* **356**, 2105 (1998).
5. J. Görres et al., *Phys. Rev. C* **51**, 392 (1995).
6. P. Bertrand et al., *17th International Conf. on Cyclotrons and their applications*, Tokyo, 2004.
7. R. Anne et al., *Nucl. Instr. and Meth. A* **257**, 215 (1987).
8. V. Z. Golberg et al., *Phys. At. Nucl.* **60**, 1061 (1997).
9. F. de Oliveira Santos et al., *Eur. Phys. J. A* **24**, 237 (2005).
10. A. Lane, and R. Thomas, *Rev. Mod. Phys.* **30**, 257 (1958).
11. E. Berthoumieux et al., *Nucl. Instr. and Meth. B* **55**, 136–138 (1998).
12. G. Audi et al., *Nucl. Phys. A* **729**, 337 (2003).
13. D. Tilley et al., *Nucl. Phys. A* **564**, 1 (1993).
14. F. de Oliveira Santos et al., *Phys. Rev. C* **55**, 3149 (1997).
15. R. Chatterjee et al., *Nucl. Phys. A* **764**, 528 (2006).
16. K. Langanke et al., *Z. Phys. A* **324**, 3 (1986).
17. K. Langanke et al., *Z. Phys. A* **324**, 147 (1986).
18. L. V. Grigorenko et al., *Phys. Rev. C* **72**, 015803 (2005).
19. N. G. Kelkar et al., *Phys. Rev. C* **70**, 024601 (2004).

# Quartetting in fermion matter and alpha particle condensation in nuclear systems

P. Schuck[*], Y. Funaki[†], H. Horiuchi[**], G. Röpke[‡], A. Tohsaki[**] and T. Yamada[§]

[*]*Institut de Physique Nucléaire, CNRS, UMR8608, Orsay, F-91406, France*
*Université Paris-Sud, Orsay, F-91505, France*
[†]*The Institute of Physical and Chemical Research (RIKEN), Wako, Saitama 351-0198, Japan*
[**]*Research Center for Nuclear Physics (RCNP), Osaka University, Ibaraki, Osaka 567-0047, Japan*
[‡]*Institut für Physik, Universität Rostock, D-18051 Rostock, Germany*
[§]*Laboratory of Physics, Kanto Gakuin University, Yokohama 236-8501, Japan*

**Abstract.** The fameous Hoyle state ($0_2^+$ at 7.654 MeV in $^{12}$C) is identified as being an almost ideal condensate of three $\alpha$-particles, hold together only by the Coulomb barrier. It, therefore, has a $^8$Be-$\alpha$ structure of low density. Transition probability and inelastic form factor together with position and other physical quantities are correctly reproduced without any adjustable parameter from our two parameter wave function of $\alpha$-particle condensate type. The possibility of the existence of $\alpha$-particle condensed states in heavier $n\alpha$ nuclei is also discussed.

**Keywords:** Quartetting, $\alpha$-particle condensation
**PACS:** PACS numbers; 05.30.Fk

## INTRODUCTION

Quantum condensation of particles is one of the most amazing phenomena of many body systems. Striking well known examples are superconducting metals and superfluid $^4$He. Also nuclei are superfluid. However, in nuclei the most tightly bound cluster is not a pair but a quartet. Therefore, what about $\alpha$-particle condensation in nuclei? The only nucleus which in its ground state has a pronounced $\alpha$-cluster structure is $^8$Be. In Fig. 1(a) we show the result of an exact calculation with a realistic $N$-$N$ interaction for the density distribution in the laboratory frame, whereas in Fig. 1(b) we see the same in the intrinsic, deformed frame where in addition the question has been asked where to find the second $\alpha$-particle when the first is placed at a given position. So we see that the two $\alpha$'s are $\sim 4$ fm apart giving raise to a very low average density $\rho \sim \rho_0/3$ as seen on Fig. 1(a) where $\rho_0$ is the nuclear saturation density. $^8$Be is also a very large object with an rms radius of $\sim 3.7$ fm to be compared with the nuclear systematics of $R = r_0 A^{1/3} \sim 2.44$ fm. Definitely $^8$Be is a rather unusual nucleus. One may ask the question what happens when one brings a third $\alpha$-particle alongside of $^8$Be. We know the answer: the 3-$\alpha$ system collapses to the ground state of $^{12}$C which is much denser than $^8$Be and cannot accommodate with its small rms radius of 2.4 fm three $\alpha$-particles barely touching one another. One nevertheless may ask the question whether the dilute three $\alpha$ configuration $^8$Be-$\alpha$ may not form an isomeric or excited state of $^{12}$C. This will be the main subject

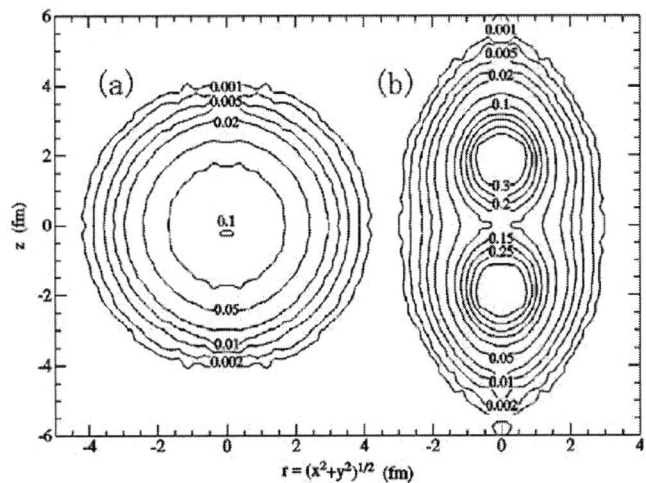

**FIGURE 1.** Contours of constant density (taken from [1]), plotted in cylindrical coordinates, for $^8$Be($0^+$). The left side (a) is in the "laboratory" frame while the right side (b) is in the intrinsic frame.

of our considerations.

## $\alpha$-CONDENSATE STATES IN SELF-CONJUGATE 4n-NUCLEI

We now will show that $\alpha$-particle condensation most likely exists in n$\alpha$ nuclei around energies of $\alpha$-particle break up thresholds. We want to give the demonstration here that, due to the existence of ample experimental data, we have identified at least one nucleus where such an $\alpha$-particle condensed state exists and then we discuss the indications and the likelyhood that such states very naturally also are present in other nuclei and that it may be a quite general phenomenon in nuclear systems.

The nucleus we want to draw our attention to is $^{12}$C. We, indeed, will give strong arguments that the $0_2^+$ state at 7.654 MeV in $^{12}$C is a state of $\alpha$-particle condensate nature.

First, it should be noticed that the $0_2^+$ state in $^{12}$C is actually, as $^8$Be, unstable and situated about 300 keV above the three $\alpha$-break up threshold.

This state only is stabilised by the Coulomb barrier. It has a width of 8.7 eV and a corresponding lifetime of $7.6 \times 10^{-17}$ s. As well known, this state is of extreme astrophysical importance concerning the synthesis of $^{12}$C in the universe and its existence was predicted in 1953 by the astrophysicist F. Hoyle [2] and shortly after discovered by W. A. Fowler in 1957 [3]. It is also well known that this Hoyle state, as it is called now, is a notoriously difficult state for any nuclear theory and for example the most modern no-core shell model calculations predict the $0_2^+$ state in $^{12}$C to occur at around 17 MeV, that is more than two times its actual value [4]. This fact alone may tell us that the

Hoyle state must have a very unusual structure and it is easy to understand that, should it indeed have a loosely bound three $\alpha$-particle structure, a shell model type of calculation would have great difficulties to reproduce its properties. However, about 30 years ago, two Japanese physicists, M. Kamimura [5] and K. Uegaki [6], with their collaborators, almost simultaneously achieved to reproduce the Hoyle state from a microscopic theory, i.e. employing a twelve nucleon wavefunction together with a Hamiltonian containing an effective nucleon-nucleon interaction. Their works, at that time, probably did not attract the wide spread attention they deserved and the true importance of their achievement is eventually only recognised now. Both authors started from practically the same ansatz for the $^{12}$C wavefunction, having the following three $\alpha$-cluster structure: $\langle \vec{r}_1...\vec{r}_{12}|^{12}C\rangle = \mathcal{A}[\chi(\vec{\mathcal{R}},\vec{s})\phi_1\phi_2\phi_3]$ where $\phi_i$ is an intrinsic $\alpha$-particle wavefunction of prescribed Gaussian form, i.e. $\phi = \exp([(\vec{r}_1-\vec{r}_2)^2 + (\vec{r}_1-\vec{r}_3)^2 + ...]/b^2)$ where the size parameter is adjusted to get the rms value of the free $\alpha$-particle radius right. $\chi(\vec{\mathcal{R}},\vec{s})$ is the yet to be determined three body wavefunction for the center of mass motion of the three $\alpha$'s with $\vec{\mathcal{R}}$ and $\vec{s}$ the corresponding Jakobi coordinates. The unknown function $\chi$ was then determined via a GCM [6] and RGM [5] calculation using the Volkov I and Volkov II nucleon-nucleon forces which well reproduce $\alpha$-$\alpha$ phase shifts. The precise solution of this complicated three body problem was, 30 years back, a truely pioneering work. The results were up to expectation. The position of the Hoyle state as well as other properties like inelastic form factor and transition probability successfully reproduced the experimental data. Other states of $^{12}$C below and around the energy of the Hoyle state were also successfully described. It also was already recognised that the three $\alpha$'s in the Hoyle state form sort of gas like state, a feature which had already been pointed out by H. Horiuchi [7] prior to the works of [5] and [6] using the orthogonality condition model (OCM) [8]. All of these authors also concluded from their studies that the linear chain state of three $\alpha$-particles, as this was postulated by Morinaga many years back [9], had to be excluded.

Though, as already said, the afore mentioned authors all had already stressed the somewhat $\alpha$-gas like nature of the Hoyle state, eventually two important aspects were missed at that time. First comes the fact that, because all three $\alpha$'s move in identical $S$-wave orbits, this forms an $\alpha$-condensate state, albeit not in the macroscopic sense. Second is that the complicated three body wave function can be replaced by a structurally and conceptually very simple microscopic three $\alpha$ wave function of the condensate type which has practically 100 percent overlap with the previous ones [10]. We now shortly want to describe this condensate wave function.

For this we make an analogy to the Cooper pair BCS wave function of ordinary pairing. The latter wave function can be written in position space as

$$\langle \vec{r}_1...\vec{r}_N|\text{BCS}\rangle = \mathcal{A}[\phi(\vec{r}_1,\vec{r}_2)\phi(\vec{r}_3,\vec{r}_4)...\phi(\vec{r}_{N-1},\vec{r}_N)] \tag{1}$$

where $\phi(\vec{r}_1,\vec{r}_2)$ is the Cooper pair wave function, including spin and isospin, which is being determined variationally by the well known BCS equations. As before $\mathcal{A}$ is the antisymmetriser. The condensate character of the BCS ansatz is born out by the fact that we have a product of $N/2$ times the same pair wave function $\phi$. Formally it now is a simple matter to generalise (1) to $\alpha$-particle condensation. We write

$$\langle \vec{r}_1...\vec{r}_N|\Phi_{n\alpha}\rangle = \mathcal{A}[\phi_\alpha(\vec{r}_1,\vec{r}_2,\vec{r}_3,\vec{r}_4)\phi_\alpha(\vec{r}_5,..,\vec{r}_8)...\phi_\alpha(\vec{r}_{N-3},..,\vec{r}_N)] \tag{2}$$

where $\phi_\alpha$ is the wave function common to all condensed $\alpha$-particles. Of course, in general, the variational solution for $\phi_\alpha(\vec{r}_1,..,\vec{r}_4)$ from (2) is extraordinarily more complicated than to find the Cooper pair wave function of (1). However, in the case of the $\alpha$-particle and for relatively light nuclei, the complexity of the problem can be reduced dramatically. This stems from the fact that, as was already recognised by the authors in [5],[6], an intrinsic wave function of the $\alpha$-particle of Gaussian form with only the size parameter $b$ to be determined, is an excellent variational ansatz (see above). In addition, and here resides the essential and crucial novelty of our wave function, even the center of mass motion of the various $\alpha$-particles can very well be described by a Gaussian wave function with, this time, a size parameter $B \gg b$ to account for the motion over the whole nucleus. We therefore write

$$\phi_\alpha(\vec{r}_1,\vec{r}_2,\vec{r}_3,\vec{r}_4) = e^{-\vec{R}^2/B^2} \phi(\vec{r}_1-\vec{r}_2,\vec{r}_1-\vec{r}_3,...) \quad (3)$$

where $\vec{R} = (\vec{r}_1+\vec{r}_2+\vec{r}_3+\vec{r}_4)/4$ is the c.m. coordinate of one $\alpha$-particle and $\phi(\vec{r}_1-\vec{r}_2,...)$ is the same intrinsic $\alpha$-particle wave function of Gaussian form as already used in [5],[6] and written out above. Of course, in (2) the center of mass $\vec{X}_{cm}$ of the three $\alpha$'s, i.e. of the whole nucleus, should also be eliminated what is easily achieved by replacing $\vec{R}$ by $\vec{R}-\vec{X}_{cm}$ in each of the $\alpha$ wave functions in (2). The $\alpha$-particle condensate wave function (2) with (3), proposed in [11] and henceforth called THSR-wavefunction, now depends only on two parameters, $B$ and $b$. The expectation value of the microscopic Hamiltonian

$$\mathscr{H}(B,b) = \langle\Phi_{n\alpha}(B,b)|H|\Phi_{n\alpha}(B,b)\rangle/\langle\Phi_{n\alpha}|\Phi_{n\alpha}\rangle \quad (4)$$

can be evaluated and the corresponding two dimensional energy surface quantised in using the two parameters $B$ and $b$ as Hill-Wheeler coordinates.

Before coming to the results, let us discuss the THSR- wave function a little more. The inocuously looking ansatz (2) with (3) is actually more subtle as it might seem. One should realise that it contains two limits exactly: if $B = b$, then (2) boils down to a standard Slater determinant with harmonic oscillator wave functions with oscillator length $b$ as the single variational parameter. This holds because (3), with $B = b$, becomes a product of four identical Gaussians and the antisymmetrisation creates all the necessary $P$, $D$, etc. harmonic oscillator wave functions automatically [11]. On the contrary, when $B \gg b$, the density of $\alpha$-particles is very low and in the limit $B \to \infty$, the average distance between $\alpha$-particles is so large that the antisymmetrisation between $\alpha$'s can be neglected, i.e. the operator '$\mathscr{A}$' in front of (2) can be taken off. Our wave function then becomes an ideal gas of independent condensed $\alpha$-particles, i.e. a pure product state of $\alpha$'s! On the other hand, in realistic cases, the antisymmetriser $\mathscr{A}$ can not be neglected and the evaluation of the expectation values in (4) becomes an analytical (but non-trivial) task. For the Hamiltonian in (4) we took the one of [12] with an effective nucleon-nucleon force of the Gogny type whose parameters have been adjusted to $\alpha$-$\alpha$ scattering phase shifts about 15 years back. It also leads to very reasonable properties of nuclear matter. Our theory is therefore free of any adjustable parameter. The energy landscapes $\mathscr{H}(B,b)$ for various $n\alpha$ nuclei are interesting by themselves [13] but for brevity not shown here.

**TABLE 1.** Comparison of the binding energies, rms radii ($R_{\text{r.m.s.}}$), and monopole matrix elements ($M(0_2^+ \to 0_1^+)$) for $^{12}$C given by solving Hill-Wheeler equation based on (2) and by RGM [5]. Volkov No. 2 force as the effective two-nucleon force is adopted in the two cases, for which the $3\alpha$ threshold energy is calculated to be $-82.04$ MeV.

|  |  | condensate w. f. (H. W.) | RGM [5] | Exp. |
|---|---|---|---|---|
| $E$ (MeV) | $0_1^+$ | $-89.52$ | $-89.4$ | $-92.2$ |
|  | $0_2^+$ | $-81.79$ | $-81.7$ | $-84.6$ |
| $R_{\text{r.m.s.}}$ (fm) | $0_1^+$ | 2.40 | 2.40 | 2.44 |
|  | $0_2^+$ | 3.83 | 3.47 |  |
| $M(0_2^+ \to 0_1^+)$ (fm$^2$) |  | 6.45 | 6.7 | 5.4 |

## RESULTS FOR FINITE NUCLEI

As already mentioned above, the wave function constructed from the Hill-Wheeler equation based on (2),(3),(4), has practically 100 percent overlap with the ones in [5],[6], once the same Volkov force is used [10]. It is, therefore, not astonishing that we also get very similar results to theirs. For $^{12}$C we obtain two eigenvalues: the ground state and the Hoyle state. Theoretical values for positions, rms values, transition probabilities, compared to the data, are given in Table 1. From the comparison of the rms radii we see that the volume of the Hoyle state is a factor 3 to 4 larger than the one of the ground state of $^{12}$C. This is the aspect of dilute gas state we were talking about above. Constructing an $\alpha$-particle density matrix $\rho(\vec{R},\vec{R}')$, in integrating out of the total density matrix all intrinsic $\alpha$-particle coordinates, we find in diagonalising this density matrix that the corresponding $0S$ $\alpha$-particle orbit is occupied to 70 percent by the three $\alpha$-particles [14], [15]. This is a huge percentage, underlining the almost ideal $\alpha$-particle condensate aspect of the Hoyle state. In this regard one should remember that superfluid $^4$He has only 10 percent of the particles in the condensate! Let us also mention that for the ground state of $^{12}$C the $\alpha$-particle occupations are equally shared between $0S$, $0D$, and $0G$ orbits, thus invalidating a condensate picture for the ground state. Please notice that also the ground state energy of $^{12}$C is reasonably reproduced by our theory.

Let us now discuss the, to our mind, most convincing feature that our description of the Hoyle state is the correct one. As the authors of [5], we reproduce very accurately the inelastic form factor $0_1^+ \to 0_2^+$ of $^{12}$C. This is shown in Fig. 2. The agreement with experiment is as such already quite impressive. Additionally, however, we made the following study shown in Fig. 2. We artificially varied the extension of the Hoyle state and studied the influence on the form factor. We found that the overall shape of the form factor only varies little, for instance in what concerns the position of the minimum. On the contrary, we found a strong dependence on the absolute magnitude of the form factor and in Fig. 2 we also plot the variation of the height of the first maximum of the inelastic form factor as a function of the percentage change of the rms radius of the Hoyle state [17]. It can be seen that a 20 percent increase of the rms radius decreases the maximum by a factor of two! This strong dependence of the magnitude of the form factor makes

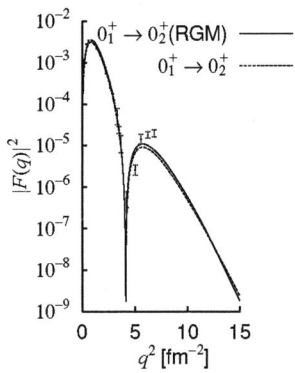
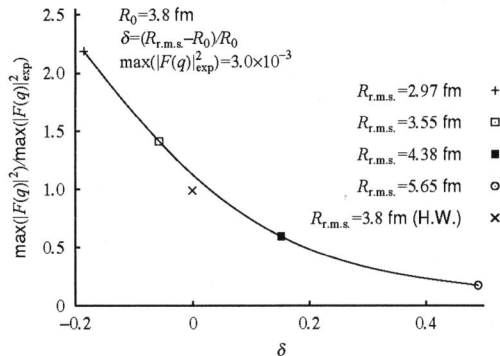

**FIGURE 2.** Present result of the inelastic form factor compared with experiment. RGM result corresponds to ref. [5] (left panel). The ratio of the value of maximum height, theory versus experiment, for the inelastic form factor, i.e. $\max|F(q)|^2 / \max|F(q)|^2_{\exp}$, is plotted as a function of $\delta$, which is defined as $\delta = (R_{r.m.s.} - R_0)/R_0$. $R_{r.m.s.}$ and $R_0$ are the rms radii corresponding to the wave function (2) and the one obtained by solving the Hill-Wheeler equation based on (2), respectively (right panel).

us firmly believe that the agreement with the actual measurement is practically a proof that our calculated wide extension of the Hoyle state corresponds to reality.

The Hoyle state can be considered as the ground state of the $\alpha$-particle condensate. Exciting one $\alpha$-particle out of the condensate and putting it into the $0D$ orbit reproduces the experimentally measured position of the $2_2^+$ state in $^{12}$C. Without going into details, we also state that the width of this state is correctly reproduced [18]. It is tempting to imagine that the $0_3^+$ state which experimentally is almost degenerate with the $2_2^+$ state is obtained by lifting one $\alpha$-particle into the $1S$ orbit. First theoretical studies [19] indicate that this view might indeed be true. However, its width ($\sim$ 3MeV) is very broad what makes a theoreticla treatment rather delicate and further investigations are necessary to validate this picture. At any rate, it would be very satisfying, if the the triplet of states, i.e. $0_2^+, 2_2^+, 0_3^+$, could all be explained from the $\alpha$-particle point of view, since those three states are precisely the ones which can not be explained within a (no core) shell model approach [4].

In conclusion, in what concerns $^{12}$C, we think we have accumulated enough facts to become convinced that the Hoyle state is, indeed, what one can call an $\alpha$-particle condensate state, being aware of the fact that 'condensate' for only three particles constitutes a certain abuse of the word. We, however, should remember in this context that also in the case of nuclear Cooper pairing, only a few Cooper pairs are sufficient to obtain clear signatures of superfluidity in nuclei!

What about $\alpha$-particle condensation in heavier nuclei? Of course, once one accepts the idea that the Hoyle state is essentially a state of three free $\alpha$-particles hold together only by the Coulomb barrier, it is hard to believe that analogous states should not also exist in heavier $n\alpha$ nuclei like $^{16}$O, $^{20}$Ne, $^{24}$Mg, .... At least our calculations systematically always show a $0^+$-state close to the $\alpha$-particle disintegration threshold. For example in

$^{16}$O we obtain three $0^+$-states: the ground state at $E_0 = -124.8$ MeV (experimental value: $-127.62$ MeV), a second state at excitation energy $E_{0_2^+} = 8.8$ MeV and a third one at $0_3^+ = 14.1$ MeV. The threshold in $^{16}$O is at 14.4 MeV. Unfortunately the experimental situation in $^{16}$O is by far not so complete as the one in $^{12}$C. For example no transition probability measurement of $0^+$-states around threshold in $^{16}$O nor inelastic form factors do exist. Recently Wakasa [20] identified a new $0^+$-state at 13.5 MeV in $^{16}$O which is the 5-th $0^+$-state. There are indications that it might be the $\alpha$-condensate state [21].

An intersting question is how many $\alpha$'s can maximally be in a self bound $\alpha$-gas state. For answering this question, a schematic investigation using an effective $\alpha$-$\alpha$ interaction in an $\alpha$-gas mean field calculation of the Gross-Pitaevsky type was performed. Our estimate yields [22] a maximum of ten $\alpha$-particles which can be held together in a condensate. However, a couple of extra neutrons may stabilise larger condensates.

Another interesting idea concerning $\alpha$-particle condensates was put forward by von Oertzen and collaborators [23, 24]. Adding more and more $\alpha$-particles to e.g. the $^{40}$Ca core, one sooner or later will arrive at the $\alpha$-particle drip. Therefore it may need little further excitation energy to shake loose further $\alpha$-particles, so that an $n\alpha$-condensate could be created on top of an inert $^{40}$Ca core. Similar ideas also have been advanced by Ogloblin [25] who imagines a three $\alpha$-particle condensate on top of $^{100}$Sn and earlier by Brenner and Gridnev who think having detected gaseous $\alpha$-particles in $^{28}$Si and $^{32}$S on top of an inert $^{16}$O core [26].

## REMARKS AND CONCLUSION

In conclusion, we see that the idea of $\alpha$-particle condensation in nuclei has already triggered a lot of new works and ideas inspite of the fact that so far strong identification of such a state only exists in $^{12}$C. However, the possibility of the existence of a completely new nuclear phase where $\alpha$-particles play the role of quasi-elementary constituents is surely fascinating and hopefully many more $\alpha$-particle states will be detected in the near future.

Let us end with some general remarks. Strongly bound $\alpha$-particles exist in nuclear physics because there are four different fermions, i.e. protons and neutrons with both spin up/down and roughly equal pairwise attraction among the four possibilities. One can also see it in a mean field picture where all four nucleons can occupy the lowest 0S-state whereas had we only e.g. neutrons, i.e. one species, two out of the four nucleons would have to be put in the next 0P-shell what is energetically very penalising. It is, therefore, conceivable that a similar situation with respect to quartetting could be created with fermions in traps, if one captured them in four different magnetic substates. Under the condition that they all attract them with more or less equal strength, one could create the very interesting situation of quartet condensation on a macroscopic scale. It would certainly be promising to investigate such experimental possibilities.

# REFERENCES

1. R. B. Wiringa, S. C. Pieper, J. Carlson, and V. R. Pandharipande, Phys. Rev. C **62**, 014001 (2000).
2. F. Hoyle, D. N. F. Dunbar, W. A. Wenzel, W. Whaling, Phys. Rev. 92, 1095 (1953)
3. C. W. Cook, W. A. Fowler, C. C. Lauritsen, T. B. Lauritesen, Phys. Rev. 107, 508 (1957)
4. P. Navrátil, J. P. Vary, and B. R. Barrett, Phys. Rev. Lett. **84**, 5728 (2000); Phys. Rev. C **62**, 054311 (2000); B. R. Barrett, B. Mihaila, S. C. Pieper, and R. B. Wiringa, Nucl. Phys. News, **13**, 17 (2003).
5. Y. Fukushima and M. Kamimura, *Proc. Int. Conf. on Nuclear Structure*, Tokyo, 1977, ed. T. Marumori (Suppl. of J. Phys. Soc. Japan, **44**, 225 (1978)); M. Kamimura, Nucl. Phys. A **351**, 456 (1981).
6. E. Uegaki, S. Okabe, Y. Abe, and H. Tanaka, Prog. Theor. Phys. **57**, 1262 (1977); E. Uegaki, Y. Abe, S. Okabe, and H. Tanaka, Prog. Theor. Phys. **59**, 1031 (1978); **62**, 1621 (1979).
7. H. Horiuchi, Prog. Theor. Phys. **51**, 1266 (1974); **53**, 447 (1975).
8. S. Saito, Prog. Theor. Phys. **40** (1968); **41**, 705 (1969); Prog. Theor. Phys. Suppl. **62**, 11 (1977).
9. H. Morinaga, Phys. Rev. **101**, 254 (1956); Phys. Lett. **21**, 78 (1966).
10. Y. Funaki, A. Tohsaki, H. Horiuchi, P. Schuck, and G. Röpke, Phys. Rev. C **67**, 051306(R) (2003).
11. A. Tohsaki, H. Horiuchi, P. Schuck, and G. Röpke, Phys. Rev. Lett. **87**, 192501 (2001).
12. A. Tohsaki, Phys. Rev. C 49, 1814 (1994).
13. A. Tosaki, H. Horiuchi, P. Schuck, and G. Röpke, *Proc. of the 8th Int. Conf. on Clustering Aspects of Nuclear Structure and Dynamics*, Nara, Japan, 2003, ed. K. Ikeda, I. Tanihata and H. Horiuchi (Nucl. Phys. A **738**, 259 (2004)).
14. T. Yamada, P. Schuck, Eur. Phys. J. A, 26, 185 (2005).
15. H. Matsumura and Y. Suzuki, Nucl. Phys. A **739**, 238 (2004).
16. I. Sick and J. S. McCarthy, Nucl. Phys. A **150**, 631 (1970); A. Nakada, Y. Torizuka and Y. Horikawa, Phys. Rev. Lett. **27**, 745 (1971); and 1102 (Erratum); P. Strehl and Th. H. Schucan, Phys. Lett. **27B**, 641 (1968).
17. Y. Funaki, A. Tohsaki, H. Horiuchi, P. Schuck and G. Röpke, Eur. Phys. J. A **28**, 259 (2006).
18. Y. Funaki, H. Horiuchi, A. Tohsaki, P. Schuck and G. Röpke, Eur. Phys. J. A, 24, 321 (2005).
19. C. Kurakowa, K. Kato, Phys. Rev. C 71, 021301 (2005).
20. T. Wakasa, private communication.
See also http://www.rcnp.osaka-u.ac.jp/annurep/2002/sec1/wakasa2.pdf
21. Y. Funaki et al, RCNP workshop, April, 2006, Osaka, to appear in Mod. Phys. Lett. A, World Scientific.
22. T. Yamada, P. Schuck, Phys. Rev. C **69**, 024309 (2004).
23. Tz. Kokalova, N. Itagaki, W. von Oertzen, and C. Wheldon, Phys. Rev. Lett. **96**, 192502 (2006).
24. W. von Oertzen et al, Eur.Phys.J.A 29, 133 (2006)
25. A. A. Ogloblin et al, Proceedings of the International Nuclear Physics Conference, Peterhof, Russia, June 28-July 2, 2005.
26. M. W. Brenner et al, Proceedings of the International Conference "Clustering Phenomena in Nuclear Physics", St. Petersburg, published in 'Physics of Atomic Nuclei (Yadernaya Fizika), 2000.

# Studies on exotic nuclei by proton-induced direct reaction at GSI and FAIR

O.A. Kiselev

*Gesellschaft für Schwerionenforschung, D-64291 Darmstadt, Germany*
*Institut für Kernchemie, Johannes Gutenberg Universität Mainz, D-55128 Mainz, Germany*

**Abstract.** The proton-induced direct reactions like elastic, quasi-elastic scattering and knock-out at intermediate energies and inverse kinematics are the most powerful classical methods for obtaining spectroscopic information on the structure of unstable exotic nuclei. Few elastic scattering experiments performed at GSI with the gaseous and liquid hydrogen targets provided the most precise data on a nuclear matter distribution and a halo-core structure of the neutron-rich He and Li isotopes. The measured differential cross sections have been also used for probing density distributions as predicted by various microscopic theories. The comparison of the data with the latest calculations will be shown. The description of the recent experiment with proton-rich $^8$B and neutron-rich Be isotopes is presented.

The experimental conditions at the future facility FAIR will provide unique opportunities for nuclear structure studies on nuclei far off stability, and will allow to reach new regions in the chart of nuclides of high interest for nuclear structure and astrophysics. In particular, predicted luminosity will allow for the investigation of direct reactions with stored and cooled radioactive beams at internal H, He, etc. targets of the storage ring NESR. This technique enables high resolution measurements down to very low momentum transfer and provides a gain in luminosity from accumulation and recirculation of the radioactive beams. In order to explore the experimental conditions for measurements planned at EXL/FAIR setup, a first attempt exploring experimentally the feasibility of its concept has been recently made. A detector setup was installed at the ESR storage ring at GSI, Darmstadt. A $^{136}$Xe beam was interacting to an internal hydrogen gas-jet target. The detector setup had all the basic ingredients as foreseen by EXL collaboration. A set of scattering reactions has been studied and the overall performance of the setup demonstrated the feasibility of the EXL experimental approach.

**Keywords:** Direct reactions, Nuclear matter distribution, Exotic nuclei, Elastic scattering, Storage rings
**PACS:** 29.20.Dh, 21.10.Gv, 24.50.+g, 25.70.Bc, 25.40.Cm

## MOTIVATION

One of the most powerful classical methods for obtaining spectroscopic information on the structure of nuclei is the investigation of light-ion induced direct reactions. The study of such reactions, as for example elastic and inelastic proton-, deuteron-, $\alpha$-scattering, or one- and few-nucleon transfer reactions, or charge exchange and knock-out reactions, etc. contributed in the past substantially to our present knowledge about stable nuclei. Before the availability of radioactive ion beams, this method was limited to stable or very long-lived nuclei, which allow to produce targets. A use of good-quality secondary radioactive ion beams, now available at several major accelerator facilities, like GANIL, JINR, MSU, ISOLDE, RIKEN and GSI, enabled to extend such studies on exotic nuclei by using the method of inverse kinematics.

Despite the experimental challenge in performing such experiments with relatively low secondary beam intensities, such investigations with radioactive beams have opened new territories of nuclei in the nuclear chart to be explored. Within the last two decades a number of experiments on direct reactions with radioactive beams have been performed (see Ref. [1] for an overview). One of the most outstanding results was the discovery of so-called "halo"-structure, which is related to the fact that in some light neutron-rich nuclei located near the neutron drip line there is a widely extended low-density distribution of loosely bound valence neutrons surrounding a compact distribution of the majority of nucleons. The investigation of reaction cross sections for some key reactions, such as total absorption cross sections, as well as cross sections for break-up reactions, knock-out reactions, transfer reactions, elastic and inelastic scattering contributed considerably to the discovery and to our present knowledge about halo nuclei [2],[3]. It allowed to deduce their ground state properties, such as the size and the radial shape of their nuclear matter density distribution, their single particle (few body) structure, as well as to determine resonant states at drip line for halo nuclei. Good example is a recent study of $^7H$ isotope, lying far beyond the drip line [4].

## ELASTIC AND QUASI-ELASTIC PROTON SCATTERING IN INVERSE KINEMATICS - A TOOL TO STUDY EXOTIC NUCLEI

The size of nuclei and the radial shape of the distribution of nuclear matter and charge are fundamental properties of nuclei, and therefore of high interest for various fields in nuclear physics. Accurate experimental data on the moments of the charge and matter distributions are of particular interest for the understanding of nuclear structure and for probing theoretical model descriptions of nuclei. Over the years a large variety of exper-

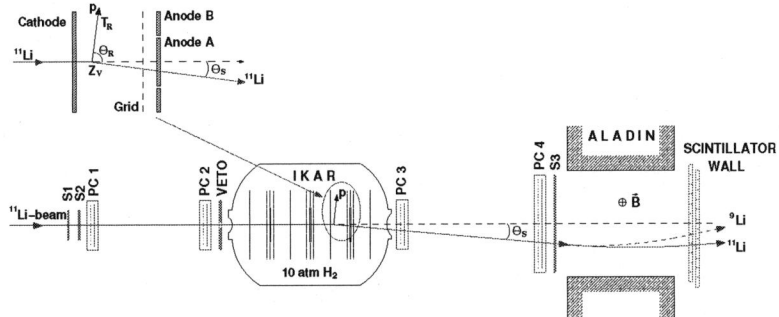

**FIGURE 1.** Schematic view of the experimental setup for small-angle elastic proton scattering on exotic nuclei in inverse kinematics. The central part shows the $H_2$-filled ionization chamber IKAR which serves simultaneously as a gas target and a detector system for recoil protons.

imental methods were developed, using leptonic probes (as electrons, muons, etc.) for investigating nuclear charge distributions, and hadronic probes (as protons, $\alpha$-particles, pions, etc.) for exploring the distributions of nuclear matter (see ref. [5]). While all these methods were applied successfully for many years for the study of stable nuclei, the investigation of the size and radial shape of exotic nuclei has become a new and exciting field of research. Proton-nucleus elastic scattering at intermediate energies around 700 - 1000 MeV is known to be a method well established for obtaining accurate nuclear matter distributions [6]. This method was applied at GSI, Darmstadt for the first time for the investigation of exotic nuclei by using the technique of inverse kinematics. The advantage of choosing protons at intermediate energies as probes is mainly due to the capability of the Glauber multiple-scattering theory of accurately describing the scattering process in this energy region, which, by using as input the proton-nucleon amplitudes from free proton-nucleon scattering data, permits to correlate the differential elastic scattering cross sections and the matter distributions of the composite nuclei in an unambiguous way. The proton scattering in the region of small momentum transfer is particularly sensitive to the nuclear matter radius and the halo structure of nuclei. Differential cross sections for elastic proton scattering at small momentum transfer were measured at GSI Darmstadt at energies around 700 MeV/u in inverse kinematics for the neutron-rich helium isotopes $^6$He, $^8$He [6, 7], and more recently for the neutron-rich lithium isotopes $^8$Li, $^9$Li, and $^{11}$Li [8]. All these experiments have been performed by using external targets. The experimental method (see Fig. 1) is based on the high-pressure hydrogen-filled ionization chamber IKAR, which serves simultaneously as gas target and recoil proton detector. The use of this dedicated active-target technique allows detecting the low energy recoil protons with sufficient angular and energy resolution. In addition projectile scattering angles were measured precisely with the tracking detectors and the scattered particles were identified by a magnetic rigidity analysis. For deducing nuclear matter distributions, differential cross sections calculated with the aid of the Glauber multiple scattering theory, using various parametrisations for the nucleon density distributions as input, were fitted to the experimental cross sections. Example for

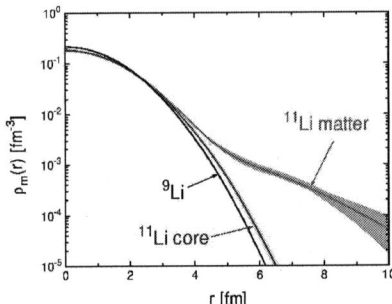

**FIGURE 2.** Nuclear core and total matter density distributions in $^{11}$Li. The shaded areas represent the envelopes of the matter and core density variations within the model parametrizations applied, superimposed by the statistical errors.

the radial distributions of total nuclear matter and core matter, deduced from these data, is displayed in Fig. 2 for the halo nucleus $^{11}$Li together with the deduced nuclear matter distribution for the $^9$Li nucleus, supposed to be a respective core. A $^{11}$Li exhibits the by far most pronounced halo structure known up to now. This is also reflected in the deduced nuclear matter radius $R_m = 3.42(11)$ fm being compatible with a radius of $^{208}$Pb and almost twice larger than those for $^8$He - $R_m = 2.49(4)$ fm.

Recently, a novel experimental approach has been accomplished with the aim to deduce the differential $^{6,8}$He proton elastic scattering cross sections at a higher momentum transfer close to the first diffraction minimum. The major difference with respect to the previous experiments was that instead of the active gaseous target a liquid hydrogen target was used, combined with a proton recoil detector [9]. The deduced values of the nuclear matter radii obtained from the analysis of this experiment were consistent with the previous measurements and confirmed the existence of an extended neutron halo in these nuclei. The nuclear charge radius of $^6$He has been recently measured for the first time using the method of isotope shift based on laser spectroscopy technique [10]. It was found to be 2.054(14) fm, and the corresponding value for the point-proton radius is 1.912(18) fm. The obtained value is in good agreement with the present value of $R_{core}$ = 1.97(9) fm (expected when assuming to have an $\alpha$-particle core + 2n halo structure) that assures the consistency of both independent measurements. The core size of $^8$He has been found to be $R_{core}$ = 1.86(8) fm and within the phenomenological approach, the nuclear matter density has been parameterized as an $\alpha$-particle core and four valence neutrons as well as a $^6$He core and two valence neutrons. Both parameterizations described the experimental cross sections with the same quality. A possible explanation is that a ground state of $^8$He is a mixture of these two configurations. This fact is supported by the analysis of the quasi-elastic scattering of $^8$He on protons, measured also during the same experiment [11]. Precise data on the differential cross sections may provide

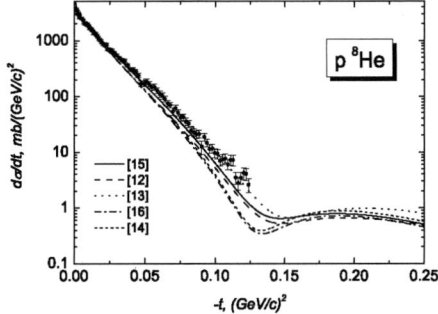

**FIGURE 3.** Experimental differential cross section versus the four momentum transfer squared -t for p$^8$He and comparison with calculations based on predictions of various theoretical models for the nuclear matter density.

a sensitive test for theoretical predictions on nuclear matter density distributions. Density distributions obtained from various theoretical approaches: relativistic mean field calculations [12], microscopic cluster model using the refined resonating group method [13], microscopic quantum Monte Carlo calculations [14], variational Monte Carlo [15], Fermionic Molecular Dynamics [16]. The cross sections p$^{6,8}$He elastic scattering have been calculated using nuclear matter density distributions. Fig. 3 shows the comparison of the data with the latest calculations for $^8$He. The best agreement has been obtained using densities from [13] and [15]. A similar comparison has been also made for p$^6$He. In this case the best agreement between the experimental data and theory was achieve with densities from [14] and [16].

Very recently, an active-target technique has been successfully applied for measurement of proton elastic scattering on with proton-rich $^8$B and neutron-rich $^{10,11,12,14}$Be isotopes. The experimental setup was similar to one used before with substantial improvement of a position resolution of the tracking devices. The preliminary analysis of $^8$B and $^{14}$Be cross sections confirms the existence of the extended matter distribution in both nuclei. This observation for the case in $^8$B is of particular interest due to it is a first proton halo known up to now.

## PERSPECTIVES FOR REACTION STUDIES AT FAIR

The next generation radioactive beam facilities will tremendously improve qualitatively and quantitatively our research potential on the physics of exotic nuclei far off stability. Such facilities, as for example the new FAIR facility at GSI, Darmstadt, will - among others - provide two major achievements, which are essential for reaction experiments with exotic nuclei far off stability [17]. Firstly, secondary beam intensities will be superior by orders of magnitude compared to those presently available, and secondly, new experimental concepts and highly advanced instrumentation will allow coping with the low production cross sections for exotic beams far off stability. For obtaining the necessary information about a structure of nuclei, there is a large variety of light-ion induced direct reactions. The most important ones are the following:

a) Elastic scattering, such as (p,p), ($\alpha$, $\alpha$), gives access to nuclear potentials and to the size and radial shape of nuclei. Especially elastic proton scattering at intermediate energies around 700 -1000 MeV/u may be effectively used for the investigation of the size and radial shape of skin and halo structures of nuclei far off stability.

b) Inelastic scattering, such as (p,p'), ($\alpha$, $\alpha$'), etc. allows to study the nuclear multipole response. Of particular interest is the investigation of low lying collective states, new collective (soft) modes, as well as electric and magnetic giant resonances, which will provide information on the evolution of the nuclear shape and collective motion over the entire chart of nuclides, on nuclear deformation parameters, B(E2)-values, transition densities, as well as the isospin dependence of giant resonances.

c) Charge exchange reactions, such as (p,n), ($^3$He,t), are excellent tools to investigate the spin-isospin response of nuclei. The determination of the strength of Gamow-Teller resonances from the cross sections of such reactions near $\theta_{cm} = 0^0$ will be of crucial importance for the calculation of stellar weak-interaction rates in supernova environments.

d) One- or few-nucleon transfer reactions such as (p,d), (p,t), (d,$^3$He), (p,$^3$He), (d,p), etc. will provide new and important information for nuclear structure, as well as for nuclear astrophysics.

f) Knock-out reactions, such as (p,2p), (p,pn), etc. are the effective tools to explore the ground state configurations and the single-particle structure of nuclei close to the drip line, and to investigate halo structures via the study of momentum distributions of the knocked-out nucleons.

A part of the FAIR facility, dedicated for the experiments with stored radioactive ion beams, will include a double ring system (see Fig. 4) consisting of a Collector Ring (CR) and the New Experimental Storage Ring (NESR) for accumulation, storage and cooling of the large emittance fragment beams provided by the new superconducting two-stage fragment separator (Super-FRS). Its design aims for a maximum fragment

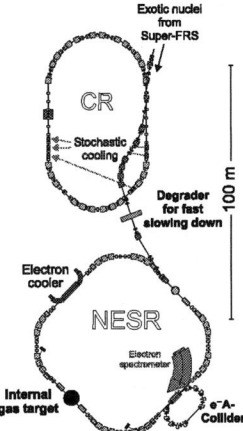

**FIGURE 4.** Schematic view of the storage ring branch of the radioactive beam facility FAIR at GSI Darmstadt.

transmission from the Super-FRS to the storage rings, and therefore will, together with the gain factor of 100 in primary beam intensities, provide a total gain factor of about $10^4$ in the intensity of secondary beams. Concerning the investigation of direct reactions with radioactive beams, this performance will provide sufficient luminosity in many cases to allow the application of new and promising concepts, such as the use of internal $^1H$, $^2H$, $^3He$, $^{34}He$, etc. targets at the NESR for nuclear reaction studies (EXL setup [18]), and the use of the intersecting electron ring (e-A-Collider) [19] for performing elastic and inelastic electron scattering experiments. The EXL setup allows in general performing similar reaction studies as at external target experiments, it has few unique features that fulfil the following considerations:

the most important information is contained in most cases (except for knock-out reactions) in the region of low momentum transfer, t demands for the conditions of inverse kinematics to detect rather low energetic target-like light reaction products with sufficient energy and angular resolution, thus limiting the maximum possible target thickness and requiring in some cases new and challenging detection schemes;

the experience obtained in reaction experiments with radioactive beams in previous experiments has already shown that it will be of big advantage, to ensure a complete kinematical characterization of each reaction event, that means to detect kinetic energy, angle and to identify each target-like recoil particle in coincidence with each beam-like reaction product. Only such an overdetermination allows for a clean and background-free identification of individual reaction channels [20].

The apparatus foreseen being installed at the internal target of the NESR storage cooler ring is displayed in Fig. 5. It includes a Si-detector array for recoiling target-like reaction products, completed by gamma-ray and slow-neutron detectors, as well as forward detectors for fast ejectiles and an in-ring spectrometer for the detection of beam-like reaction products. Whereas the design of the forward detectors and in

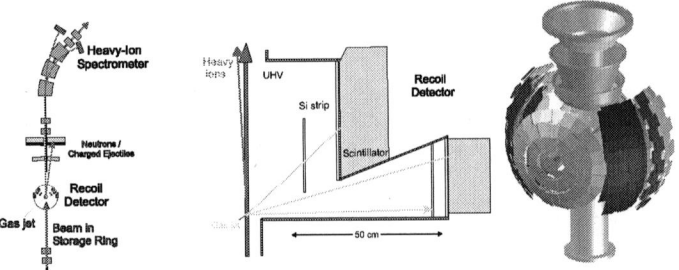

**FIGURE 5.** Schematic view of the EXL detection systems. Setup built into the NESR storage ring and a target-recoil silicon detector surrounding the internal gas-jet target.

ring spectrometers are based on technology already currently available, the design and construction of a highly-efficient, universal recoil and gamma detector system will be one of the most challenging tasks of the present research project. In particular, the detector components need to fulfil strong demands concerning angular and energy resolutions, energy threshold, dynamic range, granularity, and vacuum compartibility, that are partly not available from standard detection systems.

## EXL FEASIBILITY STUDY AT ESR

In order to prove a feasibility of the EXL setup, a first test experiment was performed at the Experimental Storage Ring ESR in GSI, Darmstadt. The experiment was performed using a beam of $^{136}$Xe at energy 350 MeV/u interacting with an internal hydrogen gas-jet target. Detectors representing all the major detector systems of EXL are shown at Fig.

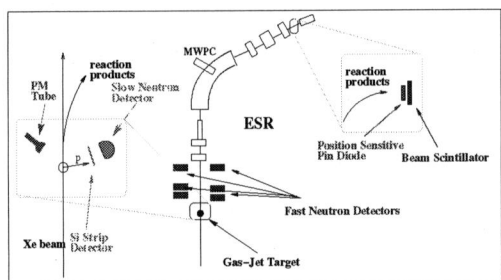

**FIGURE 6.** Experimental setup for the test experiment at ESR.

6. An UHV capable single-sided Si strip detector was mounted in the vacuum chamber close to the internal gas-jet target to detect the recoiled protons. Two walls of organic scintillators with iron converters have been installed at distances of 2.5 and 4 meters from the target. They were used to detect the fast ejectiles (i.e. neutrons and light charged particles). A scintillator and a position sensitive p-i-n silicon diode with a thickness of 300 $\mu$m were installed in a moveable vacuum pocket driven in and out of the beam tube

after the first dipole magnet of ESR. They have been used for identification and fast timing of the beam-like heavy ions. A multi-wire proportional chamber (MWPC) was used as a luminosity monitor, detecting $^{136}$Xe ions deflected out of central orbit due to atomic charge-exchange. The luminosity has been also measured by a photomultiplier

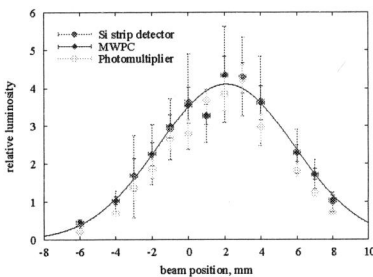

**FIGURE 7.** Target profile of the gas jet as observed from measurements with 3 different luminosity monitors by scanning the beam over the target.

installed near the target, detecting UV light produced by the interaction of the heavy ions with the hydrogen atoms. The Si strip detector served as a third luminosity monitor using the fact that the small angle proton (i.e. Coulomb) elastic scattering cross section is well known. The setup was completed with a scintillator for detecting slow neutrons from (p,n) reactions. During the experiment the beam was scanned over the target. The target profiles obtained with different detectors are consistent with each other (Fig. 7). The obtained profile reflects the overlap between the target and the beam. The size of the jet target was determined to be 8.5±0.2 mm (FWHM). The absolute luminosity has been deduced with the help of a current transformer and reached $(6 \pm 2) \times 10^{27}$ cm$^{-2}$s$^{-1}$. The recoiled protons from the elastic scattering are identified using their

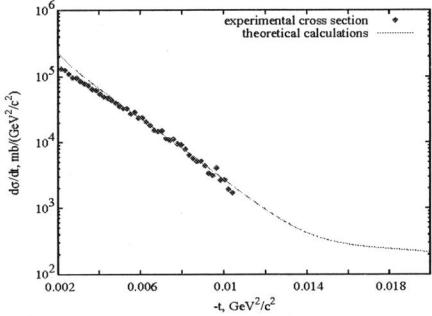

**FIGURE 8.** p$^{136}$Xe elastic scattering differential cross section versus the four momentum transfer squared.

energy loss and position in the Si strip detector. The obtained preliminary cross section is in good agreement with the theoretical one, calculated with the help of the Glauber approach, similar to the one used for the analysis of proton elastic scattering data obtained at external targets [6] (Fig. 8). This demonstrates the possibility to measure elastic scattering and to deduce nuclear matter distributions of exotic nuclei using the EXL setup. Other reaction channels, like (p,n), (p,pn), (p,2p) could be identified using correlations between the detectors for fast ejectiles or the detector for slow neutrons and the detectors for beam-like heavy ions [21].

## CONCLUSIONS

The investigation of light-ion induced direct reactions performed with radioactive beams in inverse kinematics provides important information about a nuclear structure. The method of intermediate energy elastic proton scattering has already been demonstrated for neutron-rich light nuclei to be an effective means for obtaining detailed information on the radial shape and size of halo nuclei, and will therefore be one of the key reactions of the experimental program at the planned NESR facility, allowing to extend such investigations to a wide range of medium heavy and heavy nuclei. A first test experiment with prototype detectors for the EXL setup was performed. An absolute luminosity of the order of $10^{27}$ cm$^{-2}$s$^{-1}$ was reached with the internal hydrogen target. Different methods for luminosity monitoring, applicable for EXL, were tested. Coincidences between the different detector systems were observed and the proton elastic scattering cross section is measured. This confirms the feasibility of the EXL setup, which is a very important milestone towards scattering experiments with exotic nuclei at the NESR.

## REFERENCES

1. I. Tanihata, (ed.), *Nucl. Phys. A* **693**, 1 (2001).
2. P. Egelhof, *Act. Phys. Pol. B* **30**, 487 (1999).
3. P.G. Hansen et al., *Nucl. Phys. A* **693**, 133 (2001).
4. M. Caamano et al., Proceedings of the Radioactive Nuclear Beams Conference, Cortina d'Ampezzo, Italy (2006), *to be published in Eur. Phys. J. A*.
5. J.W. Negele et al., *Adv. Nucl. Phys.* **19**, 1 (1989).
6. G.D. Alkhazov et al., *Nucl. Phys. A* **712**, 269 (2002).
7. S.R. Neumaier et al., *Nucl. Phys. A* **712**, 247 (2002).
8. A.V. Dobrovolsky et al., *Nucl. Phys. A* **766**, 1 (2006).
9. O.A. Kiselev et al., *Eur. Phys. J. A* **25**, s01 215 (2005).
10. L.-B. Wang et al., *Phys. Rev. Lett.* **93**, 142501 (2004).
11. L.V. Chulkov et al., *Nucl. Phys. A* **759**, 43 (2005).
12. S. Typel and H.H. Wolter, Proceedings of the Int. Workshop XXVI on Gross Properties of Nuclei and Nuclear Excitations, Hirschegg, Austria, GSI, Darmstadt, (1998) p. 69; S. Typel and H.H. Wolter, *Nucl. Phys. A* **656**, 331 (1999).
13. J. Wurzer and H.M. Hofmann, *Phys. Rev. C* **55**, 688 (1997); J. Wurzer and H.M. Hofmann, private communication.
14. B.S. Pudliner et al., *Phys. Rev. C* **56**, 1720 (1997).
15. S. Karataglidis et al., arXiv:nucl-th/0206050 v2 (2004); K. Amos et al., *Adv. in Nucl. Phys.* **25**, 275 (2000).
16. T. Neff and H. Feldmeier, *Nucl. Phys. A* **738**, 357 (2004).
17. Conceptual Design Report: An international accelerator facility for beams of ions and antiprotons, Gutbrod, H.H., Gross, K.D., Henning, W.F. and Metag, V. (eds.), GSI Darmstadt, Germany (2001).
18. http://www-linux.gsi.de/$\sim$wwwnusta/tech-report/05-exl.pdf.
19. http://www-linux.gsi.de/$\sim$wwwnusta/tech-report/04-elise.pdf.
20. P. Egelhof et al., *Physica Scripta*, **T104**, 151 (2003).
21. S. Ilieva et al., Proceedings of the Radioactive Nuclear Beams Conference, Cortina d'Ampezzo, Italy (2006), *to be published in Eur. Phys. J. A*.

# New Study of Reaction Cross Sections and the Nucleon Density Distribution

M. Fukuda, M. Takechi[†], M. Mihara, R. Matsumiya, K. Matsuta, T. Minamisono[‡], T. Ohtsubo[§], T. Izumikawa[¶], S. Momota[∥], T. Suzuki[††], T. Yamaguchi[††], S. Nakajima[††], K. Kobayashi[††], K. Tanaka[‡‡], T. Suda[‡‡], S. Sato[§§], M. Kanazawa[§§] and A. Kitagawa[§§]

*Dept. Phys., Osaka Univ., Toyonaka, Osaka 560-0043, Japan*
[†]*RCNP, Osaka Univ., Ibaraki, Osaka 567-0047, Japan*
*Department of Physics, Osaka University, Toyonaka, Osaka 560-0043, Japan*
[‡]*Fukui University of Technology, Fukui, 910-8505, Japan*
[§]*Department of Physics, Niigata University, Niigata 950-2102, Japan*
[¶]*RI Center, Niigata Univeristy, Niigata 951-8510, Japan*
[∥]*Kochi University of Technology, Kami, Kochi 782-8502, Japan*
[††]*Department of Physics, Saitama University, Saitama 338-3570, Japan*
[‡‡]*RIKEN, Wako, Saitama 351-0106, Japan*
[§§]*National Institute of Radiological Sciences, Chiba 263-8555, Japan*

**Abstract.** Reaction cross sections ($\sigma_R$) for $^{12}$C beam on $^9$Be, $^{12}$C, and $^{27}$Al targets and for $^{11}$Be beam on $^9$Be target were precisely measured systematically in the intermediate energy range. It is indicated that these all $\sigma_R$ data were successfully reproduced by a modified Glauber calculation, which is expected to be applied for the determination of density distributions of other exotic nuclei.

**Keywords:** Reaction cross section, Density distribution
**PACS:** 25.60.Dz, 24.10.-i, 21.10.Gv

## INTRODUCTION

The reaction cross section ($\sigma_R$) plays an irreplaceable role in investigations of nuclear sizes, especially for unstable nuclei. In the simplified black disk picture of nucleus-nucleus collision, we can understand easily that the reaction cross section contains information on the nuclear sizes as $\sigma_R = \pi(R_P^2 + R_T^2)$, where $R_P$ is the projectile radius and $R_T$ the target radius. Therefore in this case, if we know the $R_T$, we can extract $R_P$ from $\sigma_R$ data, or vice versa. The method using reaction cross sections to determine the nuclear sizes is quite suitable for the investigation for unstable nuclei, because we can utilize the in-flight type fragment separators for the production of exotic nuclear beams. While the above black disk picture is too much simplified, we usually use the Glauber model in the actual investigations to relate $\sigma_R$ to nuclear sizes or nucleon density distributions, which is a powerful tool especially at high energies [1].

The simplest form of the Glauber theory is the optical-limit approximation. In this approximation, the nuclear collision is considered as an incoherent sum of individual nucleon-nucleon scatterings [2, 3]. The $\sigma_R$ is then represented as Eq. (1), that includes only three physical quantities. These are the projectile density $\rho_P(r)$, the target one $\rho_T(r)$, and the nucleon-nucleon total cross sections $\sigma_{NN}$ which were already determined

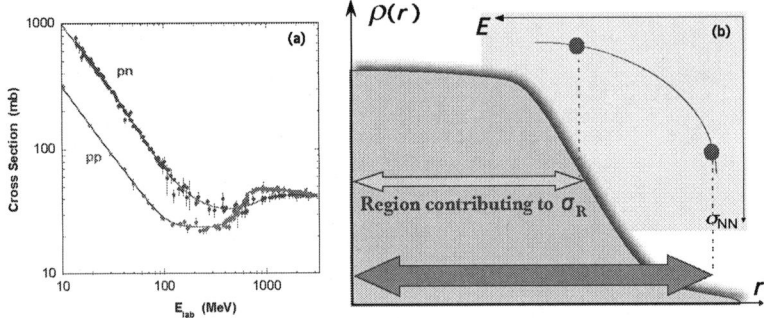

**FIGURE 1.** (a) Nucleon-nucleon total cross sections. (b) Schematic drawing of the concept of profiling nuclear surface density by changing beam energy.

precisely. Therefore in this calculation the $\sigma_R$ is calculated with only these three quantities without any free parameters.

$$\sigma_R = \int db \left[ 1 - \exp\left( \int d^2 r \sum_{ij} \sigma_{ij} \rho_z^{P_i}(r) \rho_z^{T_j}(r) \right) \right] \quad (1)$$

In Fig. 1 (a) the experimental nucleon-nucleon total cross sections are plotted as a function of energy, which shows a steep increase at lower energies as the energy decreases [4]. We can utilize this feature to extract the information of nuclear surface density as shown in Fig. 1 (b). The upper part is $\sigma_{NN}(E)$ the same as Fig. 1 (a). This shows that the $\sigma_{NN}$ is relatively small at higher energies, which means the radial region contributing to the nuclear reaction is limited to rather a large density part, inner region as indicated in the figure. On the other hand, at lower energies, $\sigma_{NN}$ is fairly large, therefore the nuclear reactions can take place even at dilute density part in the outer region. Therefore by measuring $\sigma_R$ at several different energies, we can profile the nucleon density distribution at the nuclear surface [5].

However, there was a problem that the agreement between the calculation and the experimental values of $\sigma_R$ is not so good at intermediate energies even for stable nuclei. Fig. 2 is a plot of the discrepancy between the data and the calculation. The discrepancy reaches almost 20 % at lowest energies of a few tens MeV/nucleon [6]. Therefore we should clarify the origin of this disagreement by taking precise and systematic $\sigma_R$ data for stable nuclei the densities of which are well known by electron scattering data. We also used unstable nuclei or halo nuclei, $^{11}$Be and $^8$B, the densities of which are relatively well known.

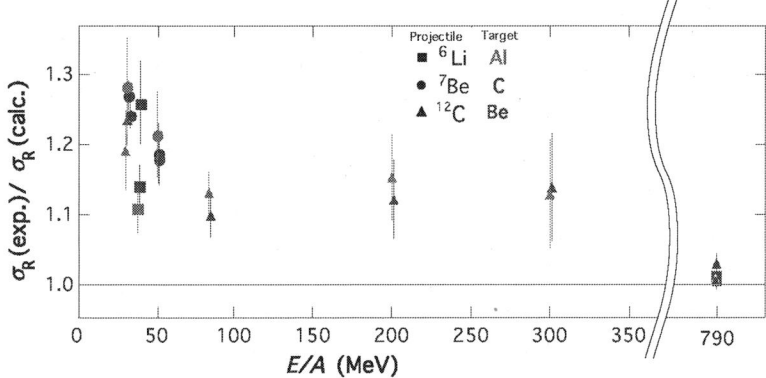

**FIGURE 2.** Experimental $\sigma_R$ divided by the optical-limit calculation plotted as a function of energy.

## EXPERIMENTS

We took the transmission method to measure the $\sigma_R$. With this method, we count the number of the objective nuclide before ($N_0$) and after ($N_1$) the target. From the attenuation of the number we can deduce the $\sigma_R$ as $\sigma_R = \frac{1}{t} ln(\frac{N_1}{N_0})$, in which $t$ is the target thickness. For this the counter after the target should have a particle identification capability. In order to correct for reactions at places other than the target, for example in the counters themselves, we carried out the target-out measurement removing the target. In this measurement the energy condition was tuned to be the same as that of the target-in measurement.

Fig. 3 shows a typical setup at HIMAC(Heavy Ion Medical Accelerator in Chiba) at NIRS(National Institute for Radiological Sciences) which is a high energy heavy ion synchrotron facility. We used the fragment separator [7] and $\Delta E$  $E$ counter telescope consisting of Si and NaI(Tl) counters. This separator is useful also for the primary beam measurements because various beam energies should be prepared for target-in and -out measurements at a lot of energy points using the degraders placed at the position of production target.

## RESULTS AND DISCUSSIONS

Fig. 4 shows a typical $\Delta E$  $E$ two dimensional spectra after the target for $^{12}C$ beam. The thick part is the no-interaction $^{12}C$ peak. The spectrum on the left-hand side is for the target-in measurement. The corresponding target-out spectrum is shown in the right-hand side. Comparing these two spectra the reaction products from the target can be clearly identified. The tail of $^{12}C$ peak corresponds to inelastic scatterings, which should be included in the reaction events. Therefore we carefully estimated the amount of the tail inside the peak using a Monte-Carlo simulation.

The experimental results for the $^{12}C$ beam are shown in Fig. 5. Comparing with the optical-limit calculation shown by the dot-dashed curves, there are some discrepancies

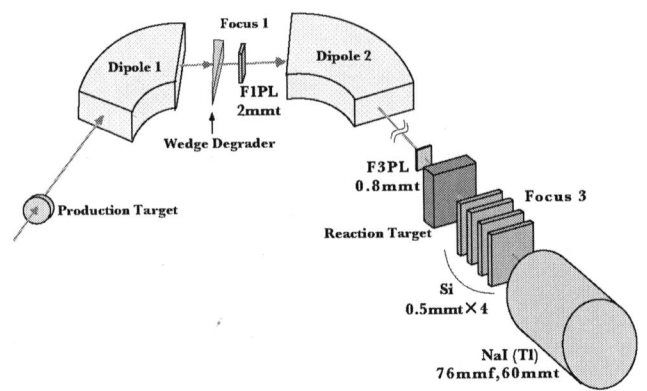

**FIGURE 3.** Typical setup at HIMAC.

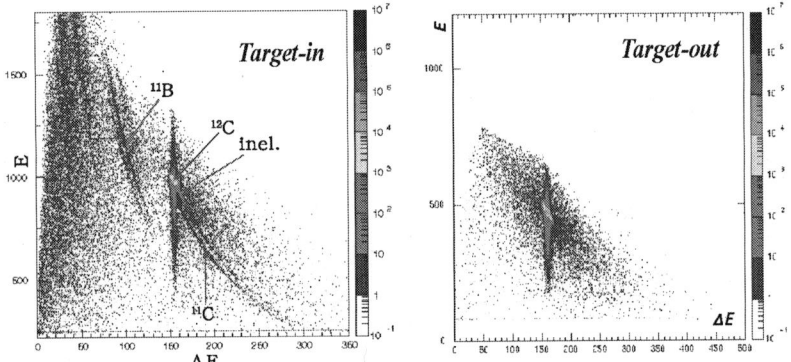

**FIGURE 4.** Typical $\Delta E$ $E$ two dimensional spectra for target-in and -out measurements.

between the calculations and the data below 200 MeV/nucleon, which are similar to all targets.

In order to improve these discrepancies, we took into account two effects. One is the multiple scattering effect, so-called Few-body effect. This is the effect of multiple scatterings of nucleons during the nuclear collision. This was appropriately taken into account by using the prescription by Ab-ibrahim and Suzuki [8]. The other is the inclusion of effect of internal motion of nucleons inside the nuclei. Including these two effects, the data can be nicely reproduced as shown by the solid curves in the figure, with almost no free parameters.

To examine this situation for unstable nuclei we used two halo nuclei, $^{11}$Be and $^{8}$B the densities of which are relatively well-known, as test probes. For these nuclei, the densities are taken from Hartree-Fock calculations by Kitagawa and Sagawa [9]. As shown in Fig. 6, the data are nicely reproduced with these densities by the same method as that for $^{12}$C.

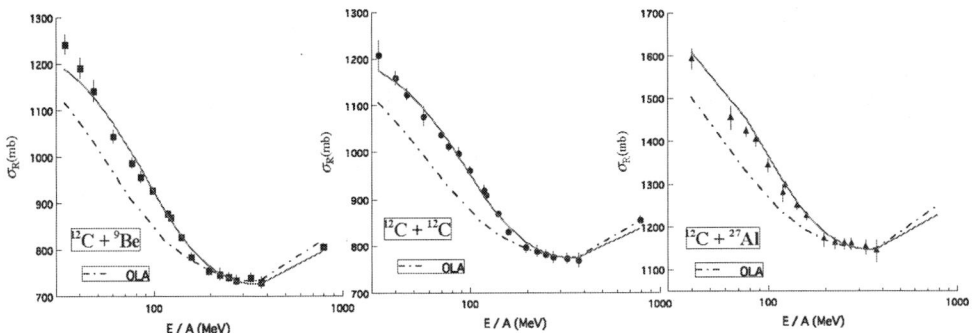

**FIGURE 5.** Experimental $\sigma_R$ compared with two Glauber calculations. One is the optical-limit (dot-dashed curves) and the other is the one with the effects mentioned in the text incorporated (solid curves) plotted as a function of energy.

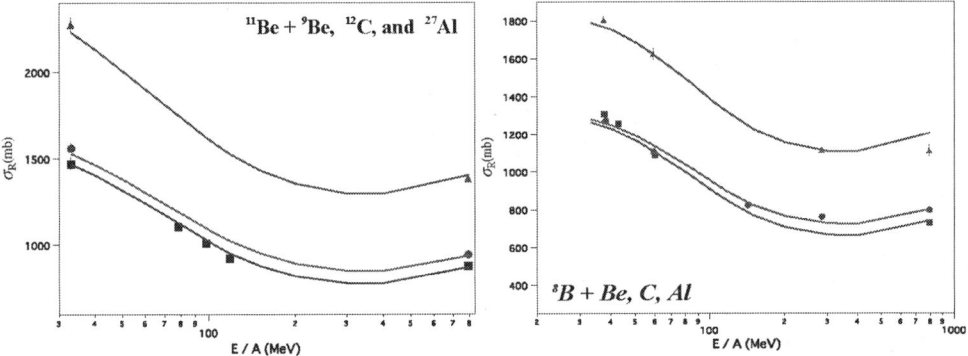

**FIGURE 6.** Experimental $\sigma_R$ for halo nuclei compared with the Glauber calculation(solid curves) described in the text plotted as a function of energy.

## SUMMARY

We systematically measured reaction cross sections for stable and unstable nuclei the densities of which are well known with a good precision in the intermediate energy range. The present data were successfully reproduced by the Glauber calculation in which the multiple scattering effect and internal motion effect were appropriately incorporated. This method is expected to be applied to a variety of unstable nuclides in order to explore the densities of exotic nuclear structures [10] and to investigate the properties of asymmetric nuclear matter.

## ACKNOWLEDGMENTS

This work was carried out as a Research Project with Heavy Ions at NIRS-HIMAC and supported by the Grant-in-Aid for Scientific Research by the Ministry of Education, Science, and Culture of Japan. It was also partly funded by the Sasakawa Scientific Research Grant from The Japan Science Society.

## REFERENCES

1. A. Ozawa, I. Tanihata, T. Kobayashi, Y. Sugahara, O. Yamakawa, K. Omata, K. Sugimoto, D. Olson, W. Christie, and H. Wieman, *Nucl. Phys. A* **608**, 63 (1996).
2. P. J. Karol, *Phys. Rev.* **C11**, 1203 (1975).
3. G. F. Bertsch, B. A. Brown, and H. Sagawa, *Phys. Rev.* **C39**, 1154 (1989).
4. P. D. Group, *Eur. Phys. J. C* **3**, 1 (1998).
5. M. Fukuda, T. Ichihara, N. Inabe, T. Kubo, H. Kumagai, T. Nakagawa, Y. Yano, I. Tanihata, M. Adachi, K. Asahi, M. Kouguchi, M. Ishihara, H. Sagawa, and S. Shimoura, *Phys. Lett.* **B268**, 339 (1991).
6. M. Takechi, M. Fukuda, M. Mihara, T. Chinda, T. Matsumasa, H. Matsubara, Y. Nakashima, K. Matsuta, T. Minamisono, R. Koyama, W. Shinosaki, M. Takahashi, A. Takizawa, T. Ohtsubo, T. Suzuki, T. Izumikawa, S. Momota, K. Tanaka, T. Suda, M. Sasaki, S. Sato, and A. Kitagawa, *Eur. Phys. J. A* **25**, 217–219 (2005).
7. M. Kanazawa, A. Kitagawa, S. Kouda, T. Nishio, M. Torikoshi, K. Noda, T. Murakami, S. Sato, M. Suda, and a. Tomitani et, *Nuclear Physics A* **746**, 393–396 (2004).
8. B. Abu-Ibrahim, and Y. Suzuki, *Physical Review C (Nuclear Physics)* **62**, 034608–12 (2000).
9. H. Kitagawa, and H. Sagawa, Private communications (1998).
10. M. Takechi, M. Fukuda, M. Mihara, R. Matsumiya, K. Matsuta, T. Minamisono, T. Ohtsubo, T. Izumikawa, S. Momota, T. Suzuki, T. Yamaguchi, S. Nakajima, K. Kobayashi, K. Tanaka, T. Suda, S. Sato, M. Kanazawa, and A. Kitagawa, *Proceedings of this symposium, "Precise studies of nucleon density distribution of $^6He$ and $^8He$"*, 2006.

# Precise Studies of Nucleon Density Distribution of $^6$He and $^8$He

M. Takechi, M. Fukuda[†], M. Mihara[†], R. Matsumiya[†], K. Matsuta[†],
T. Minamisono, T. Ohtsubo[‡], T. Izumikawa[§], S. Momota[¶], T. Suzuki[||],
T. Yamaguchi[||], S. Nakajima[||], K. Kobayashi[||], K. Tanaka[††], T. Suda[††],
S. Sato[‡‡], M. Kanazawa[‡‡] and A. Kitagawa[‡‡]

*RCNP, Osaka Univ., Ibaraki, Osaka 567-0047, Japan*
[†]*Department of Physics, Osaka University, Toyonaka, Osaka 560-0043, Japan*
*Fukui University of Technology, Fukui, 910-8505, Japan*
[‡]*Department of Physics, Niigata University, Niigata 950-2102, Japan*
[§]*RI Center, Niigata Univeristy, Niigata 951-8510, Japan*
[¶]*Kochi University of Technology, Kami, Kochi 782-8502, Japan*
[||]*Department of Physics, Saitama University, Saitama 338-3570, Japan*
[††]*RIKEN, Wako, Saitama 351-0106, Japan*
[‡‡]*National Institute of Radiological Sciences, Chiba 263-8555, Japan*

**Abstract.** Reaction cross sections ($\sigma_R$) for $^6$He and $^8$He have been measured on $^9$Be, $^{12}$C, and $^{27}$Al targets at intermediate energies. We deduced the nucleon density distribution of $^6$He and $^8$He through our data with the use of modified Glauber calculation. This method is a sensitive probe for dilute nucleon density and we successfully clarified the halo structure of $^6$He and the neutron skin-type density distribution of $^8$He.

**Keywords:** Reaction cross section, Density distribution
**PACS:** 25.60.Dz, 24.10.-i, 21.10.Gv

## INTRODUCTION

The neutron skin and halo structure of neutron-rich nuclei have been the attractive topic since the discovery of extremely large nuclear radii of those nuclei [1, 2]. $^6$He and $^8$He are nuclides most extensively studied and their extended neutron density distribution have been deduced by different methods [2, 3, 4]. However, despite of extensive investigations, details of surface density distributions have not been clarified yet and it is not clear whether $^6$He and $^8$He have halo tails or not. In this work, we precisely studied the surface nucleon density distribution of $^6$He and $^8$He through the measurements of reaction cross sections ($\sigma_R$) at intermediate energies. With the use of modified Glauber calculation, we successfully deduced surface dilute density distribution and found the halo structure of $^6$He and skin-type density distribution of $^8$He.

## EXPERIMENTS

All measurements were performed using the HIMAC synchrotron facility at NIRS (National Institute of Radiological Sciences). The primary beam of $^{13}$C with an energy of 230 MeV/nucleon was used. The secondary beam of $^6$He and $^8$He were produced through the projectile fragmentation process of $^{13}$C on Be production target. We used production targets of several kinds of thickness to perform the experiment at various energies in the intermediate energy region (from 30 to 120 MeV/nucleon). We used the transmission method to measure the $\sigma_R$. In the transmission method, we can directly determine the $\sigma_R$ following the equation below.

$$\sigma_R = \frac{1}{t} \ln\left(\frac{N_2}{N_1}\right) \qquad (1)$$

$N_1$ is the number of incoming particles, $N_2$ is the non-reacting events, and $t$ is the thickness of the reaction target. In Fig.1, the schematic view of the HIMAC projectile fragment separator and experimental setup are shown. The incoming particles $N_1$ were identified and counted with $\Delta E$ and $TOF$ (Time-Of-Flight) measured by plastic scintillation counters set at F1 and F3 focus points (F1PL and F3PL). We set the reaction target after the F3PL. We used Be, C, and Al targets in these experiments. After the reaction target, we set the $\Delta E$ $E$ counter telescope which consists of four Si detectors and a NaI(Tl) energy detector in order to count $N_2$. We also performed the target-out measurement to correct for reactions occurred in the detectors.

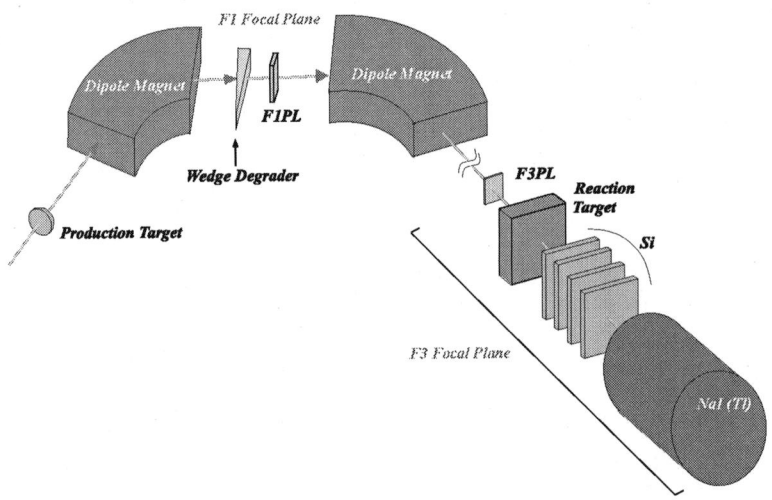

**FIGURE 1.** The schematic view of the experimental setup.

In Fig.2, we show the $\Delta E$ $E$ two dimensional spectrum after the reaction target obtained in the experiment for $^8$He. The main peak in this spectrum is $^8$He non-reacting events. We can see the clear $^6$He and $^4$He isotopes tails (see annotations in Fig.2.). In order to correctly deduced $\sigma_R$, we must carefully estimate the mixture of these isotopes and non-reacting events. In this work, a fairly qualitative estimation was done [5] with the use of Monte-Carlo simulation.

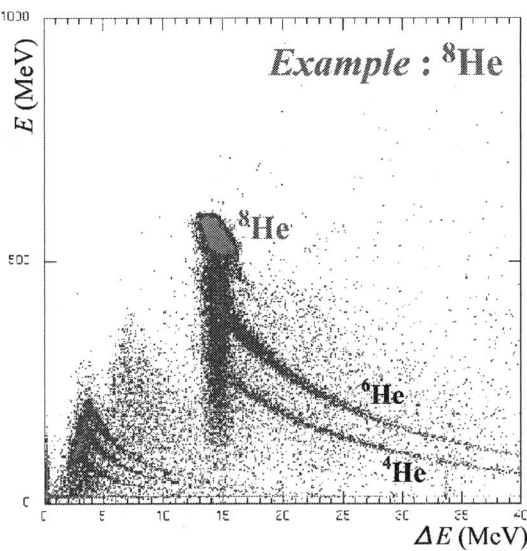

**FIGURE 2.** $\Delta E$  $E$ two dimensional spectrum obtained in the experiment for $^8$He.

## RESULTS AND DISCUSSIONS

In Fig. 3, the present $\sigma_R$ data for $^6$He and $^8$He are plotted as a function of beam energy. The data at 790MeV / nucleon are from [6]. We compare our data with the modified Glauber calculations [7]. In the Galuber calculation, we need the density distribution of $^6$He and $^8$He as input parameters. We used the currently-understood density [3] in the calculation shown with dashed-line. It can be seen that calculations performed with currently-understood densities do not agree with our data at the lower energies.

We determined the density distribution of $^6$He and $^8$He to fit the $\sigma_R$ data in Fig.3. We employed the $\chi^2$-fitting procedure [8] to deduce the density distributions. In the fitting procedure, we assumed the forms of density distributions of $^6$He and $^8$He as follows.

$$\begin{cases} \rho(r) \propto \exp\left[\frac{r^2}{b^2}\right] & ^4\text{He core} \\ \rho(r) \propto \exp\left(\frac{r^2}{c^2}\right) + X \exp(\lambda r) \; r^2 & \text{Valence neutrons} \end{cases} \quad (2)$$

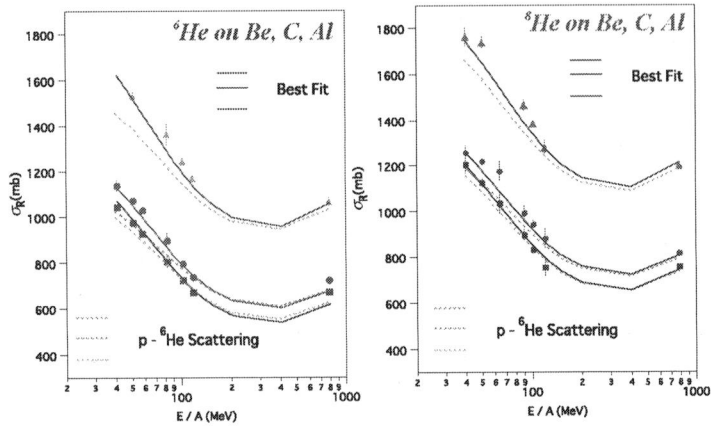

**FIGURE 3.** The $\sigma_R$ data for $^6$He and $^8$He potted as a function of the beam energy.

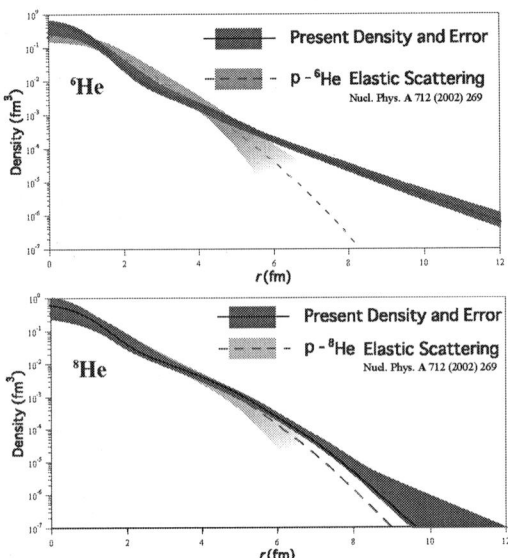

**FIGURE 4.** The present result of the nucleon density distribution for $^6$He and $^8$He compared with existing data.

It should be noted that the above assumption based on $^4$He core + valence 2 or 4 neutrons model of $^6$He and $^8$He. We show the best-fit results of density distribution in Fig.4. It is shown that our best-fit result for $^6$He has a long halo tail. On the other hand, the result for $^8$He have neutron skin-type density and is in good agreement with the existing experimental result. In Fig.3, we show the Glauber calculations performed with present best-fit densities (solid line). It can be seen that $\sigma_R$ data are reproduced well with our present densities in the lower energy region.

## SUMMARY AND FUTURE PROSPECTS

We have measured reaction cross sections for $^6$He and $^8$He at intermediate energies. Using the modified Glauber calculation, we deduced the nucleon density distribution by fitting the $\sigma_R$ data assuming model densities. The present best-fit densities clearly show the existence of halo tail for $^6$He and the neutron skin-type distribution of $^8$He. In the future, we will try model-independent analysis to deduce the nucleon density distribution.

## ACKNOWLEDGMENTS

This work was carried out as a Research Project with Heavy Ions at NIRS-HIMAC and supported by the Grant-in-Aid for Scientific Research by the Ministry of Education. It was also partly funded by the Sasakawa Scientific Research Grant from The Japan Science Society.

## REFERENCES

1. I. Tanihata *et al.*, *Phys. Lett.* **B 289**, 261 (1992).
2. I. Tanihata *et al.*, *Nucl. Phys.* **A 488**, 113c (1988).
3. G. D. Alkhazov *et al.*, *Nucl. Phys.* **A 712**, 269 (2002).
4. L. R. Gasques *et al.*, *Phys. Rev. C* **67**, 024602 (2003).
5. M. Takechi, *Doctoral thesis, Osaka university* (2006), to be pablished.
6. A. Ozawa *et al.*, *Nucl. Phys. A* **693**, 32 (2001).
7. M. Fukuda*et al.*, *New Study of Reaction Cross Sections and Nucleon Density Distributions*, in this proceedings, 2007.
8. M. Fukuda *et al.*, *Nucl. Phys. A* **656**, 209 (1999).

# Coulomb Excitation of $^{26}$Ne

D. Beaumel[1], J. Gibelin [1,2], T. Motobayashi[3], N. Aoi[3], H. Baba[3], Y. Blumenfeld[1], Z. Elekes[4], S. Fortier[1], N. Frascaria[1], N. Fukuda[3], T. Gomi[3], K. Ishikawa[5], Y. Kondo[5], T. Kubo[3], V. Lima[1], T. Nakamura[5], A. Saito[6], Y. Satou[5], E. Takeshita[2], S. Takeuchi[3], T. Teranishi[6], Y. Togano[2], A.M. Vinodkumar[5], Y. Yanagisawa[3], K. Yoshida[3]

[1] *Institut de physique nucléaire, 15 rue G.Clemenceau 91406 Orsay cedex, France*
[2] *Department of Physics, Rikkyo university, 3-34-1 Nishi-Ikebukuro, Toshima, Tokyo 171-8501, Japan*
[3] *RIKEN, Heavy-Ion science lab., 2-1 Hirosawa, Wako, Saitama 351-0198, Japan*
[4] *Institute of Nuclear Research of the Hungarian Academy of Sciences, PO Box 51, H-4001 Debrecen, Hungary*
[5] *Department of Physics, Tokyo Institute of Technology, Tokyo 152-8551, Japan*
[6] *Center for Nuclear Study, University of Tokyo, RIKEN campus, 2-1 Hirosawa, Wako, Saitama 351-0198, Japan*

**Abstract.** Coulomb excitation of the exotic neutron-rich nucleus $^{26}$Ne on a $^{208}$Pb target was measured at 58 A.MeV in order to search for low-lying E1 strength above the neutron emission threshold. Data were also taken on an Al target to estimate the nuclear contribution. The radioactive beam was produced by fragmentation of a 95 A.MeV Ar beam delivered by the RIKEN Research Facility. The set-up included a NaI gamma-ray array, a charged fragment hodoscope and a neutron wall. Using the invariant mass method in the $^{25}$Ne+n channel, we observe a sizable amount of E1 strength between 6 and 10 MeV excitation energy. By performing a multipole decomposition of the differential cross-section, a reduced dipole transition probability of $B(E1)=0.49\pm0.16$ e$^2$fm$^2$ is deduced. For the first time, the decay pattern of low-lying strength in a neutron-rich nucleus is measured.

**Keywords:** Giant resonances, Coulomb excitations
**PACS:** 24.30.Cz; 25.70.De;25.60.-t

## INTRODUCTION

Giant Resonances are a general feature of nuclei and their properties give us a handle on the collective behavior of nuclei. These modes have been extensively studied in stable nuclei over the last 50 years and the recent inception of Radioactive Ion Beam facilities opens the opportunity to extend these investigations to exotic nuclei. Far from stability new modes have been predicted to appear in the early 90's [1,2]. In particular, the dipole response of neutron-rich nuclei exhibits strength at energies lower than the standard Giant Dipole Resonance (GDR), often depicted as the oscillation of a deeply bound core against a neutron halo or skin, giving rise to a so-called pygmy resonance. Experimentally, though the experimental study of giant resonances in exotic nuclei is still in its infancy, the presence of low-lying dipole strength exhausting a sizeable amount of the Thomas-Reiche-Kuhn (TRK) sum rule in neutron-rich nuclei is now an experimental fact. It has been first revealed in light dripline nuclei in breakup reactions using high-Z targets [3]. Later on, the non-resonant nature of the dipole strength found in some light neutron-rich nuclei such as $^{11}$Be has been stated. More questionable is the nature of the low-lying dipole strength, recently observed in heavier nuclei. In the recent work performed at GSI on Oxygen [4] and Tin [5] isotopes, the experimental low-lying dipole strength distribution is extracted and compared to various theoretical predictions. Essentially, all calculations are able to predict the energy and amount of E1 strength observed, but the conclusions of the models differ concerning the microscopic structure of the states. In the case of $^{130,132}$Sn, relativistic quasi-particle random phase approximation (RRPA) calculations [6,7] predict relatively collective pygmy states, while non-relativistic quasi-particle RPA including phonon coupling involves essentially individual transitions [8]. Both approaches nevertheless agree on the exclusive contribution of neutrons to the excitations.

In the present work we investigate low-lying dipole strength in the $^{26}$Ne isotope for which an important redistribution of dipole strength as compared to the stable $^{20}$Ne is predicted by Cao and Ma [9]. In this calculation, almost 5% of the Thomas-Reiche-Kuhn (TRK) energy weighted sum rule is exhausted by a structure centered around 8 MeV. This region of energy is located between the one neutron and the two neutron emission threshold. We performed a Coulomb excitation experiment of $^{26}$Ne at intermediate energy on a lead target, and used the invariant mass method to reconstruct the B(E1) strength from the $^{26}$Ne→ $^{25}$Ne* + n channel.

After determining the strength distribution, we try to go a step further by extracting for the first time neutron branching ratios to levels in the daughter nucleus ($^{25}$Ne) of the populated pygmy states. It is established that these observables provide insight on the microscopic structure of the populated states, as long as the observed decay mode is not statistical.

## EXPERIMENTAL DETAILS

The experiment was performed at the RIKEN Accelerator Research Facility. A secondary $^{26}$Ne beam was produced through fragmentation of a 95 A.MeV $^{40}$Ar primary beam on a 2mm-thick $^{9}$Be target. The $^{26}$Ne produced fragments were separated by the RIKEN Projectile Fragment Separator (RIPS) [10]. Beam particle identification was unambiguously performed by means of the time-of-flight (TOF) and the purity was 80%. The $^{26}$Ne beam of intensity ~5.10$^3$ pps and incident energy 58 A.MeV, was tracked with two parallel-plate avalanche counters providing incident angle and hit position onto the target. 230 mg/cm$^2$ $^{nat}$Pb and 130 mg/cm$^2$ $^{27}$Al were used as reaction targets. Data obtained with aluminum target are used in the following to estimate the contribution of nuclear excitation to the data.

The outgoing charged fragments were measured using a set of telescopes placed at 1.2 m downstream of the target. They consisted of two layers (X and Y) of 500 μm single-sided silicon strip detectors (SSD) with 5 mm strips which resulted in a summed energy resolution of 1.5 MeV (FWHM). The last layer used 3mm-thick Si(Li) detectors from the charged-particle array MUST [11]. The resolution on the remaining energy (E) was 9 MeV (FWHM). Unambiguous mass and charge identification of all projectile-like fragments was obtained using the E-ΔE method.

In-beam gamma rays were detected using a 4Π gamma array DALI2 [12], which consists of 152 NaI detectors placed around the target. For 1.3 MeV gamma-rays, the measured efficiency is approximately 15% and the energy resolution is 7% (FWHM). The Doppler corrected gamma energy distribution obtained in coincidence with the $^{25}$Ne isotope allows us to identify the gamma decay from the adopted 1702.7(7) keV and 3316.4(11) keV excited states. In addition we observed a keV gamma-ray which we related to the keV excited state, only seen up to now in a transfer reaction experiment [13].

The hodoscope for neutron detection was an array of 4 layers of 29 plastic rods each, placed 3.5 m downstream of the target. Each layer was composed of 13 [2.1mx6x6cm$^2$] and 16 [1.1m x6x6 cm$^2$] rods, arranged in a shape of a cross. The total intrinsic efficiency for the detection of 60 MeV neutrons was calculated to be ~25%. Finally, 29 thin plastic rods covered the front face of the wall in order to veto charged particles as well as to provide an active beam stopper. The neutron position was determined with an error of ±3 cm and the energy, from TOF information, with a 2.5 MeV (FWHM) resolution for the neutrons of interest.

## RESULTS

A simulation of the experimental setup using the Geant 3 package [14] was performed. Using this simulation, the differential cross-section for elastic scattering of $^{26}$Ne on $^{nat}$Pb at 55 A.MeV was obtained and compared with optical model calculations based on a $^{20}$Ne+$^{208}$Pb at 40 A.MeV optical potential from [15]. For the $^{26}$Ne+$^{27}$Al reaction, we empirically generated optical potential parameters as described in [16].

Using the invariant mass method, the excitation energy of an unbound state in the $^{A}$X nucleus decaying to a state in $^{A-1}$X can be expressed by: $E^* = E_{rel} + S_n + \Sigma_i E_{\gamma i}$, where $E_{rel}$ is the relative energy between the neutron and the fragment $^{A-1}$X, $S_n$ the one neutron emission threshold, and $\Sigma_i E_{\gamma i}$ the summed energy of the gamma involved in the subsequent decay of the daughter nucleus $^{A-1}$X. The gamma detection efficiency was not high enough to apply an event-by event reconstruction technique. To extract the excitation energy spectrum, we hence used a method based on adequate subtraction of relative energy spectra built for these two kind of events.

The method can be illustrated in the schematic case where the daughter nucleus has only one excited state below its neutron threshold. The excitation energy spectrum can be decomposed in the sum of two contributions: the decay of $^{A}X$ to the ground-state (GS) of $^{A-1}X$, and the decay of $^{A}X$ to the excited state of $^{A-1}X$.

The first contribution can be obtained by subtracting from the inclusive relative energy spectrum the relative spectrum of gamma-coincidence events divided by the gamma detection efficiency, and shifted by the one neutron emission threshold $S_n$. The second contribution is simply the relative energy spectrum of gamma-coincidence events shifted by the gamma energy $E_\gamma$. Finally, the excitation energy spectrum for the $^{25}$Ne+n decay channel is the summed spectrum of the two contributions. The method was tested by simulation in this schematic case, and also in the realistic case taking into account the experimental decay scheme of $^{25}$Ne [16].

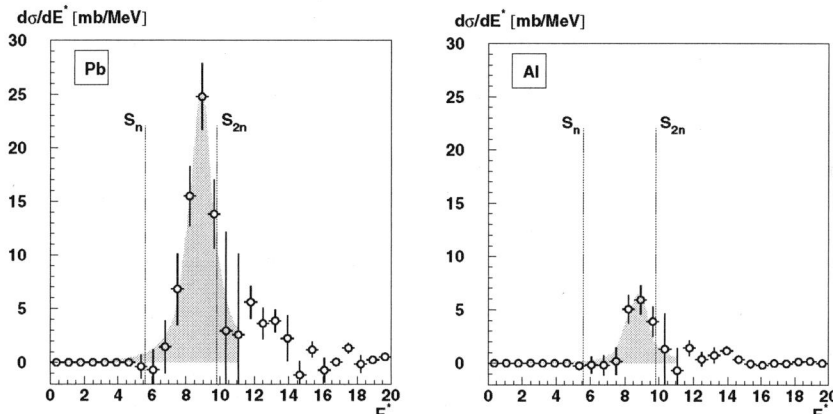

**FIGURE 1. Left:** Excitation energy distribution in $^{26}$Ne with Pb target. The shaded area is a tentative Lorentzian fit. **Right:** Same as previous but for the Al target.

The excitation energy spectra reconstructed for the Ne+n decay channel obtained with the Pb and Al targets are represented in Fig. 1. Note that above 10 MeV, the decay of $^{26}$Ne is expected to occur mainly by 2 neutron emission. Between 8 and 10 MeV, a sizable amount of cross-section is observed for both targets. In intermediate energy inelastic scattering with a heavy target such as Pb, the Coulomb dominance of the E1 excitation is well-known. The contribution of possible E2 excitation to the spectrum obtained with the lead target has been determined in a first step by using data taken with the aluminum target and the coupled channels ECIS 97 [17] code. Assuming simple collective vibration mode with equal nuclear and Coulomb deformation lengths, the E2 deformation parameters were extracted from the measured cross-section with the Al target ( $\sigma_{Al}$ = 9.1±2.3 mb). The cross section in lead was then calculated using the deformation lengths extracted in the previous step. We obtained $\sigma_{Pb}^{L=2}$ = *17.9 ± 4.3 mb*. After subtraction of the $L=2$ contribution, the resulting $\sigma_{Pb}^{L=1}$ = *48.5 ± 4.8 mb* cross section corresponds to a Coulomb deformation parameter $\beta_C$ = *0.087±0.05* which leads to B(E1) = *0.55 ± 0.05 $e^2fm^2$* via the following relation with the Coulomb radius $R_C$:

$$B(E1;0^+ \to 1^-) = (\frac{3}{4\pi} Z_p e R_C \beta_C^{L=1})^2$$

This value of reduced transition probability corresponds to 5.5±0.6% of the TRK energy weighted sum rule for an excitation energy of 9 MeV.

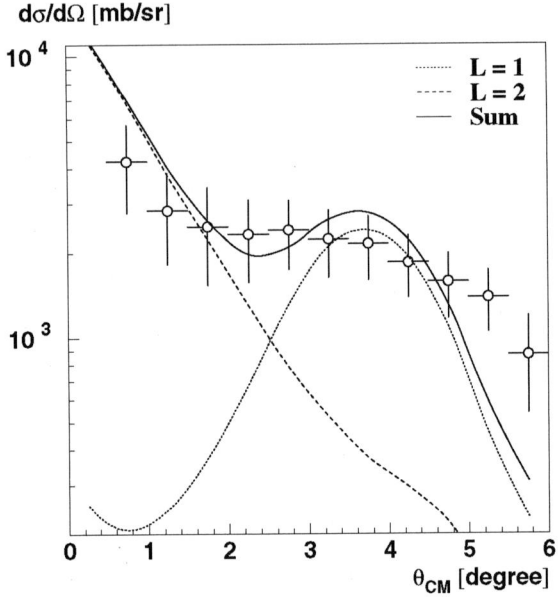

**FIGURE 2.** Result (solid line) of the multipole decomposition of the experimental differential cross-section of the structure at E*=9 MeV excited in $^{26}$Ne using the Lead target, and contributions of L=1 (dashed line) and L=2 (dotted line) multipoles.

Thanks to the high granularity and the good resolution of the present setup, it was possible to reconstruct the scattering angular distribution for $^{26}$Ne on the $^{nat}$Pb target to extract the E1 excitation by means of a multipole decomposition analysis. The *L=1* and *L≥2* angular distributions (dashed and dotted lines) were obtained from simulations based on ECIS 97 angular distribution calculated for *E*=9 MeV*. The data were fitted with a linear combination of the two distributions since L= 2 and L>2 distributions were found to exhibit similar shapes. The result of the fit gives B(E1) = 0.49 ±0.16 $e^2fm^2$ which corresponds to 4.9±1.6% of the TRK sum rule around 9MeV excitation energy. Assuming that the remaining part of the contribution is due to *L=2* excitation, we obtain B(E2↑) = 49±8 $e^2fm^2$.

The two methods thus provide very consistent results, the one using the multipole decomposition being the most direct. In the next section, we compare the results obtained from our experiment concerning the E1 transition to theoretical calculations.

## COMPARISON WITH THEORETICAL CALCULATIONS

Within the relativistic QRPA framework and the response function formalism, Cao and Ma [9] predict an E1 pygmy state centered around 8.4 MeV and exhausting 4.5% of the TRK, sum rule, close to our experimental values. Another calculation has been performed by Khan *et al.* [18] in the spherical non-relativistic QRPA framework with the effective SGII Skyrme interaction. They predict a redistribution of the strength at low energy centered around $E^* = 11.7$ MeV exhausting ~5% of the TRK sum rule. Two other preliminary calculations have been performed in the deformed QRPA framework using Gogny forces [19], and in the deformed relativistic QRPA framework [20]. Both predict a redistribution of the strength at low energy, centered around 10.7 MeV and 7.5 MeV respectively. In the former calculation, only ~1% of the TRK sum rule is exhausted. Preliminary shell-model calculations have also been performed by Nowacki *et al.* [21], in which an E1 state at 9.3 MeV is found. All these calculations agree on the presence of a structure a low excitation energy with variable percentage of TRK sum rule. But it is found that such an agreement is not reached concerning the transitions involved as well as their number.

## DECAY OF PYGMY STATES IN $^{26}$NE

Our method to reconstruct the excitation energy allows us to extract for the first time data on the decay of pygmy resonance of neutron-rich nuclei. The experimental branching ratios to several states in $^{25}$Ne are presented in Table 1. For both Pb and Al targets, the branching ratio for the decay to the ground-state of $^{25}$Ne is compatible with zero. The large difference between branching ratios obtained with the two targets proves that states of different nature have been excited. For comparison, we also performed a statistical decay calculation assuming $L=1,2,3$ emitting states using the CASCADE code[22], spins and parities of populated states in $^{25}$Ne being those listed in Table 1. The decay is not statistical, as expected in such a light nucleus. At the present time, no predictions of the direct decay of pygmy states yet exist from the previously mentioned microscopic models. Future comparisons with our data should be a strong test for these models. Qualitatively, one can already conclude that the observed decay pattern (no decay to the ground-state) exclude that the s1/2→p1/2 and s1/2→p3/2 transitions dominate, as it does in the case of the unperturbed strength calculated in [9].

**TABLE 1.** Experimental neutron branching ratios for the structure at $E^*$~9 MeV in $^{26}$Ne compared to statistical decay calculations for several multipolarities

| Final state in $^{25}$Ne | | Experiment | | Statistical decay calculation | | |
|---|---|---|---|---|---|---|
| Label | $J^\pi$ | Pb target | Al target | L=1 | L=2 | L=3 |
| GS | $1/2^+$ | $5_{-5}^{+17}$% | < 10% | 40% | 28% | 22% |
| (I) | $5/2^+ + 3/2^+$ | 66% ± 15% | $95_{-15}^{+5}$% | 55% | 67% | 75% |
| (II) | $3/2^-$ | 35% ± 9% | $5_{-5}^{+6}$% | 5% | 4% | 3% |

# CONCLUSION

We performed the Coulomb excitation of $^{26}$Ne in order to measure its low lying dipole strength below 10 MeV excitation energy, using the invariant mass method. We extracted an E1 strength value corresponding to *B(E1) = 0.49±0.16* $e^2fm^2$ at ~9MeV excitation energy. These results are compatible with various theoretical predictions. For the first time, the decay pattern of pygmy states in a neutron-rich nucleus could be measured. Future comparison of the neutron branching ratios with theoretical predictions should allow us to elucidate the microscopic structure of theses states.

# REFERENCES

1. Y. Suzuki, K. Ikeda, and H. Sato, *Progr. Theor. Phys.* 83, 180 (1990).
2. P.V. Isacker, M.A. Nagarajan, and D.D. Varner, *Phys. Rev. C* 45, R13 (1992)
3. I. Tanihata, *Progr. Part. Nucl. Phys.* 35, 505 (1995) and refs. therein.
4. A. Leistenschneider, et al., *Phys. Rev. Lett.* 86 (2001) 5442.
5. P. Adrich, et al., *Phys. Rev. Lett.* 95 (2005) 132501.
6. N. Paar, P.Ring, T.Niksicm, and D. Vretenar, *Phys. Rev. C* 67, 34312 (2003).
7. D. Vretenar, N. Paar, P.Ring, and G.A. Lalazissis, *Nucl. Phys. A* 692, 496 (2001).
8. D. Sarchi, P.F.Bortignon, and G. Colo, *Phys. Lett. B* 601, 27 (2004).
9. L.-G. Cao, Z.-Y. Ma, *Phys. Rev. C* 71 (2005) 034305.
10. T. Kubo, et al., *Nucl. Intr. Meth.* B 70 (1992) 309.
11. Y. Blumenfeld, et al., *Nucl. Instr. Meth.* A 421 (1999) 471.
12. S. Takeuchi, et al., *RIKEN Accel. Prog. Rep.* 36 (2002) 148.
13. R. H. Wilcox, et al., *Phys. Rev. Lett.* 30 (1973) 866.
14. R. Brun, et al., CERN DD/EE/84-1.
15. T. Suomijärvi, et al., *Nucl. Phys.* A 491 (1989) 314.
16. J. Gibelin, Ph.D. Thesis, Paris XI University, unpublished.
17. J. Raynal, ECIS-97 code, Unpublished.
18. E. Khan, et al., Private communication (2005).
19. S. Péru, et al., Private communication (2006).
20. P. Ring, et al., Private communication (2006).
21. F. Nowacki, et al., Private communication (2005).
22. F. Puhlhofer, *Nucl. Phys.* A 280 (1977) 267.

# New Developments for Isochronous Mass Measurements of Short-Lived Nuclei

R. Knöbel[*,†], S.A. Litvinov[*,†], B. Sun[*,**], K. Beckert[*], P. Beller[*],
F. Bosch[*], D. Boutin[*,†], C. Brandau[*], L. Chen[*,†], I.J. Cullen[‡],
C. Dimopoulou[*], A. Dolinskii[*], B. Fabian[†], H. Geissel[*,†], M. Hausmann[§],
C. Kozhuharov[*], J. Kurcewicz[*,¶], Yu.A. Litvinov[*,†], Z. Liu[‡],
M. Mazzocco[*], F. Montes[*], G. Münzenberg[*], A. Musumarra[‖],
S. Nakajima[††], C. Nociforo[*], F. Nolden[*], T. Ohtsubo[‡‡], A. Ozawa[§§],
Z. Patyk[₥], W.R. Plaß[†], C. Scheidenberger[*,†], M. Shindo[***], M. Steck[*],
T. Suzuki[††], P.M. Walker[‡], H. Weick[*], N. Winckler[*,†], M. Winkler[*] and
T. Yamaguchi[††]

[*]*Gesellschaft für Schwerionenforschung GSI, 64291 Darmstadt, Germany*
[†]*Justus-Liebig-Universität Gießen, 35392 Gießen, Germany*
[**]*School of Physics, Peking University, Beijing 100871, China*
[‡]*University of Surrey, Guildford, GU2 7XH, U.K.*
[§]*Michigan State University, East Lansing, Mi 48824, U.S.A.*
[¶]*Warsaw University, 00-681 Warszawa, Poland*
[‖]*Laboratori Nazionali del Sud, INFN Catania, Italy*
[††]*Saitama University, 338-8570 Saitama, Japan*
[‡‡]*Niigata University, Niigata 950-2181, Japan*
[§§]*University of Tsukuba, Tsukuba 305-8577, Japan*
[₥]*Soltan Institute for Nuclear Studies, 00-681 Warszawa, Poland*
[***]*University of Tokyo, Tokyo 113-0033, Japan*

**Abstract.** The combination of the in-flight separator FRS and the storage-ring ESR at GSI offers unique possibilities for high accuracy mass and lifetime measurements of bare and few-electron fragments. Operating the ESR in the isochronous mode allows for measurements of revolution frequencies of stored ions without cooling. Isochronous Mass Spectrometry (IMS) can be applied to fragments with half-lives as short as several tens of microseconds. Newly developed magnetic rigidity tagging increases the resolving power of IMS to about 500000. IMS can be used to measure masses of nuclei with rates even lower than one ion per day, a property also needed for the purpose of the ILIMA project at the future facility FAIR.

**Keywords:** Stored highly-charged ions; Exotic nuclei; Masses and half-lives
**PACS:** 21.10.Dr, 29.20.Dh

## INTRODUCTION

Masses and half-lives of exotic nuclei play an important role in modern nuclear physics and astrophysics. The investigations of the evolution of shell closures, nucleon-nucleon pairing correlations, nuclear deformations etc. are closely related to accurate measurements of binding energies of nuclides far off the valley of $\beta$-stability. The separation energies, which determine the possible decay channels, are obtained from masses of neighboring nuclei. One example of a negative two-proton separation energy is $^{45}$Fe for which the two-proton radioactivity was recently discovered [1].

One of the hot questions in nuclear astrophysics is the creation of chemical elements in stars. The nuclear masses, on the one side, settle the pathways of the nucleosynthesis processes. On the other side, the duration of these processes and, finally, the abundances of the nuclides, are determined by the nuclear half-lives.

Due to these wide applications, mass spectrometry is a rapidly developing field. Although mass and half-life measurements of exotic nuclei are performed or planned in the near future at many laboratories worldwide, it will probably be impossible to measure them for all nuclei, e.g., those involved in the r-process nucleosynthesis. Therefore, one will have to rely on the predictions of nuclear models. There is a significant progress in the theoretical calculations over the last years. For detailed information on different models the reader is referred to Refs. [2, 3]. However, the predictive power of all modern theories is not yet sufficient for most of the needs in nuclear structure and astrophysics. This is why one of the experimental tasks is to perform tests of theoretical predictions and to provide new data for model improvements.

In the following sections, we present the status of Isochronous Mass Spectrometry, which is one of the methods for direct mass measurements developed at FRS-ESR at GSI [4].

## BASIS FOR MASS SPECTROMETRY AT THE FRS-ESR

The combination of the in-flight fragment separator FRS [5] and the cooler-storage ring ESR [6] at GSI provides unique experimental conditions for studying exotic nuclei.

Products from the projectile fragmentation or fission reactions are produced by impinging primary beams with incident kinetic energies of (400-1000) A MeV on thick (1-8) g/cm$^2$ production targets. The fragments emerge from the target at energies of typically (300-400) A MeV. At these energies, the fragments are mostly fully ionized or have one to three bound electrons. After being separated in the FRS within a few hundreds of nanoseconds the ions are injected into the ESR. The revolution frequencies $f$ of ions stored in the ESR depend on their mass-to-charge ratio $m/q$ via:

$$\frac{\Delta f}{f} = -\frac{1}{\gamma_t^2} \frac{\Delta(m/q)}{m/q} + \frac{\Delta v}{v}(1 - \frac{\gamma^2}{\gamma_t^2}), \qquad (1)$$

with $\Delta f = (f_2 - f_1)/\overline{f}$ and $\Delta(m/q) = [(m/q)_2 - (m/q)_1]/\overline{(m/q)}$. $v$ represents the velocity of the ions and $\gamma$ is the relativistic Lorentz-factor. The parameter $\gamma_t$ corresponds to the transition energy and is about 1.41 for the case of the ESR. From Eq. (1) it follows that measured frequencies of stored ions can be related to their mass-to-charge ratios only if one can diminish the second term on the right side. In the experiment this can be done either by cooling the ions and thus making $\Delta v/v$ small or by injecting ions at $\gamma$ equal to $\gamma_t$. Both possibilities are realized at the FRS-ESR. The first, Schottky Mass Spectrometry [7, 8, 9], is applied to nuclides with half-lives longer than a few seconds which are needed for electron cooling [10]. The second, Isochronous Mass Spectrometry (IMS)

[11, 12, 13], has somewhat lower resolving power but is suitable for investigating ions with half-lives down to about 20 $\mu$s.

## ISOCHRONOUS MASS SPECTROMETRY (IMS)

For the IMS, the ESR is tuned in a special ion-optical mode, in which the ring is operated at the transition point ($\gamma = \gamma_t$). In this mode, the revolution times of stored ions of one nuclear species become independent of their velocity spread. One can imagine that a faster ion travels on a longer orbit while a slower ion takes a shorter one. Typically, a spread of mass-to-charge ratios accepted by the ESR is about $\Delta(m/q)/(m/q) = 13\%$. However, the isochronicity condition is only fulfilled for a small range of this mass-to-charge spectrum. This is illustrated in Fig. 1.

The revolution time is measured with a dedicated time-of-flight (TOF) detector placed inside the ring [14]. The detector consists of a thin carbon foil (17 $\mu g/cm^2$ coated with CsI) which is penetrated by the circulating ions. Secondary electrons are released from the foil at each passage of each ion. These electrons are guided by crossed electric and magnetic fields to two micro-channel-plate (MCP) detectors. Signals from the MCPs are recorded and stored on disk with a fast sampling oscilloscope (Tektronix, 15 GHz input

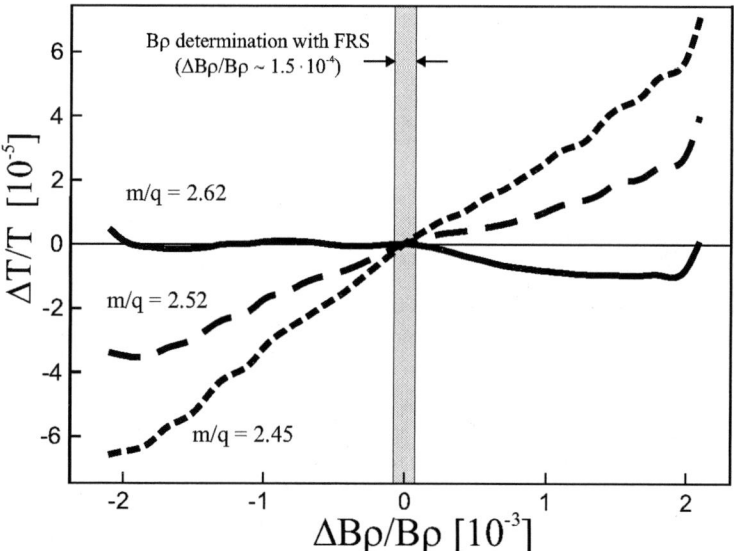

**FIGURE 1.** Isochronicity conditions for three mass-to-charge ratios. The FRS-ESR is tuned so that the primary beam $^{238}$U$^{91+}$ is isochronous (solid line). The widths of revolution time peaks for two other mass-to-charge ratios are calculated from the measured curve, assuming that particles with same magnetic rigidities have the same closed orbits. Higher resolving power is achieved if ions within a small range of magnetic rigidities are injected. The gray band illustrates the range of magnetic rigidities that was used in our last experiment. T stands for the revolution time, $B\rho$ is the magnetic rigidity, $\Delta T = (T_2 - T_1)/\overline{T}$, $\Delta B\rho = [(B\rho)_2 - (B\rho)_1]/\overline{(B\rho)}$.

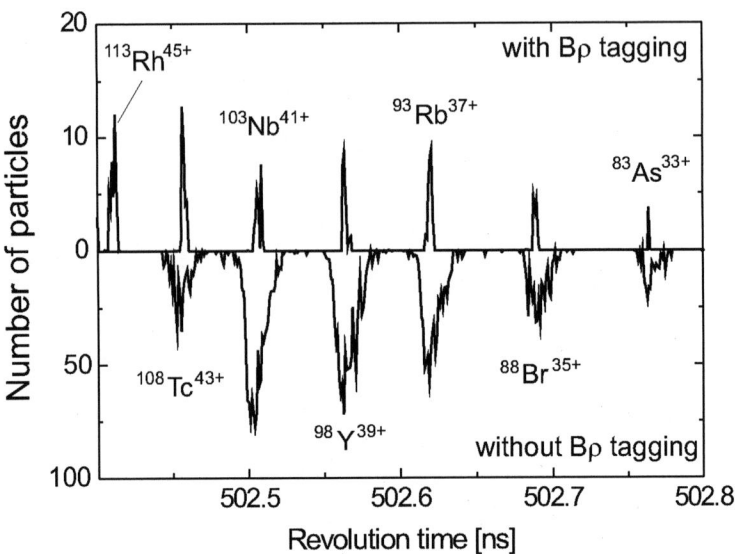

**FIGURE 2.** Lower part: TOF spectrum recorded without magnetic rigidity tagging [15]. Upper part: TOF spectrum from a very similar FRS-ESR ion-optical setting as shown in the lower panel but with magnetic rigidity tagging.

bandwidth, 40 GSamples/s on two channels) which is the prerequisite for precise time extraction.

Although, in our pioneering experiments only narrow parts of the accumulated TOF spectra with best isochronous conditions have been used, important new mass values have been measured, e.g. of relevance for the astrophysical rp-process of nucleosynthesis [11]. In a test experiment, followed by a production run, an approach to improve the IMS measurements has been proposed and successfully implemented. It has been suggested to measure the magnetic rigidity or the velocity of each stored ion [16]. By doing this, the reduction of the width of the corresponding TOF peak can be achieved also for ions further away from good isochronous conditions. In these first measurements we used the FRS as a high resolution magnetic spectrometer to collimate the beam at a dispersive focal plane. Therewith the spread of magnetic rigidities $\Delta B\rho/B\rho$ has been restricted to about $1.5 \cdot 10^{-4}$.

This $\Delta B\rho/B\rho$ range is schematically illustrated as a gray band in Fig. 1. One can see from the figure that the widths of the TOF peaks should reduce. Indeed, the obtained resolving power increased by a factor of more than two for the best isochronous particles and by at least a factor of five when moving away from good isochronicity. The experimental spectra are shown in Fig. 2. The achieved mass resolving power amounts to about 500000 and remains nearly constant over the whole TOF spectrum.

A new experiment has been performed with the goal to measure the masses of neutron-rich fission products of $^{238}$U-projectiles at the r-process path. The data analysis is underway. The high potential of the IMS has clearly been shown and the achieved

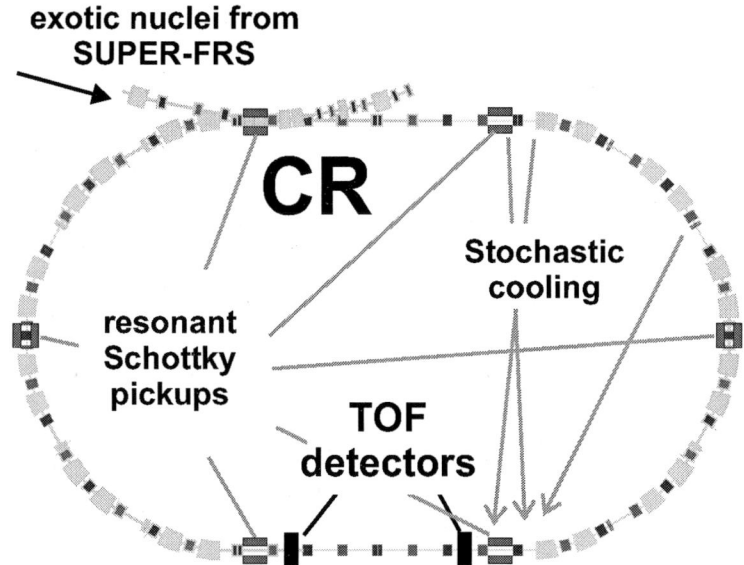

**FIGURE 3.** Layout of the collector ring CR of the future ring complex at FAIR [17]. The CR is specially designed to be operated in the isochronous mode for mass measurements of exotic nuclei provided by the new-generation fragment separator Super-FRS [19]. Two TOF detectors are foreseen for in-ring measurement of velocities of each stored ion. CR will also be equipped with stochastic pre-cooling and with a set of fast Schottky pick-ups for Schottky mass and lifetime spectrometry.

mass resolving power should significantly improve the identification of isomers.

## FUTURE PERSPECTIVES

The ILIMA project (Isomeric beams, LIfetimes, and MAsses) will continue the present program at the planned Facility for Antiproton and Ion Research (FAIR) [17]. The goals for FAIR will be on the one hand to increase the intensities to be able to go to the outskirts of the chart of nuclei which will also provide more information on the nucleosynthesis pathways. On the other hand it aims for improvements and new developments to increase the precision, the accuracy, and the sensitivity.

At FAIR, a system of dedicated storage rings [18] will replace the function of the present storage-cooler ring ESR. Fragments which will be produced and separated with the large-acceptance SUPERconducting FRagment Separator (Super-FRS) [19] will be collected in the Collector-Ring (CR) [20]. This storage ring is designed to be run in the isochronous mode and therefore will be a future tool for further mass measurements of very short-lived isotopes. The in-ring velocity measurement for each stored particle will be done by means of two TOF detectors placed on a straight section of the CR at a distance of about 34 m. The layout of the CR is shown in Fig. 3. In addition to two TOF detectors, the CR will be equipped with a set of fast Schottky pick-ups and stochastic cooling [21]. This setup will provide ideal conditions for half-life measurements, similar to the ones presently performed at FRS-ESR [22, 23, 24].

## REFERENCES

1. M. Pfützner, et al., Eur. Phys. J. A14 (2002) 279.
2. D. Lunney, J.M. Pearson, and C. Thibault, Rev. Mod. Phys. 75 (2003) 1021.
3. M. Bender, et al., Rev. Mod. Phys. 75 (2003) 121.
4. F. Bosch, et al., Int. J. Mass Spectrometry 251 (2006) 212.
5. H. Geissel, et al., Nucl. Instr. and Meth. B70 (1992) 286.
6. B. Franzke, Nucl. Instr. and Meth. B24-25 (1987) 18.
7. T. Radon, et al., Phys. Rev. Lett. 78 (1997) 4701.
8. T. Radon, et al., Nucl. Phys. A677 (2000) 75.
9. Yu. A. Litvinov, et al., Nucl. Phys. A756 (2005) 3.
10. M. Steck, et al., Nucl. Instr. and Meth. A532 (2004) 357.
11. J. Stadlmann, et al., Phys. Lett. B586 (2004) 27.
12. M. Hausmann, et al., Nucl. Instr. and Meth. A446 (2000) 569.
13. M. Hausmann, et al., Hyperfine Interactions 132 (2001) 291.
14. J. Trötscher, et al., Nucl. Instr. and Meth. B70 (1992) 455.
15. M. Matoš, Doctoral Thesis, University of Gießen, 2004.
16. H. Geissel, Yu. A. Litvinov, J. Phys. G: Nucl. Part. Phys 31 (2005) S1779-S1783.
17. An International Accelerator Facility for Beams of Ions and Antiprotons, Conceptual Design Report, GSI (2001), *http://www.gsi.de/GSI-Future/cdr/*.
18. P. Beller, et al., in Proc.: $9^{th}$ Eur. Part. Acc. Conf., Lucerne, Switzerland, 2004, 1174.
19. H. Geissel, et al., Nucl. Instr. and Meth. B204 (2003) 71.
20. A. Dolinskii, et al., Nucl. Instr. and Meth. A532 (2004) 483.
21. F. Nolden, et al., Nucl. Instr. and Meth. A532 (2004) 329.
22. H. Irnich, et al., Phys. Rev. Lett. 75 (1995) 4182.
23. Yu.A. Litvinov, et al., Phys. Lett. B573 (2003) 80.
24. T. Ohtsubo, et al., Phys. Rev. Lett. 95 (2005) 052501.

# Exotic cluster states in $^{12}$Be *via* $\alpha$-inelastic scattering

A. Saito*, S. Shimoura[†], T. Minemura**, Y. U. Matsuyama[‡], H. Baba**,
N. Aoi**, T. Gomi[§], H. Higurashi**, K. Ieki[‡], N. Imai[¶], N. Iwasa[‖],
H. Iwasaki*, S. Kanno[‡], S. Kubono[†], M. Kunibu[‡], S. Michimasa**,
T. Motobayashi**, T. Nakamura[††], H. Ryuto**, H. Sakurai**, M. Serata[‡],
E. Takeshita[‡], S. Takeuchi**, T. Teranishi[‡‡], K. Ue*, K. Yamada** and
Y. Yanagisawa**

*Department of Physics, University of Tokyo, 7-3-1 Hongo, Bunkyo, Tokyo 113-0033, Japan
[†]Center for Nuclear Study, University of Tokyo, RIKEN Campus, 2-1 Hirosawa, Wako, Saitama 351-0198, Japan
**RIKEN, 2-1 Hirosawa, Wako, Saitama 351-0198, Japan
[‡]Department of Physics, Rikkyo University, 3-34-1 Nishi-Ikebukuro, Toshima, Tokyo 175-8501, Japan
[§]National Institute of Radiological Science, 4-9-1 Anagawa, Inage, Chiba, Chiba 263-8555, Japan
[¶]Institute of Particle and Nuclear Studies, KEK, 1-1 Oho, Tsukuba, Ibaraki 305-0801, Japan
[‖]Department of Physics, Tohoku University, Aza-Aoba, Aramaki, Aoba, Sendai, Miyagi 980-8578, Japan
[††]Department of Physics, Tokyo Institute of Technology, 2-12-1 O-Okayama, Meguro, Tokyo 152-8551, Japan
[‡‡]Department of Physics, Kyushu University, 6-10-1 Hakozaki, Higashi, Fukuoka 812-8581, Japan

**Abstract.** Excited states in the neutron-rich nucleus $^{12}$Be were experimentally investigated *via* $\alpha$-inelastic scattering. The excited states in the $^{12}$Be nucleus were populated by a $^{12}$Be($\alpha,\alpha'$) reaction at 60 $A$ MeV in the inverse kinematics, and identified by measuring a $^6$He+$^6$He breakup channel in coincidence. The differential cross section and the angular correlations between the decay particles were obtained for each excitation energy from 10 MeV to 20 MeV, reconstructed by the measured momentum vectors of the two $^6$He's.

A multipole decomposition analysis based on the distorted-wave Born approximation was applied for the angular distribution of the inelastic scattering together with the angular correlation between the decay particles with respect to the directions of the incident beam and to the momentum transfer simultaneously. From the decomposed excitation energy spectra for $L$=0, 2, and 4, new excited states with $J^\pi$=0$^+$ and $J^\pi$=2$^+$ were identified, which were candidates of the band-head of a largely deformed rotational band. According to their similar energy spacings in $J$=0 and $J$=2 energy spectra, these excited states were possibly forming rotational bands with almost same large moments of inertia. The structures of the observed states are discussed with predictions of theoretical models.

**Keywords:** Cluster structure, $\alpha$-inelastic scattering, multipole decomposition analysis
**PACS:** 23.20.En Angular distribution and correlation measurements, 25.60.-t Reactions induced by unstable nuclei, 27.20.+n 6$\leq A \leq$19

# INTRODUCTION

Clustering aspects are important for understanding nuclear many-body systems as well as single-particle motions in nuclei, especially for light nuclei. The cluster structure is expected to appear at the vicinities of the respective threshold energies for the breakup into subunit nuclei. The feature of the cluster structures of so-called $4N$ nuclei is represented by the Ikeda diagram [1].

In the present study, highly excited states in $^{12}$Be above the $^{6}$He+$^{6}$He threshold were investigated *via* the $\alpha$-inelastic scattering on $^{12}$Be. The energy of the $^{12}$Be beam was chosen to $60\,A$ MeV which is the same as experiments for stable nuclei (e.g. Ref. [2]). The excitation energy of $^{12}$Be was deduced event-by-event using the invariant mass of the final state of $^{6}$He+$^{6}$He. Since the level density is rather high at this excitation energy region, the excitation energy spectrum of each multipole was decomposed by a multipole decomposition analysis (MDA) based on the distorted-wave Born approximation (DWBA) applying to the differential cross section of the inelastic scattering and the angular correlation between breakup products.

# EXPERIMENTAL PROCEDURE

The experiment was carried out at the RIKEN Accelerator Research Facility (RARF). A secondary $^{12}$Be beam was produced using the projectile fragmentation of an $^{18}$O beam at $100\,A$ MeV on a 1.48 g/cm$^2$ $^{9}$Be target. The $^{18}$O primary beam was supplied by an accelerator complex of the AVF Cyclotron and the RIKEN Ring Cyclotron. The intensity of the primary beam was 150 pnA. The secondary beam was selected using RIPS [3]. The energy of the $^{12}$Be beam was $59.9\,A$ MeV with an energy spread of $0.7\,A$ MeV at the center of the secondary target. A typical intensity of the $^{12}$Be beam was around $3.0\times10^4$ cps.

The measurement required in the present experiment were mainly divided into two parts; measurement of momentum vectors of the incident particle and outgoing parti-

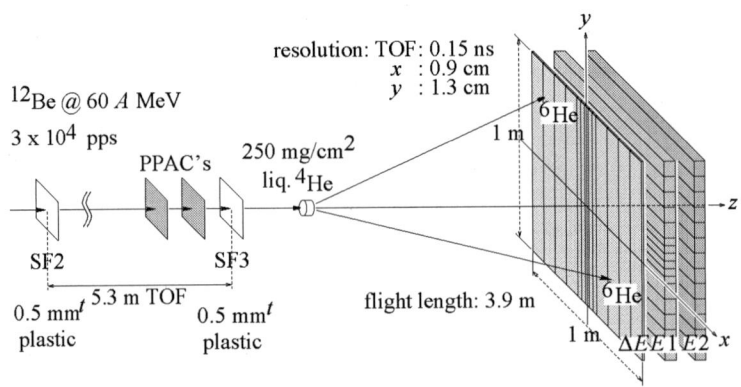

**FIGURE 1.** Schematic view of the experimental setup.

cles. Figure 1 shows a schematic view of the experimental setup. The $^{12}$Be beam was monitored by beam-line detectors consisting of two plastic scintillators and two Parallel Plate Avalanche Counters. The reaction products were measured by a hodoscope consisting of plastic scintillators. The detectors for both incident beams and scattered particles were installed in a vacuum chamber to reduce the reaction and multiple scattering in the materials along the beam line. For the measurement of $\gamma$ rays from bound states, an array of sixty-eight NaI(Tl) scintillators was also set in the air.

## DISTORTED-WAVE BORN APPROXIMATION ANALYSIS

The DWBA calculations for the $\alpha$-inelastic scattering were performed with external optical and transition potentials which are deduced by a single-folding model, in which the potentials are derived from the ground-state density. Monte Carlo simulations for acceptance of the present detector system were performed based on the calculated double-differential cross section. A multipole decomposition analysis (MDA) was applied to the experimental data.

The optical and transition potentials were obtained from the single-folding model with density dependent effective nucleon-$\alpha$ interactions in the calculation of the cross sections for $\alpha$-inelastic scattering. The optical potential $\mathscr{U}$ is given by

$$\mathscr{U}(r) = \int \mathscr{V}(|r-r'|,\rho_0(r'))\rho_0(r')dr', \quad (1)$$

where $\mathscr{V}(|r-r'|,\rho_0(r'))$ is the effective nucleon-$\alpha$ interaction, and $\rho_0(r')$ is the ground-state density. In the present DWBA calculation, a density-dependent effective nucleon-$\alpha$ interaction used in Ref. [4] is assumed. The interaction potential $\mathscr{V}$ is defined as

$$\mathscr{V}(|r-r'|,\rho_0) = -V\left(1+\beta_V\rho_0^{2/3}\right)e^{-|r-r'|^2/\alpha_V} - iW\left(1+\beta_W\rho_0^{2/3}\right)e^{-|r-r'|^2/\alpha_W}, \quad (2)$$

where $V$, $\alpha_V$, and $\beta_V$ ($W$, $\alpha_W$, and $\beta_W$) are the parameters characterizing the strengths, the ranges and the density-dependences for the real (imaginary) part, respectively. The parameters were taken from those obtained for $^{28}$Si [4] (Table 1), except for the depth of the imaginary part which was adjusted to reproduce the experimental data of the angular distribution of the $^{12}$Be($\alpha,\alpha'$)$^{12}$Be($2^+$) reaction [5].

**TABLE 1.** Parametrization of the nucleon-$\alpha$ interaction parameters.

| $V$ [MeV] | $\alpha_V$ [fm$^2$] | $\beta_V$ [fm$^2$] | $W$ [MeV] | $\alpha_W$ [fm$^2$] | $\beta_W$ [fm$^2$] |
|---|---|---|---|---|---|
| 38.0 | 3.7 | $-1.9$ | 14.5 | 3.7 | $-1.9$ |

For a state with multipolarity $L$ and excitation energy $E_x$, the radial form $\delta\mathscr{U}_L(r,E_x)$ of the transition potential are given by

$$\delta\mathscr{U}_L(r,E_x) = \int \delta\rho_L(r',E_x)\left[\mathscr{V} + \rho_0\frac{\partial\mathscr{V}}{\partial\rho_0}\right]dr', \quad (3)$$

where $\delta\rho_L(r', E_x)$ is the transition density for the considered state. The transition densities for collective excitations [6] are used in the present study.

The ground-state density was assumed to be a sum of the single particle densities calculated by the harmonic oscillator wave functions of occupied orbitals. For $^{12}$Be, the density distribution is the sum of four and eight nucleons in $s$- and $p$-orbitals, respectively.

$$\rho(r) \propto \left[1 + \frac{4}{3}\left(\frac{r}{a}\right)^2\right] \exp\left[-\left(\frac{r}{a}\right)^2\right]. \tag{4}$$

The parameter $a$ is chosen to be 1.76 fm to reproduce the matter distribution deduced from the measurement of the interaction cross section [7].

The double-differential cross section for each angular momentum transfer was obtained using the scattering amplitude for each magnetic substate, $\beta_{LM}(E_x; \Theta)$. In practice, the amplitude $\beta_{LM}$ is normalized so that the cross section for each multipole is normalized to be unity. The relation between the double-differential cross section for a transferred angular momentum $L$ and the amplitude $\beta_{LM}$ is given as the following,

$$\frac{d^2\sigma_L}{d\Omega^s d\Omega^d}(E_x; \Theta; \vartheta, \varphi) = \left|\sum_M \beta_{LM}(E_x; \Theta) Y_{LM}^*(\vartheta, \varphi)\right|^2, \tag{5}$$

where $d\Omega^s$ and $d\Omega^d$ are solid angles for $\Theta$ and for $(\vartheta, \varphi)$. One can obtain the differential cross section by integrating over the decay angles $\vartheta$ and $\varphi$. According to the diagonal properties of the spherical harmonics, The result becomes to be simplified as

$$\frac{d\sigma_L}{d\Omega^s}(E_x; \Theta) = \sum_M |\beta_{LM}(E_x; \Theta)|^2. \tag{6}$$

On the other hand, the angular correlation between the decay particles with respect to the beam direction calculated by integrating over the scattering angle $\Theta$. The angular distributions of the inelastic scattering and the one for the particle decay were included to the Monte Carlo simulation for detection efficiencies together with the effects of the angular resolutions and the acceptance of the detector system.

It is noted that the DWBA amplitudes $\beta_{LM}$'s for the inelastic scattering to a bound excited state are also used for the angular distribution of $\gamma$ rays. We tested the adequacy of the present calculation of the DWBA amplitudes by analyzing the observed $^{12}$Be$(\alpha,\alpha')^{12}$Be$(2_1^+)$ data in the same DWBA framework [5].

The multipole decomposition analysis (MDA) [8] was performed to obtain the excitation energy spectra for each transferred angular momentum. This method has been applied to the angular distributions in many studies for the giant resonances (e.g. Ref. [9]) and other reactions of stable nuclei (e.g. Ref. [10]). It was also applied to the angular distribution of decay particles from giant resonances excited by (e,e′) reaction for a fixed scattering angle of electron (e.g. Ref. [11]). In the present study, the technique of the MDA is extensively applied to all the angular distributions of the differential cross section, the angular correlation with respect to the direction of the projectile, and the one with respect to the transfered momentum simultaneously.

The experimentally obtained distributions have been fitted by means of a least square method with a coherently weighted sum of calculated transition amplitudes of the multipole components of $L = 0$, 2, and 4, weighted with fitting coefficients amplitude $\alpha_L(E_x)$, and phase $\Phi_L(E_x)$ as

$$\frac{d\sigma^\varepsilon}{d\Omega}(E_x;x) = \sum_L \alpha_L \frac{d\sigma_L^\varepsilon}{d\Omega}(E_x;x) + \sum_{L \neq L'} \sqrt{\alpha_L \alpha_{L'}} \mathrm{Re}\left(e^{-i[\Phi_L - \Phi_{L'}]} \frac{d\sigma_{LL'}^\varepsilon}{d\Omega}(E_x;x)\right), \quad (7)$$

where $d\sigma^\varepsilon/d\Omega$ is the distribution including the detector acceptance, namely,

$$\frac{d\sigma^\varepsilon}{d\Omega}(E_x;x) = \varepsilon(E_x;x) \frac{d\sigma}{d\Omega}(E_x;x), \quad (8)$$

where $x$ is $\Theta$ or $\vartheta$. The second term in Eq. 7 is the interference effects between multipoles with $L$ and $L'$. The amplitude $\alpha_L(E_x)$ and the phase $\Phi_L(E_x)$ were determined by minimizing $\chi^2$ value. The phase $\Psi_0(E_x)$ is fixed to be 0 because of the fact that only the relative phases contribute the observables. Since the calculated differential cross section for each multipole are normalized to be unity, the amplitude $\alpha_L(E_x)$ is identical to $d\sigma_L/dE_x(E_x)$. The expected maximum angular momentum is assumed to be $L = 4$ for the present case. Only even-spin states can be populated since the final state is two identical spin-0 particles.

## RESULTS AND DISCUSSIONS

Figure 2 shows examples of the results of the fits. The differential cross section of the inelastic scattering, the angular correlations of decay particle with respect to the beam direction and to the momentum transfer ($d\sigma^\varepsilon/d\Omega^s(E_x;\Theta)$, $d\sigma^\varepsilon/d\Omega_b^d(E_x;\vartheta_b)$, and $d\sigma^\varepsilon/d\Omega_q^d(E_x;\vartheta_q)$, respectively) are shown in Figs. 2 (a–c), (d–f), and (g–i), respectively. The top, middle and bottom frames in Fig. 2 ((1), (2), and (3)) are the results corresponding to $E_x = 10.8$ MeV, 11.4 MeV, and 12.4 MeV, respectively (Fig. 2). The open circles represent the experimental data. The fits were simultaneously performed to all the distributions of $d\sigma^\varepsilon/d\Omega^s$, $d\sigma^\varepsilon/d\Omega_b^d$, and $d\sigma^\varepsilon/d\Omega_q^d$. The thick curves are coherent sums of all the contributions. The solid curves correspond to contributions of $L=0$, 2, and 4. The thin ones are interferences of $0\otimes 2$, $0\otimes 4$, and $2\otimes 4$. Contributions with negligibly small magnitudes are not shown. The calculated angular correlations show good agreements with the data when the contributions of the interference between different multipoles are taken into account.

Figure 3 (a) shows the excitation energy spectrum of $^{12}$Be including the result of the MDA. The corresponding energies of 10.8 MeV, 11.4 MeV, and 12.4 MeV are indicated by (1), (2), and (3), respectively.

For the $E_x = 10.8$ MeV data, the $J = 0$ component has the largest contribution. The typical characteristics of $J = 0$ component are the strong forward-peaking structure in $d\sigma^\varepsilon/d\Omega^s$ and isotropic distributions in the angular correlations. On the other hand, the data for $E_x = 11.4$ MeV is dominated by the $J = 2$ contribution, which has a flat structure at forward angles in $d\sigma^\varepsilon/d\Omega^s$ and anisotropic behaviors in angular correlations, especially in $d\sigma^\varepsilon/d\Omega_q^d$. The anisotropic structure in Fig. 2 (h) is well fitted by a coherent sum of the $J = 0$ and $J = 2$ components. By comparing Figs. 2 (d–f) and (g–i), it is clear that the angular correlation with respect to the momentum transfer ($d\sigma^\varepsilon/d\Omega_q^d$, Fig. 2 (g–i)) is quite sensitive to the multipolarity. The data for $E_x = 12.4$ MeV include certain contributions of all the multipoles.

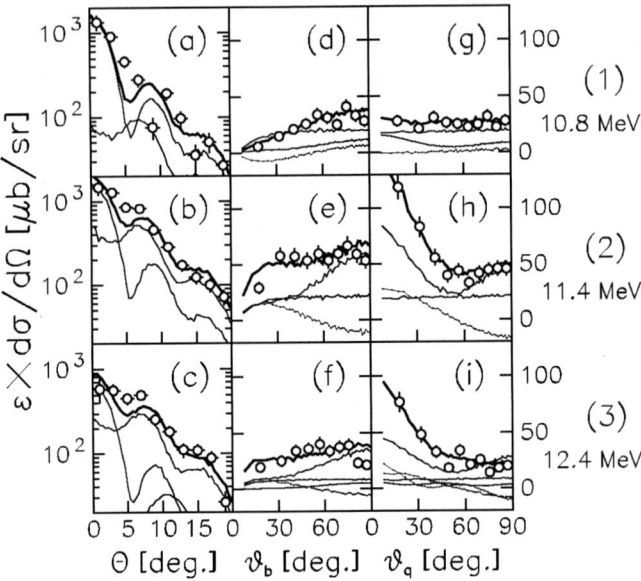

**FIGURE 2.** Examples of the angular distributions, (a–c) the differential cross section of the inelastic scattering, (d–f) the angular correlation between two $^6$He's with respect to the beam direction, (g–i) the one with respect to the momentum transfer. The open circles are the experimental data. The thick curves are the coherent sum of all the contributions. The solid curves correspond to contributions of $L=0$, 2, and 4. The thin ones are interferences of $0\otimes2$, $0\otimes4$, and $2\otimes4$. Contributions with negligibly small magnitudes are not shown.

Excitation energy spectra of $J = 0, 2$, and 4 are separately shown in Figs. 3 (b), (c), and (d), respectively. Several peaks are observed in the spectra of $J = 0$ and $J = 2$, whereas no distinctive peak is seen in the $J = 4$ spectrum. The energy spectra of $J = 0$ and $J = 2$ are fitted by Gaussians including experimental energy resolution, together with continuum components. The continuum component was calculated using the penetrabilities $P_L(E_d)$ and the density of states $\rho^{d.s.}(E_d)$, namely,

$$\frac{d\sigma_L^c(E_d)}{dE_d} \propto P_L(E_d)\rho^{d.s.}(E_d), \quad (9)$$

where $P_L$ and $\rho^{d.s.}(E_d)$ are the penetrability and the density of states, respectively. The density of states was assumed to have the form

$$\rho^{d.s.}(E_d) = \left(c_0 + c_1 E_d + c_2 E_d^2\right)^{-1}, \quad (10)$$

where $c_i$'s are free paramters in the fit.

The fit was performed simultaneously for the $J = 0$ and $J = 2$ spectra (Figs. 3 (b) and (c)). The results of the fit are indicated with thick curves. The individual components are shown with thin curves.

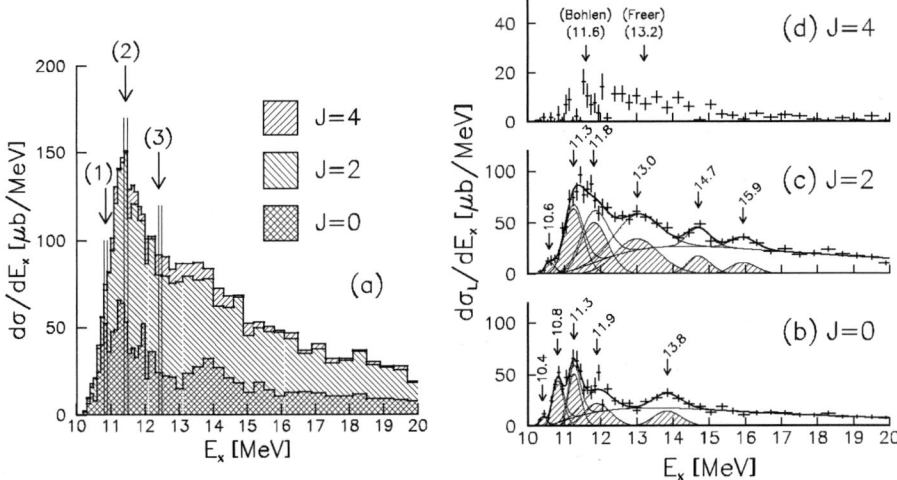

**FIGURE 3.** Excitation energy spectra of (a) total data, (b) $J = 0$, (c) $J = 2$, and (d) $J = 4$. The peaks indicated by arrows in (b) and (c) are newly observed ones in the present work. The levels indicated in (d) are taken from Refs. [12, 13].

The peaks indicated with arrows in Figs. 3 (b) and (c) are newly observed ones in the present work. The numerical values are the excitation energies in MeV. The arrows in Fig. 3 (d) are previously reported levels taken from Refs. [12, 13]. In Fig. 3, five new $J = 0$ peaks and five new $J = 2$ peaks are observed. In Fig. 3 (d), the excited states previously reported in Refs. [12, 13] are not clearly seen, possibly because of the low statistics due to a small cross section of the inelastic excitation.

The present and previously reported experimental results are compiled in Fig. 4. An interesting feature found in the comparison of the $J^\pi = 0^+$ and $2^+$ results (Figs. 3 (b) and (c)) is a similarity in energy spacings between the peaks. One may speculate these excited states are possibly classified under several groups as shown in the energy-spin systematics of $^{12}$Be (Fig. 4). The closed squares are the low-lying states. The black line is the ground-state band with $\hbar^2/2\mathscr{I} = 360$ keV. The open triangles are excited states observed in the three-neutron transfer reaction $^9$Be($^{15}$N,$^{12}$N)$^{12}$Be* [13] by assuming their spins according to the energy spacings. The dash-dotted line is a possible rotational band with $\hbar^2/2\mathscr{I} = 210$ keV. The plus symbols are the excited states observed in the $^{12}$Be→$^6$He+$^6$He reaction [12]. The closed circles are the new excited states observed in the present study. The solid line is the fit to the second lowest $0^+$ and $2^+$ states at 10.8 MeV and 11.3 MeV, respectively, which is observed in the present work, together with the $4^+$, $6^+$, and $8^+$ states at 13.2, 16.1, and 20.9 MeV, respectively, in Ref. [12]. The deduced moment of inertia correspond to $\hbar^2/2\mathscr{I} = 140$ keV. The moment of inertia is close to that of two $^6$He's touching each other ($\hbar^2/2\mathscr{I} = 165$ keV) with their radii given by $r = 1.5 \times A^{1/3}$. The dashed lines show possible rotational bands with the same moment of inertia as the solid one. As described above, the energy spacings in $J = 0$ and $J = 2$ are similar to each other. Several rotational bands with approximately same

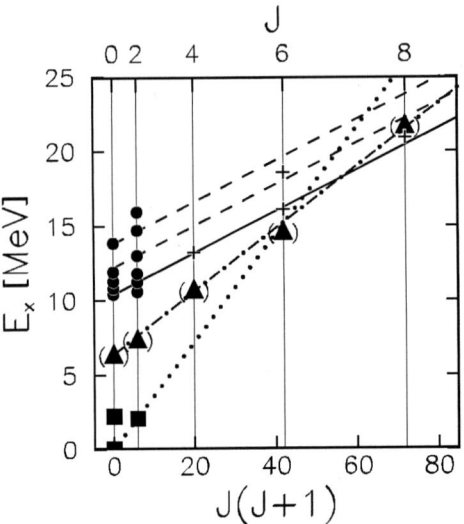

**FIGURE 4.** Energy-spin systematics of $^{12}$Be. The closed squares and the black line are low-lying states with even spins and the ground-state band. The closed circles are the present data. The plus symbols and the open triangles are the data taken from Refs. [12] and [13], respectively. The open triangles include assumptions of spins. The solid line is fitted to 10.8 MeV ($0^+$), 11.3 MeV ($2^+$), 13.2 MeV ($4^+$), 16.1 MeV ($6^+$), and 20.9 MeV ($8^+$). The dash-dotted line is fitted to the open triangles. The dashed lines correspond to possible rotational bands with $\hbar^2/2\mathscr{I} = 140$ keV, which is same as that of the solid line.

moment of inertia may exist as indicated by dashed lines in Fig. 4 which moment of inertia are the same as that for solid one. The rotational band of the 11.9 MeV $0^+$ and the 13.0 MeV $2^+$ state may be connected to the 18.6 MeV $6^+$ state with the above assumption.

The possible existence of several molecular bands is qualitatively consistent with the coupled-channels calculation [14]. The wave functions of the band members may consist of the contribution from the configurations of $^6$He($0^+$)+$^6$He($0^+$), $^6$He($0^+$)+$^6$He($2^+$), $\alpha+^8$He($0^+$), $\alpha+^8$He($2^+$), etc., with different spectroscopic amplitudes. The widths of the states depend on the spectroscopic factors. The developed $^6$He+$^6$He cluster structures are expected to appear with larger widths. The highly excited states were studied by the theoretical calculation based on the antisymmetrized molecular dynamics (AMD) and the reduced widths were calculated by Kanada-En'yo and Horiuchi [15]. The experimental widths are compared with the theoretical results of the AMD, in which low-spin members of the rotational band with developed cluster structures are predicted [15]. According to this calculation, the low-spin members of this rotational band, $0_3^+$ and $2_4^+$ states, lie at around 11 MeV and around 10 MeV, respectively. They predicted large reduced widths for the $^6$He- and $^8$He decay of the $0_3^+$ and the $2_4^+$ states. Using these theoretical values of the reduced widths, the decay widths for all the new excited states were estimated. The experimentally determined widths are comparable to or larger than the ones predicted by the AMD calculation. The newly observed excited states can be candidates of the predicted $0_3^+$ and $2_4^+$ states with novel molecular-orbital structure [15].

## SUMMARY

Cluster states in the neutron-rich nucleus $^{12}$Be were investigated *via* the $\alpha$-inelastic excitation. Several excited states with $0^+$ and $2^+$ above the $^6$He+$^6$He threshold were identified and were found to be candidates of rotational bands with developed cluster structures.

Highly excited states in $^{12}$Be above the $^6$He+$^6$He threshold were populated by the $\alpha$-inelastic scatterings. The inelastic excitations were identified by measuring the $^6$He+$^6$He breakup channel and their spectra was reconstructed by an invariant-mass method.

We extensively performed MDA based on the DWBA. The present MDA successfully reproduced both the angular distribution of the inelastic scattering and the angular correlation between the reaction products. We found clear resonant structures in the multipole-decomposed spectra.

A possibility of the existence of several rotational bands with approximately same large moment of inertia was suggested. The experimental data was compared with the theoretical calculation of AMD. The observed widths of the new excited states are comparable to or larger than the predicted ones by the AMD calculation. The new excited states may be candidates of the exotic cluster state in the AMD calculation with a novel molecular orbital structure.

## ACKNOWLEDGMENTS

The authors would like to thank RIKEN Ring Cyclotron crews for producing a nice beam during the experiment. The present study is partially supported by the Grant-In-Aid for Scientific Research of the Japan Ministry of Education, Science and Cluture under the program numbers (B) 08454069 and 11440081.

## REFERENCES

1. K. Ikeda, et al., *Prog. Theor. Phys.* **Extra Number**, 464 (1968).
2. D. H. Youngblood, et al., *Phys. Rev.* **C55**, 2811 (1997).
3. T. Kubo, et al., *Nucl. Inst. Meth. in Phys. Res.* **B70**, 309 (1992).
4. A. Kolomiets, et al., *Phys. Rev.* **C61**, 034312 (2000).
5. S. Shimoura, *Nucl. Phys.* **A738**, 162 (2004).
6. G. R. Satchler, *Nucl. Phys.* **A472**, 215 (1987).
7. I. Tanihata, et al., *Phys. Lett.* **B206**, 592 (1988).
8. M. A. Moinester, *Can. J. Phys.* **65**, 660 (1987).
9. M. Itoh, et al., *Phys. Rev.* **C68**, 064602 (2003).
10. T. Kawabata, et al., *Phys. Rev.* **C70**, 034318 (2004).
11. M. Kohl, et al., *Phys. Rev.* **C57**, 3167 (1998).
12. M. Freer, et al., *Phys. Rev. Lett.* **82**, 1383 (1999).
13. H. G. Bohlen, et al., *Proc. Int. Symp. on Exotic Nuclei 2001 (EXON2001)* p. 453 (2002).
14. M. Ito, and Y. Sakuragi, *Phys. Rev.* **C62**, 064310 (2000).
15. Y. Kanada-En'yo, and H. Horiuchi, *Phys. Rev.* **C68**, 014319 (2003).

# Excitation and Charged Particle Decay of Dipole Resonance Analogs in the $\alpha$ Clusters of $^6$Li and $^7$Li

T. Yamagata*, H. Akimune*, S. Nakayama†, M. Fujiwara**, K. Fushimi†,
M.B. Greenfield‡, K. Hara**, K.Y. Hara*, K. Hashimoto**, K. Ichihara†,
H. Ikemizu*, K. Kawase**, M. Kinoshita*, Y. Matsui†, K. Nakanishi**,
M. Ohta*, M. Sakama§, M. Tanaka¶, H. Utsunomiya*, N. Warashina* and
M. Yosoi**

*Department of Physics, Konan University, Kobe 658-8501, Japan
†Department of Physics, University of Tokushima, Tokushima 770-8502, Japan
**Research Center for Nuclear Physics, Osaka University, Osaka 567-0047, Japan
‡Department of Physics, International Christian University, Tokyo 181-8585, Japan
§School of Health Science, University of Tokushima, Tokushima 770-8509, Japan
¶Kobe Tokiwa College, Kobe 654-0838, Japan

**Abstract.** We investigated charged-particle decay from the high-lying dipole resonances at $E_x$=24 MeV in $^6$He and at 18 MeV in $^7$He via the $^{6,7}$Li($^7$Li,$^7$Be $x$) reactions at 455 MeV and at 0°. $d$- and $t$-decay, and $t$-decay was observed as the dominant decay modes for these resonances in $^6$He and $^7$He, respectively. The yield ratios observed for charged particle decay were qualitatively consistent with those expected from the assumption that the resonances are the analogs of the dipole resonances in the $\alpha$ cluster in $^6$Li and $^7$Li, but could not be explained by simple phase space calculations.

## INTRODUCTION

Clusters in nuclei play an important role in nuclear structure and nuclear reactions. Since clusters in nuclear systems are weakly bound, and spatially localized subsystems composed of strongly correlated nucleons [1], we can expect two types of excitations in the clustering nuclei. One is the excitation due to an inter-cluster relative motion. A typical example has been observed as the rotational excitation of a clustering nucleus [2]. The other is the excitation due to an intrinsic excitation of the cluster itself. Its existence has been suggested in recent years [3]. Such a cluster excitation is very interesting as a new concept of nuclear excitation.

$^6$Li and $^7$Li are the well established nuclei such as their ground states have the $\alpha$ clustering structure. Here, the dipole resonance (DR) consisting of the isovector giant dipole resonance (GDR) and spin dipole resonance (SDR) is the most significant excitation mode in a free $\alpha$ particle ($^4$He). So we searched for the excitation of the $\alpha$ cluster in $^6$Li and $^7$Li by observing the DR's.

The first evidence for the excitation of the $\alpha$ clusters in $^6$Li and $^7$Li was obtained in the $^{6,7}$Li($^7$Li,$^7$Be) reactions in which the high-lying DR's were observed in $^6$He and $^7$He at excitation energies of $E_x$~24 MeV and 18 MeV, respectively [4]. Similar DR's were

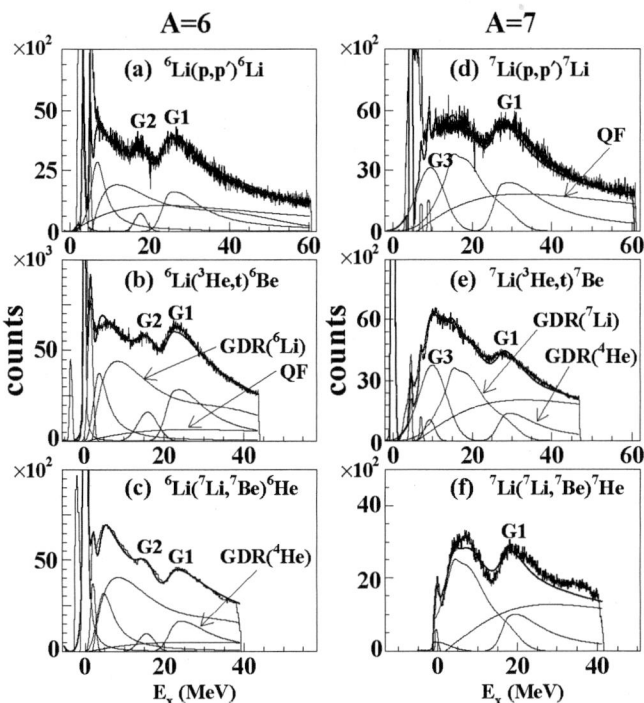

**FIGURE 1.** The spectra for the reactions of (p,p') at $\theta_L=8°$, ($^3$He,t) at 2.7°, and ($^7$Li,$^7$Be) at 0°. The resonances denoted by G1's have been proposed being the dipole resonances in the $\alpha$ clusters of $^{6,7}$Li in Ref. [5]. Thin solid lines show the peak fitting results.

also observed in $^{6,7}$Li via the (p,p') reactions and in $^{6,7}$Be via the ($^3$He,t) reactions with nearly the same reaction Q-values as the DR's in $^6$He and $^7$He [5]. Figure 1 shows the resonances (denoted by G1's) observed in these reactions. These resonances are located much higher excitation energies than the excitation energies for the GDR in $^6$Li ($E_x=12$ MeV) and $^7$Li ($E_x=17$ MeV) [6].

We had compared the excitation energies, widths, excitation cross sections for the dipole resonances (DR's) with those for the DR in $^4$He observed in the (p,p') reaction [7] and for the GDR observed in the ($\gamma$,n) reactions [6, 8]. Based on these results we had suggested that the resonances in $^{6,7}$Li were consistent with the DR in the $\alpha$ clusters of $^{6,7}$Li, and that the resonances in $^{6,7}$He and $^{6,7}$Be were consistent with the analogs of the DR's [4, 5, 7]. Since the idea for the DR excitation of the $\alpha$ clusters is not yet well established, it is very important to confirm this interpretation from different viewpoints.

In the present work, we investigated charged particle decay from the high-lying DR's in $^6$He and $^7$He at $E_x\sim 24$ and 18 MeV, respectively. Figure 2 shows the threshold energies for the particle decay channels in $^6$He and $^7$He. The DR's are located at higher excitation energies than the thresholds for the two or three particle emission. We supposed that the DR's can be excited via the charge exchange reactions on the $\alpha$

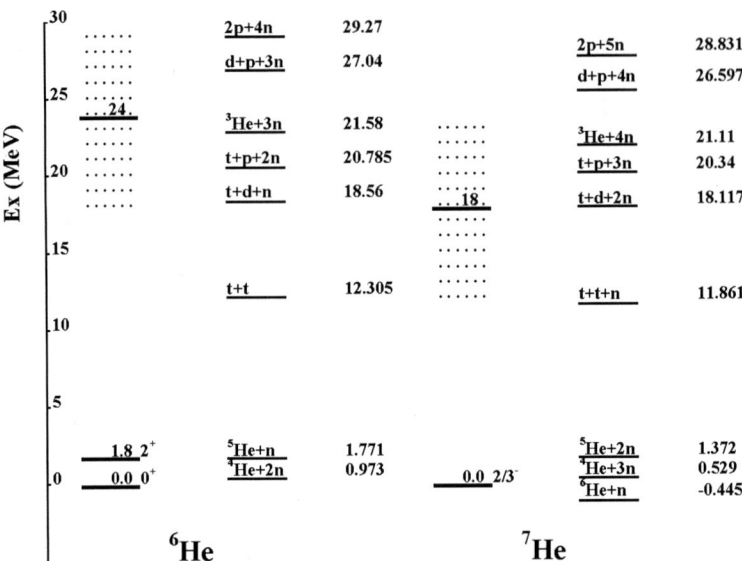

**FIGURE 2.** The threshold energies for the decay channels in $^6$He and $^7$He. The location and widths for the resonances of the $\alpha$ cluster excitation in $^{6,7}$He are indicated.

clusters in $^6$Li and $^7$Li leaving the $d$ cluster in $^6$Li and $t$ cluster in $^7$Li as spectators for the reactions [3]. The excitation of $\alpha$ clusters forms $^4$H* in the ($^7$Li,$^7$Be) reactions. $^4$H* is the resonant system of $n+t$. Therefore, $^6$He and $^7$He nuclei involving the DR's of the $\alpha$ cluster may have components of $d+n+t$, and $t+n+t$, respectively. Thus we expect, if the DR's are due to the DR analogs in the $\alpha$ clusters of $^6$Li and $^7$Li, the dominant charged decay-products would be $d$ and $t$ in the $^6$Li($^7$Li,$^7$Be) reaction, and $t$ in the $^7$Li($^7$Li,$^7$Be) reaction. If, for these relevant decay modes, the coincidence yields measured in the experiment are much larger than the statistical calculation, it should be strong evidence for that the DR's are due to the excitation of the $\alpha$ cluster.

## EXPERIMENT

The experiment was carried out by using the 455-MeV $^7$Li beams from the ring cyclotron at the Research Center for Nuclear Physics (RCNP), Osaka University. Experimental setup was very similar to the previous work [9]. Targets used were self-supporting metallic foils of enriched $^6$Li (95.4%) and $^7$Li (99.5%) with thicknesses of 0.6 and 0.7 mg/cm$^2$, respectively. The targets were tilted by 45° with respect to the beam direction to minimize the energy loss of the decay particles in the target.

$^7$Be emitted from the reactions were analyzed by using the magnetic spectrograph "Grand Raiden" located at $\theta_L = 0°$, and were detected with the focal plane detector system consisting of two multi-wire drift chambers backed by a $\Delta E$-$E$ plastic scintillator telescope [10]. The aperture of the entrance slits of the spectrograph was ±20 mr

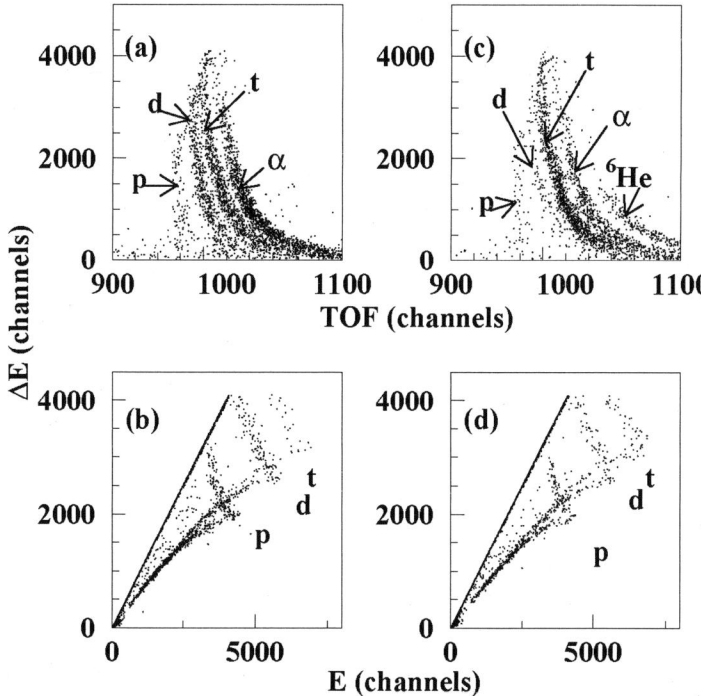

**FIGURE 3.** The two-dimensional scatter-plots of the coincidence events for identification of decay particles. The upper and lower figures show scatter-plots for $\Delta E$ vs. TOF, and $\Delta E$ vs. $\Delta E+E$, respectively, for the $^6$Li($^7$Li,$^7$Be $x$) reaction (a) and (b), for the $^7$Li($^7$Li,$^7$Be $x$) reaction (c) and (d).

horizontally and $\pm 30$ mr vertically. After passing through the target, the beams were stopped with a Faraday cup located inside the spectrograph. Typical values of energy resolution were 800 keV, which was mainly due to the energy spread of the incident beam and the $^7$Be-particle excitation ($E_x$=0.43 MeV). Since the momentum acceptance of the spectrograph is 5%, the reaction spectra for an excitation-energy region from 5 to 43 MeV was measured in the coincidence experiment.

Charged particles emitted from $^{6,7}$He excited via the ($^7$Li,$^7$Be) reactions were detected by using eight Si detectors (SSD's) telescopes, each of them consisting of a 500-$\mu$m $\Delta E$ and 300-$\mu$m $E$ detectors with an aperture of 3.8 cm$^2$. The SSD's were positioned at the angles between $\theta_L$= 90° and $\theta_L$= 160° at intervals of 10° and all at a distance of 30 cm from the target. A time of flight (TOF) technique was utilized for identification of the low energy particles which stopped in the $\Delta E$ detectors. Figures 3 (a) and (c) show the typical two-dimensional scatter-plots of TOF vs. $\Delta E$. Particle identification for the particles penetrating the $\Delta E$ detectors were done with the $\Delta E$-$E$ method. Figures 3 (b) and (d) show two-dimensional scatter plot of $\Delta E$ vs. $\Delta E+E$. It is clearly seen that each particle was well separated in both methods.

Since the threshold energy due to the noise discrimination level of the detectors could

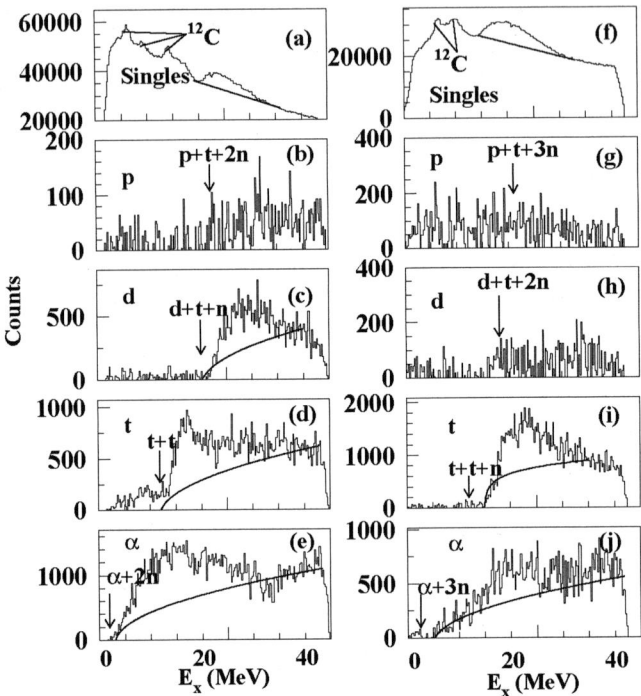

**FIGURE 4.** The left figures show the $^6$Li($^7$Li,$^7$Be) spectra for (a) singles, coincidence with (b) $p$, (c) $d$, (d) $t$, and (e) $\alpha$ particles. The right figures show the $^7$Li($^7$Li,$^7$Be) spectra for (f) singles, coincidence with (g) $p$, (h) $d$, (i) $t$, and (j) $\alpha$ particles. The arrows indicate the threshold energies for each particle emission. The coincidence events in every SSD's have been summed and accidental coincidence events have been subtracted. Contaminant peaks due to $^{12}$C are indicated in singles spectra. The solid lines show assumed continuum shapes.

be set as low as possible in the present setup, e.g., about 0.1 MeV, the missing number of low energy particles due to the detector discrimination level remained small, at most 3%. A small contribution from the $^{12}$C contamination was recognized in the singles spectra. The amount of the $^{12}$C contribution was checked by measuring the spectra with a $^{12}$C target and was confirmed to be negligibly small in the coincidence spectra.

## RESULTS

Figures 4(a) and 4(f) show singles spectra for the ($^7$Li,$^7$Be) reactions on $^6$Li and $^7$Li, respectively. The spectra observed in the present study are very similar to those observed in the previous work [4, 5, 9, 11]. The resonances observed at $E_x \sim$ 24 MeV in $^6$He and at 18 MeV in $^7$He have been assigned to the analog of the $\alpha$-cluster excitation in $^6$Li and $^7$Li, respectively [4, 5].

Figures 4(b)-4(e) and 4(g)-4(j) show the coincidence spectra. Noticeable events are

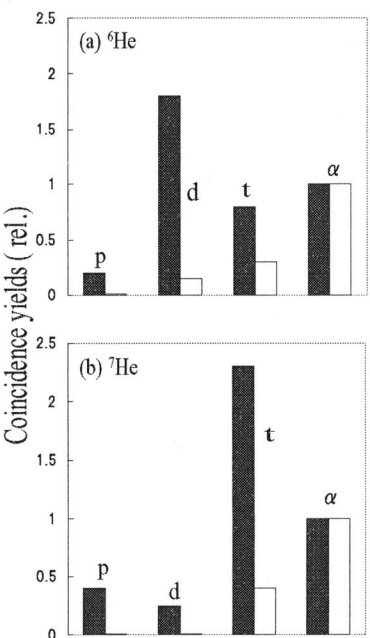

**FIGURE 5.** Preliminary results of the decay yields from the DR's in the $\alpha$ clusters of $^6$He (a) and $^7$He (b). The closed and open rectangles show the experimental yields and phase space calculations, respectively by normalizing the $\alpha$ decay yields to unity.

seen in the coincidence with $x=d$, $t$, and $\alpha$ in the $^6$Li($^7$Li,$^7$Be $x$) spectra, while coincidence events are very small for $p$. A peak seen in the coincidence spectrum for the $^6$Li($^7$Li,$^7$Be $t$) reaction at $E_x$=18 MeV is due to binary $t$-decay from the $t+t$ resonance in $^6$He [9, 11]. On the other hand, in the $^7$Li($^7$Li,$^7$Be $x$) spectra coincidence events with $x=t$ and $\alpha$ are seen to be large, and those with $p$ and $d$ are very small.

In order to deduce the yield ratios for the decay channels, we should separate the resonance contributions from the underlying continua. Since the shapes for the continua are not well known, we simply assume these shapes as smooth curves, as shown by the solid lines in Fig. 4. Excess yields from the assumed continua integrated in the region of the DR's of the $\alpha$ clusters are shown in Fig. 5, where decay yields are renormalized by assuming the yields for $\alpha$ decay to be unity. It is noted that since we have detected only one decay particle, the relative yields in Fig. 5 are not exclusive, and are not corrected for the decay multiplicities.

To understand the observed features of decay modes of the resonances, we compared the measured decay yields to the calculated ones. We calculated the decay yields with the assumption that the yields were in proportion to the volumes of the phase space for the decay particles at the final reaction channels. The detail of this calculations will be published in a forthcoming paper. The results of the calculations normalizing the $\alpha$ decay yield are shown in Fig. 5.

## DISCUSSION AND CONCLUSION

We measured charged particle decay from the DR's at $E_x$=24 MeV and 17 MeV in $^6$He and $^7$He via the ($^7$Li,$^7$Be) reactions. Since the lowest threshold energies correspond to decay into $\alpha+n$'s, the largest decay yields for the charged particles are expected to be $\alpha$ decay in the calculations for both $^6$He and $^7$He. On the other hands in the measured results, the $d$ and $t$ decay modes show the largest coincidence yields for $^6$He and $^7$He, respectively. In $^6$He, $t$ decay is also observed to be large. It is noted that the measured yields for $p$ decay in $^6$He and $p$ and $d$ decay in $^7$He are not distinguished from the contribution of the continua. The observed coincidence yield ratios are not consistent with the calculations.

The specific feature of the observed yields are qualitatively understood as follows: The ground states of $^6$Li and $^7$Li have the cluster structures of $d+\alpha$ and $t+\alpha$, respectively. The DR's observed in $^6$He and in $^7$He at $E_x$=24 MeV and 17 MeV, respectively are excited via the reaction process in which the $\alpha$ clusters themselves are excited as $\alpha \rightarrow {}^4\text{H}^*$ in both $^6$Li and $^7$Li. On the other hand, the $d$ and $t$ clusters in $^6$Li and $^7$Li, respectively remain as the spectators during the reaction process [3]. Since the produced $^4\text{H}^*$ is the resonant system of $t+n$, the DR's would have the components of $d+t+n$ and $t+t+n$ for $^6$He and $^7$He, respectively. $d$ and $t$ from $^6$He and $t$ from $^7$He can be emitted as the dominant decay products because the DR's are located above the thresholds. This interpretation is very consistent with the idea of the $\alpha$ cluster excitation. Therefore, the present experimental results are a strong evidence for the DR's in $^6$He and $^7$He being the $\alpha$ cluster excitations.

## ACKNOWLEDGMENTS

These experiments were performed at the Research Center for Nuclear Physics, Osaka University under the Program Nos. E184 and E190. The authors are grateful to the staff of the RCNP cyclotron for their support. This work was supported partly by the Grant-in-Aid (No. 16540257) by the Japan Ministry of Education, Culture, Sports, Science, and Technology.

## REFERENCES

1. K. Ikeda, J. Phys. Soc. Jpn. **58**, 277 (1989) Suppl.
2. H. Horiuchi, J. Phys. Soc. Jpn. **58**, 7 (1989) Suppl.
3. S. Nakayama *et al.*, Prog. Theor. Phys. Suppl. **146**, 603 (2002).
4. S. Nakayama *et al.*, Phys. Rev. Lett. **87**, 122502 (2001).
5. T. Yamagata *et al.*, Phys. Rev. C **69**, 044313 (2004).
6. B.L. Berman and S.C. Fultz, Rev. Mod. Phys. **47**, 713 (1975).
7. T. Yamagata *et al.*, Phys. Rev. C **74**, 014309 (2006).

8. D.V. Webb *et al.*, in Proc. Int. Conf. on Photonuclear Reactions and Applications. Asilomar, ed. B.L. Berman, Lawrence Livermore Laboratory, vol. 1, p. 149 (1973).
9. T. Yamagata *et al.*, Phys. Rev. **C 71**, 064316 (2005).
10. M. Fujiwara *et al.*, Nucl. Instrum. Methods Phys. Res., **A 422**, 484 (1999).
11. H. Akimune *et al.*, Phys. Rev. **C 67**, 051302 (2003).

# t+t clustering states in He-isotopes

S. Aoyama* and K. Arai[†]

*Integrated Information Processing Center, Niigata University, Niigata 950-2181, Japan
aoyama@cc.niigata-u.ac.jp
[†] Division of General Education, Nagaoka National College of Technology, Niigata 940-8532, Japan

**Abstract.** $t+t$ clustering states in He isotopes are investigated by using two theoretical approaches. We obtained many t+t resonances near to the t+t threshold with a RGM type approach. Further, a role of the $t+t$ cluster component in the ground state is examined with AMD triple-S, allowing the wider configuration space containing simultaneously the "$t+t$+valence neutrons" structure and "$^4$He+valence neutrons" structure. We understand the importance of the t+t component even for the ground state.

**Keywords:** cluster; neutron-rich nuclei; resonance
**PACS:** 21.30.Fe,21.60.Cs,27.20.+n

## INTRODUCTION

Neutron rich He-isotopes have been taken much attentions because of their exotic structures such as the neutron halo which is observed in $^6$He [1]. Various theoretical and experimental studies have proved that the ground and low-lying excited states have "$^4$He+valence neutrons" structures. For excited states, we have a question whether they are the core+valence neutron state or the t+t clustering state. Very recently, Akimune et al. discovered such the excited t+t states in $^6$He [2]. We studied the t+t excited resonance with an RGM type approach [3].

It is believed that the ground states of He-isotopes are essentially described by the $^4$He+valence neutrons model. But even in the most simple case as $^6$He, we can not reproduce the binding energy by the $^4$He+$n$+$n$ model. And several group suggested that the $t+t$ channel gives a certain contribution for reproducing the binding energy of the $^6$He ground state[4, 5]. However, it has not been solved whether $^6$He is a special case in the He-isotopes because of difficulties of the numerical calculation for many-body systems. Recently, we have developed a new method which is called AMD triple-S. And we analyzed the t+t clustering effect [6]. In this report, we will review the essential points of these recent studies [3, 6], and the detail of model and method will not be given because they are given in our previous publications [3, 6] and references in there.

## RESULTS

First, $t+t$ elastic scattering phase shifts for negative parity states in $^6$He with four different interactions are shown in FIGURE 1. In this calculation, we employed the two-body microscopic cluster model according to the RGM [3]. In this study, we used four different effective $N$-$N$ interactions. First interaction is the Minnesota potential ($u$=0.98)[7] with the spin-orbit term of Reichstein and Y. C. Tang [8] and the tensor term of Heiss and Hackenbroich[9]. This Minnesota pot. with the tensor term was employed by Csótó for the studies of the light nuclei[10]. Second interaction is by Csótó and Lovas[11] which was used to calculate the ground state of $^6$Li. Third is by Furutani et al.[12] and fourth is by Mertelmeier and Hofmann[13]. These are denoted by MN, CL, FU, and MH in FIGURE 1, respectively.

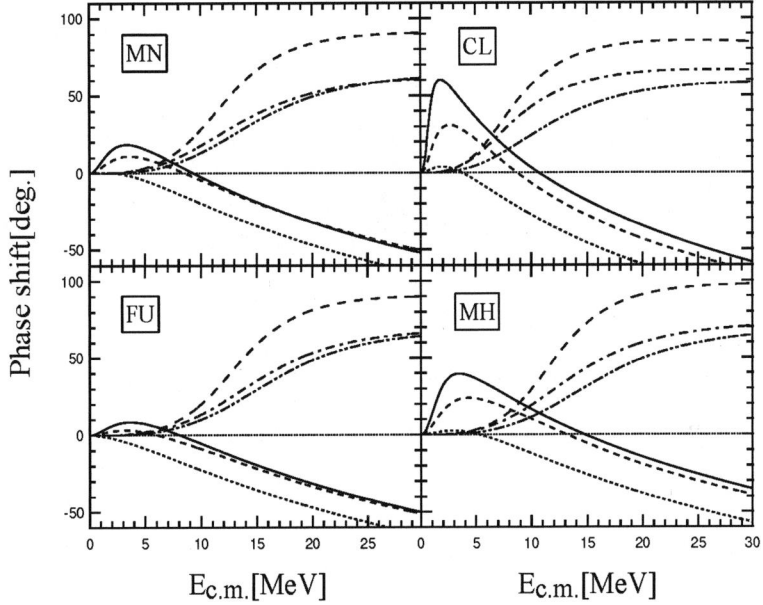

**FIGURE 1.** $t$+$t$ elastic scattering phase shifts for negative parity states in $^6$He.

As shown in FIGURE 1, the negative parity ($P$- and $F$-wave) phase shifts suggest an existence of several broad resonances. The order of the phase shifts in both waves is not sensitive at all to the choice of the effective $N$-$N$ interaction. In $P$-wave states, every interaction indicates very broad resonances just above the threshold. The $0^-$ state shows most prominent peak among the $P$-wave phase shifts and the $2^-$ resonance could lie close to the $0^-$ resonance but with much broader width. As for the $1^-$ states, $t+t$ relative interaction is much less attractive than in other two states and a peak of the calculated phase shifts does not reach more than 5 degree. Therefore it is difficult to conclude the existence of the $1^-$ resonance right above the $t+t$ threshold in present calculation. Note that this order ($0^-$, $2^-$, and $1^-$) of the $P$-wave splitting is decisively attributed to the tensor force between the two clusters as is seen in order of the $P$-wave low energy phase shifts of $N+N$[4, 10]. Without the tensor force, the order of the $P$-wave splitting is naturally $2^-$, $1^-$, and $0^-$ due to the spin-orbit force but the tensor force gives more attractive effect in the $0^-$ state and more repulsive effect in the $1^-$ state.

Next, in FIGURE 2, the calculated binding energies of He-isotopes are shown. The solid line indicates the binding energies calculated with the coupled channel model, in which the {$t+t$+valence neutrons}-channel is coupled to the {$^4$He+valence neutrons}-channel. On the other hand, the dashed line represents the binding energies calculated with the single {$^4$He+valence neutrons} channel. Here, we used a method which is called AMD triple-S. Since the detail are given in our previous papers and references in there [14, 15], we skip the explanation. As a nucleon-nucleon interaction, the Volkov No.2 potential (M=H=0.125) [16] is used for the central-force part of the $N$-$N$ interaction, and the G3RS potential is employed for the spin-orbit part with $V_0 = 2000$ MeV [17]. This choice of the interactions is the same as in our previous studies [14, 15].

The difference between the two lines comes from the {$t+t$ +valence neutrons} components. It is clearly seen that the binding energies are properly lowered by including the t+t channel. For $^6$He, Csótó [4] and Arai et al. [5] pointed out it. We also confirm it in heavier He-isotopes. As for $^8$He, the energy gain $\Delta E$ is 0.64 MeV. By comparing the value for $^6$He, $\Delta E$ is 0.35 MeV, the two extra neutrons from $^6$He enhance the $t+t$ clustering effect. As for $^{10}$He, you can think that the presence of many valence neutrons (six neutrons) may enhance further the $t+t$ clustering effect (i.e., on $^6$He and $^8$He), but at the same time the p-shell of neutrons is completely filled in this nucleus. This means that the core of the nucleus may favor the spherical shape. As seen from FIGURE 2, the calculated energy gain for $^{10}$He is obtained as $\Delta E$=0.42 MeV. This result suggests that the distortion effect is still present though it seems that the shell closure effect stops the increasing trend of the energy gain $\Delta E$ at $^{10}$He.

## SUMMARY

In summary, we reported several results which is related with t+t clustering. One is the excited t+t resonance around the t+t threshold with the RGM type approach [3]. Another is the importance of the t+t clustering effect in the ground states of He-isotopes by using AMD triple-S [6]. We concluded that the t+t clustering is very important for understanding for the He-isotopes.

## ACKNOWLEDGMENTS

This work was supported in part by Grant-in-Aid for Scientific Research (Grant No 17740137) from the Ministry of Education, Science and Culture, and also by the grant

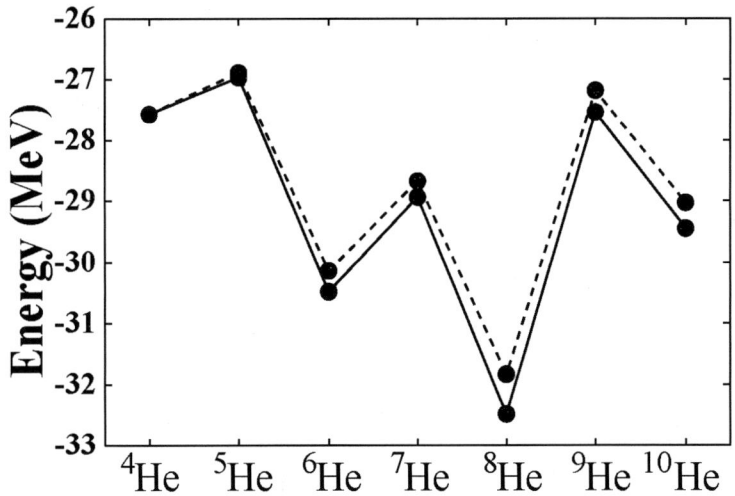

**FIGURE 2.** The calculated binding energies of the He-isotopes for the single channel case and the coupled channel case.

for the international collaboration between Japan and Belgium ( JSPS and FNRS).

## REFERENCES

1. I. Tanihata, J. Phys. **G22**, 157 (1996).
2. H. Akimune *et al.*, Phys. Rev. C **67**, 051302 (2003); S. Nakayama *et al.,ibid* **69**, 041304 (2004); T. Yamagata *et al.,ibid.* **71**, 064316 (2005).
3. K. Arai, K. Kato and S. Aoyama, Phys. RevC**74**, 034305(2006).
4. A. Csótó, Phys. Rev. **C48**, 165 (1993).
5. K. Arai, Y. Suzuki and R.G. Lovas, Phys. Rev. **C59**, 1432(1999).
6. S. Aoyama, N. Itagaki and M. Oi, Phys. RevC**74**, 017307(2006).
7. D. R. Thompson, M. LeMere, and Y. C. Tang, Nucl. Phys. **A286**, 53 (1977).
8. I. Reichstein and Y. C. Tang, Nucl. Phys. **A158**, 529 (1970).
9. P. Heiss and H. H. Hackenbroich, Phys. Lett. **30B** 373 (1969).
10. A. Csótó, R. G. Lovas and A. T. Kruppa, Phys. Rev. Lett. **70**, 1389 (1993); A. Csótó, H. Oberhummer, and R. Pichler, Phys. Rev. C **53**, 2366 (1996). A. Csótó and G. M. Hale, Phys. Rev. C **55**, 2366 (1997).
11. A. Csótó and R. G. Lovas, Phys. Rev. C **46**, 576 (1992).
12. H. Furutani, H. Horiuchi, and R. Tamagaki, Prog. Theor. Phys. **62**, 981 (1979).
13. T. Mertelmeier and H. M. Hofmann, Nucl. Phys. **A459**, 387 (1986).
14. N. Itagaki, A. Kobayakawa and S. Aoyama, Phys. Rev. **C68**, 054302 (2003).
15. N. Itagaki, S. Aoyama, S. Okabe, and K. Ikeda, Phys. Rev. **C70**, 054307 (2004).
16. A.B. Volkov, Nucl. Phys. **74**, 33 (1965).
17. R. Tamagaki, Prog. Theor. Phys. **39**, 91 (1968).

# Spectroscopy of $^9$He, Quasi-free Scattering $^6$He+$^4$He

R. Wolski,[1,2] M.S. Golovkov,[1] L.V. Grigorenko,[1] A.S. Fomichev[1]
A.V. Gorshkov,[1] S.A. Krupko,[1] A.M. Rodin,[1] S.I. Sidorchuk,[1]
R.S. Slepnev,[1] S.V. Stepantsov,[1] G.M. Ter-Akopian,[1]
A.A. Korsheninnikov,[3,4] E.Yu. Nikolskii,[3,4] E.A. Kuzmin,[4]
B.G. Novatskii,[4] D.N. Stepanov,[4] S. Fortier,[5] P. Roussel,[6] and W. Mittig[6]

[1]*Flerov Laboratory of Nuclear Reactions, JINR, Dubna, Ru-141980 Russia*
[2]*The Henryk Niewodniczański Institute of Nuclear Physics PAN, Kraków, Poland*
[3]*RIKEN, Hirosawa 2-1, Wako, Saitama 351-0198, Japan*
[4]*RRC "The Kurchatov Institute", Kurchatov sq. 1, 123182 Moscow, Russia*
[5]*Institute de Physique Nucleaire, IN2P3-CNRS, F-91406 Orsay, France*
[6]*GANIL, BP 5027, F-14076 Caen Cedex 5, France*
wolski@nrmail.jinr.ru

**Abstract.** Low lying states of the unbound $^9$He system were investigated by means of the (d,p) reaction. A cryogenic deuterium target was bombarded by a secondary $^8$He beam. Complete kinematics measurements were accomplished. The observed $^9$He states are broad and overlapping. Their spins and parities have been unambiguously determined. Data from a $^6$He($\alpha$,2$\alpha$)2n experiment have been analyzed in terms of a quasi-free scattering approach. A large effect of the 2n final state interaction on the internal $^6$He momentum distribution has been demonstrated.

**Keywords:** neutron drip line, transfer reactions, angular correlations, impulse distributions.
**PACS:** 21.10.-k, 24.50.+g, 25.55.Hp, 25.70.Ef.

## INTRODUCTION

One could distinguish two aspects in using radioactive beams for physical experiments. The first one is associated with the investigation of interaction mechanisms and the structure of beam nuclei and their subsystems. A major part of physics studied by fragmentation processes at high beam energies belongs to that class of experiments. Another aspect implies the use of radioactive beams for producing nuclear systems which are even farer than the beam nuclei from the stability, even including unbound systems. Due to the inherent transparent mechanism and not too low cross-sections the transfer reactions with radioactive beams represent the most suitable way to reach that mass region.

In this paper the results of two recent experiments, which could exemplify both kinds of studies with radioactive beams, are presented. The $^9$He spectrum was investigated by means of the (d,p) reaction with a $^8$He beam and the momentum distribution of the α-particle core in $^6$He was measured in (α,2α) quasi-free scattering of a $^6$He beam on a helium gas target. Both studies were performed at the Flerov Laboratory of Nuclear Reactions (FLNR), JINR, Dubna.

The paper is divided on two parts each for one experiment. A description of the secondary beams preparation is presented as a part of the introduction.

## *Production and Diagnostics of Secondary Beams for Experiments*

A 34 MeV/nucleon $^{11}$B primary beam accelerated by the U-400M cyclotron at a typical intensity of 4pµA bombarded a 370 mg/cm$^2$ Be production target. The ACCULINNA fragment separator [1] was used to separate either $^8$He or $^6$He beam from other products and deliver the secondary to a position where a physical target was installed in the scattering chamber. The time of flight (ToF) and trajectory measurements were provided for each incoming particle resulting in the projectile kinetic energy and coordinates of the interaction point on the targets. ToF was given by the difference of time signals from two plastic scintillators placed on a 785cm flight base. The overall time resolution was 0.8 ns. The beam tracking was done by two multi-wire proportional chambers installed 26 cm and 80 cm upstream of the target. Each chamber had two planes of mutually perpendicular wires with a 1.25 mm pitch. The beam energy was 25 MeV/nucleon in the middle of the target. The energy spread of the beam, angular divergence and the position spread on the targets were 8.5%, 0.23° and about 5.4 mm, respectively.

# **Investigation of $^9$He Structure by the (d,p) Reaction**

## *Introduction*

The $^9$He nucleus is unbound in respect to the decay into the $^8$He+n channel. The first experimental investigation of $^9$He states has been done by the double charge exchange reaction of negative pions on a $^9$Be target [2]. Several narrow resonances were found in this study albeit with a low statistics and a large background. The presence of low lying narrow $^9$He states has been confirmed in the multi-nucleon exchange reactions: $^9$Be($^{14}$C,$^{14}$O)$^9$He and $^9$Be($^{13}$C,$^{13}$O)$^9$He [3]. Conclusions drawn from these observations have been generally accepted as a well established one. In the next experiment on $^9$He a search for the 2s state in this nucleus has been done in the fragmentation reaction of $^{11}$Be projectile nuclei [4]. In order to explain the $^8$He+n invariant mass spectrum obtained the authors proposed the existence of a 2s virtual state with an upper limit for the scattering length a<−10 fm. Another source of data on the low lying $^9$He states could be provided by a recent study of $^8$He+p resonance scattering [5]. In this study the $^9$Li (T=5/2) states, which are isobaric analogue of the $^9$He, were excited. One has to mention that the existing interpretation of the $^9$He spectrum from [2, 3] is in contradiction with the expectation of single particle strength for some of the states in question [6].

In the present study of $^9$He we used a single nucleon transfer reaction. Such reactions are known to be a proper tool to test single particle strength in nuclei. A complete kinematics measurement has been exploited. The advantage of this technique had been demonstrated by us earlier in the investigation of the $^3$H(t,p)$^5$H reaction [7].

## (d,p) Experiment

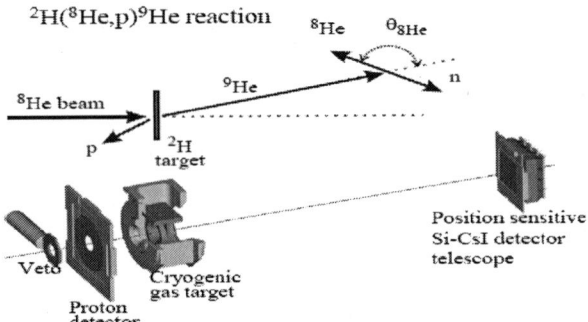

**FIGURE 1.** Scheme of kinematics for the (d,p) reaction and the experimental setup.

Secondary $^8$He beam with a typical intensity of 2 x $10^4 s^{-1}$ was focused on a cryogenic target filled with deuterium at a 1020 mPa pressure and cooled down to 25 K [8]. The target was 4 mm thick and had two 6 μm stainless steel windows of 30 mm in diameter. The experimental setup and schematic diagram of the reaction kinematics are shown in Fig. 1. Due to a peculiarity of the inverse kinematics for the (d,p) reaction the forward CM angle protons are emitted with low energy in backward direction in LAB. These slow protons were detected by a single annular, double-sided, 300 μm silicon detector with a 28mm central hole. The active area of this detector, limited by the inner and outer diameters of 32 mm and 82 mm, respectively, was segmented in 16 rings on one side and 16 sectors on the other. The detector was installed 100 mm upstream the target. The proton detection energy threshold of around 1.2 MeV corresponded to the efficiency fall for $^9$He excitation energy at above 4 MeV with cut-off energy of about 6 MeV. The $^8$He nuclei resulting from the $^9$He decay were emitted from the target in a narrow angular cone around the beam direction. These nuclei, as well as non interacting $^8$He projectiles were detected in a Si-CsI telescope mounted in air, behind the exit hole of the scattering chamber. The exit hole having 82 mm diameter was closed by a 125 μm maylar foil. The Si-CsI telescope consisted of two 60×60 mm$^2$, 1mm thick silicon detectors and an array of 16 CsI crystals coupled to the photodiode readouts. The Si detectors, each having 32 strips, provided the vertical and horizontal positions and energy loss signals. The sixteen 1.5×2×2 cm$^3$ CsI crystals were arranged in a 4×4 wall placed just behind the Si detectors. The thickness of the CsI crystals was enough to stop the $^8$He nuclei and to fetch their energy signals. Energy resolution provided by the CsI detector array came to 2% thus allowing the particle identification. The telescope was placed 50 cm downstream the target. That distance was a compromise between the requirements of the angular resolution and detection efficiency achieved for the $^8$He nuclei originating from the $^9$He decay. The total energy resolution in term of the $^9$He excitation energy was estimated by Monte-Carlo simulation giving a value of 0.8 MeV.

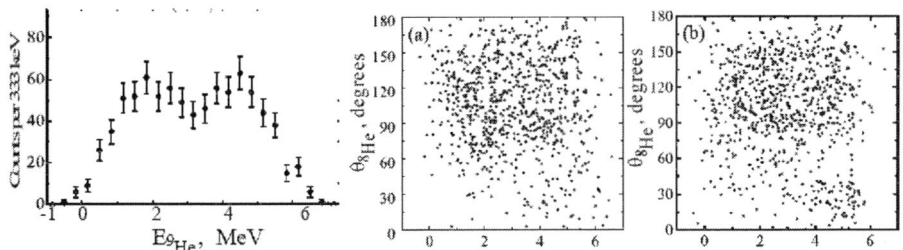

**FIGURE 2.** Experimental results obtained for the d($^8$He,p)$^9$He reaction. Left panel: $^9$He excitation energy spectrum from p+$^8$He coincidence events, (a) panel: the $^8$He angular distribution in CM of $^9$He for the investigated $^9$He energy range, (b) panel: simulation assuming $s_{1/2}$, $p_{1/2}$, $d_{5/2}$ sequence of the $^9$He states.

Data were taken at a condition of coincidence between the proton and $^8$He detectors. Additionally, events triggered by every 256-th beam nucleus were stored in order to monitor the beam parameters and estimate the experimental efficiency. The beam integral was equal to 2 x 10$^{10}$. A comparable value was collected in a run made with the evacuated target cell.

## Results and Discussion

Experimental results are presented in Fig. 2. The left panel of Fig. 2 shows the missing mass spectrum of the $^9$He extracted from the proton kinetic energy spectrum gated by the coincidence with $^8$He nuclei detected by the forward angle telescope. The coincidence was necessary to discriminate the reaction in question from others proton emission processes, like evaporation from the window material nuclei. The narrow states known from the literature are absent in the obtained spectrum. Two rather broad bumps could be seen instead. A prominent feature of the data is an apparent backward-forward asymmetry observed in the angular correlation of the $^9$He decay into the $^8$He+n channel (see the right panel in Fig. 2). The $^8$He nuclei are emitted preferably in the backward direction in CM of $^9$He. In order to explain such a phenomenon an interference of different parity states is required. At very low $^9$He energy only the *s-p* interference is possible. Since the asymmetry persists in the whole $^9$He energy range with a change in the angular correlation shape above 3 MeV indicating the appearance of a higher angular momentum wave, the assumption of *p-d* interference is unavoidable above 3 MeV of the $^9$He excitation.

It is reasonable to assume that at small angles at which the protons were detected, i.e. 3-7 deg. in CM, the orbital angular momentum is transferred with zero projection on the quantization axis which is very close to the beam axis in our case. For zero spin particles such spin alignment of the produced states leads to a very selective picture for the angular correlations. In the discussed case, due to neutron spin=1/2, the full alignment of the $^9$He states means the presence of the total spin projection: μ=+1/2 and μ=-1/2 only.

All resulting effects of asymmetry or symmetry caused by the interference of $^9$He states, on condition of full spin alignment, are collected in Table 1. One can see in Table 1 that the observed asymmetry in Fig. 2 uniquely selects the $s_{1/2}$, $p_{1/2}$, $d_{5/2}$

**TABLE 1.** The interference effect of the backward-forward asymmetry (symmetry) for overlapping $^9$He states at the full alignment in the (d,p) reaction. "No" - means the lack of the interference.

| States of $^9$He | $S_{1/2}$ | $P_{1/2}$ | $P_{3/2}$ | $D_{3/2}$ | $D_{5/2}$ |
|---|---|---|---|---|---|
| $S_{1/2}$ | xxx | Asymmetry | No | Symmetry | No |
| $P_{1/2}$ |  | xxx | Symmetry | No | Asymmetry |
| $P_{3/2}$ |  |  | xxx | Asymmetry | No |
| $D_{3/2}$ |  |  |  | xxx | Symmetry |
| $D_{5/2}$ |  |  |  |  | xxx |

sequence for the low-lying $^9$He states. This could be treated as the first experimental determination of spins and parities of these states.

A qualitative description of the data has been accomplished by a simple $l$ dependent potential model for the three states. A square well was taken with a radius of 3 fm consistent with the $1.4A^{1/3}$ dependence. The energies and weights of the $p_{1/2}$, $d_{5/2}$ states are given in the spectrum shown in Fig. 2, left panel. The high-energy shoulder of the spectrum for the $d_{5/2}$ resonance (above 5 MeV, see Fig. 2, left panel) is deformed by the efficiency fall. Therefore, the parameters of this resonance are determined less precisely. The phase shift between the $p_{1/2}$ and $d_{5/2}$ amplitudes is constrained by the experimental asymmetry. The $s_{1/2}$ contribution in the spectrum is characterized by its weight and scattering length $a$ (since this is not a resonance but a virtual state). These 2 parameters appeared to be not independent in the present analysis of the data, thus only a lower limit was obtained for the scattering length. Measurements for a larger range of the $^9$He energy and with increased resolution for the low-lying states are needed to refine the parameters for the $d_{5/2}$ and $s_{1/2}$ states, respectively. Following parameters for the investigated $^9$He quasi-bound $s_{1/2}$ state and for $p_{1/2}$ and $d_{5/2}$ resonances have been eventually proposed:

$s_{1/2}$    $a > -20$ fm
$p_{1/2}$,   $E = 2.0(0.2)$ MeV,   $\Gamma = 2$ MeV
$d_{5/2}$    $E \geq 4$ MeV,      $\Gamma > 1$ MeV

The Monte Carlo simulation made with the above parameters for the angular correlation is shown in Fig. 2, right panel.

There are several important conclusions from the study made:

1. Existence of a $s_{1/2}$ structure close to the $^9$He decay threshold has been confirmed.
2. The $p_{1/2}$, and $d_{5/2}$ resonances are broad and overlapping.
3. $^9$He resonances could be described in a single particle approach implying the $^8$He as a "good" core in the $^9$He structure.
4. All existing predictions for the $^{10}$He structure should be recalculated with new properties of the $^9$He found. Since the later is a subsystem of the former, narrow resonances for $^{10}$He seem to be rather unlikely to observe.

# Quasi Free Scattering Reaction $^6$He($\alpha$,2$\alpha$)2n

## *Introduction*

Quasi-free scattering (QFS) is a 3-body reaction which could be treated as a 2-body elastic scattering on a cluster bound in one of the collision partner. The process is recognized as a tool for nuclear structure investigation. Indeed, if one considers, for example, the $^6$He($\alpha$,2$\alpha$)2n reaction as a quasi-free scattering one, i.e. a single step $\alpha$-$\alpha$ collision going without distortions in the entrance and exit channels, the momentum conservation would yield the $\alpha$-particle internal momentum $p_\alpha$ in $^6$He

$$p_\alpha = p_1 + p_2 - (2/3)p_0 \qquad (1)$$

where: $p_0$ is the initial momentum of $^6$He, $p_1$ and $p_2$ are the momentum vectors of the two $\alpha$ particles.

Energy conservation gives the spectator excitation energy, in our case the two-neutron system. Thus kinematics alone provides valuable structure data in a model independent way. However, the assumption that distortions do not show up in the entrance and exit channels, i.e. the choice of the plane wave impulse approximation (PWIA) is hardly acceptable, except for the case when a rather high energy beam is used. QFS reactions induced by proton and $\alpha$ particle beams have been studied extensively. Usually the spectator was either a stable nucleus or a resonance. It was shown that the distorted wave impulse approximation (DWIA) [8] is more reliable for extracting data on spectroscopic factors.

The QFS reaction $^6$Li($\alpha$,2$\alpha$)d has been investigated at $\alpha$ particle energies around 100 MeV [9]. The authors used in their analysis both the plane and distorted wave approximations. The both methods lead to similar results for the momentum distribution of $\alpha$ cluster bound in the $^6$Li nucleus which were claimed to be in accord with a two-body $^6$Li wave function given by theory.

It was tempting to check whether QFS approach could be applied to examine wave functions of nuclei close to the drip line. The $^6$He($\alpha$,2$\alpha$)2n reaction was chosen for such test because the $^6$He three-body wave function is well established in theory [10]. In this particular case the spectator is unbound. Such a situation becomes typical for QFS studied with deep-line nuclei. Similar situation occurred in the $^6$Li($\alpha$,2$\alpha$)pn and $^6$Li(p,2p)pn reactions studied in Refs. [11,12] in coplanar geometry and with detectors taking small solid angles. This resulted in some limitations for data taking.

## *Experiment, Results and Discussion*

The secondary beam of $^6$He nuclei of intensity of about $2 \cdot 10^4$ s$^{-1}$ bombarded helium target cooled down to a temperature of about 16 K. The target thickness was $2 \cdot 10^{20}$ cm$^{-2}$. Two $\alpha$ particles outgoing from the target in angular ranges of $\pm(15°-55°)$ were detected in coincidence by two identical $\Delta E_1$-$\Delta E_2$-E telescopes. The telescopes were installed symmetrically in respect to the beam direction. They provided X&Y

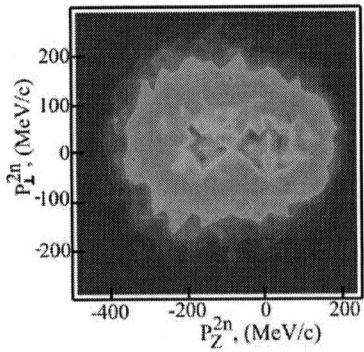
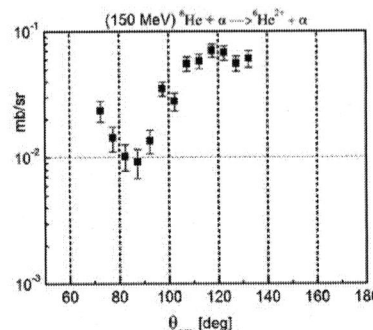

**FIGURE 3.** Left panel: the distribution of CM of neutron pair momentum $p_{2n}$ in respect to the CM of projectile. Right panel: the angular distribution for the reaction $^4$He($^6$He,$\alpha$)$^6$He($2^+$), see text.

coordinates and the energy of the reaction products. Each telescope consisted of two thin Si strip detectors (70 μ, 50×50 mm$^2$, 16 strips and 1000μ, 61×61 mm$^2$, 32 strips) for $\Delta E_1$, $\Delta E_2$ measurements, and a 6.2 mm thick Si(Li) detector of a large area, ~ 60 mm of diameter for E measurement. Knowing the energies and positions of the both coincident α particles we could calculate the relative momentum value for the two unobserved neutrons $p_{n-n}$ and the momentum vector $p_{2n}$ of their CM. A rather low background was observed in a run done with the empty target.

The $^4$He($^6$He,2α)2n reaction leads to four particles in the exit channel. One could expect that several processes, like the $^6$He inelastic scattering, the 1n and 2n transfer, etc., contribute in the kinematics region chosen for the observation of QFS events. The bulk of α-α coincidence events is shown in Fig. 3, left panel. The data are presented as a two-dimension correlation plot $p_{2n}^X$ vs. $p_{2n}^Z$ observed in the $^6$He CM for the two projections of the neutron pair CM momentum. Two partially overlapping loci are seen in Fig. 3, left panel. The first one is centered at the $^6$He CM, as is expected for the QFS reaction. The second one is centered in vicinity of $p_{2n}^z \approx -200$ MeV/c. This value roughly corresponds to the CM momentum of the total system α+α+n+n.

The lack of complete kinematics data made impossible the identification of reaction channels hampering the observation of QFS. However, one such channel, the inelastic scattering to the first $2^+$ state of $^6$He was clearly seen in the data. A relevant angular distribution is shown in Fig. 3, right panel. The inelastic scattering data could be of interest itself.

In order to further suppress non QFS contributions events showing large α-2n relative energy: $E_{\alpha 1-2n}>10$ MeV and $E_{\alpha 2-2n}>10$ MeV were eliminated. In that way, we tried to eliminate from all events those which can be attributed to the $^6$He excitation in continuum. Another constrain was introduced for the α-α CM scattering angle $\theta_{\alpha-\alpha}$.

A better separation of the two loci seen in Fig. 3, left panel, was obtained taking $60° > \theta_{\alpha-\alpha} > 120°$. We assumed that the remaining events of non QFS origin could be treated in a phase–space approximation.

To extract structural properties of the $^6$He nucleus the data were analyzed in terms of PWIA with the usual factorization of the transition matrix. The α-α double differential

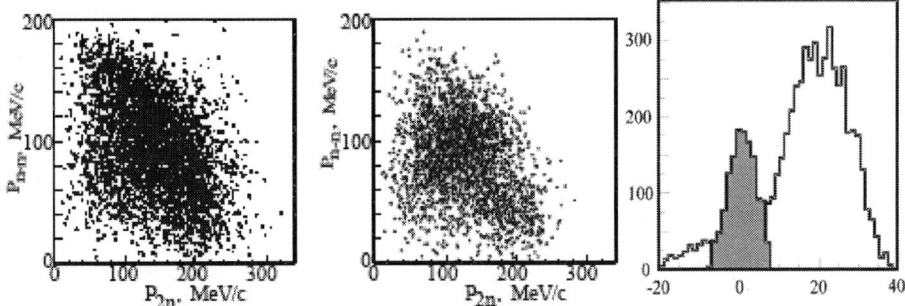

**FIGURE 4.** Left and middle panels: Distribution of events as a function of neutron momenta ($p_{2n}$ $p_{n-n}$) correlation, experimental data in the left panel, simulation in the middle panel, see text. Right panel: missing mass (in MeV) spectrum of triton spectator from QFS $^6$He$(\alpha,t\alpha)t$ reaction, the shadow area corresponds to the ground state of the spectator, see text.

cross section actually measured in the QFS experiment was expressed as a product of the structural and the dynamical parts. The last is responsible for α-α interaction which is approximated by the free (on-energy-shell) α-α elastic cross section.

$$\frac{d\sigma^{QFS}}{d\Omega_{\alpha-\alpha}} \propto S^2(\vec{p}_{n-n},\vec{p}_{2n})\frac{d\sigma^{on-shell}}{d\Omega_{\alpha-\alpha}} F_{PS} dE_{n-n} dE_{\alpha-\alpha} d\Omega_{2\alpha-2n} d\Omega_{n-n} \quad (2)$$

where $F_{PS} = \sqrt{E_{n-n}E_{\alpha-\alpha}(E_0 + Q - E_{n-n} - E_{\alpha-\alpha})}$ is the phase space factor taking into account the energy conservation law, $E_0=E_{n-n}+E_{\alpha-\alpha}+E_{2\alpha-2n}-Q$ is the total CM energy of the α+α+n+n system, $S(\vec{p}_{n-n},\vec{p}_{2n})$ is the spectral function which contains structural information and represents the probability density for the 2n system to be found in a final state characterized with momentum vectors $\vec{p}_{n-n}$ and $\vec{p}_{2n}$. Subscript 2α-2n relates to the motion of the CM of the α-α pair in respect to the CM of the two neutrons. The α-α elastic scattering cross section was calculated for the relative energy $E_{\alpha-\alpha}$ between the two detected α particles i.e. taking the post energy prescription for the α-α collision energy.

The spectral function is an overlap integral of the $^6$He three-body WF [10] with a wave function of *n-n* final state interaction (FSI) in the impulse representation.

$$S(\vec{p}_{n-n},\vec{p}_{2n}) = \int d\vec{r}_{n-n} d\vec{r}_{2n} \psi^*_{n-n}(\vec{p}_{n-n},\vec{r}_{n-n}) e^{-i\vec{p}_{2n}\vec{r}_{2n}} \psi_{6He}(\vec{r}_{n-n},\vec{r}_{2n}) \quad (3)$$

where: $\psi_{6He}(\vec{r}_{n-n},\vec{r}_{2n})$ is the three-body WF of $^6$He [10], and $\psi_{n-n}(p_{n-n},r_{n-n})$ is the two-body WF of the n-n system obtained with a realistic interaction potential. For vanishing n-n potential the spectral function converges to the Fourier transform of the $^6$He wave function in coordinate representation.

Having calculated the structural function and borrowing the α-α elastic cross section from literature data, a MC simulation of the QFS experiment was performed with all experimental conditions included. One can see that the simulated 2n momentum correlation, Fig. 4, middle panel, is similar to that obtained in the experiment, see Fig. 4, left panel. Indeed, the experimental and simulated widths and average momentum values of the $p_{n-n}$ and $p_{2n}$ distributions obtained as the projections to the axes made for the $p_{n-n}$ vs. $p_{2n}$ correlation plots are close to each other.

Other kinematics quantities extracted from the data, and not shown here, were compared with those generated by the simulation. A general agreement was observed, however, some contribution from the four-body phase-space was necessary to introduce.

The three-body $^6$He WF exhibits two characteristic bumps at (large $p_{n-n}$ and small $p_{2n}$) and vice-versa. This specific feature is not seen in the data. The experimentally obtained distribution, as seen in Fig. 4, is highly modified in respect to the source WF known for $^6$He and such peculiarity is not observed. This is mainly due to the $n$-$n$ FSI in the exit channel. One could expect that QFS studied at higher beam energies would be better separated from other processes and less affected by FSI.

It is worth to mention for completeness that another QFS has been observed in the colliding system $^6$He+$\alpha$. Events of triton - $\alpha$-particle coincidence have been recorded. These events show all characteristic phenomena of the scattering of $\alpha$ particle on a triton cluster bound in the $^6$He nucleus. Excitation energy spectrum obtained in this case for the spectator triton is shown in Fig. 4, right panel (in horizontal axis the missing mass energy is given in MeV, and the count number is presented in the vertical axis). A large background is present and energy resolution is not high. Nevertheless, the ground state of triton is clearly seen at zero excitation energy.

## ACKNOWLEDGMENTS

The work has been partially supported by the Russian Foundation for Basic Research grant 96 and 05-02-16404 and 05-02-17535, by INTAS grants 03-51-4496 and 03-54-65545, by IN2P3-Pologne grant 02-106.

## REFERENCES

1. A.M. Rodin et al., *Nucl. Instrum. Methods* B **204**, 114-118 (2003).
2. K. Seth et al., *Phys. Rev. Letters* **58**, 1930 (1987).
3. H.G. Bohlen et al., *Prog. Part. Nucl. Phys.* **42**, 17 (1999).
4. L. Chen et al., *Phys. Letters* **B 505**, 21 (2001.
5. G.V. Rogachev et al., *Phys. Rev.* **C 67**, 041603R (2003).
6. F.C. Barker, *Nucl. Rhys.* **A 741**, 42 (2004).
7. M.S. Golovkov et al., *Phys. Rev.* **C 72**, 064612 (2005).
8. N.S. Chant and P.G. Roos, *Phys. Rev.* **C 15**, 57 (1977).
9. A. Okihana et al., *Nucl. Phys.* **A 549**, 1-11 (1992).
10. B.V. Danilin et al., *Yad. Fiz.* **48**, 1206 (1988)
11. R.E. Warner et al., *Nucl. Phys.* **A 503**, 161 (1989) .
12. R.E. Warner et al., *Phys. Rev.* **C 72**, 2143 (1990).

# NUCLEAR ASTROPHYSICS (NAP)

# Cross Sections For The Production Of Residual Nuclides At Medium Energies: Status And Recent Experimental and Theoretical Progress

## Rolf Michel

*Zentrum für Strahlenschutz und Radioökologie, Leibniz Universität Hannover,
Am Kleinen Felde 30, D-30167 Hannover, Germany*

**Abstract.** The production of stable and radioactive residual nuclides by medium-energy protons and neutrons is of importance for many fields of basic and applied sciences ranging from astrophysics over space and environmental sciences, medicine, accelerator technology, space and aviation technology to accelerator driven transmutation of nuclear waste and energy amplification. In this paper, the up-to-now achievements, new developments and still open questions in the understanding of nuclide production at medium energies are discussed emphasizing the production of cosmogenic nuclides in extraterrestrial matter and pointing out some references to recent investigations related to transmutation studies.

**Keywords:** cross sections, cosmogenic nuclides, accelerator-driven technologies.
**PACS:** 96.50.S, 25.40.Sc

## NUCLIDE PRODUCTION AT MEDIUM ENERGIES

The term "nuclear reactions at medium energies" is not unambiguously defined. Here, it is understood as reactions at energies above 50 MeV and below 10 GeV. In this energy region, nuclear reactions are no longer dominated by the compound nucleus in statistical equilibrium and not yet dominated by production of elementary particles others than nucleons. At these energies, pre-equilibrium reactions, spallation- and fragmentation-reactions occur which all necessarily end up in a phase of statistical equilibrium and de-excitation via evaporation or fission at low excitation energies.

It is characteristic for nuclear reactions at medium energies that a wide variety of secondary particle, such as protons, neutrons and light charged particles, are produced and that in thick and extended targets substantial secondary particle fields built up and contribute themselves to the production of residual nuclides. Thus, the production rate $P_i$ of a nuclide $i$ in a sample with a chemical composition $c_s = (c_{s1},...,c_{sm})$ irradiated at a depth $d$ in the target with a bulk chemical composition $c_b = (c_{b1},...,c_{bn})$ and a target shape parameter $R$ (describing size and geometry) is given by

$$P_i(d,R,\mathbf{c}_s,\mathbf{c}_b) = N_L \cdot \sum_j \frac{c_{s,j}}{A_j} \cdot \sum_k \int_0^\infty \sigma_{i,j,k}(E_k) \cdot \frac{\partial J_k}{\partial E_k}(E_k,d,R,\mathbf{c}_b) dE_k \quad (1)$$

with $k$ denoting the different types of primary and secondary particles and $N_L$ and $A_j$ being Avogadro's number and the mass number of target element $j$. The bulk chemical composition $c_b$ determines the transport of primary and secondary particles and their resulting spectra inside the target. The sample chemical composition $c_s$ is the vector of chemical abundances of those elements contributing to the production of a nuclide in a particular sample of the irradiated target. With equation 1, the nuclide production at medium energies is strongly dependent on the actual irradiation scenario that means on type and energy of the radiation and on geometry and composition of the targets.

Production of residual nuclides by nuclear reactions at medium energies occurs in nature as well as man-made. Cosmic radiation and accelerators provide the necessary particle energies to induce such reactions. The residual nuclides produced have found manifold applications ranging from astrophysics over space and environmental sciences, medicine, accelerator technology, space and aviation technology to accelerator driven transmutation of nuclear waste and energy amplification.

In this paper, some achievements, new developments and still open questions in the understanding of nuclide production at medium energies are discussed taking the production of cosmogenic nuclides in extraterrestrial matter as example and giving references to recent investigations related to transmutation studies.

## COSMOGENIC NUCLIDES IN EXTRATERRESTRIAL MATTER

A wide variety of stable and radioactive cosmogenic nuclides is produced by solar (SCR) and galactic cosmic ray (GCR) interactions with extraterrestrial matter such as cosmic dust, meteoroids, and lunar and planetary surface materials. In these materials, stable cosmogenic nuclides can be observed as positive isotopic anomalies for rare gas isotopes where the small amounts of cosmogenic origin are not mimicked by high natural abundances. Cosmogenic radionuclides can either be observed by their decay, e.g. $^3$H, $^{44}$Ti, $^{26}$Al, $^{37}$Ar, $^{56}$Co, $^{22}$Na, $^{55}$Fe, and $^{60}$Co, or in case of long-lived radionuclides such as $^{10}$Be, $^{14}$C, $^{26}$Al, $^{36}$Cl, $^{41}$Ca, $^{53}$Mn, $^{60}$Fe, $^{59}$Ni, and $^{129}$I by accelerator mass spectrometry (AMS), by radiochemical neutron activation analysis ($^{53}$Mn) or by conventional rare gas mass spectrometry as for $^{39}$Ar and $^{81}$Kr. The longest-lived cosmogenic radionuclide, $^{40}$K, can only be observed in special matrices with low potassium contents such as iron meteorites by mass spectrometry. See refs. [1, 2] for reviews.

The production rate of cosmogenic nuclides by SCR interactions can be calculated straight forward for a given set of spectral parameters since production by secondary particles can be neglected because of the relatively low energies of SCR particles and since the depth dependent SCR spectra can be unambiguously calculated for an actual irradiation geometry considering electronic stopping and nuclear attenuation of the primary particles in an irradiated body [3].

For the interaction of GCR particles with matter the situation is much more complicated. Since energies up to 10 GeV/A have to be taken into account, secondary particles, in particular neutrons, become important and the depth scale on which GCR interactions occur extends to several hundreds of g/cm$^2$ in depth. Thus, GCR particles penetrate the earth's atmosphere and their interactions extend several meters into the surface of planets without gas envelope, asteroids and meteoroids.

The specific activity $A_i$ of a cosmogenic radionuclide $i$ with a decay constant $\lambda_i$ in a sample of a meteorite with a production rate $P_i$ depends of the exposure time $T_{exp}$ as well as of its terrestrial residence time $T_{terr}$

$$A_i(T_{exp}, T_{terr}) = P_i \cdot \left(1 - e^{-\lambda_i \cdot T_{exp}}\right) \cdot e^{-\lambda_i \cdot T_{terr}} \tag{2}$$

while the concentration of a stable cosmogenic nuclide $C_i$ just is increasing with the exposure time

$$C_i(T_{exp}) = P_i \cdot T_{exp} \tag{3}$$

By combining the observed abundances of stable and radioactive nuclides one can determine the exposure as well as the terrestrial ages of meteorites provided that the respective production rates are known. In recent years, purely physical models [4 - 6] were extremely successful to calculate production rates *ab initio* on the basis of spectra of primary and secondary cosmic ray particles resulting from Monte Carlo calculations and of the cross sections of the underlying nuclear reactions.

For the interpretation of cosmogenic nuclides, a cross section database has been established in a series of thin-target and thick-target experiments during more than two decades. The database covers the target elements C, N, O, Mg, Al, Si, Ca, Ti, V, Mn, Fe, Co, Ni, Cu, Rb, Sr, Y, Zr, Nb, Mo, Rh, Ag, In, Te, Ba, and La for proton-induced reactions. In 1997, the database of cosmochemically relevant cross sections of p- and α-induced reactions covered nearly 550 target/product combinations with nearly 22,000 cross sections [7 - 9]. More recent analyses covered the production of special long-lived radionuclides such as $^{14}$C [10], $^{36}$Cl [11 - 13], $^{41}$Ca [14], $^{53}$Mn [15], and $^{129}$I [16] and of stable rare gas isotopes [17, 18]. In addition, measurements of cosmochemically relevant cross sections were increasingly performed during recent years by other groups; see [19] for a review and references. For neutron-induced reactions, the respective excitation functions were derived by unfolding of thick-target production rates [20 - 22]. These thick-target experiments for the first time realistically simulated the isotropic irradiation of small bodies in space and allowed to bench-mark the model calculations of cosmogenic nuclide production in extraterrestrial matter [4 - 6].

These physical model calculations [4 - 6], validated by thick-target simulation experiments [20 - 24] yield the production rates of cosmogenic nuclides as function of the size of meteoroid and the depth of a sample in it; see the production rates for $^{21}$Ne in Fig. 1. The production rates for different cosmogenic nuclides do not have the same depth- and size-dependence due to differing energies necessary to produce them from their relevant target elements and to the energy-dependence of the excitation functions of the individual contributing nuclear reactions. The different size- and depth-dependences can be used to constrain the pre-atmospheric radius and the depth of a sample inside a meteoroid; see the comparison of $^{10}$Be and $^{26}$Al production rates and the dependence of $^{10}$Be and $^{26}$Al production rates on the $^{22}$Ne/$^{21}$Ne and $^{3}$He/$^{21}$Ne production rate ratios in Figure 1. In particular, the $^{22}$Ne/$^{21}$Ne and $^{3}$He/$^{21}$Ne production rate ratios have turned out to be reliable depth and size indicators since the respective isotopic ratios can be measured with accuracies of a few percent.

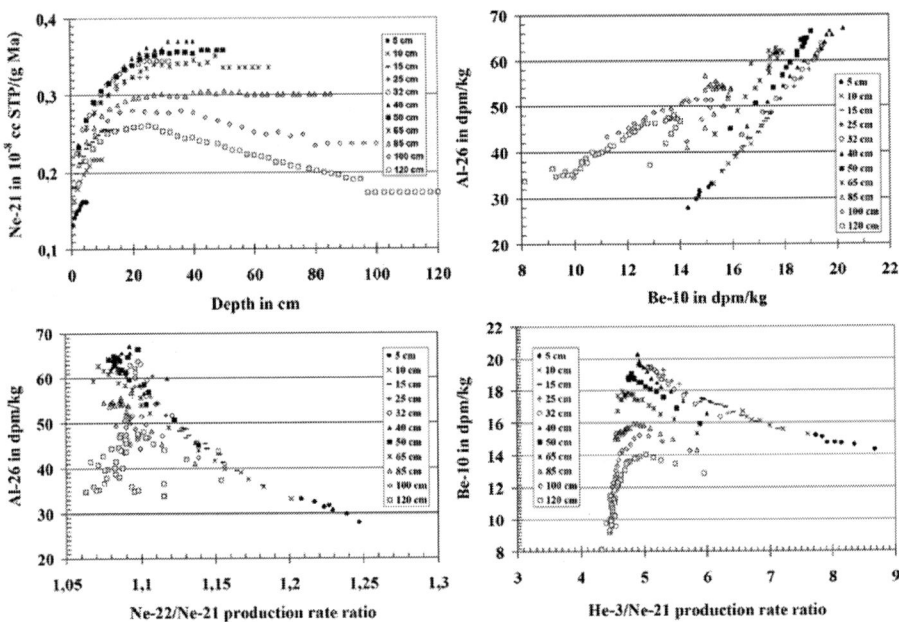

**FIGURE 1.** Production rates of $^{21}$Ne (upper left) as functions of meteoroid radii and depth inside the meteoroids, $^{10}$Be versus $^{26}$Al production rates (upper right) and production rates of $^{26}$Al (lower left) and $^{10}$Be (lower right) as functions of $^{22}$Ne/$^{21}$Ne and $^{3}$He/$^{21}$Ne production rate ratios, respectively.

Today's physical model calculations reproduce the observed $^{22}$Ne/$^{21}$Ne and $^{3}$He/$^{21}$Ne production rate ratios in meteorites within the experimental uncertainties (Figs. 2 and 3). The differences in the depth and size dependences are a delicate balance of proton- and neutron-induced production modes as can be seen for the $^{3}$He/$^{21}$Ne production rate ratio in Figure 3. In general, the physical model calculations exhibit accuracies better than 10 % and 3 % when predicting production rates of cosmogenic nuclides and production rate ratios, respectively. On the basis of these model calculations long-term spectra of solar and galactic cosmic ray particles were determined [4, 25] and the observed abundances of cosmogenic nuclides in meteorites can be interpreted with respect to their exposure histories. The success of physical models in calculating cosmogenic nuclide production rates in extraterrestrial matter such as meteorites and lunar samples mainly rests on highly reliable experimental cross sections of the underlying nuclear reactions.

As described in detail elsewhere [1] the interpretation of cosmogenic nuclides in extraterrestrial matter within the framework of physical model calculations yields information about the intensity and spectral distribution of the solar and galactic cosmic radiation in the past. Based on the observed depth profiles of $^{10}$Be, $^{26}$Al, and

[53]Mn in the lunar surface and in meteorites it can be concluded that the solar and galactic cosmic radiation was constant during the last 10 million years with long-term averaged spectra well within the range of spectra observed during the last decades [4, 25].

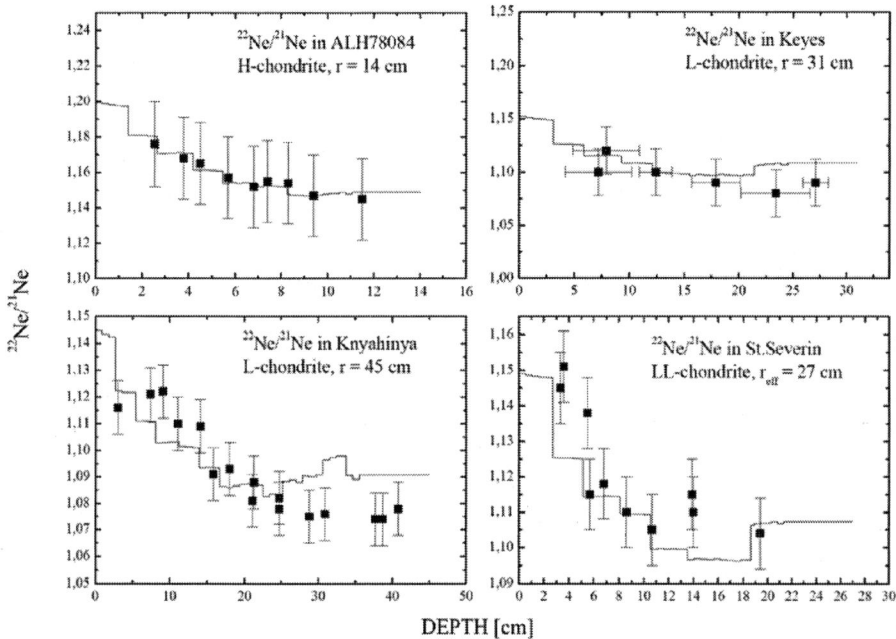

**FIGURE 2.** Depth profiles of $^{22}$Ne/$^{21}$Ne isotopic ratios in selected stony meteoroids [5].

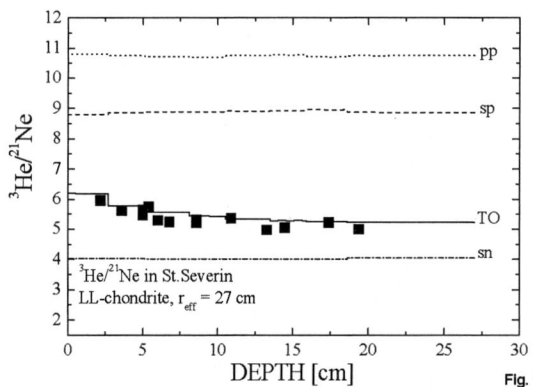

**FIGURE 3.** $^{3}$He/$^{21}$Ne production rate ratios in the LL-chondrite St. Severin [5].

Today, in a one-gram sample of a meteorite all relevant radioactive and stable cosmogenic nuclides can be measured. The physical model calculations yield then the information about the exposure and terrestrial ages as well as on the size of the meteorite and the depth of a sample in it. Moreover, in combination with mineralogical data they allow identifying paired meteorite finds. Figure 3 gives examples of exposure ages of H-chondrites found 1988 in the Allan Hills/Antarctica, of terrestrial ages of stony meteorites found in Antarctica and in Sahara and of the pre-atmospheric radii of these meteorites.

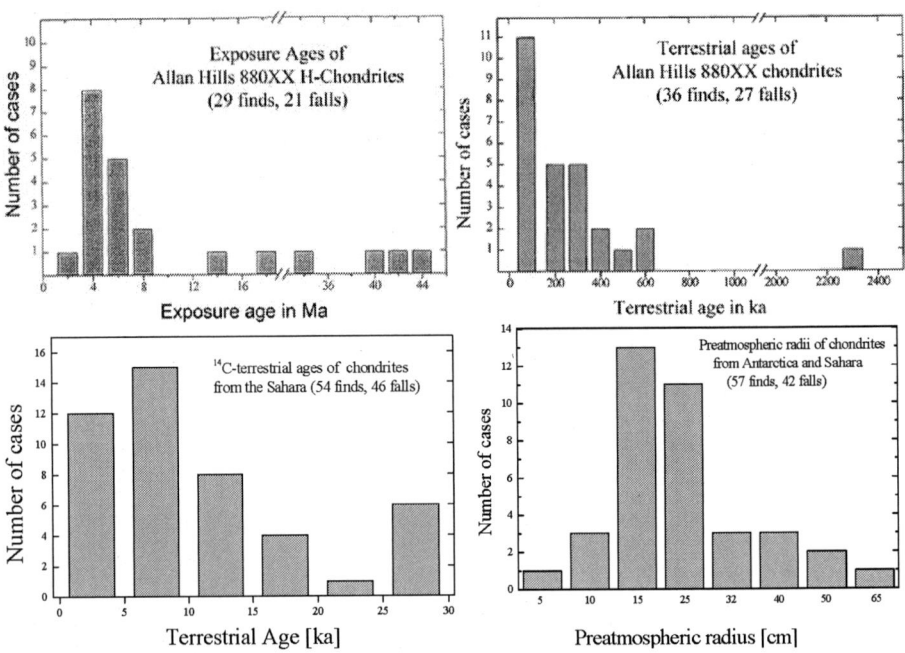

**FIGURE 4.** Distribution of exposure ages Allan Hills 880XX H-chondrites (upper left), of terrestrial ages of Allan Hills 880XX H-chondrites (upper right), of terrestrial ages of chondrites from Sahara (lower left), and of pre-atmospheric radii of chondrites from Antarctica and Sahara (lower right).

Systematic meteorite searches in Antarctica and in several hot deserts did not only yield the usual meteorites originating from the asteroid belt, but also extraordinary specimen whose origin must be attributed to Moon and Mars. These lunar and Martian meteorites are of particular interest with respect to the collision history of small bodies in the solar system. The interpretation of their observed cosmogenic nuclide record has demonstrated the complexity of this collision history [10, 26 - 33]. The achieved reliability of physical model calculations has made them a successful standard tool of cosmochemistry and –physics. However, there are still open problems as e.g. a lack of consistency in the modeling of the production of cosmogenic nuclides by medium-energy particles and by low-energy neutron capture [34 - 35]. Moreover, the physical modeling of *in-situ* production of cosmogenic nuclides in the solid Earth's surface has not yet achieved the same maturity as that in extraterrestrial matter and is a topic of intense investigation.

# NUCLEAR DATA FOR ACCELERATOR DRIVEN SYSTEMS

For accelerator technologies, production of residual nuclides at medium energies has to be modeled in order to describe the radioactive inventories of the spallation targets, the activation of accelerating structures, of the beam pipes and windows and of shielding materials, cooling materials and ambient air. Radionuclide inventories will determine the final disposal costs of spallation targets and will decide whether the burn-up of nuclear waste can be counterweighted or not by the creation of other activation products. Moreover, medium-energy cross sections of fission products and actinides are needed. Finally, the production of residual nuclides will cause chemical alteration of the irradiated components and, in particular, production of light complex particles such as $^2$H, $^3$H, $^3$He, and $^4$He will cause considerable material damage.

Integral excitation functions for the production of residual nuclides are basic quantities for the calculation of radioactive inventories of spallation targets in spallation neutron sources and in accelerator-driven devices for energy amplification or for transmutation of nuclear waste. Due to the large range of relevant target elements and the vast amount of product nuclides it will not be possible to measure all the cross sections needed. Consequently, one will have to rely widely on models and codes to calculate the required cross sections and validation of such calculations will be a high priority issue. Since previous experiences with predictions of such excitation functions were not satisfying [36], two new code systems, namely TALYS [37] and INCL4+ABLA [38, 39], were developed within the HINDAS (*High and Intermediate Energy Nuclear Data for Accelerator-Driven Systems*) project [40, 41]. Experimental investigations of the HINDAS project provided consistent sets of nuclear data of all types to allow for comprehensive tests of these models; see ref. [41] for details.

Also our experimental investigations were recently extended to heavy target elements such as Ta, W, Pb, Bi [42 - 46], and U [47] within the EC project HINDAS. For the target element lead a comprehensive set of excitation functions published recently was completed by AMS-measurements of cross sections for the production of the long-lived radionuclides $^{10}$Be, $^{36}$Cl, and $^{129}$I, [48, 49]. For natural uranium, cross sections for the production of residual nuclides are available for p-energies from 21 MeV to 69 MeV [47] and at 600 MeV [50]. Further spin-offs of the HINDAS project were mass spectrometric measurements of cross sections for the production of stable and radioactive isotopes of He, Ne, Ar, Kr, and Xe from natural lead [51].

Together with not yet published cross sections for the target elements Rb, Mo, Rh, Ag, In, Te, and La, the published cross sections for the production of residual nuclides for cosmochemically relevant target elements, and those measured within the HINDAS project for Fe, Ta, W, Au, Pb, Bi, and U, our consistent data base now contains data for nearly 1,500 nuclear reactions and more than 25,000 cross sections. Most of the data are available in EXFOR.

The phenomenology and energy dependence of the excitation functions exhibits clearly distinguishable reaction modes like multi-fragmentation, three different types of medium-energy asymmetric and symmetric fission, deep spallation, pre-equilibrium reactions and classical compound nucleus reactions. It allows for a systematic survey on the phenomenology of excitation functions and for comprehensive tests of models and codes describing medium-energy nuclear reactions.

Our investigations made use of classical kinematics and, therefore, they are confined to residual nuclides with usually at least a few hours' half-lives which mostly reveal a cumulative production due to the decay of short-lived progenitors. These investigations are complimentary to those using inverse kinematics where all the primary residuals can be studied at certain energy points but not over the whole energy range desired, e.g. [52].

Contrary to their importance for applications, the availability of cross sections for the production of residual nuclides by neutron-induced reactions above 30 MeV is scarce and theoretical predictions are not yet reliable enough. Recent investigations opened an opportunity to improve this situation. Facilities with quasi mono-energetic neutrons produced by the $^7$Li(p,n)$^7$Be reaction are now available, e.g. at UCL/Louvain La Neuve [53] and TSL/Uppsala [54 - 56], with sufficiently high neutron fluxes to allow activation experiments in which such production cross sections can be determined. However, cross sections cannot be directly calculated from the experimental data since the neutrons used are just "quasi-monoenergetic" with only about 30 to 50% of the neutrons in the high-energy peak with a width of a few MeV. Therefore, neutron cross sections have to be determined by unfolding from the experimental response integrals determined in a series of irradiation experiments with different neutron energies. A series of such experiments was performed with proton energies between 36.4 and 178.8 MeV at UCL and TSL covering the target elements C, N, O, Mg, Al, Si, Fe, Co, Ni, Cu, Ag, Te, Pb, and U [57]. The results obtained so far [58] demonstrate the feasibility of this method. They show for all target elements significant differences between the neutron- and proton-induced productions of residual nuclides emphasizing the necessity for further studies.

## CONCLUSION

Much progress has been made during recent years. For proton-induced reactions a wealth of experimental cross sections exists. The production of cosmogenic nuclides in extraterrestrial matter is well understood. The HINDAS project has very much improved our understanding of radionuclide production at medium energies. However, there are still open problems such as the theoretical description of the production of intermediate mass fragments and products of deep spallation. For neutron-induced reactions we are just at the beginning, but respective irradiation facilities and experimental methods exist to solve this problem. In spite of the progress made in the theoretical understanding of nuclide production at medium energies, the statement remains valid: "if accuracy matters one has to rely on measurements".

## REFERENCES

[1] R. Michel, *Radiochim. Acta* **87** (1999) 47 – 73.
[2] R. Michel and S. Neumann, *Proc. Indian Acad. Sci. (Earth Planet. Sci.)* **107** No. 4 (1998) 441 – 457.
[3] R. Michel et al., *Earth Planet. Sci. Letters*, **59** (1982) 33 – 48, Erratum: *ibid*, **64** (1983) 174.
[4] R. Michel et al., *Nucl. Instr. Meth. Phys. Res.* **B113**, (1996) 434 – 444.
[5] I. Leya et al., *Meteoritics & Planetary Science* **35** (2000) 259 – 286.
[6] I. Leya et al., *Meteoritics & Planetary Science* **36** (2001) 1547 – 1561.
[7] R. Michel et al., *Nucl. Instr. Meth. Phys. Res.* **B103** (1995) 183 – 222.

[8] R. Michel et al., *Nucl. Instr. Methods in Phys. Res.* **B129** (1997) 153 – 193.
[9] H.-J. Lange et al., W., *Appl. Rad. Isotop.* **46** (1994) 93 – 112.
[10] U. Neupert, Langlebige kosmogene Radionuklide in Meteoriten aus heißen und kalten Wüsten, Ph.D. thesis, Universität Hannover (1996).
[11] Th. Schiekel et al., *Nucl. Instr. Meth. Phys. Res.* **B114** (1996) 91 – 119.
[12] Th. Schiekel et al., *Nucl. Instr. Meth. Phys. Res.* **B113** (1996) 484 – 489.
[13] F. Sudbrock et al., in: G. Reffo et al. (eds.), *Conf. Proc. Vol. 59 „Nuclear Data for Science and Technology"*, Società Italana di Fisica, Bologna, (1997) 1534 – 1536.
[14] Ch. Schnabel et al., *Nucl. Instr. Meth. Phys. Res.* **B223-224** (2004) 812 – 816.
[15] S. Merchel et al., *Nucl. Instr. Meth. Phys. Res.* **B172** (2000) 144 – 151.
[16] Ch. Schnabel et al., in: G. Reffo et al. (eds.), Conf. Proc. Vol. 59 „Nuclear Data for Science and Technology", Società Italana di Fisica, Bologna, (1997) 1559 – 1561.
[17] E. Gilabert et al., *Nucl. Instr. Meth. Phys. Res.* **B145** (1998) 293 – 319.
[18] I. Leya et al., *Nucl. Instr. Meth. Phys. Res.* **B145** (1998) 449 – 458.
[19] J.M. Sisterson et al., *Nucl. Instr. Meth. Phys. Res.* **B123** (1997) 324 – 329.
[20] I. Leya et al., *Meteoritics & Planetary Science* **35** (2000) 287 – 318.
[21] I. Leya et al., *Meteoritics & Planetary Science* **39** (2004) 367 – 386.
[22] E. Gilabert et al., *Meteoritics & Planetary Science* **37** (2002) 951 – 976.
[23] R. Michel et al., *Nucl. Instr. Meth. Phys. Res.* **B16** (1986) 61 – 82.
[24] R. Michel, *Nucl. Instr. Meth. Phys. Res.*, **B42**, (1989) 76 – 100.
[25] R. Michel, in M. Arnould et al. (eds.) Tours Symposium on Nuclear Physics III, AIP (1998) 447 – 456.
[26] P. Scherer et al., *Meteoritics & Planetary Science* **32** (1997) 769 – 773.
[27] A. Bischoff et al., *Meteoritics & Planetary Science* **33** (1998) 1243 – 1257.
[28] Th. Stelzner et al., *Chem. Erde* **57** (1997) 297 – 309.
[29] L. Schultz et al., *Meteoritics & Planetary Science* **33** Supplement (1998) A138.
[30] Th. Stelzner et al., *Meteoritics & Planetary Science* **34** (1999) 787 – 794.
[31] S. Merchel et al., in: L. Schultz et al. (eds.), Workshop on Extraterrestrial Materials from Cold and Hot Deserts (2000) LPI Contribution 997, Lunar and Planetary Institute, Houston, 53 – 56.
[32] M. Pätsch et al., *Meteoritics & Planetary Science* **35** (2000) A124 – A125.
[33] I. Leya et al., *Meteoritics & Planetary Science* **36** (2001) 1479 – 1494.
[34] D. Kollár et al., *Meteoritics & Planetary Science* **41** (2006) 375 – 389.
[35] Ch. Schnabel et al., *Meteoritics & Planetary Science* **39** (2004) 453 – 466.
[36] R. Michel and P. Nagel, International Codes and Model Intercomparison for Intermediate Energy Activation Yields, NSC/DOC(97)-1, NEA/OECD, Paris, 1997
[37] A.J. Koning et al., in: R.C. Haight et al. (eds.) AIP Conf. Proc. **769**, Melville NY (2005) 1154 – 1159.
[38] A. Boudard et al., C., *Phys. Rev.* **C66**, Art. No. 044615 (2002).
[39] A.R. Junghans et al., *Nucl. Phys.* **A629**, 635-655 (1998).
[40] A. Koning et al., *Nucl. Science Technology, Supplement* **2** (2002) 1161 – 1166.
[41] J.P. Meulders et al., HINDAS, High and Intermediate Energy Nuclear Data for Accelerator-Driven Systems, Final Technical Report, EC Contract N°: FIKW-CT-2000-00031, Project N°: FIS5-00150, January 2005.
[42] M. Gloris et al., *Nucl. Instr. Meth. Phys. Res.* **A463** (2001) 593 – 633.
[43] R. Michel et al., *Nucl. Science Technology, Supplement* **2** (2002) 242 – 245.
[44] J. Kuhnhenn et al., *Radiochimica Acta* **89** (2001) 697 – 702.
[45] M.M.H. Miah et al., *Nucl. Science Technology, Supplement* **2** (2002) 369 – 372.
[46] R. Michel et al., in: R.C. Haight et al. (eds.) AIP Conf. Proc. **769**, Melville NY (2005) 1551 – 1554.
[47] M.A.M. Uosif et al., in: R.C. Haight et al. (eds.) AIP Conf. Proc. **769**, Melville NY (2005) 1547 – 1550.
[48] D. Schumann et al., in: R.C. Haight et al. (eds.) AIP Conf. Proc. **769**, Melville NY (2005) 1517 – 1520.
[49] D. Schumann et al., *Nucl. Instr. Meth. Phys. Res.* **A562** (2006) 1057 – 1059.
[50] J. Adam et al., in: R.C. Haight et al. (eds.) AIP Conf. Proc. **769**, Melville NY (2005) 1043 – 1046.
[51] I. Leya et al., *Nucl. Instr. Meth. Phys. Res.* **B229** (2005) 1 – 23.
[52] Enqvist et al., *Nucl. Phys.* **A686** (2001) 481 – 524.
[53] H. Schuhmacher et al., *Nucl. Instrum. Methods Phys. Res.* A421 (1999) 284 – 295.
[54] H. Condé et al., *Nucl. Instr. Meth. Phys. Res.* **A292** (1990) 121 – 128.
[55] S. Neumann et al., Conf. Proc. Vol. 59 „Nuclear Data for Science and Technology", G. Reffo et al. (Eds.), Società Italana di Fisica, Bologna (1998) 379 – 383.
[56] S. Pompe t al., in: R.C. Haight et al. (eds.) AIP Conf. Proc. **769**, Melville NY (2005) 780 – 783.
[57] W. Glasser et al., *Nucl. Science Technology, Supplement* **2** (2002) 373 – 376.
[58] R. Michel et al., in: R.C. Haight et al. (eds.) AIP Conf. Proc. **769**, Melville NY (2005) 861 – 864.

# Spallation nucleosynthesis by accelerated charged-particles in stellar envelopes of magnetic stars

S. Goriely[1]

*Institut d'Astronomie et d'Astrophysique*
*Université Libre de Bruxelles*
*Campus de la Plaine, CP 226*
*1050 Brussels – Belgium*

**Abstract.** Recent observations have suggested the presence of radioactive elements, such as Tc, Pm and $84 \leq Z \leq 99$ elements at the surface of the chemically-peculiar magnetic star HD 101065, also known as Przybylski's star. The peculiar $35 < Z < 82$ abundance pattern of HD 101065 has been so far explained by diffusion processes in the stellar envelope. However, such processes cannot be called for to explain the origin of short-lived radio-elements. The large magnetic field observed in Ap stars can be at the origin of a significant acceleration of charged-particles, mainly protons and $\alpha$-particles, that in turn can by interaction with the stellar material modify the surface content.

The present contribution explores to what extent the irradiation process corresponding to the interaction of the stellar material with energetic particles can by itself only explain the abundances determined by observation at the surface of the chemically peculiar star HD 101065, as well as other chemically peculiar star. Due to the unknown characteristics of the accelerated particles, a purely parametric approach is followed, taken as free parameters the proton and $\alpha$-particle flux amplitude and energy distribution as well as the time of irradiation. The specific simulations considered here can explain many different observational aspects. In particular, it is shown that a significant production of $Z > 30$ heavy elements can be achieved. The most attractive feature of the irradiation process is the significant production of Tc and Pm, as well as actinides and sub-actinides.

**Keywords:** Spallation reaction, Chemically peculiar stars, Nucleosynthesis
**PACS:** 25.40.Sc, 97.10.C,97.30.Fi

## INTRODUCTION

Chemically peculiar (CP) stars are spectral type A stars that have chemical abundances differing significantly from that of other stars of similar classification. They exhibit a remarkable variety of elemental enhancements and depletions in their photospheres, sometime 5 to 6 orders of magnitudes different than found in the sun. The two major types of CP stars are Am stars that have no detectable magnetic field and Ap stars that have magnetic fields with large-scale structure.

Rapidly oscillating Ap (roAp) stars are a subgroup of magnetic Ap stars which oscillate with non-radial, low order acoustic *p*-modes with the axis of oscillation aligned with the axis of the magnetic field [1]. The mechanisms responsible for the excitation of roAp stars and other physical parameters that distinguish them from non-pulsating

---

[1] S.G. is FNRS Research Associate

CP stars remains an open question. The comparison between roAp and non-roAp star surface composition do not reveal large abundance differences. So far, the CP star abundances have been almost uniquely explained on the basis of diffusion processes, i.e diffusive segregation of ionic and isotopic species resulting from a balance between radiative and gravitational forces within the atmosphere and sub-atmospheric levels of warm stars.

Recent observations have suggested the presence of short-lived radioactive elements, such as Tc, Pm and $84 \leq Z \leq 99$ elements at the surface of the CP roAp star HD 101065, also known as Przybylski's star. Such an observation can in no way be explained by diffusion processes. Observations of CP-stars in the X-ray regime show evidence that high energy phenomena are associated with, if not produced by, some CP stars. As shown by [2], Ap stars are 3-4 times more likely to be associated with X-ray sources than normal A stars, so that high-energy events can be expected in the vicinity of (if not on) many Ap stars. The large magnetic field observed in Ap stars can be at the origin of a significant acceleration of charged-particles, mainly protons and $\alpha$-particles, that in turn can by interaction with the stellar material modify the surface content.

The purpose of the present work is to study to what extent the irradiation process resulting from the interaction of the stellar material with energetic particles can by itself only explain the abundances determined by observation at the surface of the roAp star HD 101065. This includes the particularly large abundances of the heavy elements with $35 < Z < 82$ by 3-4 dex [3], so far explained by diffusion processes [4], but also the possible presence of radioactive elements, such as Tc, Pm and $84 \leq Z \leq 99$ elements that have been identified by [5, 6]. No effort is done at the present stage to explain the possible mechanisms responsible for the particle acceleration. As already discussed, such particles could be locally accelerated, but they could also come from an external source.

## MODELLING

To describe the changes in abundance of the nuclei as a result of the irradiation process, a nuclear reaction network including all reactions of relevance is used. All nuclei with $0 \leq Z \leq 102$ and located between the proton drip line and the neutron-rich side of the valley of stability are included in the network. The chosen set of nuclear species are then coupled by a system of differential equations corresponding to all the reactions affecting each nucleus, i.e. mainly proton, $\alpha$ and neutron captures, $\beta$- and $\alpha$-decays, as well as spontaneous fission decay. The rate of change of the molar fraction $Y_{(Z,A)}$ of a nucleus $(Z,A)$ with charge number $Z$ and mass number $A$ can be written as

$$\begin{aligned}\frac{dY_{(Z,A)}}{dt} = \; & - Y_{(Z,A)} \Phi_p <\sigma>_p + \sum_k Y_k \Phi_p <\sigma>_{p,k}^{(Z,A)} \\ & - Y_{(Z,A)} \Phi_\alpha <\sigma>_\alpha + \sum_k Y_k \Phi_\alpha <\sigma>_{\alpha,k}^{(Z,A)} \\ & - Y_n \rho N_a <\sigma v>_n^{(Z,A)} Y_{(Z,A)} + \sum_k Y_n \rho N_a <\sigma v>_{n,k}^{(Z,A)} Y_k \end{aligned}$$

$$- \lambda_{\beta^-}^{(Z,A)} Y_{(Z,A)} - \lambda_{\beta^+}^{(Z,A)} Y_{(Z,A)} + \lambda_{\beta^-}^{(Z-1,A)} Y_{(Z-1,A)} + \lambda_{\beta^+}^{(Z+1,A)} Y_{(Z+1,A)}$$
$$- \lambda_{\alpha}^{(Z,A)} Y_{(Z,A)} + \lambda_{\alpha}^{(Z+2,A+4)} Y_{(Z+2,A+4)} - \lambda_{sf}^{(Z,A)} Y_{(Z,A)} + \sum_k \lambda_{sf,k}^{(Z,A)} Y_k \quad (1)$$

where $\lambda_\beta^\pm$, $\lambda_\alpha$ and $\lambda_{sf}$ are the $\beta^\pm$, $\alpha$ and spontaneous fission decay rate, respectively; $N_a$ is the Avogadro number, $\rho$ the local density and $Y_n$ the neutron abundance. Note that in Eq. 1, the recycling by p-induced, $\alpha$-induced or spontaneous fission is assumed to take place by symmetric fission to one unique species. $<\sigma>_p$ and $<\sigma>_\alpha$ are the total effective cross sections for proton and $\alpha$-captures averaged over the energy distribution $\Phi(E)$, i.e for the proton case

$$<\sigma>_p = \frac{1}{\Phi_p} \int_0^\infty \Phi(E) \sigma_p(E) dE \quad (2)$$

where $\sigma_p(E)$ is the $E$-dependent total proton capture cross section and $\Phi_p = \int_0^\infty \Phi(E) dE$ the proton flux amplitude (see below). Similarly, $<\sigma>_{p,k}^{(Z,A)}$ is the partial effective cross sections for proton capture on a nucleus $k$ leading to the residual $(Z,A)$ nucleus.

In the present work, a special attention is paid to the abundance of the neutrons emitted by the spallation process. We assume that the so-produced neutrons are thermalized on timescales much shorter than the typical capture timescales. An upper value for the thermalization time can be estimated by

$$\tau \simeq \frac{1}{N_a \rho \sigma v_{th}} \frac{\sqrt{2}}{\sqrt{2}-1} \quad (3)$$

where $v_{th}$ is the thermal velocity. Assuming the matter to be essentially made of H, the H – neutron collision cross section amounts to about $\sigma \simeq 20$ b, so that for a surface temperature of $T \simeq 7000$ K and $\rho = 10^{-7}$ g/cm$^3$, $\tau = 2$ s which justifies the fast thermalization of the neutrons. In this case, their energy spectrum follows a traditional Maxwell-Bolztmann distribution at a temperature $T$. The neutron capture rates $<\sigma v>_n$ (where $v$ is the relative target-neutron velocity) are estimated by the Hauser-Feshbach reaction code MOST [7]. In Eq. 1, $<\sigma v>_{n,k}^{(Z,A)}$ corresponds to either the $(n,\gamma)$, or $(n,p)$ or $(n,\alpha)$ reaction rate leading to the formation of the $(Z,A)$ nucleus.

The energy spectrum of the accelerated particles is assumed to remain constant in time. Different energy distributions were considered, but only one is presented here. It corresponds to a energy-constant distribution in the $0 \leq E \leq E_{max}$ energy range. The impact of such a choice and a comparison with the more traditional power law proposed by [8] is discussed in [9].

The proton and $\alpha$-particle capture cross sections $\sigma(E)$ are either taken from experiments (essentially for light species [10]) or estimated with the Talys nuclear reaction code [11] which takes into account all types of direct, pre-equilibrium and compound mechanisms to estimate the total reaction probability as well as the competition between the various open channels. The cross sections are estimated at energies up to 200 MeV. The calculation includes single-particle (nucleons and alpha) as well as multi-particle emissions and fission. All the experimental information on nuclear masses, deformation and low-lying states spectra is considered, whenever available. If not, global nuclear level formulas, and nucleon and $\alpha$-particle optical model potentials are considered to determine the excitation level scheme and the particle transmission coefficients, respectively. Due to the large number of open channels at high energies, typically about 30 different (p,$x$n $y$p $z\alpha$) reaction types corresponding to the emission of $x$ neutrons, $y$ protons, and $z$ $\alpha$-particles, need to be taken into account per target nucleus. This implies that the reaction network includes 5000 different species with $0 \leq Z \leq 102$ and some 250000 proton and $\alpha$-particle capture reactions. Finally, note that the initial abundance distribution is assumed to be the solar one.

No effort is done in the present work to explain the possible mechanisms responsible for the particle acceleration. The energy spectrum, the proton and $\alpha$-particle fluxes, as well as the irradiation time are taken as free parameters. It can however be kept in mind that in the specific case of solar flares typical proton fluences are of the order of $10^{17-18}$cm$^{-2}$, while early solar system models call for fluences of about $10^{23}$cm$^{-2}$ to explain the origin of now-extinct short-lived radionuclides known to have been present in the early solar system [12]. If HD 101065 is not a T-Tauri star, its putative companion could possibly be one. In addition, the proton fluence can also be expected to be increased by a factor as high as $10^6$ if the protons are guided along the magnetic field lines. On the other hand, it can also be of interest to apply such spallative model to the jet particles emitted, for example, in the case of bipolar supernova explosions [13]. In such a situation, at a radius of 1000 km, the particles can reach fluences of about $10^{37}$ cm$^{-2}$ and a speed of about $c/6$, i.e some 10 MeV per nucleon. The interaction of such jet particles (initially protons which would recombine into $\alpha$-particles at later time during the expansion) either with the supernova envelope or a low-density medium at large distances represent another extreme case that could be considered by the present model, but will not be discussed here.

## RESULTS

In the above-described model, the parameter space spanning the different possible characteristics of the accelerated particles is extremely vast. For this reason, we will restrict ourselves to present the results for a limited number of cases and leave a broader discussion for future publication [9]. Only proton fluxes $\Phi_p = 10^{-4}$ mb$^{-1}$s$^{-1}$ and a $\alpha$-to-proton flux ratio $\Phi_\alpha/\Phi_p = 0.1$ typical of solar flares will be considered. The energy spectra is a constant distribution with $E_{min} = 0$ and $E_{max} = 10$ MeV.

Quite generally, in the case of small fluxes, the proton and $\alpha$ captures are slower than the $\beta^+$ decays, so that the nuclear flow follows a path along the valley of $\beta$-stability on the neutron-deficient side. For larger values of $\Phi_p$ and $\Phi_\alpha$, neutron-deficient nuclei are produced significantly and many unstable nuclei are produced up to the proton-drip line.

For the adopted energy distribution, the low energy component of the projectile flux is responsible for radiative captures, while the high energy one favours a large number of transfer reactions. In particular, the (p,$x$n $y$p $z\alpha$ reactions are responsible, at least at large fluences for a significant production of neutrons which through their radiative neutron captures tend to bring the flow back to the valley of $\beta$-stability. Such secondary neutrons may play a crucial role, especially if we are concerned by a possible production of

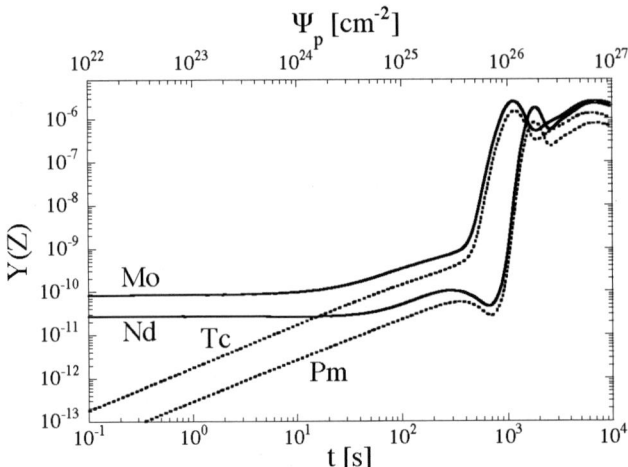

**FIGURE 1.** Evolution with time t, or equivalently proton fluence $\Psi_p$, of the Tc and Pm molar fractions as well as their neighbour element Mo and Nd, respectively. The abundance evolution results from the irradiation by energetic particles corrresponding to $\Phi_p = 10^{-4}$ mb$^{-1}$ s$^{-1}$, $\Phi_\alpha/\Phi_p = 0.1$, and $E_{max} = 10$ MeV.

actinides. No p- or $\alpha$-captures could possibly lead to the synthesis of actinides or sub-actinides due to the low $\mu$-second $\alpha$-unstable island on the $85 \leq Z \leq 89$ p-rich side. Only a nuclear path on the neutron-rich side could be taken to produce transuranium significantly. The neutron abundance is followed with the same time-dependent equation as in Eq. 1. It is of particular interest to note that the abundance distributions obtained with Eq. 1 is rather *independent* of the local density $\rho$. Test calculations were performed for densities ranging between $\rho = 10^{-9}$ to 1 g/cm$^3$, and systematically lead to a constant product $\rho Y_n$, which in first approximation can be expressed by

$$\rho Y_n = \frac{\sum_{k,x} Y_k x [\Phi_p <\sigma>^x_{p,k} + \Phi_\alpha <\sigma>^x_{\alpha,k}]}{\sum_k Y_k N_a <\sigma v>_n} . \qquad (4)$$

The resulting abundance distributions are therefore unaffected by changes in the density. In other words, the impact of the neutron captures remains rather insensitive to the depth within the stellar surface at which the irradiation process takes place. All calculations shown here have been obtained for a local density of $\rho = 10^{-7}$ g/cm$^3$.

In Fig. 1, we show the evolution of the two radio-elements Tc and Pm as well as their neighbour elements Mo and Nd, respectively, as a function of time, or equivalently the proton fluence $\Psi_p = \int_0^t \Phi_p dt$. Minimum values of $10^{23}$ cm$^{-2}$ are required to produce Tc ad Pm significantly. In Fig. 2, the full elemental abundance distribution is compared with the solar and HD 101065 ones for four different irradiation times. If fluences of the order $10^{26-27}$ cm$^{-2}$ can be achieved, not only globally the abundances of the elements heavier

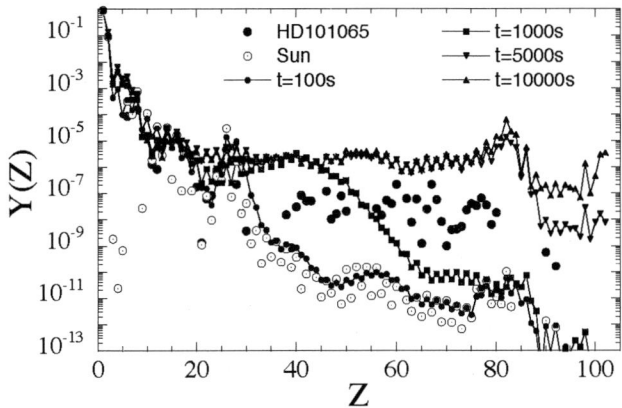

**FIGURE 2.** Abundance distribution resulting from the irradiation process with $\Phi_p = 10^{-4}$ mb$^{-1}$ s$^{-1}$, $\Phi_\alpha/\Phi_p = 0.1$ and $E_{max} = 10$ MeV at four different irradiation times. For comparison, the HD 101065 abundances determined by Cowley et al. [3] and the solar abundances taken as initial composition are also plotted.

than iron can be increase by 5 orders of magnitude, but also the neutron flux becomes large enough to bridge the $N > 126$ $\alpha$-unstable region between Bi and Ra and produce actinides as high as $Z \simeq 100$ in very large amount. Obviously, such high fluences can hardly be expected in normal stellar flares and the relevance of such a scenario must be confirmed by more realistic astrophysics models.

It could be noted in Fig. 2 that the abundance pattern is rather flat, and indeed with the dominant capture of charged-particles with energies up to 10 MeV per nucleon, all nuclear structure effects are smoothed out. The high-fluence abundance distribution is consequently quite different from the observed one. Some structure effects are, however, recovered if, at the end of the irradiation, neutrons are still largely present and keep on participating to the nucleosynthesis.

If we now assume that the material irradiated for a given time $\tau_{irr}$ is mixed (in a way or another) with unaffected surounding layers of solar-type composition, or equivalently that only part of the observed material has been irradiated, the final abundances can be estimated by

$$Y^{mix}_{(Z,A)} = f_{mix} Y^{irr}_{(Z,A)} + (1 - f_{mix}) Y^{\odot}_{(Z,A)} \qquad (5)$$

where $f_{mix}$ is an arbitrary dilution factor. For the specific parameter sets considered in the case of Fig. 3, we see that globally the observed abundances can be remarkably explained, including the presence of Tc, Pm and transuranium. Even the isotopic ratio of $^6$Li/$^7$Li=0.2 is relatively close to the value of 0.3 derived from observation [14], although the Li abundance is here overestimated by about 1 dex.

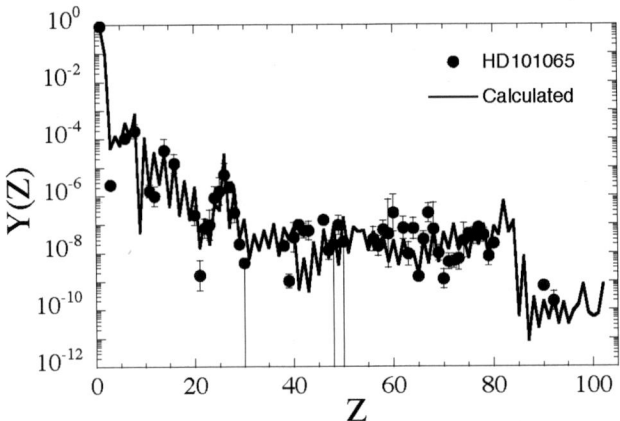

**FIGURE 3.** Abundance distribution resulting from the mixing of solar abundances with irradiated material. The fluxes are characterized by $\Phi_p = 10^{-4}$ mb$^{-1}$ s$^{-1}$, $\Phi_\alpha/\Phi_p = 0.1$, $E_{max} = 10$ MeV and an irradiation timescale of $\tau_{irr} = 5000 s$. The irradiation takes place up to a time $t = 10^4$ s before being mixed with the solar composition. The dilution factor is set to $f_{mix} = 0.02$. The full dots correspond to the HD 101065 abundance determination of Cowley et al. (2000) [3].

Finally, it should be noted that all abundances shown so far correspond to nuclei produced at the end of the irradiation and many of them are consequently unstable. After the irradiation and the freeze-out of the neutron captures, most of these unstable nuclei decay back to the valley of $\beta$-stability by $\beta^\pm$ and $\alpha$-decays. In particular, the abundance distribution shown at a time $t = 10^4$ s in Fig. 3 can still be slightly modified by such decays, but without any major change in our present conclusions.

## CONCLUSION

The present study aims at exploring spallation nucleosynthesis as a possible explanation of the peculiar abundances observed at the surface of HD 101065. Due to the unknown characteristics of the accelerated particles that could be held responsible for the nucleosynthesis process, a purely parametric approach is followed, taken as free parameters the proton and $\alpha$-particle flux amplitude and energy distribution, the time of irradiation and the possible mixing with unaffected material. We have shown that at least the specific simulation shown here with large fluences tend to be compatible with many different observational aspects. In particular, a significant production of $Z > 30$ heavy elements by radiative and spallation reactions can be achieved, and most of all a significant synthesis of radio-elements, such as Tc, Pm but also actinides up to at least $Z \simeq 100$. This requires, however, fluences by far larger than what would be expected in normal or even extreme stellar flares. The possible existence of such irradiation events need to be confirmed by hydrodynamics simulations, but most of all by spectroscopic observations through the detection of short-lived radio-elements. If such observations clearly indicate the need to introduce large-fluence irradiations, theoretical astrophysics will face a new challenge, relatively similar as the one faced by the r-process nucleosynthesis, i.e identifying the relevant astrophysical sites in which such required extreme conditions could be met.

The present study did not aim to explain the observed pattern in all details but rather to show to what extent spallative nucleosynthesis only could possibly and globally explain the surface composition of the chemically peculiar Ap star HD 101065. Improvements still need to be brought to the parametric modelling, in particular a better description of the neutron capture rate at low temperatures and a full survey of the large parameter space characterizing the property of the accelerated particles. Realistic models describing the acceleration processes and the possibility to find the large fluences required need to bring their share in making this scenario a viable one. But most of all, observation spectroscopy need to confirm the presence of short-lived radio-elements at the surface of CP stars and hopefully in a close future provide abundance determination for these clear tracers of nuclear activity at the surface or in the nearby surroundings.

## REFERENCES

1. D.W. Kurtz, Mon. Not. R. Astron. Soc. **200**, 807 (1982)
2. P. Padovani, in Proc. IAU symposium Nr 224, (eds. Zverko et al.), 485 (2004)
3. C.R. Cowley, T. Ryabchikova, F. Kupka, D.J. Bord, G. Mathys and W.P. Bidelman, Mon. Not. R. Astron. Soc. **317**, 299 (2000)
4. G. Michaud, Astroph. J. **160**, 641 (1970)
5. C.R. Cowley, W.P. Bidelman, S. Hubrig, G. Mathys and D.J. Bord, Astron. Astrophys. **419**, 1087 (2004)
6. V.F. Gopka, A.V. Yushchenko, A.V. Shavrina, D.E. Mkrtichian, A.P. Hatzes, S.M. Andrievsky and L.V. Chernysheva, in Proc. IAU symposium Nr 224, (eds. Zverko et al.), 734 (2004)
7. M. Arnould and S. Goriely, Nucl. Phys. **A777**, 157 (2006)
8. D. Clayton, E. Dwek and S.E. Woosley, Astroph. J. **214**, 300 (1977).
9. S. Goriely, Astron. Astrophys., submitted (2007)
10. J.P. Meyer, Astroph. J. Suppl. **7**, 417 (1972).
11. A.J. Koning, S. Hilaire and M.C. Duijvestijn, Proc. Int. Conf. on Nuclear Data for Science and Technology (eds. C. Haight et al.) AIP Conference Vol. 769, p. 1154, 2005.
12. I. Leya, A.N. Halliday and R. Wieler, Astroph. J. **594**, 605 (2003).
13. K. Maeda and K. Nomoto, Astroph. J. **598**, 1163 (2003)
14. A.V. Shavrina, N.S. Polosukhina, Ya. V. Pavlenko *et al.*, Astron. Astrophys. **409**, 707 (2003)

# Accelerated Particle Properties in Solar Flares from Gamma-Ray Line Observations

J. Kiener*, V. Tatischeff*, G. Weidenspointner[†], M. Gros** and A. Belhout[‡]

*CSNSM, IN2P3-CNRS and Univ Paris-Sud, 91405 Campus Orsay, France
[†]Centre d'Etude Spatiale des Rayonnements, 9 av. du Colonel Roche, 31028 Toulouse, France
**DSM/DAPNIA/Service d'Astrophysique, CEA Saclay, 91191 Gif-sur-Yvette, France
[‡]USTHB, Faculté de Physique, BP 32, El Alia, 16111 Bab Ezzouar, Algiers, Algeria

**Abstract.** In solar flares with strong $\gamma$-ray emission, several prominent narrow lines can usually be observed emerging from a broad structure in the several hundred keV to several MeV region. They are produced by interactions of energetic protons, $^3$He and $\alpha$-particles with the most abundant isotopes in the solar atmosphere. The underlying continuum is due to electron and positron bremsstrahlung and to the superposition of broad lines from accelerated heavy ions as well as numerous narrow nuclear lines which are too weak to be resolved individually. The properties of the accelerated particles in the October 28, 2003 flare observed with the gamma-ray spectrometer SPI onboard INTEGRAL were deduced by comparison of prominent narrow line fluence ratios, the line shapes of the 4.4 MeV $^{12}$C and 6.1 MeV $^{16}$O lines and the total nuclear line emission in the framework of a thick target interaction model.

**Keywords:** solar impulsive flares, nuclear gamma-ray lines
**PACS:** 25.10.+s, 95.55.Ka, 96.60.qe

## INTRODUCTION

Solar flare $\gamma$-ray emission is induced by interactions of accelerated particles within the solar atmosphere. Below several hundred keV, the emission is mostly due to electron bremsstrahlung where the differential photon flux usually follows a power law in energy. At higher photon energies one often observes an additional component superposed on the power-law continuum which is mainly from nuclear $\gamma$-ray line emission and sometimes becomes the dominant process in the several MeV range. This emission can be fairly complex with observations showing the presence of strong and relatively narrow lines and an underlying broad structure probably composed of strong broad lines and a quasi-continuum of numerous unresolved weak lines.

The most prominent narrow lines are due to the prompt deexcitation of low-lying levels in the abundant species $^{12}$C, $^{16}$O, $^{20}$Ne, $^{24}$Mg, $^{28}$Si and $^{56}$Fe which are excited by collisions with energetic protons, $^3$He and $\alpha$-particles. Other important lines are from $\alpha$ + $^4$He reactions leading to the production of $^7$Li and $^7$Be in their first excited states, from neutron capture on hydrogen and from positron annihilation. Their relative intensities can be used to extract information about the composition of the ambient medium (e.g. [1]).

The prompt lines are slightly Doppler-broadened and energy-shifted in solar flares due to the recoil of the emitting nucleus after excitation in the collision with the light ion. The energy shift and the broadening do usually not exceed a few percent of the line energy, its details may, however, contain information about the energetic particle composition and their directional distribution with respect to the line-of-sight. A comparison of calculated line shapes with observed ones has already been done for several flares observed by SMM [2] [3]. The conclusions on accelerated particle distributions were however limited by the modest resolution of the scintillation detectors.

Since the launch of RHESSI and INTEGRAL in 2002, both equipped with high-resolution Ge detectors, several flares with prominent narrow lines susceptible to detailed line-shape analysis were observed. We will present in the first chapter such an analysis of the solar flare of October 28, 2003 observed by INTEGRAL/SPI with emphasis on the shape of the 4.438 MeV line of $^{12}C$ and the 6.129 MeV line of $^{16}O$.

The same prompt lines can also be produced in reverse kinematics by interactions of accelerated heavy ions with hydrogen and helium. The velocity of the emitting nuclei is much higher in this case leading to broader lines. Observations of these broad lines would provide a direct access to the heavy-ion content of the energetic particles interacting within the solar atmosphere. However, the extraction of those lines is particularly difficult due to their large width and the superposition with the numerous unresolved transitions from high-lying nuclear levels which are excited in the nuclear collisons of the accelerated light and heavy ions. The second chapter describes our approach to model the total $\gamma$-ray line emission including the prominent narrow and broad lines and the quasi-continuum of weak lines.

## LINE SHAPES IN THE LABORATORY AND IN SOLAR FLARES

First calculations of the spectral shape of prompt emission lines were performed by Ramaty & Crannell [4] and further developed by Ramaty et al. [5] and Murphy, Kozlovsky & Ramaty [6]. These calculations were based on empiric parameterizations of differential reaction cross sections and population amplitudes for the excited nuclear levels.

Werntz, Lang & Kim [7] and Lang & Werntz [8] introduced nuclear reaction model calculations to predict line shapes produced in proton and $\alpha$-particle inelastic scattering. Kiener, de Séréville & Tatischeff [9] used new accelerator data for the 4.438 MeV line of $^{12}C$ and extensive optical model calculations to refine the line shape calculations for inelastic scattering off $^{12}C$ and to deduce values for a simple parameterized model of proton and $\alpha$-particle induced spallation of $^{16}O$.

The 4.438 MeV $\gamma$ ray of $^{12}C$ is probably the best candidate for line-shape analysis of prompt emission lines observed in solar flares. It is often the most intense prompt line and has a relatively large width of typically hundred keV, which is easily resolved by modern high-resolution $\gamma$-ray detectors. Nuclear reaction calculations are not very complex for this line because the emitting $2^+$, 4.439 MeV level in $^{12}C$ is the only particle-bound excited state in this nucleus. This reduces greatly its population by cascade transitions from higher-lying levels, which can be safely neglected in the calculations.

The main nuclear reaction mechanism is proton inelastic scattering off $^{12}C$, except for very soft energetic particle spectra, where $\alpha$-particle inelastic scattering may dominate. Further significant reactions are proton and $\alpha$-particle induced spallation of $^{16}O$. Other reactions like spallation of $^{14}N$ or still heavier abundant species like $^{20}Ne$ do not

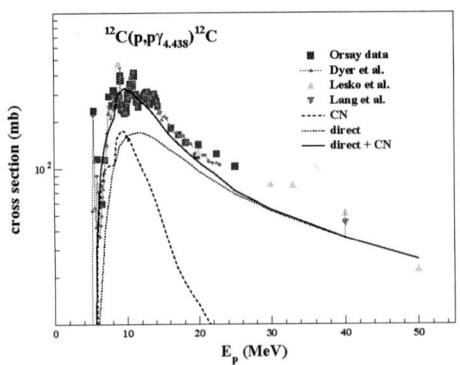 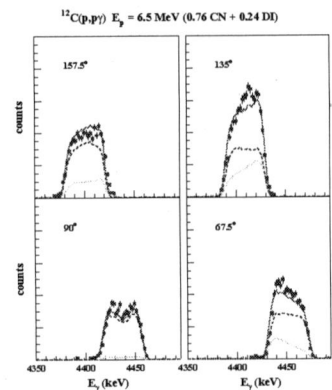

**FIGURE 1.** Left: Cross section excitation function for the 4.438 MeV $\gamma$-ray production by proton inelastic scattering off $^{12}$C. Besides the data from Orsay, experimental data of Dyer et al. [12], Lesko et al. [13] and Lang et al. [14] are shown. Right: Measured line shapes of the 4.438 MeV $\gamma$-ray of $^{12}$C produced by a 6.5 MeV proton beam incident on a thin carbon foil. The laboratory angles of the $\gamma$-ray detectors with respect to the beam direction are indicated. Dashed and dotted lines in both figures show the calculations for the compound nucleus component and the direct reaction mechanism, respectively. The sum of both is shown by the solid lines.

contribute significantly. As an example, we present in the following the line shape calculations for proton inelastic scattering.

Figure 1 shows the cross section excitation function for $^{12}$C(p,p$\gamma_{4.438}$)$^{12}$C as measured in two experiment campaigns at the tandem accelerator of IPN Orsay [10, 11] together with other literature data. One can distinguish two energy regions for the two main reaction mechanisms at work for this particular reaction. The peaks present at energies below about 15 MeV are due to isolated resonances in the compound nucleus $^{13}$N. At higher energies, the smoother behaviour indicates the dominance of the direct reaction mechanism. Calculations for the compound nucleus mechanism (CN) with the Hauser-Feshbach formalism and the direct mechanism with the coupled-channels formalism (CC) illustrate their relative importance as a function of energy. The sum of both contributions describes reasonably well the magnitude of the cross section and its average energy dependence.

While at energies above 15 MeV CC calculations for the inelastic scattering reactions reproduce fairly well measured 4.438 MeV line shapes in the laboratory, the contribution of CN resonances must be added below that energy. We show an example at $E_p$ = 6.5 MeV which corresponds to an excitation energy of 7.94 MeV in the compound nucleus $^{13}$N, close to the maximum of the broad ($\Gamma \approx$ 1.5 MeV) 3/2$^+$ resonance at 7.90 MeV. For this resonance the only possible incoming orbital angular momentum is l = 2 and the orbital angular momenta of the outgoing proton + $^{12}$C$^{\star}_{2^+,4.439}$ are l = 0, 2 (l = 4 can be neglected due to small Coulomb + centrifugal barrier transmission). Angular distributions and correlations between the outgoing proton and the emitted $\gamma$-ray are completely fixed by angular momentum algebra (see e.g. Ferguson [15]) and one only has to tune the relative contributions of $l = 0$ and $l = 2$ proton emission and the ratio of CN and direct reaction (see Fig. 1).

This study will be extended to all measured energies in the Orsay experiments for both proton and $\alpha$-particle scattering off $^{12}C$ which provide a neat energy coverage of the CN resonance region. So far line-shape analyses of solar flares were done with less elaborated methods (see [9]) with estimated theoretical uncertainties on the line profiles of the order 20-40%. This applies also to the analysis of the solar flare presented hereafter.

One of the strongest observed flares occured on October 28, 2003, the X-ray flux starting at 9:41 UT at solar coordinates 18E20S as observed by the GOES satellites. The $\gamma$-ray flux as detected by INTEGRAL started around 11:02 UT and lastet for about 15 minutes. Photon spectra taken with the $\gamma$-ray spectrometer SPI onboard INTEGRAL showed the presence of a very strong line at 2.2 MeV from neutron capture on hydrogen and strong lines at 4.4 MeV from $^{12}C$ deexcitation and 6.1 MeV from $^{16}O$ deexcitation. Details of the observation can be found in [16] [17].

For both deexcitation lines SPI observed a total of 500-700 photons, corresponding to fluences of about 130-140 photons/cm$^2$. The good statistics, together with the low background and excellent energy resolution of the Ge telescope of SPI (2-3 keV at 1.33 MeV) allowed a detailed analysis of the line shapes. We calculated the theoretical line profiles assuming a thick target interaction model for the interaction of the accelerated protons and $\alpha$-particles within the solar atmosphere using standard solar abundances of C, O and Ne and another set with reduced C/O. We included also broad line production by accelerated heavy ions in inverse kinematics. The particle energy spectra were taken as unbroken power-law distributions from reaction threshold to 500 MeV per nucleon for all species. The power-law index s was extracted from the fluence ratio of the 2.2 MeV line to the 4.4 and 6.1 MeV deexcitation lines to be $3 \leq s \leq 4$ [18].

Theoretical line shapes were compared to observed ones by varying systematically the energetic $\alpha$-particle to proton ratio $\alpha/p$ and the angular distribution of the energetic particles around the flare axis defined as the normal to the solar surface at the flare location. For each set of parameters a normalization parameter for the 4.4 + 6.1 MeV line intensity was fitted. Best fits for different power-law indices s and different time periods pointed all to downward-directed angular distributions and $\alpha/p \approx 0.1$ in agreement with solar energetic particle (SEP) observations in interplanetary space [19] and. An example for the time period with strongest prompt line emission is shown in Fig. 2.

For the angular distributions we used among others results of the model of Murphy et al. [20] for particle transport in a magnetic loop including pitch-angle scattering on MHD irregularities. The chromospheric portion of the loop where most interactions take place is taken perpendicular to the solar surface. Converging field lines there lead to magnetic mirroring of the energetic particles. The parameter $\lambda$ in this model characterizes the mean-free path of the particles against pitch-angle scattering scaled to the half-length of the chromospheric portion of the loop. $\lambda = 30$ gives a relatively narrow downward-directed distribution of the interacting particles due to saturated pitch-angle scattering while $\lambda = 300$ gives a broader distribution approaching a fan beam parallel to the solar surface due to the longer mean-free path and therefore increasing importance of magnetic mirroring. $\lambda = 300$ could be excluded at the 99.7% level in combined fits of the 4.4 and 6.1 MeV line shapes. Similar conclusions could be drawn in the analysis of

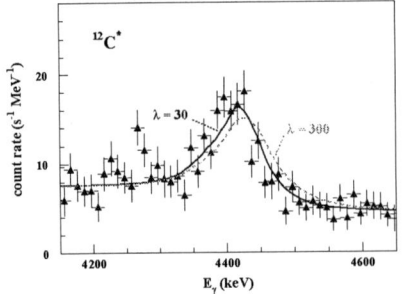
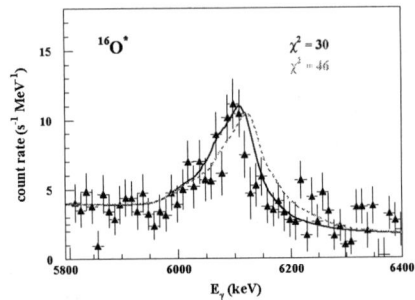

**FIGURE 2.** Line shapes of the 4.4 MeV $^{12}$C and 6.1 MeV $^{16}$O $\gamma$-rays from the solar flare of Oct. 28, 2003 as observed by INTEGRAL/SPI. Continuous lines show calculated line shapes with best-fit $\alpha/p$ and a predominantly downward-directed angular distribution ($\lambda = 30$) of energetic particles, while dashed lines show the best fit with a broader angular distribution ($\lambda = 300$). The indicated $\chi^2$'s on the right figure are for the combined fit of the 4.1 and 6.1 MeV line with a total of 51 d.o.f. The small values of the reduced $\chi^2$ can be explained by the theoretical uncertainties of the line shape calculations which have been added for conveniance in the fits to the statistical errors of the data points.

two other recent solar flares observed with SPI [21].

## TOTAL NUCLEAR $\gamma$-RAY LINE EMISSION IN SOLAR FLARES

Additionally to the above mentionned lines, two other lines of $^{16}$O at 6.9 and 7.1 MeV and the deexcition lines of the first excited states of $^{20}$Ne, $^{24}$Mg and $^{28}$Si could be identified above the $2\sigma$-significance level. The residual structure between $\approx$ 0.6 - 8.5 MeV (effective detection range of SPI for solar flare photons, for details see [17]) is due to bremsstrahlung of accelerated electrons, strong broad lines and a quasi-continuum of weak nuclear lines. For hard energy spectra, bremsstrahlung of energetic positrons and electrons from pion decay may also be important in the region above several MeV.

Share & Murphy [22] made an attempt to extract the broad lines from accelerated heavy-ion interactions in solar flare spectra recorded by SMM. They found indication that the accelerated Fe/C ratio in these events exceeded the ambient Fe/C by a factor of $\approx$ 5 which was consistent with enhancements of the Fe/C ratio observed in SEP fluxes from interplanetary space after impulsive-type solar flares [19]. It was, however, noted that a substantial uncertainty remained because of the weak-line quasi-continuum component, whose strength and shape was relatively uncertain.

There is only very sparse experimental information on this component whose importance for solar flare spectra was first discussed in [5]. Since then there was, if any, only poor progress related to this problem. We decided in a first step to rely entirely on calculations, taking profit of the development of modern global nuclear reaction codes like TALYS [23] and Empire [24]. Both include the major nuclear reaction mechanisms at energies below $\approx$ 250 MeV per nucleon and use comprehensive librairies of nuclear structure data.

We made calculations with TALYS for proton, $^3$He and $\alpha$-particle induced reactions with 19 different nuclei from $^4$He to $^{58}$Ni, representing more than 99% of the total number of target nuclei (except H) of the solar atmosphere. For the calculated $\gamma$-ray spectrum, we included transitions from up to the first 25 known experimental levels. This amounted in total to $\approx$ 5800 different explicit $\gamma$-ray transitions in the target nuclei and reaction products. Higher-lying levels and their deexcitations were treated in the continuum approach with standard level-density models and $\gamma$-ray strength functions. Above 250 MeV per nucleon, we took the results of the calculation at 250 MeV. This should be a reasonable approximation since reactions at these energies make less than 10% of the total prompt $\gamma$-ray production for s $\geq$ 3.0.

The $\gamma$-ray spectrum was then calculated with the thick target interaction model. We used coronal abundances for the composition of the interaction region and throughout an accelerated $^3$He/$\alpha$ ratio of 0.5, which is a typical value observed in SEP from impulsive flares [19]. For the accelerated electron bremsstrahlung continuum we used an unbroken power-law. However, because we had no data below 500 keV, we could neither extract the spectral index s($\gamma e^-$) nor the normalization. We explored therefore the range 2.5 $\leq$ s($\gamma e^-$) $\leq$ 3.5 and treated the normalization as a free parameter. We also took into account the continuum emission from the bremsstrahlung of positrons and electrons produced in the decay of charged pions, which we calulated with the formalism given in [25]. The production of pions was normalized to that of the narrow 4.4 MeV line fluence.

We calculated spectra for 3 different combinations of the spectral index s and the $\alpha$/p ratio determined in [18] which fit the observed fluence ratio $\Phi_{2.2}/\Phi_{4.4+6.1}$. Thereof we could exclude the combination s = 4 and $\alpha$/p = 0.5, because it overproduced the 3.56 MeV line of $^6$Li, mainly produced by the $^4$He($^3$He,p)$^6$Li reaction. This agrees with the line-shape analysis, where an $\alpha$/p ratio of 0.5 with s = 4 could be excluded at the 99.7% level. Similarly, s = 3 and $\alpha$/p = 0.02 produced too much high-energy positrons and electrons from pion decay, their bremsstrahlung flux exceeding the total observed $\gamma$-ray emission for E $\geq \approx$ 6 MeV.

A satisfactory reproduction of the spectrum was observed with s = 3.5 and $\alpha$/p = 0.1, and energetic particle composition similar to the SEP impulsive flare composition of Reames ([19], table 9.1) (see Fig. 3). Strong enhancements of accelerated heavy ions ($>$ 3) with respect to this composition could be excluded because it resulted in too strong continuum emission above 2 MeV. Unfortunately, more stringent limits could not be set because of the uncertainty on the electron bremsstrahlung component.

With the exception of s = 4 and $\alpha$/p $\approx$ 0.5, excluded by the line-shape analysis the calculations all underestimate the observed spectrum just below the 2.2 MeV neutron-capture line. This may be explained by Compton scattering of the 2.2 MeV $\gamma$ rays in the solar atmosphere, which would be compatible with downward directed beams resulting in neutron capture in relatively deep layers.

We would also like to point out the strong feature from $\alpha$ + $^4$He reactions around 450 keV which is very interesting for extracting the flux and angular distribution of the energetic $\alpha$-particles. It is shown by the dotted line in Fig. 3 (left panel), where one can observe several other strong lines (dashed line), mainly of $^{23}$Na and $^{23}$Mg deexcitation which are superposed. In particular the 440 keV line of $^{23}$Na is strongly produced by proton and $\alpha$-particle induced spallation of $^{24}$Mg and fusion-evaporation reactions of $\alpha$-

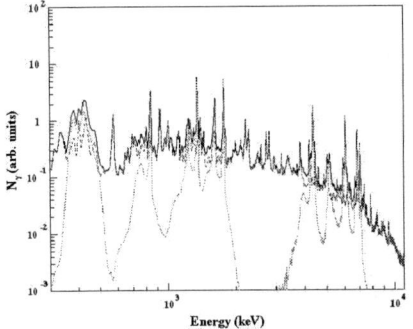
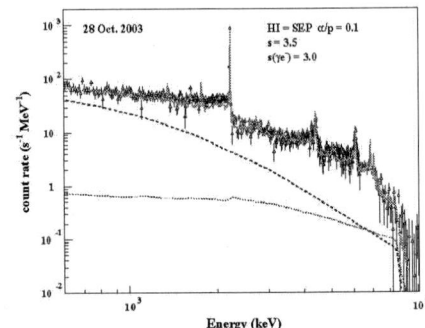

**FIGURE 3.** Left panel: Calculated prompt nuclear γ-ray line spectrum with s = 3.5, $\alpha/p$ = 0.1, $^3$He/$\alpha$ = 0.5, coronal abundances for the ambient medium and energetic particle composition from impulsive flare SEP observations. Dotted lines show contributions of 17 strong prominent lines, the contribution of other lines is indicated by the dashed line. Right panel: Spectrum of the Oct. 28, 2003 solar flare with calculated spectrum including bremsstrahlung from accelerated electrons (dashed line) and leptons from pion-decay (dotted line). The prompt nuclear γ-ray line emission was normalized to reproduce the observed fluence of the narrow 4.4 MeV line.

particles with $^{20}$Ne. For $\alpha/p$ = 0.1 and s = 3.5, $\alpha$ + $^4$He reactions contribute only about 50% of the total flux between ≈ 300 and 500 keV. In solar flare spectra, this structure is superposed with the 3γ-continuum from positron annihilation. It requires thus a careful line-shape analysis of the line complex to infer the intensity and the angular distribution of the accelerated $\alpha$-particles. This should be possible in solar flare spectra observed by RHESSI, at least for flares with high $\alpha/p$ ratios.

## CONCLUDING REMARKS

We have shown the potential of line-shape analysis applied to solar flare spectra to constrain the properties of the accelerated particles interacting in the solar atmosphere. The prominent narrow 4.4 and 6.1 MeV lines of the solar flare of Oct. 28, 2003 were investigated with this method in the framework of a thick-target interaction model. The spectral index of the accelerated particles, and the $\alpha/p$ ratio could be determined.

We could furthermore determine the angular distributions of the interacting particles, which favored relatively narrow downward-directed distributions. The theoretical line-shape calculations which have been employed suffer, however, still from approximations. They could be improved by explicitly taking into account isolated CN resonances as we showed in fits of measured 4.4 MeV line shapes in an experiment at the Orsay tandem accelerator.

The total prompt nuclear γ-ray line emission is also sensitive to accelerated particle properties. Using results of a preliminary calculation with a global nuclear reaction code to which were added bremsstrahlung continuums from accelerated electrons and pion-decay leptons reproduced fairly well the observed spectrum. The derived composition of the fast particles interacting in the solar atmosphere is compatible with the average measured composition in interplanetary space of SEP from impulsive flares. The calculations should be systematically confronted with existing data to improve the reliability of the different models employed in the nuclear reaction codes and extended to energies above 250 MeV.

It would be very interesting to obtain in the future solar flare $\gamma$-ray spectra extended to several hundreds of MeV. This would include the bump around 65 MeV from $\pi^0$ decay, whose shape depends strongly on the energy spectrum and the angular distribution of the accelerated particles. A full spectral coverage from several keV to several GeV may be possible for flares at the beginning of the next solar cycle when the satellite GLAST will be launched in 2007 and if RHESSI and INTEGRAL continue their mission for some more years.

## ACKNOWLEDGMENTS

We like to thank M. Harris for interesting discussions on solar flares and S. Goriely for his advice and discussions on nuclear reaction codes.

## REFERENCES

1. R. J. Murphy, R. Ramaty, B. Kozlovsky, and D. V. Reames, *Astrophys. J.* **371**, 793 (1991).
2. G. H. Share, and R. J. Murphy, *Astrophys. J.* **485**, 409 (1997).
3. G. H. Share, R. J. Murphy, J. Kiener, and N. de Séréville, *Astrophys. J.* **573**, 464 (2002).
4. R. Ramaty, and J. C. Crannell, *Astrophys. J.* **203**, 766 (1976).
5. R. Ramaty, B. Kozlovsky, and R.E. Lingenfelter, *Astrophys. J. Suppl. Ser.* **40**, 487 (1979).
6. R.J. Murphy, B. Kozlovsky, and R. Ramaty, *Astrophys. J.* **331**, 1029 (1988).
7. C. Werntz, F. L. Lang, and Y. E. Kim, *Astrophys. J. Suppl. Ser.* **73**, 349 (1990).
8. F. L. Lang, and C. W. Werntz, in AIP Conf. Proc. 232, Gamma-Ray Line Astrophysics, P. Durouchoux & N. Prantzos (New York:AIP), 445 (1991)
9. J. Kiener, N. de Séréville, and V. Tatischeff, *Phys. Rev.* **C 64**, 025803 (2001).
10. J. Kiener, M. Berheide, N. L. Achouri, A. Boughrara, A. Coc, A. Lefebvre, F. de Oliveira Santos, and Ch. Vieu, *Phys. Rev.* **C 58**, 2174 (1998).
11. A. Belhout, J. Kiener, A. Coc et al., to be submitted to *Phys. Rev. C*
12. P. Dyer, D. Bodansky, A. G. Seamster, E. B. Norman, and D. R. Maxson, *Phys. Rev.* **C 23**, 1865 (1981).
13. K. T. Lesko, E. B. Norman, R.-M. Larimer, S. Kuhn, D. M. Meekhof, S. G. Crane, and H. G. Bussell, *Phys. Rev.* **C 37**, 1808 (1988).
14. F.L. Lang, C.W. Werntz, C.J. Crannell, J.L. Trombka, and C.C. Chang, *Phys. Rev.* **C 35**, 1214 (1987).
15. A. J. Fergusen, "Angular Correlation Methods in Gamma-Ray Spectroscopy", North-Holland Publishing Company, Amsterdam, 1965
16. M. Gros, V. Tatischeff, J. Kiener et al., in *Proc. 5th INTEGRAL Workshop*, Munich, Germany, 16-20 February 2004, ed. V. Schönfelder, G. Lichti & C. Winkler, ESA SP-552, 669
17. J. Kiener, M. Gros, V. Tatischeff, and G. Weidenspointner, *Astronomy&Astrophysics* **445**, 725 (2006).
18. V. Tatischeff, J. Kiener, and M. Gros, in *Proc. of the 5th Rencontres du Vietnam, "New Views on the Universe"*, edited by L. Celnikier, Y. Giraud-Heraud, and J. Tran Thanh Van (The Gioi Publishers, Hanoi, 2005), 137
19. D. V. Reames, *Space Science Rev.* **90**, 413 (1999).
20. R. J. Murphy, X.-M. Hua, B. Kozlovsky, and R. Ramaty, *Astrophys. J.* **351**, 299 (1990).
21. M.J. Harris, V. Tatischeff, J. Kiener, M. Gros, and G. Weidenspointner, *Astronomy&Astrophysics*, in print

22. G. H Share and R. J. Murphy, in *Proc. XXVI International Cosmic Ray Conference*, Salt Lake City, USA, 17-25 August 1999, ed. D. Kieda, M. Salamon, and B. Dingus
23. A. J. Koning, S. Hilaire, and M. C. Duijvestijn, "TALYS: Comprehensive nuclear reaction modeling", *Proceedings Int. Conf. on Nuclear Data for Science and Technology - ND2004*, Sep. 26 - Oct. 1, 2004, Santa Fe, USA, AIP Conference Proceedings, Vol. 769, 1154
24. M. Herman et al., *Proceedings Int. Conf. on Nuclear Data for Science and Technology - ND2004*, Sep. 26 - Oct. 1, 2004, Santa Fe, USA, AIP Conference Proceedings, Vol. 769; available at URL: http://www.nndc.bnl.gov/empire-2.18
25. R. J. Murphy, C. D. Dermer, and R. Ramaty, *Astrophys. J. Suppl. Ser.* **63**, 721 (1987)

# Neutrino Probes of Galactic and Extragalactic Supernovae

Shin'ichiro Ando

*California Institute of Technology, Mail Code 130-33, Pasadena, CA 91125*

**Abstract.** Neutrinos are a messenger of extreme condition inside a supernova core and a new-born neutron star. Since current ground-based detectors have potential to detect $\sim 10,000$ neutrinos from supernova at the galactic center, they could tell us lots of important physics. It includes: explosion mechanism, shock wave propagation, core temperature, and gravitational binding energy, as well as neutrino properties as elementary particle. In addition to the galactic supernova neutrino burst, one can still learn about them with diffuse supernova neutrino background, which is also soon to be detected. We review current situation from both points of view, and discuss prospects for future neutrino astrophysics.

**Keywords:** supernovae; neutrinos; diffuse background; elementary particle physics
**PACS:** 97.60.Bw; 98.70.Vc; 13.15.+g; 14.60.Pq; 95.85.Ry

## 1. INTRODUCTION

A core-collapse supernova explosion is one of the most spectacular events in astrophysics, and it attracts a great deal of attention from many physicists and astronomers. It also produces a number of neutrinos; 99% of its gravitational binding energy is transformed to neutrinos. Therefore, neutrinos play an essential role in supernovae. For example, neutrino interaction might be affecting the core-collapse dynamics, by causing the observed kick velocity of newly born neutron stars (see, e.g., Ando [1], and references therein).

However, how core-collapse supernovae explode still remains one of the greatest mysteries in astrophysics. Nuclear fusion reactions in the core of a massive star produce progressively heavier elements until a Chandrasekhar mass of iron is formed, and electron degeneracy pressure cannot support the core under the weight of the stellar envelope. The core collapses until it reaches nuclear densities and neutrino emission begins; then an outgoing bounce shock should form, unbinding the envelope and producing the optical supernova. While this is successful in nature, in most numerical supernova models, the shock stalls, so that the fate of the entire star is to produce a black hole (after substantial neutrino emission), but no optical supernova.

Since the gravitational energy release transferred to neutrinos, about $3 \times 10^{53}$ erg, is $\sim 100$ times greater than the required kinetic energy for the explosion, it is thought that neutrino emission and interactions are a key diagnostic or ingredient of success. However, very little is directly known about the total energies and temperatures of the neutrino flavors. The $\simeq 20$ events from SN 1987A were only crudely consistent with expectations for $\bar{\nu}_e$, and gave very little information on the other flavors. It is thus essential to collect more supernova neutrino events to understand the supernova physics more precisely. More concretely, we are eager to know what the explosion mechanism is, what the initial conditions of new-born protoneutron stars would be. Neutrinos potentially give us the answers (or at least implications) for these questions, since they are the direct messenger of the collapsing cores unlike photons. Furthermore, we can learn neutrino properties with supernova neutrinos, and this would be fascinating for the particle physics community. With the progressive results by many experiments, we learned a lot about neutrino physics and most of them were quite surprising, but still, only very little is known about the neutrino sector. Supernova neutrinos can play a unique role for us to learn more about neutrinos because of the extreme and (therefore) unique conditions of the core, from which neutrinos are produced.

At present, several large-volume underground detectors are running. Super-Kamiokande (SK), which is at the Kamioka mine in Japan, is the representative Čerenkov detector, filled with 50-kton pure water (32-kton fiducial volume for the burst mode and 22.5-kton for other modes). In addition to SK, there are many detectors of various types; e.g., Sudbury Neutrino Observatory (SNO; heavy water Čerenkov detector) and KamLAND (scintillator) both having $\sim$ 1-kton fiducial volume. With these detectors, we have made remarkable progress on the neutrino physics especially about flavor mixing, or neutrino oscillation, using the atmospheric, solar, and reactor neutrino data. Furthermore, as a fascinating target of these detectors, a supernova neutrino burst has been considered by many researchers for many years. If a core-collapse supernova explosion occurs in the galaxy (e.g., at 10 kpc from the Earth), SK is expected to detect about 10,000 neutrinos (mainly $\bar{\nu}_e$ events), and SNO and KamLAND would also detect several hundreds of events (mainly $\nu_e$ and $\bar{\nu}_e$ for SNO, and $\bar{\nu}_e$ for KamLAND). Because the extreme physical conditions realize inside the supernova core such as very high density and magnetic fields, and neutrinos escape directly from the core unlike photons, then they would be a unique probe of these environments. We note that the order of 10,000 events would be enough for statistically significant analysis.

On the other hand, it is not strange at all even if there occurs *no* supernovae in our galaxy in the near future (while detectors are running). This is because the supernova rate is estimated to be a few per century. Even under such discouraging situation, we can still learn physics through supernova neutrinos, using *diffuse neutrino background* from cosmological supernovae. In terms of statistics, this is not competitive to the galactic burst; nevertheless, it is important because it could be the only method to probe supernova core directly with neutrinos. Furthermore, this could tell us what the *average* neutrino emission spectrum would be, which would turn out to be the average condition of supernova core. This could also provide unique method for particle astrophysics and cosmology including test of long range neutrino lifetime and probe of cosmic star formation history.

The aim of this contribution is to reveal the fundamental physics with supernova neutrinos, ranging from supernova physics such as explosion mechanisms to neutrino properties as the elementary particle. More precisely, we focus on future prospects. This is partly because we still do not have any data of supernova neutrinos except for $\sim$ 20 events from SN 1987A detected by Kamiokande-II [2] and IMB [3]; the analysis of these $\sim$ 20 data has been performed and several implications for supernova and neutrino physics have been discussed by many researchers. To what extent we can learn from the future supernova neutrino observation is the main thrust, and we visit this point by studying two different categories—galactic and cosmological supernovae—in the following discussions.

## 2. IMPLICATIONS FROM THE GALACTIC SUPERNOVA NEUTRINO BURST

In this section, we discuss several possible implications from the future galactic supernova neutrino burst as well as the existing neutrino data from SN 1987A. Using the currently working detectors such as SK, we can obtain the distributions of energy, time, and direction of incident neutrinos, from each of which we can derive useful information as detailed below.

### 2.1. Supernova neutrino spectrum

#### *2.1.1. Temperature and binding energy of protoneutron stars*

It is an intriguing question what the temperature and gravitational binding energy of the newly born protoneutron stars are. That information then implies the mass and radius of the star, telling us the equation of state of very high-dense nuclear matter. Since the expected number of events are $\sim 10,000$, we can precisely construct the incident neutrino energy distribution, which reflects the physical condition of the protoneutron stars, such as temperature. Furthermore, by analyzing time evolution of the detected spectrum, we can follow the evolution of the temperatures and radius of the protoneutron star. It is also worth mentioning the result of supernova neutrino burst from SN 1987A obtained by Kamiokande-II and IMB. Many researchers analyzed these data, and obtained some implications for the supernova parameters (see, e.g., Refs. [4, 5, 6], and references therein).

#### *2.1.2. Neutrino mixing parameters*

Neutrinos undergo flavor mixing while they propagate the stellar envelope, in some characteristic ways. As the result, the neutrino spectrum (mainly $\bar{\nu}_e$ in the case of SK) at the Earth is quite different from the original one. Since we know some mixing parameters very precisely, we can reconstruct the original spectrum from the obtained one to obtain the information on physical parameters of the supernova.

Another possibility is to derive the information on unconstrained mixing parameters such as $\theta_{13}$ and mass hierarchy from the obtained neutrino spectrum [7, 8, 9]. Especially if the mass hierarchy is inverted, the complete conversion between $\bar{\nu}_e$ and $\bar{\nu}_{\mu,\tau}$ could occur, enhancing the high-energy tail of the spectrum predicting large number of events. In the case of normal mass hierarchy, on the other hand, SK has little sensitivity to the value of $\theta_{13}$ since it mainly captures $\bar{\nu}_e$. In this case, SNO plays an important role, by efficiently collecting $\nu_e$ events, whose mixing scheme strongly relies on $\theta_{13}$. Therefore, a combined analysis of the SK and SNO (and other) data would become essential in evaluating the mixing parameters as well as the quantities regarding supernova physics. We can solve the degeneracy between models using both the $\nu_e$ and $\bar{\nu}_e$ signals, although the statistics would not be great especially with SNO.

Since supernovae would be highly magnetized, we could also test the neutrino magnetic moment. This effect causes a similar effect as flavor oscillation, but it also changes spin, e.g., $\bar{\nu}_e \leftrightarrow \nu_\mu$ conversion. This spin-flavor conversion in the context of supernova neutrinos have been studied in Ref. [10, 11, 12] (see also references therein).

## 2.2. Time evolution of neutrino luminosity

### 2.2.1. Explosion mechanism

From the neutrino luminosity curve, we study the dynamical evolution of the central remnant such as neutron stars. If the explosion proceeds by the delayed mechanism, we could identify the matter accretion phase, where neutrino luminosity becomes higher due to the shock stall and fall back of the matter. On the other hand, if the explosion is driven by the prompt mechanism without any shock stalls, then there should not be any luminosity enhancement; the protoneutron star cooling phase should start immediately [13]. We can also follow the evolution of the protoneutron stars by observing the late-time luminosity curve. In the case of the burst at 10 kpc, we would have enough statistics to follow the evolution precisely.

### 2.2.2. Shock wave propagation

Although neutrinos can be directly emitted from the core, we might be able to infer the shock wave propagation through the stellar envelope (at far outer region). This is because the neutrino mixing probability could change as the shock changes density profiles at the resonance region, and that signature might appear in the neutrino spectrum and luminosity curve (see, e.g., Ref. [14, 15, 16], and references therein). Furthermore, if the reverse shock forms when the supersonically expanding neutrino-driven wind collides with the slower earlier supernova ejecta that gives rise to a "double dip" feature in the number of events and average neutrino energy as a function of time [15]. These features are observable in the $\bar{\nu}_e$ signal for an inverted and in the $\nu_e$ signal for a normal neutrino mass hierarchy, provided $\sin^2\theta_{13} \gg 10^{-5}$. This is because the higher resonance, which can be affected by the shock waves, would be relevant only in these cases.

### 2.2.3. Black hole formation

Black hole formation may be triggered by mass accretion or a sudden change in the high-density equation of state. If this transition occurs during the neutrino luminosity is still high, then the neutrino signal from the supernova will be terminated abruptly (the transition takes $< 0.5$ ms). The properties and duration of the signal before the cutoff are important measures of both the physics and astrophysics of the cooling protoneutron star. Discussing more concretely, if the black hole formation occurs while the neutrino luminosity is still as high as $10^{52}$ erg s$^{-1}$ (for the first $\sim 1$ s), the rate just before the cutoff (hereafter $t_{\rm BH}$) is about 1500 s$^{-1}$ at SK, in the case of supernova at 10 kpc. Even if it occurs at the late phase, when the luminosity is one order of magnitude smaller, the rate is $\simeq 150$ s$^{-1}$. After $t_{BH}$, the rates are zero. The 0.5 ms duration of the cutoff can be disregarded, which should contain about 0.4 events in the early case and about 0.04 events in the late case. Since these are fewer than 1, the cutoff can be considered to be sharp [17].

## 2.3. Determination of the supernova direction

Since neutrinos can immediately escape from the supernova core, contrary to the optical signal that is caused by the shock breakout from the photosphere, their detection tells us the upcoming optical explosion in advance by about a couple of hours. Therefore, it would be essential to ask how well the supernova can be located by its neutrino signal. Elastic scattering events ($\nu e^- \to \nu e^-$) strongly peaks at the forward direction and should help locate the supernova. Using Monte Carlo simulations, we found that the supernova direction can be determined with the accuracy of about $10°$ ([18]; see also, Refs. [19, 20]) in the case of supernova occurring at 10 kpc. This information would be informed to the astronomical community through the Supernova Early Warning System (SNEWS; [21]).

## 3. DIFFUSE NEUTRINO BACKGROUND FROM COSMOLOGICAL SUPERNOVAE

Because supernova explosions have occurred very commonly in both the past and present universe, tracing the cosmic star formation rate (SFR), they should have emitted a great number of neutrinos, which now make a diffuse background; we refer this as supernova relic neutrinos (SRNs). Involved physics in SRN ranges quite widely— from cosmic SFR and supernova physics to neutrino properties as elementary particles. Therefore, detecting SRNs or even setting limits on their flux can give us quite useful and unique implications for various fields of astrophysics, cosmology and particle physics. Detectability of SRNs in various detectors and its implications have been discussed in many theoretical papers from various points of view (see Ref. [22] for a review), about which we discuss in the following.

Flux estimation requires models of neutrino spectrum emitted from each supernova explosion and cosmic SFR. Furthermore, the effect of neutrino oscillation should now be taken into account appropriately. In addition to a flux estimation as precise as possible, a detailed discussion of background events, which hinder the SRN detection, is essential. A stringent observational upper limit on the SRN flux is obtained by the SK group [23] and its value is just above the recent theoretical predictions [24, 22]. In order to make the SRN detection more likely, a promising method was proposed by Beacom and Vagins [25]. Their basic idea is to dissolve GdCl$_3$ into the water Čerenkov detectors, which greatly reduces the background events if it is applied to the currently working or proposed future detectors such as SK, Hyper-Kamiokande (HK) and Underground Nucleon Decay and Neutrino Observatory (UNO). Therefore, we are now at an exciting stage, where SRNs might soon be detected actually, and be used to obtain several implications for various fields of astrophysics as a unique and complementary method to usual observations of the light.

Before moving on to details, we briefly mention another possibility. In the intermediate regime between galactic and cosmological scales, megaton detectors could also detect supernova neutrinos from relatively nearby galaxies (< 10 Mpc) with a rate of $\sim 1 \text{ yr}^{-1}$ [26]. This could also help our understanding on supernova neutrinos together with galactic and cosmological signals.

## 3.1. Flux and event rate of supernova relic neutrinos

The cosmic SFR is being progressively established and is inferred from rest-frame UV, NIR Hα, and FIR/submillimeter observations. In the local universe, all studies show that the comoving SFR monotonically increases with $z$ out to a redshift of at least 1. The core-collapse supernova rate can be obtained either through the SFR, or by the direct number count of supernova. These different approaches give a consistent result, and therefore, we can trust them. Here we use the SFR model obtained by the GALEX satellite [27], and convert it to the supernova rate by assuming the Salpeter initial mass function.

As for the original supernova neutrino spectrum, here we use the result of numerical simulation by Ref. [13]. The effect of neutrino oscillation is also included, assuming the case of normal mass hierarchy.

Figure 1(a) shows the SRN flux as a function of neutrino energy. The flux of solar neutrinos, reactor, and atmospheric neutrinos, which becomes background events for SRN detection, is shown in the same figure. The SRN flux peaks around 5 MeV and exponentially decreases above that energy. The energy range in which we are more interested is high-energy regions such as $E_\nu > 19.3$ MeV and $E_\nu > 11.3$ MeV, because as discussed below, the background events are less critical and the reaction cross section increases as $\propto E_\nu^2$. In such a range, the SRN flux is found to be 5.5 cm$^{-2}$ s$^{-1}$ ($E_\nu > 11.3$ MeV) and 1.1 cm$^{-2}$ s$^{-1}$ ($E_\nu > 19.3$ MeV), for our model.

With the SK detector, whose fiducial volume is 22.5 kton, we might be able to detect SRNs. Furthermore, much larger water Čerenkov detectors such as HK and UNO are being planned. The expected event rates at such detectors are shown in Fig. 1(b) in units of (22.5 kton yr)$^{-1}$ MeV$^{-1}$; with SK, it takes a year to obtain the shown SRN spectrum. The event rate integrated over various energy ranges is 5.5 (22.5 kton yr)$^{-1}$ for $E_e > 10$ MeV and 2.4 (22.5 kton yr)$^{-1}$ for $E_e > 18$ MeV. This clearly indicates that if the background events that hinder the detection are negligible, the SK has already reached the required sensitivity for detecting SRNs.

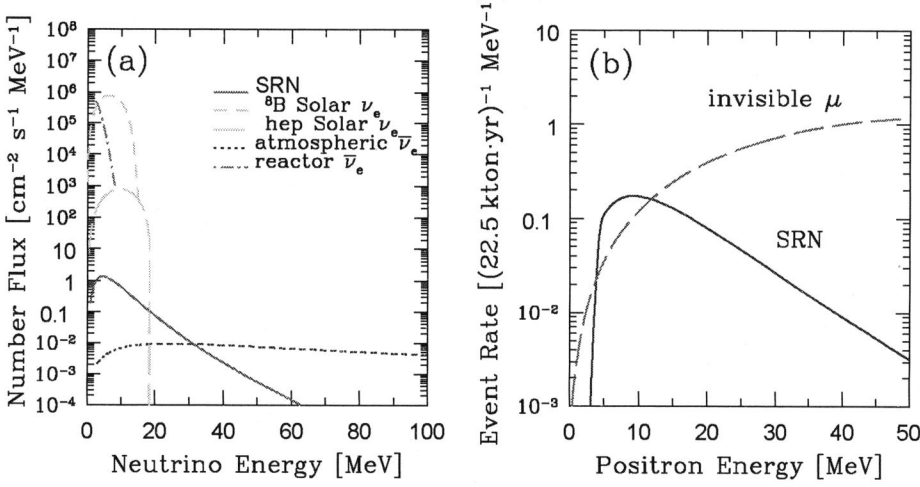

**FIGURE 1.** (a) Number flux of SRNs compared with other background neutrinos. (b) Event rate of SRNs and invisible muon decay products. From Ref. [22].

### 3.2. Observational upper limit and implications

The background events are atmospheric [28, 29] and solar neutrinos [30], antineutrinos from nuclear reactors [31], spallation products induced by cosmic-ray muons and decay products of invisible muons (e.g., [32]). We show in Figs. 1(a) and (b) the flux and event rate of these background events.

The most stringent upper limit on the SRN flux is obtained by the observation for 1496 days at the SK detector [23]. This limit is obtained by the statistical analysis including these background events from atmospheric neutrinos and invisible muons, and is $< 1.2$ cm$^{-2}$ s$^{-1}$ over the energy region of $E_\nu > 19.3$ MeV (90% CL). Comparing with the prediction for the same energy range, we find that the current SK limit is *just above* the prediction. This strong constraint motivated many theoretical studies [33, 34, 35, 24, 36, 37, 38, 39, 40] and has been translated into constraints on various quantities. The shaded region of Fig. 2(*left*) corresponds to the SK flux limit [23], reinterpreted in terms of the time-integrated luminosity $L_\nu$ and the spectrum average energy $\langle E_\nu \rangle$ for $\bar{\nu}_e$ [39]. Most numerical simulations predict around point D shown in the same figure, and hence, the current limit is encouragingly close to this "canonical" point.

In the near future, sensitivity of water Čerenkov detectors for the SRN detection would be significantly improved by the promising technique proposed recently [25]. The basic idea is the same as the delayed coincidence technique actually adopted at SNO or KamLAND, but GdCl$_3$ is dissolved into the pure-water of SK (or other future detectors), which enables us to actively identify $\bar{\nu}_e$ by capturing neutrons produced by the $\bar{\nu}_e p \to e^+ n$ reaction. Owing to this proposal, the range 10–30 MeV would

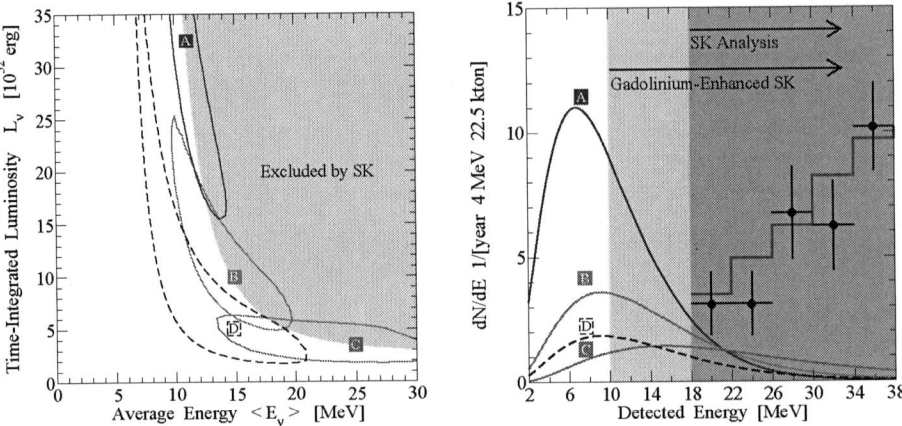

**FIGURE 2.** *Left:* Joint limits on the time-integrated luminosity $L_\nu$ and the spectrum average energy $\langle E_\nu \rangle$ for $\bar\nu_e$ due to the SRN non-detection (shaded region). Contours are possible 90% C.L. measurements of the emission parameters for each point A–D after 5 years running of a gadolinium-enhanced SK detector. *Right:* SRN detection spectra for points A–D (solid and dashed curves), efficiency-corrected SK data (points with error bars), and detector background (solid steps). From Ref. [39].

be an energy window because we can positively distinguish the $\bar\nu_e$ signal from other backgrounds such as solar neutrinos ($\nu_e$), invisible muon events and spallation products. Figure 2(*right*) eloquently tells us how promising this proposal is for the SRN detection. SK is so far sensitive only to the dark shaded region of Fig. 2(*right*) above 18 MeV due to high backgrounds at lower energies. With the addition of gadolinium, these backgrounds in the range 10–18 MeV would be removed, and that shown would be reduced by a factor $\sim 5$, opening up the light shaded region for analysis. As the result, we can distinguish three typical models A–C quite well, as shown as contours in Fig. 2(*left*). We note that these models give almost the same yield at the current SK above 18 MeV. In addition, a model supported by many simulations (point D) can also be probed quite well with five-year data.

## ACKNOWLEDGMENTS

The author acknowledges support from Sherman Fairchild Foundation at Caltech.

## REFERENCES

1. S. Ando, *Phys. Rev. D* **68**, 063002 (2003).
2. K. Hirata, et al., *Phys. Rev. Lett.* **58**, 1490 (1987).
3. R. M. Bionta, et al., *Phys. Rev. Lett.* **58**, 1494 (1987).
4. B. Jegerlehner, F. Neubig, and G. Raffelt, *Phys. Rev. D* **54**, 1194–1203 (1996).
5. M. Kachelriess, R. Tomas, and J. W. F. Valle, *JHEP* **01**, 030 (2001).
6. A. Mirizzi, and G. G. Raffelt, *Phys. Rev. D* **72**, 063001 (2005).
7. A. S. Dighe, and A. Y. Smirnov, *Phys. Rev. D* **62**, 033007 (2000).

8. K. Takahashi, and K. Sato, *Prog. Theor. Phys.* **109**, 919 (2003).
9. C. Lunardini, and A. Y. Smirnov, *J. Cosmol. Astropart. Phys.* **06**, 009 (2003).
10. S. Ando, and K. Sato, *Phys. Rev. D* **67**, 023004 (2003).
11. S. Ando, and K. Sato, *Phys. Rev. D* **68**, 023003 (2003).
12. S. Ando, and K. Sato, *J. Cosmol. Astropart. Phys.* **0310**, 001 (2003).
13. T. Totani, K. Sato, H. E. Dalhed, and J. R. Wilson, *Astrophys. J.* **496**, 216 (1998).
14. K. Takahashi, K. Sato, H. E. Dalhed, and J. R. Wilson, *Astropart. Phys.* **20**, 189 (2003).
15. R. Tomàs, M. Kachelriess, G. G. Raffelt, A. Dighe, H.-T. Janka, and L. Scheck, *J. Cosmol. Astropart. Phys.* **0409**, 015 (2004).
16. G. L. Fogli, E. Lisi, A. Mirizzi, and D. Montanino, *J. Cosmol. Astropart. Phys.* **0504**, 002 (2005).
17. J. F. Beacom, R. N. Boyd, and A. Mezzacappa, *Phys. Rev. D* **63**, 073011 (2001).
18. S. Ando, and K. Sato, *Prog. Theor. Phys.* **107**, 957 (2002).
19. J. F. Beacom, and P. Vogel, *Phys. Rev. D* **60**, 033007 (1999).
20. R. Tomàs, D. Semikoz, G. G. Raffelt, M. Kachelriess, and A. S. Dighe, *Phys. Rev. D* **68**, 093013 (2003).
21. K. Scholberg (1999), `astro-ph/9911359`.
22. S. Ando, and K. Sato, *New J. Phys.* **6**, 170 (2004).
23. M. Malek, et al., *Phys. Rev. Lett.* **90**, 061101 (2003).
24. L. E. Strigari, M. Kaplinghat, G. Steigman, and T. P. Walker, *J. Cosmol. Astropart. Phys.* **0403**, 007 (2004).
25. J. F. Beacom, and M. R. Vagins, *Phys. Rev. Lett.* **93**, 171101 (2004).
26. S. Ando, J. F. Beacom, and H. Yuksel, *Phys. Rev. Lett.* **95**, 171101 (2005).
27. D. Schiminovich, et al., *Astrophys. J.* **619**, L47 (2005).
28. G. D. Barr, T. K. Gaisser, P. Lipari, S. Robbins, and T. Stanev, *Phys. Rev. D* **70**, 023006 (2004).
29. M. Honda, T. Kajita, K. Kasahara, and S. Midorikawa, *Phys. Rev. D* **70**, 043008 (2004).
30. J. N. Bahcall, *Neutrino Astrophysics*, Cambridge University Press, 1989.
31. C. Bemporad, G. Gratta, and P. Vogel, *Rev. Mod. Phys.* **74**, 297 (2002).
32. S. Ando, K. Sato, and T. Totani, *Astropart. Phys.* **18**, 307 (2003).
33. M. Fukugita, and M. Kawasaki, *Mon. Not. R. Astron. Soc.* **340**, L7 (2003).
34. S. Ando, and K. Sato, *Phys. Lett. B* **559**, 113 (2003).
35. S. Ando, *Phys. Lett. B* **570**, 11 (2003).
36. S. Ando, *Astrophys. J.* **607**, 20 (2004).
37. L. E. Strigari, J. F. Beacom, T. P. Walker, and P. Zhang, *J. Cosmol. Astropart. Phys.* **0504**, 017 (2005).
38. J. F. Beacom, and L. E. Strigari, *Phys. Rev. C* **73**, 035807 (2006).
39. H. Yüksel, S. Ando, and J. F. Beacom, *Phys. Rev. C* **74**, 015803 (2006).
40. C. Lunardini, *Phys. Rev. D* **73**, 083009 (2006).

# Heavy element nucleosynthesis in jets from collapsars

Shin-ichirou Fujimoto*,†, Masa-aki Hashimoto**, Kei Kotake‡ and Shoichi Yamada§,¶

*Department of Electronic Control, Kumamoto National College of Technology, Kumamoto 861-1102, Japan
†Institut d'Astronomie et d'Astrophysique, Université libre de Bruxelles, CP226 Boulevard du Triomphe, B-1050 Brussels, Belgium
**Department of Physics, School of Sciences, Kyushu University, Fukuoka 810-8560, Japan
‡Division of Theoretical Astronomy, National Astronomical Observatory Japan, 2-21-1, Osawa, Mitaka, Tokyo, 181-8588, Japan
§Science & Engineering, Waseda University, 3-4-1 Okubo, Shinjuku, Tokyo 169-8555, Japan
¶Advanced Research Institute for Science & Engineering, Waseda University, 3-4-1 Okubo, Shinjuku, Tokyo 169-8555, Japan

**Abstract.** We investigate nucleosynthesis in collapsars, based on long-term, magneto-hydrodynamic simulations of a rapidly rotating massive star of $40 M_\odot$ during the core collapse. We have calculated detailed composition of magnetically driven jets ejected from the collapsars, in which the magnetic fields before the collapse, are uniform and parallel to the rotational axis of the star and the magnitudes of the fields, $B_0$, are $10^{10}$ G or $10^{12}$ G. We follow the evolution of chemical composition up to about 4000 nuclides inside the jets from the collapse phase to the ejection phase through the jet generation phase with use of a large nuclear reaction network. We find that the $r$-process successfully operates in the jets from the collapsar of $B_0 = 10^{12}$ G, so that U and Th are synthesized abundantly. Abundance pattern inside the jets is similar to that of $r$-elements in the solar system. Furthermore, we find that $p$-nuclei are produced without seed nuclei: not only light $p$-nuclei, such as $^{74}$Se, $^{78}$Kr, $^{84}$Sr, and $^{92}$Mo, but also heavy $p$-nuclei, $^{113}$In, $^{115}$Sn, and $^{138}$La, can be abundantly synthesized in the jets. The amounts of $p$-nuclei in the ejecta are much greater than those in core-collapse supernovae (SNe). In particular, $^{92}$Mo, $^{113}$In, $^{115}$Sn, and $^{138}$La deficient in the SNe, are significantly produced in the ejecta. On the other hand, in the jets from the collapsar of $B_0 = 10^{10}$ G, the $r$-process cannot operate and $^{56}$Ni, $^{28}$Si, $^{32}$S, and $^4$He are abundantly synthesized in the jets, as in ejecta from inner layers of Type II supernovae. An amount of $^{56}$Ni is much smaller than that from SN 1987A.

**Keywords:** Accretion, accretion disks — nuclear reactions, nucleosynthesis, abundances — stars: supernovae: general — MHD — methods: numerical — gamma rays: bursts
**PACS:** 26.30

## INTRODUCTION

A major fraction of elements heavier than iron are considered to be produced through a rapid neutron capture process ($r$-process) in an explosive event [e.g., 1, and references therein]. Recent observations of metal poor stars (MPSs) have shown that $r$-elements have been produced in an early phase of metal enrichment of the Galaxy [e.g., 2]. An abundance profile of $r$-elements in some MPSs has similar patterns of the $r$-elements in the solar system. Although various sites of $r$-process have been proposed, all the proposed scenarios have some deficiencies [e.g., 3, and references therein].

On the other hand, any neutron capture processes cannot produce the $p$-nuclei that is 35 neutron deficient stable nuclei with mass number $A \geq 74$. The nuclei are considered to be synthesized by sequences of $(\gamma, n)$ photodisintegrations of $s$-nuclei ($s$-process seeds) processed during helium core burning in a massive star. The scenario of the synthesis of the $p$-nuclei via the sequential photodisintegrations, or $p$-process, is proposed in the oxygen/neon layers of highly evolved massive stars [4, 5]. The scenario, however, has a conspicuous shortcoming [6]; some $p$-nuclei, such as $^{92}$Mo, $^{94}$Mo, $^{96}$Ru, $^{98}$Ru and $^{138}$La, are underproduced compared with those in the solar system.

During collapse of a massive star greater than 35-40$M_\odot$, stellar core is considered to promptly collapse to a black hole [7, 8]. When the star has sufficiently high angular momentum before the collapse, an accretion disk is formed around the hole and jets are shown to be launched from the inner region of the disk near the hole through magnetic processes [9, 10]. Gamma-ray bursts (GRBs) are expected to be driven by the jets. This scenario of GRBs is called a collapsar model [11]. For accretion rates greater than $0.1 M_\odot \text{s}^{-1}$, the accretion disk is so dense and hot that nuclear burning is expected to proceed efficiently. In fact, an innermost region of the disk related to GRBs becomes neutron-rich through electron capture on nuclei [12, 13, 14]. Nucleosynthesis inside outflows from the neutron-rich disk has been investigated with steady, one-dimensional models of the disk and the outflows [12, 14, 15, 16]. Not only neutron-rich nuclei [14] but $p$-nuclei [16] are shown to be produced inside the outflows. However, abundances of the outflows are shown to highly depend on electron fractions of the outflows. The electron fractions are expected to change during the generation phase of the outflows near the base of the outflows. Nevertheless, the electron fraction of the outflow at the base is assumed to be that of the disk at the base [14, 15, 16]. Therefore, in order to evaluate change in the electron fractions of the outflows and to examine nucleosynthesis in a collapsar, we need non-steady, multi-dimensional simulations of collapsars from the collapsing phase of the star to the ejection phase of the outflows from the star.

In the present paper, in order to investigate heavy element synthesis in collapsars, we have calculated detailed composition of magnetically-driven jets ejected from the collapsar, based on long-term, magnetohydrodynamic (MHD) simulations of a rapidly rotating, magnetized massive star of $40M_\odot$ during core collapse, which has been performed in Fujimoto et al. [9] recently. We follow evolution of the electron fraction and abundances of about 4000 nuclides from the collapse phase to the ejection phase through the jet generation phase, with the aid of a large nuclear reaction network.

In §2 we briefly describe a numerical code for MHD calculation of the collapsing star, initial conditions of the star, and properties of jets from the star. In §3, we firstly present Lagrangian evolution of ejecta through the jets and a large nuclear reaction network, in which spontaneous and $\beta$-delayed fission is taken into account. Then, we proceed abundances of the ejecta through the jets from collapsars.

## MHD SIMULATIONS OF COLLAPSARS

We carry out Newtonian MHD calculations of the collapse of a rotating massive star of $40M_\odot$. The core of the star is assumed to be collapsed to a black hole promptly.

Calculations are performed over the region other than the central region, smaller than 50km, of the star. Fluid is freely absorbed through the inner boundary of 50km, which mimics a surface of the black hole with the mass of $M$. The mass is initially set to be that of the central region of the progenitor $\leq$ 50km ($0.001 M_\odot$) and is continuously increased by the mass of infalling gas to the region during the calculation.

To calculate the structure and evolution of the collapsing star, we solve the Newtonian MHD equations,

$$\frac{D\rho}{Dt} + \rho \vec{\nabla} \cdot \vec{v} = 0, \tag{1}$$

$$\rho \frac{D\vec{v}}{Dt} = -\nabla P - \rho \nabla (\Phi + \Psi) + \frac{1}{4\pi} \left( \vec{\nabla} \times \vec{B} \right) \times \vec{B}, \tag{2}$$

$$\rho \frac{D}{Dt} \left( \frac{e}{\rho} \right) = -P \vec{\nabla} \cdot \vec{v} - L_\nu, \tag{3}$$

$$\frac{\partial \vec{B}}{\partial t} = \vec{\nabla} \times \left( \vec{v} \times \vec{B} \right), \tag{4}$$

$$\triangle \Phi = 4\pi G \rho, \tag{5}$$

where $\rho, P, \vec{v}, e, \Phi, \Psi, \vec{B}, L_\nu$ are the mass density, the pressure other than the magnetic pressure, the fluid velocity, the internal energy density, the gravitational potential of fluid, the gravitational potential of the central object, the magnetic field and the neutrino cooling rate, respectively. We denote the Lagrange derivative as $\frac{D}{Dt}$.

The numerical code for the MHD calculations employed in this paper is based on the ZEUS-2D code [17]. We have extended the code to include a realistic EOS [18] based on the relativistic mean field theory [19]. For lower density regime ($\rho < 10^5$ g/cc), where no data is available in the EOS table with the Shen EOS, we use another EOS, which includes contributions from an ideal gas of nuclei, radiation, and electrons and positrons with arbitrary degrees of degeneracy [20]. We carefully connect two EOS at $\rho = 10^5$ g/cc for physical quantities to vary continuous in density at a given temperature [9].The use of the realistic equation of state is important for nucleosynthesis. This is because it enable us to evaluate temperatures of fluid preciously.

We consider neutrino cooling through electron-positron pair capture on nuclei, electron-positron pair annihilation, and nucleon-nucleon bremsstrahlung. The total neutrino cooling rate is evaluated with a simplified neutrino transfer model based on the two-stream approximation [21], with which we can treat the optically thin and thick regime on neutrino reaction, approximately. We ignore resistive heating, whose properties are highly uncertain, not as in Proga et al. [10]. We note that the change in the electron fraction is ignored in the MHD calculations (or $DY_e/Dt = 0$).

Spherical coordinates, $(r, \theta, \phi)$ are used in our simulations and the computational domain is extended over 50km $\leq r \leq$ 10000km and $0 \leq \theta \leq \pi/2$ and covered with 200($r$) $\times$ 24($\theta$) meshes, with which we can resolve a fastest growing mode of MRI for models, whose magnetic fields are greater than $10^{10}$G before the core collapse. We assume the fluid is axisymmetric and the mirror symmetry on the equatorial plane. We mimic strong gravity around the black hole in terms of the pseudo-Newtonian potential [22]:

$$\Psi = -\frac{GM}{r - r_g},\qquad(6)$$

where $G$ is the gravitational constant and $r_g = 2GM/c^2$ is the Schwarzschild radius.

We set the initial profiles of the density, temperature and electron fraction to those of the spherical model of a $40M_\odot$ massive star before the collapse [23]. The radial and azimuthal velocities are set to be zero initially, and increase due to the collapse induced by the central hole and self-gravity. The computational domain is extended from the iron core to an inner oxygen layer. The boundaries of the silicon layers between the iron core and the oxygen layers are located at about 1800 km ($1.88M_\odot$) and 3900km ($2.4M_\odot$), respectively. We adopt an analytical form of the angular velocity $\Omega(r)$ of the star before the collapse:

$$\Omega(r) = \Omega_0 \frac{R_0^2}{r^2 + R_0^2},\qquad(7)$$

as in previous works of collapsars [24] and SNe [25, 26]. Here $\Omega_0$ and $R_0$ are parameters of our model. We consider a set of $(\Omega_0, R_0)$ = (10 rad/s, 1000km) or R models (cases with rapidly rotating core) in [9]. For the set of $\Omega_0$ and $R_0$, the maximum specific angular momentum is about $10^{17}$ cm$^2$/s, which is comparable to that of the Keplerian motion at 50km around a black hole of $3M_\odot$. Therefore the centrifugal force can be larger than the gravitational force of the central black hole and the formation of disk like structure is expected near the hole.

Initial magnetic field is assumed to be uniform, parallel to the rotational axis, and weak elsewhere ($\beta = 8\pi P/B^2 \gg 1$). In the present paper, we consider two cases with $B_0 = 10^{10}$ and $10^{12}$ G. It should be noted that the magnetic pressure is much smaller than the other pressure before the collapse even if the initial magnetic field, $B_0$, is equal to $10^{12}$ G. In brief, we consider two collapsars, or R10 and R12 in [9], whose $B_0 = 10^{10}$ and $10^{12}$ G, respectively, in the present study.

We briefly describe results of our MHD calculation [9]. Lagrangian evolution of physical quantities inside the jets, which is important for nucleosynthesis, is shown in §3.1 in detail. We find that the jets can be magnetically driven from the central region of the star along the rotational axis; After material reaches to the black hole with high angular momentum of $\sim 10^{17}$cm$^2$s$^{-1}$, a disk is formed inside a surface of weak shock, which is appeared near the hole due to the centrifugal force and propagates outward slowly. The magnetic fields, which are dominated by the toroidal component, are chiefly amplified due to the wrapping of the field inside the disk and propagate to the polar region along the inner boundary near the black hole through the Alfvén wave. Eventually, the jets can be driven by the tangled-up magnetic fields at the polar region near the hole at $t = 2.58$ and $0.20$s for R10 and R12, respectively. The velocities of the jets are at most 10% of the velocity of light.

# NUCLEOSYNTHESIS IN A COLLAPSAR

We examine nucleosynthesis in collapsars, based on the results of our MHD simulations. We firstly show Lagrangian evolution of ejecta through jets in detail and then proceed nucleosynthesis inside the jets from the collapsars of R10 and R12.

## Lagrangian evolution of ejecta through jets

In order to calculate chemical composition of material inside jets, we need Lagrangian evolution of physical quantities, such as density, temperature, and, velocity of the material. We adopt a tracer particle method [27] to calculate the Lagrangian evolution of the physical quantities from the Eulerian evolution obtained from our MHD calculations of the collapsars. For R10 (R12), particles are initially placed from a Si-rich layer (the Fe core) to an inner O-rich layer, or 2000 (800)km to 10,000 (5000)km. The numbers of the particles in a layer are weighted to the mass in the layer. We find that 21 (59) particles can be ejected via the jets for R10 (R12) with the 50,000 (1000) particles, with which we can follow ejecta through the jets appropriately.

We have performed MHD calculation until $t = t_f$ ($t_f = 2.62$ and $0.36$s for R10 and R12, respectively). We need much longer the Lagrangian evolution of the jets for nucleosynthesis calculation, or $\gg 1$ s. However, we cannot follow Eulerian evolution during such a long time with the MHD calculation due to our limited computational time and environment. In addition, we also need much larger computational domain to follow the longer evolution, which requires additional computational time. Hence, after $t_f$, we assume that particles are adiabatic and expand spherically and freely. Therefore, velocity, position, density, and temperature of a particle are set to be $v(t) = v_0$, $r(t) = r_f + v(t)(t - t_f)$, $\rho(t) = \rho_f(r_f/r(t))^3$, $T(t) = T_f(r_f/r(t))$, respectively, where $v_0$ is constant in time and set to be $v_f$. Here $v_f, r_f, \rho_f$, and $T_f$ are velocity, position, density, and temperature of each particle at $t_f$, respectively.

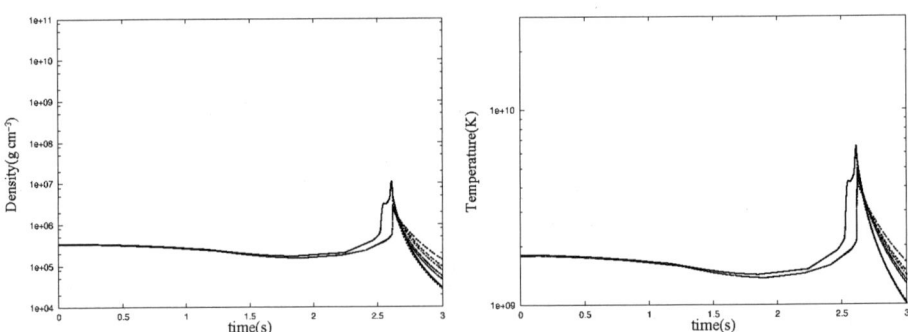

**FIGURE 1.** Lagrangian evolution of densities (left panel) and temperatures (right panel) of particles for R10.

Figures 1 and 2 show the evolution of density and temperature for R10 and R12, respectively. Initial density and temperature of a particle is set to be those of the pre-supernova in a layer where the particle is initially located. As the particle falls near the hole, the density and temperature increase. For R10, as a result of relatively low densities and temperatures, electron capture on protons cannot operate inside the jets so that the jets cannot be neutron-rich. On the other hand, for R12, some particles become enough high densities ($> 10^{10}$ g/cm$^3$) and high temperatures ($> 10^{10}$ K) for protons to capture electrons efficiently and can be neutron-rich. For the most neutron-rich particle of $Y_e = 0.10$, the density and temperature stay at high levels during $t = 0.1$-$0.2$s because of convective motion near the black hole. Here $Y_e$ is the electron fraction of the particle at $T = 10^9$ K. The motion is important for decrease in the electron fraction.

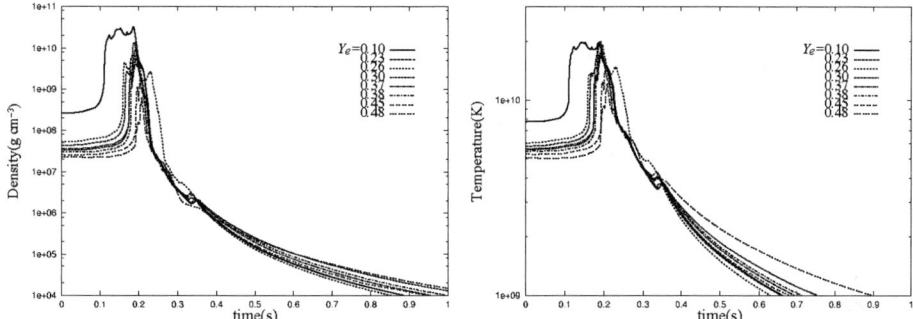

**FIGURE 2.** Same as Fig.1 but for R12.

## Nuclear Reaction Network

As shown in Figure 2 (right panel), the ejecta can attain to high temperature greater than $9 \times 10^9$K near the black hole, where material is in nuclear statistical equilibrium (NSE). The abundances of the material in NSE can be expressed with simple analytical expressions [1, 28], which are specified by $\rho$, $T$, and the electron fraction as in Fujimoto et al. [13]. The electron fraction changes through electron and positron capture on nuclei. It should be emphasized that changes in the electron fraction through neutrino interactions can be ignored because the material is transparent for neutrinos even in the most dense part of the computational domain [9].

In the relatively cool regime of $T < 9 \times 10^9$K, NSE breaks and the chemical composition is calculated with a nuclear reaction network. The network includes about 4000 nuclei from neutron and proton up to Fermium, whose atomic number, $Z$, is 100 [see NETWORK B of Table 1 in 29]. The networks contain reactions such as two and three body reactions, various decay channels, and electron-positron capture. The experimental masses and reaction rates are adopted if available or theoretical masses and rates are taken into account otherwise [see 3, 29, for detailed nuclear data]. Moreover, spontaneous and $\beta$-delayed fission is taken into accounts in the network. We note that empirical formula [eq.(5) in 30], in which the distribution of fission yields is asymmetric, is adopted about decay products through the fission. However, for R10, in which the temperatures are lower than $7 \times 10^9$K and the jets are not neutron-rich, the abundances are calculated with a small nuclear reaction network, which includes 463 nuclei from neutron and proton up to Kr and fission is not taken into accounts [14].

## Abundances of ejecta through the jets from the collapsars

Once we obtain the density and temperature evolution of a particle, we can follow abundance evolution of the particle during the infall and ejection, through post-processing calculation using the nuclear reaction network presented in the previous subsection. The ejected particles are initially located in the oxygen-rich layers and the iron core for R10 and R12, respectively. As the particles infall near the black hole, the particles become hotter than $9 \times 10^9$K for R12, above which material in the particles is in NSE. We therefore switch to the NSE code from the large nuclear reaction network to calculate abundance change of the particle. On the other hand, during the ejection, the temperature of the particles decreases and the particle becomes cooler than $9 \times 10^9$K. We switch back to the reaction network from the NSE code for the post-processing calculation of the particle composition. We perform the post-processing calculation of nucleosynthesis inside the jets until $\sim 10^{10}$ yr. Therefore, almost all unstable nuclei has been decayed to their corresponding stable nuclei. After we calculate the abundances of all the particles, we can integrate particle abundances weighted by their masses to obtain the chemical composition of the jets.

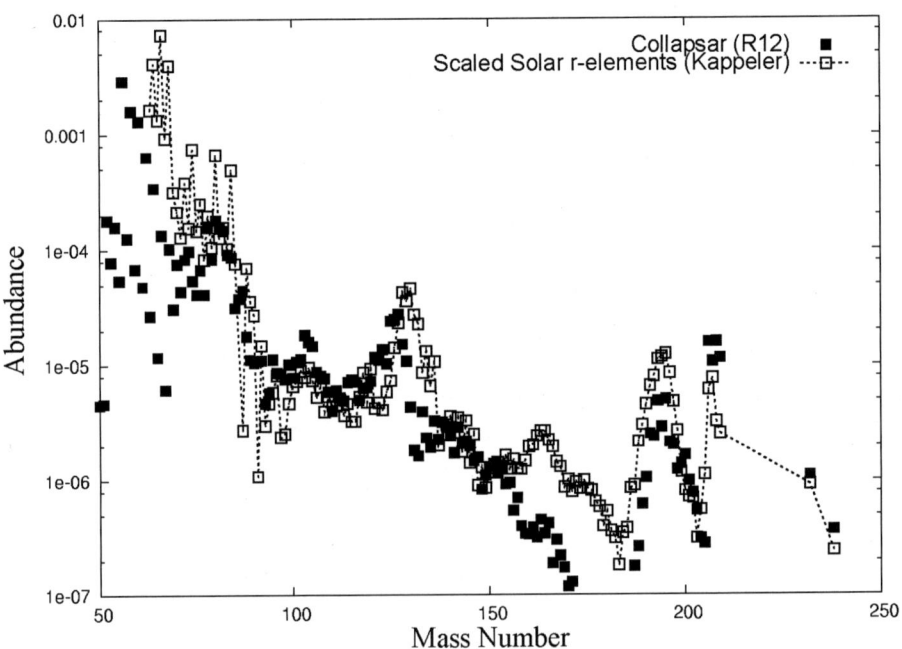

**FIGURE 3.** Abundances of the jets from a collapsar (R12) after decays.

For R10, $\alpha$-rich freezeout takes place in a hot particle with $T_{\mathrm{peak}} > 5 \times 10^9$K, where $T_{\mathrm{peak}}$ is the peak temperature of the particle. The particle has abundant $^4$He and $^{56}$Ni, which decays to $^{56}$Fe, after the ejection from the collapsar. While a cool particle with $T_{\mathrm{peak}} < 5 \times 10^9$K becomes abundant in $\alpha$ elements, such as $^{28}$Si and $^{32}$S, through complete O- and incomplete Si-burning. The ejecta for R10 is chiefly composed from $^{56}$Fe, $^{28}$Si, $^{32}$S and $^4$He. The ejected amount of $^{56}$Ni is much smaller than that from SN 1987A, which is estimated to be $\sim 0.07 M_\odot$, due to a small amount of the ejecta less than $10^{-3} M_\odot$ through the jets.

For R12, various heavy nuclei are produced inside the jets through $r$-process because of the neutron-richness of the jets, in which the electron capture on protons efficiently operates. Figure 3 shows abundances of the jets for R12 after decays. The abundances of $r$-process elements in the solar system are also shown in Figure 3, where the abundances are scaled in the abundance of $^{153}$Eu. We find that the abundances of the jets have a profile similar to that of the solar r-elements. Heavier nuclei are produced inside a particle with lower $Y_e$, or a more neutron-rich particle. We note that U and Th can be abundantly synthesized in the jets, in particular, in the particle of $Y_e = 0.1$, in which neutron capture operates enough fast to produce neutron-rich nuclei near the neutron-drip line and nuclei with mass number $A$ greater than 200 are abundantly synthesized. A fraction of the heavy nuclei with $A > 200$ decay to U and Th through sequences of $\beta$ and $\alpha$ decays.

We also find that an interesting feature on the abundances of the jets for R12; $p$-nuclei are significantly synthesized in the ejecta in spite of neutron-richness of the ejecta. Figure 4 shows the mass fractions of $p$-nuclei abundantly produced in the ejecta. The fractions are normalized by those in the solar system [31]. We also show mass fractions of $p$-nuclei from core-collapse SNe averaged over an initial mass function [6]. Not only light $p$-nuclei, such as $^{74}$Se, $^{78}$Kr, $^{84}$Sr, and $^{92}$Mo, but also heavy $p$-nuclei, $^{113}$In, $^{115}$Sn, and $^{138}$La, are produced in the ejecta abundantly. The abundances of $p$-nuclei in the ejecta are much greater than those in core-collapse SNe [6]. It should be emphasized that $^{92}$Mo, $^{113}$In, $^{115}$Sn, and $^{138}$La, which are deficient in core-collapse SNe [6], are significantly produced in the ejecta through the jets from the collapsar of R12.

## REFERENCES

1. Qian, S. 2003, Prog. Part. Nucl. Phys. 50, 153
2. Truran, J. W., Cowan, J. J.,Pilachowski, C. A., & Sneden, C. 2002, PASP, 114, 1293
3. Fujimoto, S., Kotake, K., Yamada, S., & Hashimoto, M. 2006b, preprint, Astro-ph/0602460 (ApJ accepted)
4. Arnould, M. 1976, A&A, 46, 117
5. Woosley, S. E., & Howard, W. M., 1978, ApJS, 36, 285
6. Rayet, M., Arnould, M., Hashimoto, M., Prantzos, N., & Nomoto, K. 1995, A&A, 298, 517
7. Woosley, S. E., & Weaver, T. A. 1995, ApJS, 101, 181
8. Heger, A., Fryer, C. L., Woosley, S. E., Langer, N., & Hartmann, D. H. 2003, ApJ, 591, 288
9. Fujimoto, S., Kotake, K., Yamada, S., Hashimoto, M., & Sato, K. 2006a, ApJ, 644, 1040
10. Proga, D., MacFadyen, A. I., Armitage, P. J., & Begelman, M.,C. 2003, ApJL, 599, 5
11. Woosley, S. E. 1993, ApJ, 405, 273
12. Pruet, J., Woosley, S. E., & Hoffman, R. D. 2003a, ApJ, 586, 1254

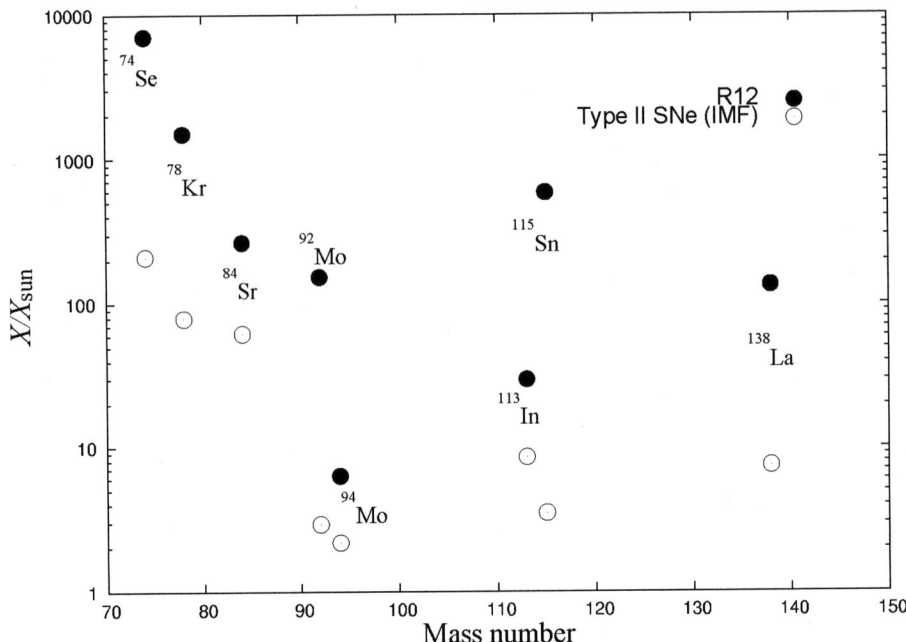

**FIGURE 4.** Mass fractions of *p*-nuclei abundantly produced in the ejecta through the jets. The fractions are normalized by those in the solar system [31]. IMF averaged abundances of *p*-nuclei from core-collapse SNe [6] are shown with open circles.

13. Fujimoto, S., Hashimoto, M., Arai, K., & Matsuba, R. 2003, *Origin of Matter and Evolution of the Galaxies 2003* ed. M. Terasawa et al., pp.344-353 (Singapore: World Scientific).
14. Fujimoto, S., Hashimoto, M., Arai, K., & Matsuba, R. 2004, ApJ, 614, 817
15. Pruet, J., Thompson, T. A., & Hoffman, R. D. 2004, ApJ, 606, 1006
16. Fujimoto, S., Hashimoto, M., Arai, K., & Matsuba, R. 2005, Nucl. Phys. A758, 47
17. Stone, J. M., & Norman, M. L. 1992, ApJS, 80, 791
18. Kotake, K., Sawai, H., Yamada, S., & Sato, K. 2004, ApJ, 608, 391
19. Shen, H., Toki, H., Oyamatsu, K., & Sumiyoshi, K. 1998, Nucl. Phys. A., 637, 435
20. Blinnikov, S. I., Dunina-Barkovskaya, N. V., & Nadyozhin, D. K. 1996, ApJS, 106, 171
21. Di Matteo, T., Perna, R., & Narayan, R. 2002, ApJ, 579, 706
22. Paczyńsky, B., & Wiita, P. J. 1980, A&A, 88, 23
23. Hashimoto, M. 1995, Prog. Theor. Phys. 94 663.
24. Mizuno, Y., Yamada, S., Koide, S., & Shibata, K. 2004, ApJ, 606, 395
25. Kotake, K., Yamada, S., Sato, K., Sumiyoshi, K., Ono, H., & Suzuki,H. 2004, Phys. Rev. D., 609, 124004
26. Yamada, S., & Sawai, H., 2004, ApJ, 608, 907
27. Nagataki, S., Hashimoto, M., Sato, K., & Yamada, S. 1997, ApJ, 486, 1026
28. Clayton, D. D. 1968, *Principles of Stellar Evolution and Nucleosynthesis* (Newyork: MacGraw-Hill).
29. Nishimura, S., Kotake, K., Hashimoto, M., Yamada, S., Nishimura, N., Fujimoto, S., & Sato, K. 2006, ApJ, 642, 410
30. Kodama, T., & Takahashi, K. 1975, Nucl. Phys. A239 489.
31. Anders, E., & Grevesse, N. 1989, Geochim. Cosmochim. Acta 53, 197

# Low-energy nuclear reactions and the alpha-nucleus optical potential: where do we stand?

P. Demetriou and M. Axiotis

*Institute of Nuclear Physics, NCSR "Demokritos", 15310 Athens, Greece*

**Abstract.** Recent efforts to develop an accurate and reliable $\alpha$-nucleus optical potential at low energies are presented. In view of the advent of new data on $\alpha$ elastic scattering and $\alpha$ radiative-capture reactions, the global semi-microscopic $\alpha$-nucleus potential is revisited and compared with the updated database. Needs for improvements are discussed.

Keywords: Nuclear reactions: Elastic scattering: optical model: Statistical model: p process
PACS: 24.10.Ht: 24.60.Dr: 25.60.Bx: 26.30.+k: 26.50.+x

## INTRODUCTION

The p process of nucleosynthesis is the mechanism responsible for the production of the stable neutron-deficient nuclei heavier than iron that are observed in the solar system. The most favoured scenarios proposed for the p process involve the photodisintegration of intermediate and heavy elements at high temperatures (2–3 billion degrees Kelvin) that can be achieved only in the deep O-Ne-rich layers of massive stars during their pre-supernova or supernova phases [1]. During the photodisintegration process, $(\gamma,n)$, $(\gamma,p)$, $(\gamma,\alpha)$ reactions and their inverse processes compete with one another and with $\beta$ decays. Other proposed sites include the C-rich zones of Chandrasekhar-mass white dwarfs exploding as Type Ia supernovae [2, 3] or exploding sub-Chandrasekhar mass white dwarfs on whcih He-rich material has accreted [4]. In the latter case, recent studies have shown the importance of $\alpha$-radiative captures on top of the standard photodisintegrations, leading to the so-called $\alpha$p process, as well as of (n,p) reactions made possible by high neutron densities that revive the nuclear flow towards higher-mass nuclei (pn process).

Given the above-mentioned scenarios, p-process calculations involve an extended network of almost 20000 reactions on 2000 nuclei. A very small fraction of these reaction cross sections can or have been measured in the laboratory, so the calculations rely largely on predictions of the statistical model. Although the predictions for nucleon-nucleus reactions seem to be well understood and adequately described, the situation is far more bleak for $\alpha$-induced reactions. At the p-process temperatures the $\alpha$-particle energies are typically of a few MeV. At these energies, the Hauser-Feshbach cross sections are governed by the $\alpha$-particle transmission coefficient which is extracted from an appropriate optical model potential (OMP). The latter remains to date poorly known at low energies mainly due to the lack of relevant data on $\alpha$ elastic scattering and $\alpha$-induced reactions at such low energies close to or even below the Coulomb barrier.

Considerable effort has been devoted in recent years to improve our knowledge of the $\alpha$-nucleus optical potential at low energies of astrophysical relevance. A brief account of recent developments is given in Section 2. The only existing global $\alpha$-optical model potential is compared with new data and is further improved in Section 3. Some conclusions are presented in Section 4.

## WHERE DO WE STAND?

The main obstacle in developing reliable and precise $\alpha$-nucleus OMPs at low astrophysically-relevant energies is the lack of experimental data. The imaginary part of the OMP is known to depend strongly on energy at energies below the Coulomb barrier. This means that for astrophysics applications the determination of the OMP parameters need to focus on experimental information at energies as close as possible to the astrophysically-relevant energies. OMP parameters are usually derived from the analysis of elastic scattering angular distributions. However, for $\alpha$-particle energies well below the Coulomb barrier the elastic scattering cross section is non-diffractive and dominated by the Rutherford component. Only very few precise experimental data exist on $\alpha$-elastic scattering at energies relevant to astrophysics and these data were taken into account by Demetriou et al. [5] in the determination of a global $\alpha$ OMP at low energies. As far as $\alpha$-induced reactions and $\alpha$-emission reactions are concerned, the cross sections at sub-Coulomb energies are very small due to the Coulomb barrier penetration, and as a result only a few measurements have been performed and can be found in the literature (see [5] and Refs. therein). Since the publication of the first global $\alpha$ OMP at low energies [5], considerable effort has been made to extend the database of $\alpha$ elastic scattering and reaction cross sections. The new measurements performed since 2003 are presented in Table 1.

**TABLE 1.** New data published since 2002 and used to re-adjust the improved OMP III

| Reaction | Nucleus | Energy | Reference | |
|---|---|---|---|---|
| $(\alpha,\alpha)$ | $^{106}$Cd | 15,17,19. | Kiss et al. [6] | |
| $(\alpha,\alpha)$ | $^{112,124}$Sn | 14.4,19.5 | Galaviz et al. [7] | |
| $(\alpha,\gamma)$ | $^{72}$Ge, $^{91,92}$Zr, $^{92}$Mo | 5-12 MeV | Harissopulos [8] | prelim. |
| " | $^{104}$Pd,$^{116,118}$Sn | " | " | data |
| $(\alpha,\gamma)$ | $^{106}$Cd | 8-12 MeV | Gyürky et al. [9] | |
| $(\alpha,p)$ | " | " | " | |
| $(\alpha,n)$ | " | " | " | |
| $(\alpha,\gamma)$ | $^{107}$Ag | 7-11 MeV | Baglin et al. [10] | |
| $(\alpha,\gamma)$ | $^{112}$Sn | 8-11 MeV | Özkan et al. [11] | |
| $(\alpha,\gamma)$ | $^{63}$Cu | 6-9 MeV | Basunia et a;. [12] | |

On the theoretical front, the precise measurements of elasti scattering angular distributions at energies close to the Coulomb barrier shown in Table 1, have also allowed the determination of precise OMPs specific to the nuclei and energies measured. Nevertheless, these local $\alpha$ OMPs fail to reproduce the $\alpha$-induced reaction data when the latter are available for the same nucleus. This is not surprising since elastic scattering and nuclear reactions probe different regions of the nuclear potential. It is well known

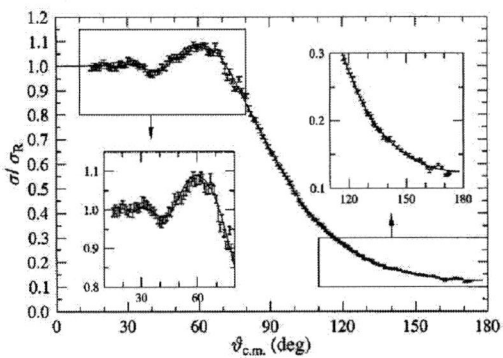 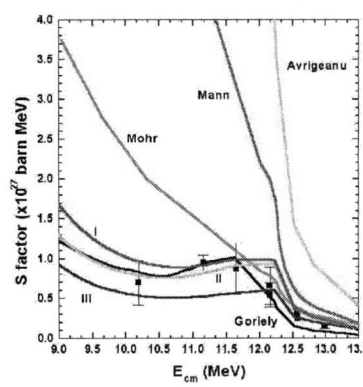

**FIGURE 1.** Elastic $\alpha$ scattering on $^{144}$Sm at 20 MeV and the S factor of the $^{144}$Sm$(\alpha,\gamma)$ reaction using the local OMP of Ref. [13] (label Mohr).

that the elastic scattering cross sections are more sensitive to the surface part of the nuclear wavefunction while the reaction cross sections are also sensitive to the inner part of the wavefunction. Determining the parameters of an OMP from elastic or reaction data only does not guarantee that the same OMP will be able to describe both sets of data. This is evident in the case of $^{144}$Sm nucleus, where the local double-folding $\alpha$ OMP determined by Mohr et al. [13] from $\alpha$-scattering cross sections failed to reproduce the $^{144}$Sm$(\alpha,\gamma)^{148}$Gd cross sections [14] at the lowest energies as can be seen in Fig. 1.

Other approaches towards a precise $\alpha$-nucleus OMPs at low energies relevant to astrophysics involve the development of a semi-local $\alpha$ OMP by Avrigeanu et al. [15] for nuclei in the A$\sim$100 mass region. The real part of the potential is again based on the double-folding method whereas the imaginary part is described by a Woods-Saxon functional. The potential also takes into account the dispersive corrections to the real potential. The parameters of this OMP were determined from fitting low- and medium-energy $\alpha$ elastic-scattering cross sections for nuclei in the above-mentioned mass region. However, when the derived potential was applied to the description of (n,$\alpha$) reaction cross sections on several Mo isotopes, it failed to give a satisfactory result. The authors found they had to introduce an additional free parameter in the density function used in the double-folding integral in order to reproduce both elastic scattering and (n,$\alpha$) reaction data on the Mo isotopes [16]. This new parametrization has yet to be tested on other nuclei besides the four Mo isotopes studied in Refs. [15, 16]. Also worth mentioning, is the semi-local $\alpha$ OMP of Rauscher [17] based on Woods-Saxon functionals and determined on $(\alpha,\gamma)$ and (n,$\alpha$) cross sections for nuclei in the A$\sim$140 mass region. The key feature of this OMP is the strong energy dependence of the depth of the imaginary part of the potential, which was deemed necessary to reproduce the reaction data at the lowest energies.

All these very important efforts to determine precise $\alpha$-nucleus OMPs at low energies, highlight the difficulties inherent in such an endeavor and stress the necessity to include both scattering and reaction data when adjusting the parameters of the OMP. In this direction, and in view of the fact that astrophysics applications require input of cross-sections for an enormous number of nuclei covering an extensive part of the nuclear chart, an attempt was made to develop a global $\alpha$-nucleus OMP, over the whole mass region, that would be able to reproduce all existing data on elastic scattering and reactions at astrophysically relevant energies [5]. This $\alpha$ OMP as well as recent improvements will be presented in the following section.

## GLOBAL SEMI-MICROSCOPIC ALPHA-NUCLEUS OMP

Three different global $\alpha$-nucleus OMPs were determined in Ref.[5] as a result of exploring the importance of surface versus volume effects, and the additional constraints imposed by applying the dispersive relations between the real and imaginary parts. Of the three OMPs, OMP III is by far the most complete as it includes all three features mentioned above, i.e. volume plus surface contributions in the imaginary potential, as well as dispersive contributions arising from the dispersive relation. The details of the potentials can be found in Ref. [5].

A comparison of the results of OMP III with elastic scattering cross sections and radiative capture cross sections on several nuclei is shown in Figs. 2-3. As can be seen in the figures, OMP III overall gives a very good description of all the data, including the new data that have been measured since 2003 and were not included in the parameter fit. At energies beyond the neutron-emission threshold reflect the uncertainties in the cross section calculations are associated with other nuclear ingredients as well, such as the nucleon-nucleus OMPs, nuclear level densities and $\gamma$-ray strenght functions. Once the different possible reaction channels open and compete with one another, the HF cross section depends not only on the $\alpha$-nucleus OMP but on all the other nuclear properties mentioned previously that descirbe the various reaction channels. For this reason, it is not always straightforward to draw conclusions on the $\alpha$ OMP from comparisons with data in these energy regions. Furthermore, there seem to be very few data available in the mass regions A$\sim$100 and A$\sim$200. Certainly, the recent measurements of [8] have provided considerable information for the former as will be seen in the attempt to improve OMP III in this mass region, however no data exist for the latter.

In more detail, as far as the elastic scattering data are concerned, OMP III tends to underpedict the backward-angles for most of the Mo, Cd, and Sn isotopes shown in Fig. 2 (dashed lines). As for the $(\alpha,\gamma)$ cross sections, OMP III seems to slightly overestimate the new data of $^{63}$Cu and underestimate the new data of $^{92}$Mo as well as the old data of $^{144}$Sm at energies below the neutron threshold. An attempt to improve these predictions by modifying certain parameters of OMP III is presented in the following section.

### Improvements

An important requirement that should be satisfied by any OMP is the dispersive relation linking the imaginary part of the potential to the real part. This requirement

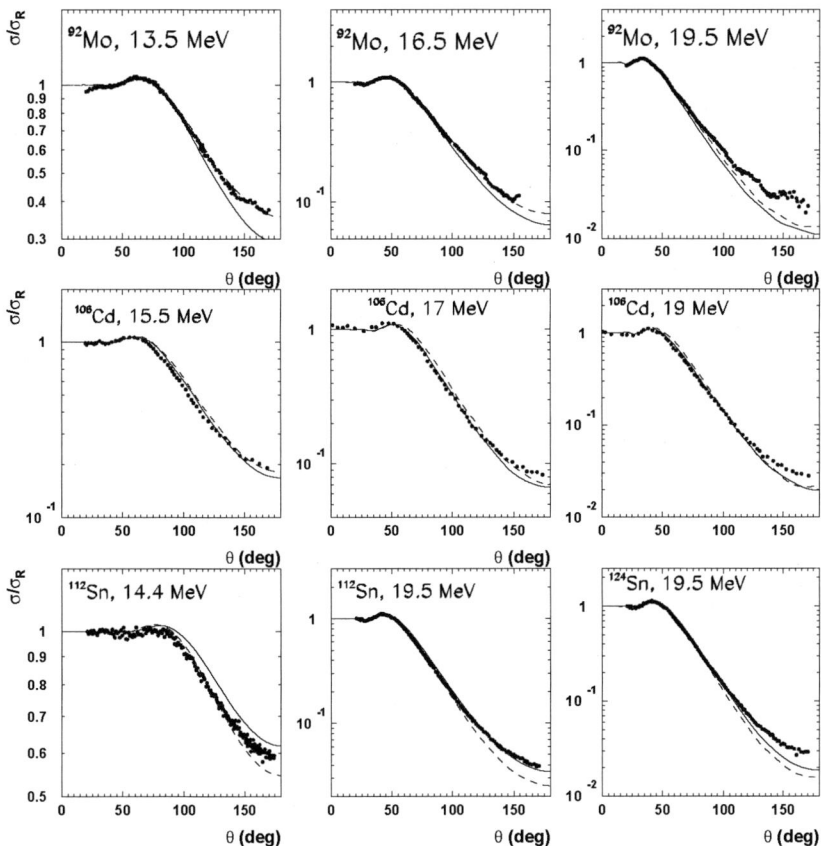

**FIGURE 2.** Elastic $\alpha$ scattering on $^{92}$Mo, $^{106}$Cd, $^{112,124}$Sn compared with the predictions of OMP III [5] (solid line) and the improved version of the present work (dashed line). Details about the data are presented in Table 1.

essentially ensures that flux is emitted after the reaction takes place. The dispersive relation imposes additional constraints on the free parameters of the potential. In this improved version of OMP III, we calculate the dispersive correction $\Delta V(R;E)$ to the real potential by the following dispersion relation [18]

$$\Delta V(R;E) = \frac{\mathscr{P}}{\pi} \int_{E_0}^{\infty} \frac{W(R;E')}{E'-E} dE', \qquad (1)$$

with $\mathscr{P}$ denoting the Cauchy principal value and $V(R;E), W(R;E)$ are the real and imaginary potentials, respectively. Due to the uncertainties involved in the calculation of the absolute magnitude of the dispersive correction $\Delta V$ at a nuclear radius $R$ in

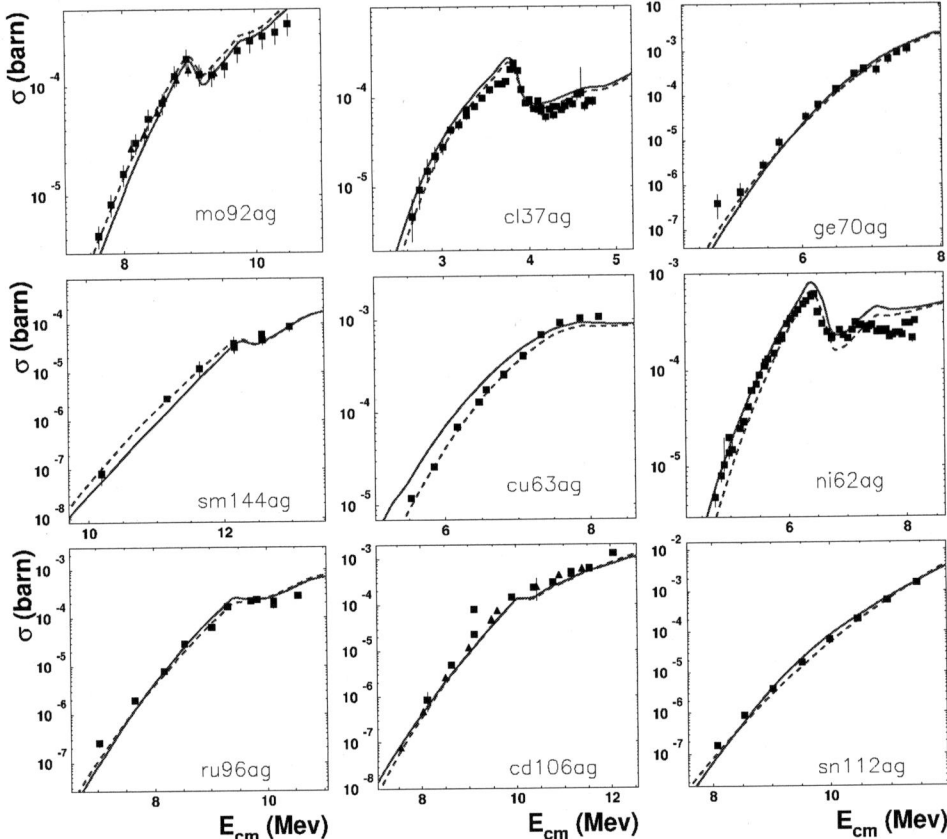

**FIGURE 3.** Cross sections for the $(\alpha,\gamma)$ reaction on several nuclei included in Ref. [5] and in Table 1. The solid lines correspond to the original OMP III [5] and the dashed lines to the improved version of this work.

the nucleus-nucleus case, we shall consider only the "subtracted dispersion relation" following Mahaux *et al* [18]

$$\Delta V_{E_s}(E) = \Delta V(E) - \Delta V(E_s) = $$
$$(E-E_s)\frac{\mathscr{P}}{\pi}\int_{E_0}^{\infty}\frac{W(E')}{(E'-E_s)(E'-E)}dE',$$

where $E_0$ is taken to be the smallest of the bound-state energies of the target-plus-projectile system. $E_s$ is a reference energy which lies in the energy domain of interest, i.e. between $E_0$ and 150 MeV. In our calculations we choose it to take the value 150 MeV. The dispersive relation integral is calculated numerically according to the prescription in Ref.[18].

The other parameters of the original OMP III modified in our attempt to improve its performance are listed below along with the new values obtained after adjusting them to reproduce the bulk of the existing old data considered in Ref. [5] and the new data included in Table 1:

$$a_W = 0.3$$
$$E^* = 0.0854 \cdot A + 1.1307 \cdot a^*$$

$$J_0 = \begin{cases} 135 - 0.644 \cdot A & A \leq 96 \\ 75 & A > 96 \end{cases}$$

$$C = \begin{cases} 2.15 - 0.165 \cdot E & E < 13 \\ 0.005 & E \geq 13 \end{cases}$$

where $a_W$ is the diffuseness of the imaginary volume term, $E^*$, $a^*$ are the parameters of the Fermi function describing the energy depemndence of the volume integral $J_I$, $J_0$ is the saturation value of the volume integral $J_I$ and $C$ is the damping coefficient of the damped surface term of the imaginary potential. All the other parameters remain the same as in OMP III of Ref. [5].

The resulting cross sections for elastic scattering and radiative capture reactions are shown in Figs. 2-3 for several nuclei. All the calculations presented in this work were performed by the reaction code TALYS [19]. The code was modified to perform calculations using transmission coefficients and reaction cross sections read from external files. This way we were able to feed into the code our semi-microscopic dooublefolding improved version of OMP III. The transmission coefficients and reaction cross sections were obtained using the optical model subroutine SCAT-2000. The other input ingredients of the HF calculations include the nucleon-nucleus OMP of Koning and Delaroche [20], the microscopic nuclear level densities of Demetriou and Goriely [21], and the $\gamma$-ray strength functions described in Ref.[19]. As can be seen in Figs. 2-3, there is notably an improvement in the description of the data, mainly due to the changes in the relative contributions of volume and surface terms and their impact on the real potential through the dispersive relations. The backward-angles of the elastic-scattering angular distributions are in agreement with the data for most of the nuclei with the exceptions of $^{92}$Mo at 19.5 MeV, and $^{112,124}$Sn at 19.5 MeV. The reaction cross sections are somewhat increased at energies below neutron threshold for $^{92}$Mo, $^{144}$Sm, and decreased for $^{63}$Cu in agreement with experiment. There is also improvement for $^{37}$Cl but some underestimation of the $^{62}$Ni cross section at low energies.

## CONCLUSIONS

Owing to the intense efforts devoted in recent years to the study of the $\alpha$-nucleus OMP at low energies, it has become clear that to obtain a global and reliable $\alpha$-nucleus OMP, it is important to compare with both elastic scattering and reaction data at low energies. An attempt to improve the only existing global double-folding $\alpha$-nucleus OMP using an updated and extended experimental database has been made. The results reflect how difficult it is to reproduce both elastic scattering and reaction data at low energies with the same global $\alpha$-nucleus OMP. More work is needed and more experimental data are required particularly in the higher mass region where they are completely lacking.

## ACKNOWLEDGMENTS

This work was funded under the Marie Curie European Reintegration Grand contract no. MERG-6-CT-2005-516789. The authors are greatly indebted to Dr. S. Harissopulos for allowing the use of unpublished data.

## REFERENCES

1. M. Arnould, *Astron. Astrophys.* **46**, 117 (1976).
2. R. A. Chevalier, *Science* **276**, 1374 (1997).
3. K. Nomoto, K. Iwamoto, N. Kishimoto, *Science* **276**, 1378 (1997).
4. M. Arnould, and S. Goriely, *Phys. Rep.* **384**, 1 (2003).
5. P. Demetriou, C. Grama, and S. Goriely, *Nucl. Phys. A* **707**, 142 (2002).
6. G. G. Kiss, Zs. Fülöp, Gy. Gyürky, et al., *Eur. Phys. J. A* **27**, 197 (2006).
7. D. Galaviz, Zs. Fülöp, Gy. Gyürky, et al., *Phys. Rev. C* **71**, 065802 (2005).
8. S. Harissopulos, *private communication*.
9. Gy. Gyürky, G. G. Kiss, Z. Elekes, et al., *arXiv:nucl-ex/0605034* (2006).
10. C. M. Baglin, E. B. Norman, R-M. Larimer, and G. A. Rech,in: Proc. Int. Conf. on Nuclear Data for Sci. Tech., Santa Fe, USA, ND2004, AIP Conference Proceedings, vol. 769, AIP, N.Y., 2004, p. 1370.
11. N. Özkan, G. Efe, R. T. Güray, et al., *Eur. Phys. J. A* **27**, 145 (2006).
12. M. S. Basunia, E. .B. Norman, H. A. Shugart, et al., *Phys. Rev. C* **71**, 035801 (2005).
13. P. Mohr, T. Rauscher, H. Oberhummer, et al., *Phys. Rev. C* **55**, 1523 (1997).
14. E. Somorjai, Zs. Fülöp, A. Z. Kiss, et al., *Astron. Astrophys.* **333**, 1112 (1998).
15. M. Avrigeanu, W. von Oertzen, A. J. M. Plompen, and V. Avrigeanu, *Nucl. Phys. A* **723**, 104 (2003).
16. M. Avrigeanu, W. von Oertzen, and V. Avrigeanu, *Nucl. Phys. A* **764**, 246 (2006).
17. T. Rauscher, *Nucl. Phys. A* **719**, 73c (2003).
18. C. Mahaux, H. Ngô, G. R. Satchler, *Nucl. Phys. A* **449**, 354 (1986).
19. A. J. Koning, S. Hilaire, and M. C. Duijvestijn, TALYS:Comprehensive nuclear reaction modeling, in: Proc. Int. Conf. on Nuclear Data for Sci. Tech., Santa Fe, USA, ND2004, AIP Conference Proceedings, vol. 769, AIP, N.Y., 2004, p. 1154
20. A. J. Koning, and J. P. Delaroche, *Nucl. Phys. A* **713**, 231 (2003).
21. P. Demetriou, and S. Goriely, *Nucl. Phys. A* **695**, 95 (2001).

# Probing universe with fast neutrons

Y. Nagai, T. Shima, A. Tomyo, M. Segawa, Y. Temma

*Research Center for Nuclear Physics, Osaka University, Ibaraki, Osaka 567-0047, Japan*

T. Ohsaki and M. Igashira

*Laboratory for Nuclear Reactors, Tokyo Institute of Technology, Meguro, Tokyo 152-8552, Japan*

**Abstract.** Fast neutrons play crucial roles as a probe to trace the history of the universe. The present paper describes our recent work on the measurement of the neutron capture and neutron inelastic scattering cross sections of various nuclei from primordial and stellar nucleostyntheses point of view, which have been carried out by developing a high sensitive measurement system. The obtained results are compared to previous works and theoretical calculations.

**Keywords:** Radiative capture; Primordial nucleosynthesis, Stellar nucleosynthesis, Rapid process, Slow process, Nuclear-cosmochronology.
**PACS:** 25.40.Dn; 25.40.Fq; 25.40.Lw; 26.30.+k; 26.50.+x; 28.20.-v; 28.20.Cz;29.40.Mc; 29.30.Kv.

## I. INTRODUCTION

A free neutron has many unique features. It decays into three unique particles, proton, electron, and neutrino, which interact differently with other particles. Because of this unique property of neutrons, detailed studies of the neutron itself, such as its half-life, the axial coupling constant, and the electric dipole moment provide crucial information on the trace of the history of universe.

Fast neutrons play an important role in the eras of primordial nucleosynthesis and stellar nucleosynthesis [1]. Predictions of primordial light element abundance, D, $^3$He, $^4$He, and $^7$Li, and their comparison with observations are a crucial test of the standard big bang cosmology [2]. However, the uncertainties remain in these predictions, and they are claimed to be dominated by the nuclear physics input from reaction cross sections relevant to the primordial nucleosynthesis, in which the radiative neutron capture on a proton is one of the key reactions. We measured the reaction cross section from 10 to about 550 keV by employing a prompt $\gamma$-ray detection method [3,4] as shown in Fig. 1 together with a recent theoretical calculation [5]. Its inverse reaction, the deuteron photodisintegration, was also studied recently by Hara, Utsunomiya et al. [6]. Experimental uncertainties of the measured cross sections were around 5-8%, while the theoretical uncertainty is claimed to be less than 3%. Further experimental efforts are required to accurately determine the cross section within an uncertainty of a few % to set stringent of the primordial $^4$He abundance and test the theoretical prediction. The $^2$H($p,\gamma$)$^3$He reaction cross section is important not only in the primordial era, but also in the proto-stars era [7], where energy generated by the reaction would have slowed down the contraction due to the gravitational force, and thus the accurate value is required to construct solid models for the proto-stars. Casella et al. have succeeded to obtain the $S$-factor by measuring the cross section [7]. Since the reaction is quite important as mentioned above, it would be worthwhile to derive the $S$-factor using its mirror reaction, the $^2\text{H}(n,\gamma)^3\text{H}$ reaction. The study would also provide important information on an electron screening effect, which appears significantly in the $^2\text{H}(p,\gamma)$ reaction at low energy [7]. The $^2\text{H}(n,\gamma)^3\text{H}$ reaction cross section, however, has not ever measured at stellar temperature.

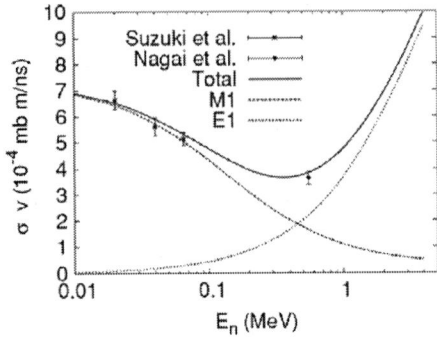

Fig. 1 The measured cross section of the $\text{H}(n,\gamma)^2\text{H}$ reaction compared to theory. $En$ is the laboratory energy of neutrons. Dashed and dotted curves are the $M1$ and $E1$ contributions to the total cross section [5].

Recently astronomers have observed elemental abundance in ultra metal deficient stars with metallicity of -3.1, which could provide crucial information on the nucleosynthesis in the earliest galactic stellar generations. The observed heavy elements abundance was found to be consistent with a scaled solar system r-process elemental distribution [8], which would mean that in the earlier stellar generations, the stellar r-process nucleosynthesis dominated the s-process one. An iron abundance in such stars is about 1000 times less than that of the solar system. However, the abundance of light elements such as C, N, O, and Ne would be abundant, and therefore these light elements would work as a neutron poison, affecting the nucleosynthetic yield of heavy elements. How much the yield is reduced depends on their neutron capture cross sections. Hence, we have been measuring the neutron capture cross sections of these light nuclei [9,10]. Some of these isotopes are shown to become a strong neutron poison even for the weak s-process nucleosynthesis of massive stars in the solar metallicity [11].

The overproduction of s-isotope $^{62}\text{Ni}$ has been a long-standing problem in the weak s-process nucleosynthesis in massive stars [12]. The $\sigma N$ value for the s-isotopes between mass 90 and 209 is constant, whereas that for below mass 90 is not. Here, $\sigma$ is the neutron capture cross section of a nucleus, and $N$ is the elemental abundance. Hence, in the weak process nucleosynthesis the individual neutron capture cross section determines overall distributions. Several years ago, Rauscher et al. calculated the nucleosynthetic yield of massive stars of solar metallicity starting from the onset of central hydrogen burning through explosion as Type II supernovae [12]. Although the calculated yields of most isotopes from mass 16 to 90 agree with the solar abundances, the isotope $^{62}\text{Ni}$ is overproduced. The neutron capture cross section of $^{62}\text{Ni}$ is shown to affect strongly the overall calculated abundance for weak s-process elements. An uncertainty of the nuclear physics input such as the neutron capture cross section of $^{62}\text{Ni}$ could be one of the reasons of the overproduction of $^{62}\text{Ni}$. Here it is interesting to note that there is a large difference between old data sets of the cross section. What could be a problem in obtaining the cross section? In $^{63}$Ni, there is two resonances near the neutron threshold; one is a sub-threshold resonance at -0.077 keV and another is a near threshold resonance at 4.5 keV. Because of a possible interference of the incident neutrons with and these threshold resonances, one can hardly apply a 1/v law to derive the cross section at stellar temperature and therefore one has to measure the cross section.

The $^{187}$Re-$^{187}$Os pair is known to be one of the good nuclear cosmo-chronometers owing to following characteristic [13]. First, $^{187}$Re is only produced by the r-process nucleosynthesis. Second, the half-life of $^{187}$Re is quite long of 42.3 Gyear. Third, since the geochemical property of Re and Os is same, they have not been strongly fractionated from each other. Hence, knowing the decay rate of $^{187}$Re one could extract information of stellar duration of r-process nucleosynthesis. However, there are problems mentioned below. First, $^{187}$Os is produced also by the stellar s-process nucleosynthesis, since $^{186}$Os is a pure s-isotope. Second, $^{187}$Os is depleted by the neutron capture reaction of the first excited state as well as the ground state of $^{187}$Os, since the energy of the first excited state is 10 keV, and therefore the first excited state could be significantly populated at stellar temperature. Hence, one has to obtain the production and depletion rates of $^{187}$Os mentioned to derive the fraction of $^{187}$Os due to the $^{187}$Re decay. The s-process component of $^{187}$Os can be obtained using the constant $\sigma$N curve. Note that $^{186}$Os is the pure s-only nucleus, and therefore the s-process abundance of $^{187}$Os, ($^{187}$Os)$_s$, can be obtained as follows.

$$(^{187}Os)_s = \sigma_{n\gamma}(^{186}Os) \times (^{186}Os)_s / \sigma_{n\gamma}(^{187}Os) \quad (1)$$

Here, ($^{186}$Os)$_s$ is the total abundance of the pure s-isotope of $^{186}$Os. Hence, the $^{187}$Os fraction due to the decay of $^{187}$Re is given by subtracting the s-process abundance of $^{187}$Os from the total abundance of $^{187}$Os as given below.

$$(^{187}Os)_{cosm.}/(^{187}Re) = \{(^{187}Os)_{tot.} - <\sigma^*_{n\gamma}(^{186}Os)> \times (^{186}Os)_s / <\sigma^*_{n\gamma}(^{187}Os)>\}/(^{187}Re) \quad (2)$$

Here, $\sigma^*_{n\gamma}$ is the neutron capture cross section at stellar environment.

Here it should be mentioned that in order to obtain information on the depletion rate via the first excited state of $^{187}$Os, the measurement of the neutron inelastic scattering cross section off the ground state of $^{187}$Os to its first excited state was suggested by Fowler [14], since the cross section would give us information on the sensitivity of the first excited state to neutrons.

So far, the neutron capture cross sections for $^{186}$Os and $^{187}$Os were measured by using hydrogen-free liquid scintillation detectors such as C$_6$D$_6$ and/or C$_6$F$_6$ [15,16]. However, there remains a discrepancy between old data sets beyond their respective statistical uncertainty, which indicates that there is a systematic uncertainty in these data. Similarly to the case mentioned, previous data on the neutron elastic as well as inelastic scattering cross sections for $^{187}$Os have large uncertainties. Hence, one could hardly use these data to accurately derive the first excited neutron capture cross section for $^{187}$Os. Note that old data were measured using a plastic scintillation counter with rather low enriched Os samples.

Under such experimental situations of the neutron capture cross sections on $^2$H, $^{62}$Ni, and $^{186,187}$Os, and the neutron inelastic as well as elastic scattering cross sections on $^{186,187}$Os, we aimed at determining these cross sections with a small systematic uncertainty using an anti-Compton NaI(Tl) spectrometer [17] and a newly developed measurement system for the $(n,n)$ and $(n,n')$ reaction cross sections of a nucleus [18] as described below.

## II. EXPERIMENTAL PROCEDURE AND RESULTS

The keV neutron capture reaction cross section of a nucleus has been measured by two different methods of direct and indirect ones. The latter method uses either Coulomb dissociation using radioactive beams and/or photodisintegration reaction using real photons. Here it should be stressed perspective about the study of direct reaction. Using the method we have been measuring the cross section of a keV neutron induced reaction of a nucleus. At the moment, we can measure such a cross section mostly on a stable target, despite a great progress one has made to measure the cross section on unstable nuclei. Since the stellar nucleosynthesis proceeds via various reactions on unstable nuclei as well as stable nuclei, one derives the cross section on unstable nuclei using theoretical models. In order to construct solid theoretical models it is naturally important to obtain crucial information on the nuclear reaction mechanism and nuclear structure as much as possible using a neutron induced reaction for stable targets (direct method). Therefore, we designed our experimental equipments so as to obtain such information [9,10,17].

Among two direct methods of activation and prompt $\gamma$-ray detection, we used the latter one. The method can be applied to all nuclei free from the half-life of a final nucleus. In addition, one can obtain not only the total but also partial cross sections, which are of vital importance to study the reaction mechanism as well as nuclear structure relevant to dominate the partial cross section. Since a partial cross section can be obtained by detecting a $\gamma$-ray from the neutron capture by a nucleus to a low-lying state, one has to use a $\gamma$-ray detector with good energy resolution. Here, it is noted that the neutron capture cross section of light nuclei is as small as a few µbarn, but the elastic scattering cross section is as large as a ten barn [9,10]. Hence, scattered neutrons from a sample nucleus would enter into a $\gamma$-ray detector, producing a large background. Therefore, many groups have been using a hydrogen-free scintillation detector such as $C_6D_6$ to get rid of this neutron related background. A $BaF_2$ detector with a large solid angle of $4\pi$ has been also used to detect all $\gamma$-rays emitted from the $(n,\gamma)$ reaction with high efficiency of 100% [19].

We have been using an anti-Compton NaI(Tl) spectrometer, which allows us to take a pulse height spectrum from the $(n,\gamma)$ reaction. A possible detection of a discrete $\gamma$-ray from the reaction to the ground state of a final nucleus indicates that we can be sure that we detect $\gamma$-rays from a sample nucleus but not a background $\gamma$-ray.

We measured the cross sections of various nuclei using pulsed keV neutrons, which were produced by the $^7$Li$(p,n)^7$Be reaction. A pulsed proton beam was provided from the 3.2 MV Pelletron accelerator at Tokyo Institute of Technology. The thus produced neutrons were captured by a sample, which was placed about 12 cm downward from the $^7$Li neutron target position. Prompt $\gamma$-rays were detected by the NaI(Tl) spectrometer, which was placed at 125 degrees with respect to the proton beam direction. Hence, γ-ray yields obtained give an angle integrated cross section for a dipole transition. A gold sample was used to normalize the neutron capture cross section of a sample, since the cross section of Au is well known within an uncertainty of 3 %.

The background subtracted γ-ray spectrum from the $^2$H$(n,\gamma)^3$H reaction is shown in Fig. 2, where one sees clearly the γ-ray from the $^2$H$(n,\gamma)^3$H reaction for the first time with a good signal to noise ratio [20].

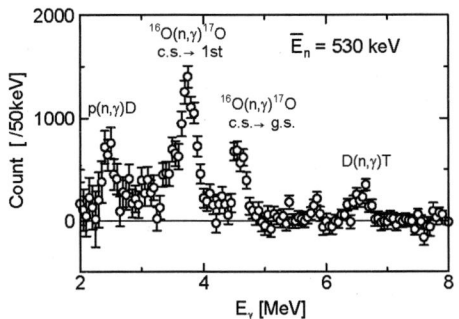

Fig. 2 Background subtracted γ-ray spectrum from the $^2$H$(n,\gamma)^3$H reaction at $En$=530 keV. Since the $^2$H$_2$O sample was used, the γ-ray from the $^{16}$O$(n,\gamma)^{17}$O reaction was also Observed [20].

The thus obtained cross section is compared to values reported previously, which was derived by extrapolating the measured thermal neutron capture cross section by deuterium and the measurement on the inverse reaction. The new data is significantly different from the previous value, by a factor two at 30 keV and about 20% at 550 keV. The new data are compared with theoretical predictions, which are based on the Faddeev scheme with single-nucleon current, meson-exchange current, and the three-nucleon force. They marginally agree with theoretical values [20].

The γ-ray spectrum from the $^{62}$Ni$(n,\gamma)^{63}$Ni reaction is shown in Fig. 3. Before the measurement it was considered that the reaction would proceed via the compound process for such a heavy nucleus and therefore one would observe γ-rays spectra characteristic to a statistical decay. What we observe is quite different from such a picture, and we see an intense γ-ray from the reaction to the ground state in $^{63}$Ni. The partial cross section corresponding to the γ-ray transition is about a half of the total cross section [21].

The Maxwellian averaged cross section thus obtained is about three times larger than the previous value. Hence, the overproduction of $^{62}$Ni is reduced by about 40% [21]. On the other hand, s-process yields in the region of mass between 60 and 90 are enhanced by about 40% due to the large capture cross section of $^{62}$Ni, which should be solved in the future study. Here it should be noted that the cross section decreases with increasing the neutron energy, and thus the neutron capture reaction by $^{62}$Ni proceeds mainly by an s-wave neutron capture but not a p-wave non-resonant capture. The present data is different from a recent result obtained by an activation method [22] by a factor two.

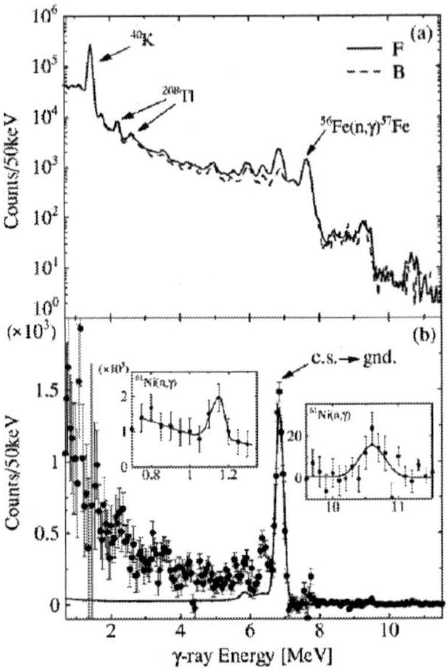

Fig. 3 Foreground (including background), and background γ-ray spectra (a), and a background subtracted spectrum (b) from the $^{62}$Ni$(n,\gamma)^{63}$Ni reaction. A strong peak at 6.8 MeV is the γ-ray from the reaction to the ground state in $^{63}$Ni [21].

Similarly to the case mentioned, we have for the first time succeeded to take the γ-ray energy spectrum from the $^{187}$Os$(n,\gamma)^{188}$Os reaction in the neutron energy range from 10 to 70 keV, where we could see clearly discrete γ-rays from the reaction feeding to low-lying states of $^{188}$Os including to the ground state [23]. The clear detection of the γ-rays to these low-lying states demonstrates the reliability of the present experiment, since these γ-rays characterize the final nucleus of $^{188}$Os. In addition partial cross sections derived using these discrete γ-rays transition strength would be quite useful to improve reliability of theoretical models.

The inelastic scattering cross section by $^{187}$Os was measured by using a newly constructed system shown in Fig. 4, in which neutrons scattered by a sample were detected by four $^{6}$Li-glass scintillation detectors with a diameter of 50 mm and a thickness of 10 mm [18]. We used neutrons from 10 to 70 keV. Hence, inelastically scattered neutrons from $^{187}$Os were detected together with elastically scattered neutrons.

It was therefore necessary to subtract the elastically scattered neutron events from $^{187}$Os from total neutron events containing background events. Here, the elastic scattering events were subtracted by measuring elastic scattering neutron events by $^{186}$Os. Note although the absolute elastic scattering cross section for $^{187}$Os may differ from that for $^{186}$Os, the energy dependence of the elastic scattering cross section for $^{187}$Os is calculated to be the same as that for $^{186}$Os by Goriely et al. [24]. The normalization in the subtraction was made by referring to neutron events above $En \geq 60$ keV. Note that neutron energy we used was in the energy range from 10 to 70 keV.

Hence, neutrons events detected in the energy range from 60 to 70 keV are only due to the elastically scattering events. The absolute value of the cross section was determined by normalizing the elastic scattering events by Carbon, which is well known with a small uncertainty of 4%. Consequently, we could determine accurately the inelastic scattering cross section for $^{187}$Os. Detailed analysis is in progress for the neutron capture cross sections and neutron elastic as well as inelastic scattering cross sections for $^{186}$Os and $^{187}$Os.

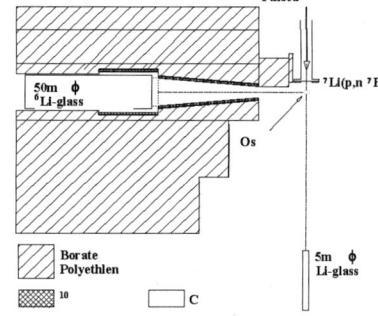

Fig. 4 Measurement system of the (n,n) and (n,n') reaction cross section of Os with a $^6$Li-glass detector. A cadmium vessel containing a boron powder was placed around a throat of tapered collimator.

## III. SUMMARY

In this paper we discussed several experiments, which have been recently carried out by our group. There are many interesting topics on experimental studies using fast neutrons, which are related to a branching point nucleus, isotope anomaly, and a new reaction mechanism. These studies would contribute to trace the history of the galaxy. Therefore in order to promote further these studies Japanese scientists are now installing new measurement systems of the fast neutron induced reaction cross section of a nucleus at the 4 MV Pelletron accelerator facility in Japan Atomic Energy Agency, and at J-PARC in Tokai, where one can expect to use neutron flux much higher than the flux available now.

## ACKNOWLEDGMENTS

We would like to thank K.Takahashi, S. Goriely, and A. Mengoni for discussions.

## REFERENCES

1. E. M. Burbidge, G. R. Burbidge, W. A. Fowler and F. Hoyle, *Rev. Mod. Phys.* 29, 547 (1957).
2. P. J. E. Peebles, *Phys. Rev. Lett.* 16, 410 (1966), H. Sato, *Prog. Theo. Phys.* 38, 1083 (1967), R. V. Wagoner, W. A. Fowler, and F. Hoyle, *Astrophys. J.* 148, 3 (1967).
3. T. S. Suzuki, Y. Nagai, T. Shima, T. Kikuchi, H. Sato, T. Kii, and M. Igashira, *Astrophys. J.* 439, L59 (1995).
4. Y. Nagai, T. S. Suzuki, T. Kikuchi, T. Shima, T. Kii, H. Sato, and M. Igashira, *Phys. Rev.* C56, 3173 (1997).
5. S. Ando, R. H. Cyburt, S. W. Hong, and C. H. Hyum, *Phys. Rev.* C74, 025809 (2006).
6. K. Y. Hara et al., *Phys. Rev.* D68, 072001 (2003).

7. C. Casella et al., *Nucl. Phys.* A706, 203 (2002).
8. .C. Sneden et al., *Astrophys. J.* 533, L139 (2000).
9. Y. Nagai et al., *Astrophys. J.* 372, 683 (1991), T. Ohsaki et al., *Astrophys. J.* 422, 912 (1994).
10. M. Igashira et al., *Astrophys. J.* 441, L89 (1995).
11. M. Rayet, and M. Hashimoto, *Astron. Astrophys.* 354, 740 (2000), S. E. Woosley, A. Heger, T. Rauscher, and R. D. Hoffman, *Nucl. Phys.* A718, 3c (2003).
12. T. Rauscher, A. Heger, R. D. Hoffman, and S. E. Woosley, *Astrophys. J.* 576, 323 (2002), F. X. Timmes, S. E. Woosley, T. A. Weaver, *Astrophys. J. Supplement.* 98, 617 (1995).
13. D. D. Clayton, *Astrophys. J.* 139, 637 (1964).
14. W. A. Fowler, *Rev. Mod. Phys.* 56, 149 (1984).
15. R. R. Winters and R. L. Macklin, *Phys. Rev.* C25, 208 (1982), R. R. Winters et al., *Astron. Astrophys.* 171, 9 (1987).
16. J. C. Browne et al., *Phys. Rev.* C23, 1434 (1981).
17. T. Ohsaki et al., *Nucl. Instr. Meth.* A425, 302 (1999).
18. M. Segawa et al., *Nucl. Instr. Meth.* A564, 370 (2006).
19. K. Wisshak et al., *Nucl. Instr. Meth.* A292, 595 (1990).
20. Y. Nagai et al., *Phys. Rev.* C74, 025804 (2006).
21. A. Tomyo et al., *Astrophys. J.* 623, L153 (2005).
22. H. Nassar et al., *Phys. Rev. Lett.* 94, 092504 (2005).
23. M. Segawa, "Measurements of the (n,γ) and (n,n') reaction cross sections on $^{186,187,189}$Os and $^{187}$Re-$^{187}$Os nuclear chronometer", Ph.D. Thesis, Osaka University, 2005.
24. S. Goriely, private communication (2006).

# Underground studies of pp and CNO

## Heide Costantini

*INFN Genova (Italy) and University of Notre Dame, IN (USA)*

**Abstract.** Cross section measurements for quiescent stellar H and He burning are hampered mainly by extremely low counting rate and cosmic background. Some of the main reactions of H-burning phase have been measured at the LUNA facility (Laboratory for Underground Nuclear Astrophysics) taking advantage of the very low background environment of the Underground Gran Sasso National Laboratory in Italy. An overview of the adopted experimental techniques is given together with the latest results on the $^{14}N(p,\gamma)^{15}O$ reaction and the status of the ongoing $^{3}He(^{4}He,\gamma)^{7}Be$ experiment.

**Keywords:** Hydrogen burning, Underground Nuclear Astrophysics, 14N(p,γ)15O, 3He(4He,γ)7Be, cross section measurements
**PACS:** 25.40.Lw-26.20.+f

## H-BURNING IN STARS

The fusion of hydrogen into helium represents the greater part of the stars life (main sequence stars) and is responsible for the prodigious luminosity of those stars. The basic concept of hydrogen burning is:

$$4p \rightarrow {}^4He + 2e^+ + 2\nu + 26.73 MeV \tag{1.1}$$

This transformation can occur through two different processes: the p-p chain and the CNO cycle.

The sequence of reactions for the p-p chain is shown in fig. 1.

If in addition to hydrogen and helium, heavier elements are present in the star's interior, a second possibility for the conversion of hydrogen into helium is offered by a reaction cycle investigated in 1938 by H. Bethe and C.F. Von Weiszäker: the CNO-cycle.

When the central temperature increases to $T \sim 15 \times 10^6$ K, carbon present in the star can react with the proton sea producing $^{13}N$, which decays to $^{13}C$ which, in turn, captures another proton.

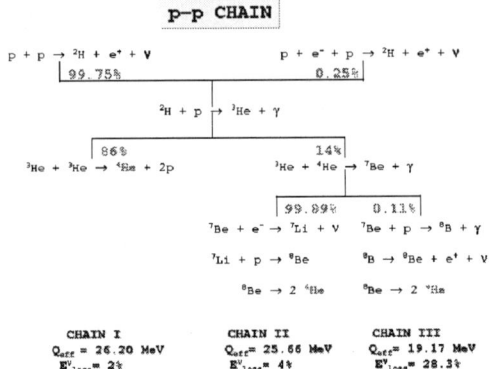

**FIGURE 1.** Scheme of the p-p nuclear reactions chain. The reactions are divided in three chains and the final result is the transformation of four protons into a helium nucleus.

As shown in equation 1.2 at the end of this first part of the CNO-I cycle, four protons are transformed in one helium nucleus, exactly as in the p-p chain and with the same Q-value:

$$^{12}C + p \rightarrow {}^{13}N + \gamma$$
$$^{13}N \rightarrow {}^{13}C + e^+ + \nu$$
$$^{13}C + p \rightarrow {}^{14}N + \gamma$$
$$^{14}N + p \rightarrow {}^{15}O + \gamma \quad (1.2)$$
$$^{15}O \rightarrow {}^{15}N + e^+ + \nu$$
$$^{15}N + p \rightarrow {}^{12}C + \alpha$$
$$^{15}N + p \rightarrow {}^{16}O + \gamma$$

The CNO cycle energy production rate increases faster with temperature than the p-p chain reaction rate.

$$\varepsilon(CNO) = \rho X Z_{CNO} T^{18}$$
$$\varepsilon(p-p) = \rho X^2 Z_{CNO} T^4 \quad (1.3)$$

where X is the hydrogen abundance and Z is the metal abundance. Therefore when the central stellar temperature exceeds $15 \times 10^6$ K, H-burning occurs mainly through the CNO cycle.

Nuclear reactions during H-burning occur between charged particles and since the typical energy of the interacting nuclei (KT~keV) is much smaller than the Coulomb barrier, nuclear reactions occur through tunneling effect.

For charged particle reaction it is therefore possible to define the cross section as:

$$\sigma(E) = \frac{S(E)}{E} e^{-2\pi\eta} \quad (1.4)$$

where the exponential term takes into account the tunneling probability and the S-factor, S(E), smoothly varying function with the energy for non resonant reactions, includes all the nuclear properties of the reaction [1].

The reaction rate can thus be expressed as:

$$\langle \sigma v \rangle = \sqrt{\frac{8}{\pi \mu}} \frac{1}{(KT)^{3/2}} \int_0^\infty \frac{S(E)}{E} e^{-2\pi\eta} e^{-E/KT} dE \qquad (1.5)$$

Where T is the stellar temperature and $\phi(E) \propto e^{-E/KT}$ is the Maxwell-Boltzmann energy distribution that determines the velocity distribution of the nuclei inside the stellar plasma in the case of non-degenerate matter like in quiescent H-burning.

The product of the two exponential terms leads to a well defined peak (the Gamow peak). For a given stellar temperature T, nuclear reactions are taking place mainly inside the Gamow peak if no resonances are present. In the case of H-burning typical energies are of the order of tens of keV (for example ~27 keV for $^{14}N(p,\gamma)^{15}O$ and 22 keV for $^3He(^3He,2p)^4He$).

One of the goal of experimental nuclear astrophysics is to measure nuclear reactions at the energies at which they take place inside stars. Typical cross sections at the Gamow peak energy for H-burning reactions are of the order of $10^{-9}$-$10^{-12}$ barn corresponding to experimental counting rate ranging from few events per day to few events per month with typical laboratory conditions. The main problem in performing these reaction measurements at surface laboratory is that the detectors are continuously bombarded by cosmic rays, that interacting with the detector, the target and the surrounding materials, create background in the detectors. The cosmic background rate is generally much larger than the reaction rate at the Gamow peak.

Therefore experimentalists measure nuclear reactions at higher energies transform the cross section into S-factor and then extrapolate the S-factor by means of different techniques (for example the R-matrix method [2]).

However extrapolations can sometimes fail, for example in the case of an unpredicted narrow resonance at low energies or in the case of contributions from a subthreshold state. One solution to overcome this problem is to perform nuclear reaction measurements in an underground laboratory where the cosmic flux is reduced by several orders of magnitude.

## THE LUNA PROJECT

The Laboratory for Underground Nuclear Astrophysics (LUNA) has been designed to measure nuclear reactions mainly of H-burning both of p-p chain and CNO cycle at energies as close as possible to the Gamow peak. It is located deep underground in the Laboratori Nazionali del Gran Sasso (LNGS) in Italy. The Gran Sasso site is protected from cosmic rays by a rock cover (1400 m thick) equivalent to 3800 m water, suppressing the flux of cosmic ray induced μ by six orders of magnitude and the neutron flux by three orders of magnitude.

During the first phase of the experiment a homemade 50 kV accelerator was operated, and the nuclear processes $^3He(^3He,2p)^4He$ and $d(p,\gamma)^3He$ [3] [4] were measured reaching for the first time the relevant astrophysical energy of the Gamow peak.

After the success of this first phase a new 400 kV accelerator was installed.

Since the main feature of an underground nuclear reaction measurement is the extremely low cross section, high beam current up to several hundreds of μA is a fundamental requirement. For the 400 kV accelerator helium and proton beams are operated at currents of approximately 500 μA for protons and 250 μA for α particles.

Measurements generally last several weeks and months and therefore long term energy stability becomes very important. Furthermore since the cross section depends exponentially from the energy (see eq. 1.4), an uncertainty of few per cent in the energy brings a very large error in the cross section determination. Therefore a good energy resolution is required. For the LUNA 400 kV accelerator a beam energy stability of 5 eV/hour has been measured and the energy spread is of the order of 70 eV [5].

## The $^{14}$N(p, γ) $^{15}$O reaction measurement

The first measurement performed at the 400 kV accelerator was the $^{14}$N(p,γ)$^{15}$O reaction .

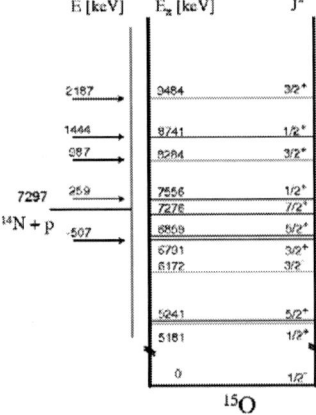

**FIGURE 2.** Level scheme of the $^{15}$O nucleus.

This capture reaction, the slowest process in the H-burning CNO cycle [1], it's of high astrophysical interest as its reaction rate influences sensitively the age determination of globular clusters [6] and the solar neutrino spectrum [7,8]. The capture cross section needs to be known down to $E_0$=30 keV (the Gamow peak in core H-burning stars), which is far below the low-energy limit of direct γ-ray measurements, i.e. the center-of-mass energy E=240 keV [9]. Thus, the data had to be extrapolated over a large energy gap leading to a substantial uncertainty for the astrophysical S-factor at zero energy, S(0). According to the data and analysis of [9], there are two major and nearly equal contributions to S(0): the direct capture (DC) to the 6.79 MeV state in $^{15}$O and the capture to the ground state (gs) in $^{15}$O (fig. 2). The latter process is enhanced due to a subthreshold resonance at $E_R$ =-507 keV, the width of which was taken as a free parameter in the fit [9]. Subsequently, the data of Schröder et al [9] were reanalyzed by [10] using an R-matrix approach. Contrary to the extrapolation by [9] for capture to the ground state, they reported a negligible contribution due to a smaller total width of the subthreshold resonance. A smaller width of the 6.79 MeV state was supported by a lifetime measurement via Doppler-shift method [11] and by a Coulomb excitation measurement [12]. The LUNA collaboration started in 2001 a reinvestigation of $^{14}N(p,\gamma)^{15}O$ studying this reaction in two different phases [13, 14,15].

The measurement of the $^{14}N(p,\gamma)^{15}O$ reaction is particularly well suited for an underground experiment since the Q-value of the reaction is 7.3 MeV.

As a matter of fact for γ-detectors, the advantage of an underground laboratory is particularly appreciated at γ-energies above 3 MeV (see fig. 3). In this energy region the dominant background source are cosmic rays and by bringing the detector underground the background rate is reduced by more than three orders of magnitude. On the other hand at γ energies below 3 MeV, the background spectrum is dominated by γ radiation coming from environmental radioactive isotopes ($^{40}K$, $^{208}Tl$, $^{214}Bi$ etc.) that are always present in the rocks surrounding the laboratory.

Radiative reaction measurements are consequently favored underground especially for high Q-value reactions.

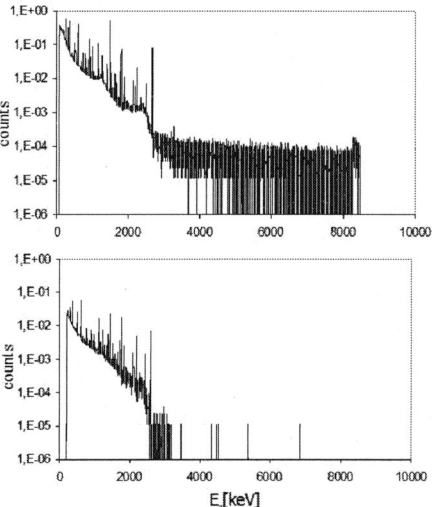

**FIGURE 3.** Background spectra taken with the 126% high purity germanium detector at surface (top spectrum) and underground with a 5 cm lead shielding (bottom spectrum). The measurement time is the same for both spectra and the counts are expressed in arbitrary units.

The goal of the first phase was to study the single γ-transitions and in particular the ground state transition. Therefore a solid target coupled with a high resolution HpGe-detector was used and it was possible to distinguish the single γ-decays. The ground state transition energy is close to the Q-value of the reaction and so a clean signal could be detected thanks to the background free energy spectra around 7 MeV (fig. 3).

In order to lower beam induced background, a careful study of different solid targets and backing materials was performed and TiN sputtered targets on a Ta backing were finally chosen. Cross section measurements with solid target setup were performed in the energy range $E_b$=140-400 keV.

Beam induced background was mainly disturbing the measurements at intermediate energies, since by decreasing the beam energy, the Coulomb barrier was sensitively affecting the cross section of the parasitic reactions.

The weak side of using a high resolution germanium detector is the relatively low detection efficiency.

Therefore to push the cross section measurements toward lower energies a second phase of the experiment was started using a nearly $4\pi$ BGO summing crystal. All the $\gamma$-cascades are summed together to a peak at $E_\gamma$=Q+$E_{cm}$ around 7 MeV where the detection efficiency is about 65% [15].

Due to the intense beam current ($I_p$=500 μA), the gas target local density along the beam path is decreased [16]. A careful study of beam heating effect was performed and the results were implemented in the final data analysis [17]. In the gas target experiment the main source of beam induced background was coming from impurities in the collimators and the beam stop.

By replacing the N gas with the inert $^4$He gas, beam induced background measurements could be performed and the obtained spectra subtracted to the $^{14}$N spectra. Again beam induced background was a major problem at intermediate energies. At the lowest measured energy ($E_b$=80 keV) the main background source in the ROI was coming from (n,$\gamma$) reactions from neutrons produced either by ($\alpha$,n) from natural radioactivity or residual cosmic μ.

The two different approaches were complementary and both took extreme advantage of the low background laboratory (see fig. 4).

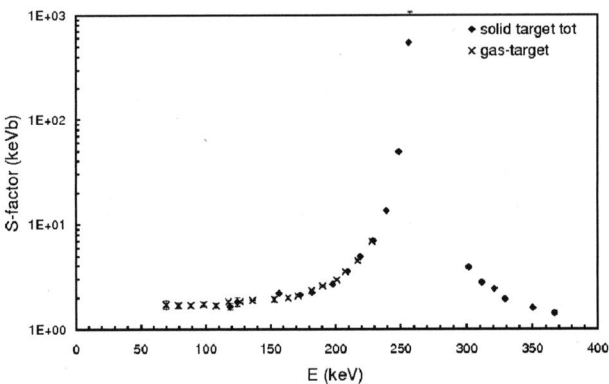

**FIGURE 4.** Comparison between the astrophysical factor obtained with the LUNA high resolution and high efficiency experiments for the $^{14}$N(p,$\gamma$)$^{15}$O reaction.

The final results from both experiments were in good agreement with new measurement by [18, 19] but differed in the weight of the contributions from the various transitions.

The extracted stellar reaction rate confirmed the conclusion of [10] that the rate has to be reduced by nearly a factor of two at low temperatures, but it is in good agreement with NACRE [20] above $T_6$=150. In conclusion, with the present determination of the reaction rates the main astrophysical consequence is the age increase of the Globular Clusters by about 1 Gyear, i.e about 14±1 Gyears depending by the metallicity of the Globular Cluster [6] and the reduction of a factor 2 of solar CNO neutrinos.

## The $^3$He($^4$He,$\gamma$)$^7$Be reaction measurement

In the case of low Q-value reactions, the advantage of an underground laboratory is not evident at first sight. Environmental background is also present underground.

Detectors can be shielded passively with proper Lead and Copper shield as on surface. However there is a big advantage in a underground laboratory.

In a surface laboratory passive shielding can be built around the detectors but above a certain thickness the shield efficiency cannot be increased by adding further shield material since cosmic $\mu$ interact with the shielding material and can create background signals in the detector.

Obviously this problem is dramatically reduced in an underground laboratory.

The $^3$He($^4$He,$\gamma$)$^7$Be has a Q-value of 1.6 MeV. This reaction is presently under study at LUNA. The $^3$He($^4$He,$\gamma$)$^7$Be reaction is one of the major source of uncertainty in determining the Boron solar neutrino flux and dominates over the present observational accuracy $\Delta\phi(B)/\phi(B)$=7% [7]. The foreseeable accuracy of the new generation solar neutrino experiments is $\Delta\phi(B)/\phi(B)$=3%. This result could illuminate about solar physics if the uncertainty on $S_{34}$ is reduced to a corresponding level. Moreover this reaction plays an important role in understanding the primordial $^7$Li abundance [21].

Past measurements, that go back to twenty years ago, have been performed using two different methods. In the first method prompt $\gamma$-rays ($E_\gamma$~1.6 and 1.2 MeV), coming from direct $\alpha$-capture, are detected, while in the second one the delayed $\gamma$s ($E_\gamma$ = 478 keV), coming from $^7$Be decay through electron capture, are counted.

A global analysis indicates that the extrapolated S(0) obtained with the activation method is sistematically 13% higher than the prompt-$\gamma$ result.

A recent activation study [22] reduces the discrepancy to 9% still not at the precision level of the $^8$B neutrino data.

The goal of the experiment at LUNA is to measure the cross section of the reaction using both techniques at the same time reducing the error on the astrophysical factor S(3,4) to 4%.

The prompt capture $\gamma$-rays are measured with an ultra-low 135% background germanium detector heavily shielded (0.3 m$^3$ of Lead and Copper) and placed at close distance to a $^3$He windowless gas target. The suppression factor obtained with the shield for γs below 2 MeV is of five orders of magnitude (see fig. 5).

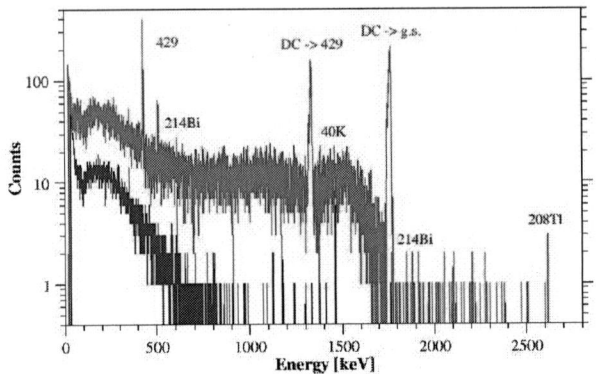

**FIGURE 5.** Comparison between the background spectrum and the reaction spectrum at $E_\alpha$= 400 keV obtained with the ultra low background 135% HpGe detector. For both spectra the measurement time was approximately 4.5 days. The suppression factor due to the passive Pb-Cu shield of the natural environmental radioactivity is of five orders of magnitude.

Besides from the rocks, environmental background can come also from all the materials surrounding the detector and from the detector itself. Therefore the target chamber and all the setup close to the detector were made of OFC copper and no welding materials have been used in the chamber assembling.

The $^7$Be nuclei are collected on the beam stop and are counted off-line by a 125% HpGe detector, completely shielded by 15 cm of lead and 10 cm of copper on each side, positioned in the low activity laboratory at LNGS. The typical beam current is about 250 μA and therefore to avoid systematic uncertainties due to beam heating effects the target density is measured through α-Rutherford scattering cross section with a silicon detector positioned inside the target chamber.

First results of the activation measurements have been obtained at $E_{cm}$=117, 148 and 169 keV [23] with a total uncertainty of 4%. The uncertainty (systematic and statistical combined in quadrature) is comparable to or lower than previous activation studies at high energy and lower than prompt-γ studies at comparable energy. The data analysis for the online γ-measurement is still in progress and the comparison with activation results will be of fundamental importance to understand previous discrepancies between the two methods.

When the $^3$He($^4$He,γ)$^7$Be reaction measurement is finished, LUNA will start to study the $^{25}$Mg(p,γ)$^{26}$Al reaction.

It's the slowest reaction of the Mg-Al cycle. The $\beta^+$ decay of $^{26}$Al$_{gs}$ to the excited state of $^{26}$Mg gives rise to a 1.8 MeV γ-ray, one of the most important line for γ-astronomy [24]. The level scheme of $^{26}$Al is very complicated and a lot of resonances of low intensities are present in the astrophysical energy region. The measurement of the weak low energy resonances will be performed at LUNA coupling a high efficient

$4\pi$ BGO summing crystal with a high purity $^{25}$Mg solid target. First test of target purity and stability has been performed.

With the completion of the $^{25}$Mg(p,$\gamma$)$^{26}$Al measurement, LUNA will end the current scientific approved program.

In September 2005 a working group has been formed inside the LUNA collaboration with the goal to determine a list of reaction of astrophysical relevance that could be studied at the 400 kV accelerator and for which an underground approach represents a clear advantage. An other goal of the working group is to investigate the importance and the possibility of the installation of a higher energy accelerator that could be used for the study of He-burning key nuclear reactions

## REFERENCES

1. C. Rolfs and W.S. Rodney: *Cauldrons in the cosmos, University of Chicago Press* (1988)
2. A.N. Lane and R.G. Thomas, *Rev. Mod. Phys.* 30(1958)257
3. C. Casella et al, *Nucl. Phys. A* 706 (2002) 203-216
3. R. Bonetti et al.,*Phys. Rev. Lett.* 82(1999)5205
5. A. Formicola et al., *Nucl. Inst. Meth. A*, 507(2003)609
6. G. Imbriani et al, *A&A*, 420(2004)625
7. J. N. Bahcall and M. H. Pinsonneault, *Phys. Rev. Lett.* 92(2004)121301
8. S. Degl'Innocenti et al., *Phys. Lett. B* 590(2004)13
9. U. Schröder et al, *Nucl. Phys. A* 467(1987)240
10. C. Angulo and P. Descouvemont, 2001, *Nucl. Phys. A* 690(2001)755
11. P.F. Bertone et al, *Phys. Rev. Lett.* 87(2001)152501
12. K. Yamada et al, *Phys. Lett. B* 579(2004)265
13. A. Formicola et al, *Phys.Lett.B*, 591(2004)61-68
14. G. Imbriani et al, *Eur.Phys.Journal A*, 25(2005)455-466
15. A. Lemut et al., *Phys. Lett. B* 634(2006)483-487
16. J. Görres et al., *Nucl. Inst. Meth. A*, 177(1980)295
17. D. Bemmerer et al., In Press in *Nucl. Phys. A*
18. A.M. Mukhamedzhanov et al, *Phys. Rev. C* 67(2003)065804
19. R.C. Runkle et al.,*Phys. Rev. Lett.* 94(2005)082503
20. C. Angulo et al, *Nucl. Phys. A* 656(1999)3
21. A. Coc et al., *Astrophys. J.* 600(2004)544
22. B.S. Nara Singh et al., *Phys. Rev. Let.* 93(2004)262503
23. D. Bemmerer et al, *Phys. Rev. Let.* 97(2006)122502
24. R. Diehl et al. *A&A*, 298(1995)445.

# Bound Electron Screening Corrections to Reactions in Hydrogen Burning Processes

Sachie Kimura* and Aldo Bonasera*,†

*Laboratori Nazionali del Sud, INFN, via S.Sofia 62, 95125 Catania, Italy
†Libera Università Kore, Enna, Italy

**Abstract.** We estimate the bare astrophysical S-factor($S_b(E)$) in the PP-chains, through the polynomial expression with the adiabatic enhancement factor by the electron screening. The obtained $S_b$ is significantly different from the simple extrapolation from high energy data, however $S_b$ at the zero incident energy is in agreement with the results of one of recent R-matrix analysis.

**Keywords:** Astrophysical S-factor, Electron screening in the laboratories
**PACS:** 29.87.+g

## INTRODUCTION

The series of reactions which convert hydrogen into helium on stellar site is known as the proton-proton chains. It is a key to understand the evolution of the stars. These reactions are measured at laboratory energies and are then extrapolated to thermal energies [1], because of their small cross sections at such low energies. This extrapolation is done by introducing the astrophysical S-factor:

$$S(E) = \sigma(E) E e^{2\pi \eta(E)}, \qquad (1)$$

where $\sigma(E)$ is the reaction cross section at the incident center-of-mass energy $E$ and $\eta(E) = Z_T Z_P \alpha \sqrt{\frac{\mu c^2}{2E}}$, denoting the atomic numbers and the reduced mass of the target and the projectile as $Z_T$, $Z_P$, $\mu$, respectively. The exponential term in the equation represents the Coulomb barrier penetrability. Since one has factored out the strong energy dependence of $\sigma(E)$ due to the barrier penetrability, S-factor could be approximated by a polynomial expression. In laboratory experiments, the targets are usually in gas or solid state. There the experimental data of the S-factor in the low energy region show large enhancement to the extrapolation from high energy data for many reactions [2]. This enhancement is, usually, attributed to the screening by the bound electrons around the target and is discussed in terms of a constant potential shift(screening potential $U_e$). A puzzle had been that the experimentally observed enhancements are systematically and significantly larger than that within the adiabatic limit.

On this issue, the dynamical effect to this problem has been studied by Caltech group [3]. They followed time evolution of the atomic wave function in the classical allowed region by solving time dependent Hartree-Fock equation and evaluated the screening potential. Their results suggest that the screening potential approaches to the adiabatic limit as the incident energy becomes lower. The influence of the tunneling phenomenon to this problem has been studied, as well [4]. And, there, the screening potential could go over the "conventional" adiabatic limit, only in the case there is excited state component at the classical turning point. The modification to the adiabatic limit is, nevertheless, too small to explain large discrepancies of all the reactions. We have examined the problem using molecular dynamics approach with constraints [5, 6], to see the effect of the fluctuations in our previous studies [7, 8]. The obtained average enhancement factors, again, do not exceed the adiabatic limit, there are events which give larger enhancemnt factors than that in the adiabatic limit. There are other attempts [2, 9, 10] to explain the mechanism to get such a large enhancement, over the adiabatic limit, however none of them is affirmative to the screening potential which goes over the adiabatic limit.

In this connection, we mention that the difficulty lies in the determination of the bare S-factor. Remember that the bare S-factor is usually determined by extrapolations from high energy data and then the screening potential is determined by taking the ratio of the data and extrapolated S-factor. Instead, Barker [11] performed the fit including whole data using either a polynomial(quadratic or cubic) or R-matrix determining parameters simultaneously and obtained more consistent values of screening potentials to the adiabatic limit for some reactions. Experimentally, Catania group tried to extract the bare cross section using the Trojan Horse method(THM) [12, 13, 14].

In this paper, we determine the bare S-factors($S_b(E)$) of the reactions, especially, in hydrogen burning process through the polynomial expression with the screening enhancement within the adiabatic limit. The results are compared with the extrapolated $S_b(E)$ by polynomial expressions from high energy data. The obtained $S_b$ at the zero incident energy are compared with the results by the R-matrix analyses [15].

## ENHANCEMENT FACTOR IN THE ADIABATIC LIMIT

In order to discuss the enhancement quantitatively, we determine the enhancement factor:

$$f_e = \frac{\sigma(E)}{\sigma_0(E)}, \qquad (2)$$

in terms of the real cross section $\sigma(E)$ and the bare cross section $\sigma_0(E)$. If one assumes that the effect of the electron screening can be represented by the constant shift $U_e$(screening potential) of the potential barrier, the enhancement factor is approximated by,

$$U_e \sim \frac{E}{\pi \eta(E)} \log f_e. \qquad (3)$$

The $U_e$ can be estimated easily in two limiting cases. One is the case where the inter-nuclear velocity is much higher than that of electrons velocity, i.e. at the sudden limit. Within this limit the electron wave function is frozen during the reaction. In the opposite case where the inter-nuclear motion is much slower than electrons motion, the bound electrons follow the motion of nuclei adiabatically. Within this adiabatic limit The screening potential is expressed by the difference of the binding energies between the initial target atom($BE_T$) and the united atom($BE_{UA}$) which is formed during the reaction.

$$U_e^{(AD)} = BE_T - BE_{UA} \qquad (4)$$

**TABLE 1.** Reactions in PP-chains, it's minimum incident energy in the center-of-mass system measured so far and enhancement factor within the adiabatic limit at the minimum energy.

| reactions | $E_{min}$ [keV] | $f_e^{(AD)}(E_{min})$ |
|---|---|---|
| H$(p,\beta^+\nu_e)$D | | |
| D$(p,\gamma)^3$He | 2.52 | 1.07 |
| $^3$He$(^3$He$,2p)^4$He | 20.76 | 1.22 |
| $^3$He$(\alpha,\gamma)^7$Be | 107.2*, 127.† | 1.02 |
| $^7$Be$(e^-,\nu_e)^7$Li | | |
| $^7$Li$(p,\alpha)^4$He | 12.7, 10.** | 1.18 |
| $^7$Be$(p,\gamma)^8$B | 115.6 | 1.01 |

\* prompt-$\gamma$ method
† activation method
\*\* indirect THM

The screening potential within this limit gives the upper limit.

## BARE S-FACTORS OF PP-CHAIN REACTIONS

A list of reactions in the PP-chains are shown in Table 1. The first reaction H$(p,\beta^+\nu_e)$D involves the $\beta$-decay and has too small cross section to be measured experimentally. Its S-factor is calculated from first principle [16]. We, therefore, concentrate on the other 5 reactions except the electron capture reaction $^7$Be$(e^-,\nu_e)^7$Li. In the table the minimum incident energies, measured so-far, for each reaction are also shown. For three reactions D$(p,\gamma)^3$He, $^3$He$(^3$He$,2p)^4$He and $^7$Li$(p,\alpha)^4$He cross sections have been measured already including the low energy region. The other two reactions are proton or $\alpha$ capture reactions which have even smaller cross sections. The S-factor of the reaction $^3$He$(\alpha,\gamma)^7$Be has been re-determined with high precision recently by detecting $\gamma$-ray from $^7$Be [17]. Its S-factor in the low energy region is extrapolated from high energy data by R-matrix fitting. The reaction $^7$Be$(p,\gamma)^8$B involves unstable nuclei. The S-factor of this reaction has been determined by means of the direct capture reaction [18, 19] and the coulomb dissociation method [20] and it is one of questions under discussion that there is a discrepancy between the results by two methods [19, 21].

The previous studies of the electron screening effect suggest us that the enhancement factor cannot be over the adiabatic limit. We, therefore, adopt the enhancement factor within the adiabatic limit:

$$f_e^{(AD)} = e^{\pi\eta(E)\frac{U_e^{(AD)}}{E}}, \qquad (5)$$

and determine the bare S-factor by fitting the experimental data of the reactions. The fit of the experimental data is performed by assuming a polynomial expression for bare S-factor:

$$S(E) = S_b(E) \cdot f_e^{(AD)}; \qquad S_b(E) = S_b(0) + S_1 E + S_2 E^2 + S_3 E^3 \qquad (6)$$

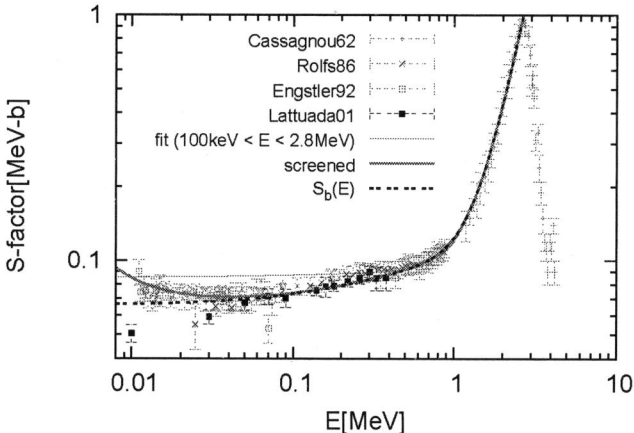

**FIGURE 1.** S-factor for the reaction $^7\text{Li}(p,\alpha)^4\text{He}$ as a function of the incident center-of-mass energy. The experimental points are from [22](Cassagnou62), from [23](Rolfs86), from [24, 25](Engstler92) and from [13](Lattuada01).

**TABLE 2.** Fitting parameters of the reaction $^7\text{Li}(p,\alpha)^4\text{He}$ in the high energy region(the first row) and in the whole energy region(the second row).

|  | $S_b(0)$[MeVb] | $S_1$[b] | $S_2$[MeV$^{-1}$b] | $S_3$[MeV$^{-2}$b] |
|---|---|---|---|---|
| $S_b(E)$ | $0.080 \pm 0.008$ | $0.04 \pm 0.03$ | $-0.06 \pm 0.03$ | $0.067 \pm 0.006$ |
| $S_b(E) \cdot f_e^{(AD)}$ | $0.066 \pm 0.002$ | $0.08 \pm 0.01$ | $-0.10 \pm 0.01$ | $0.076 \pm 0.004$ |

## PARTICULAR REACTIONS

## $^7\text{Li}(p,\alpha)^4\text{He}$

In Fig. 1 the S-factor of the reaction $^7\text{Li}(p,\alpha)^4\text{He}$ from several direct measurements are shown with error bars. Extracted S-factor data by THM are especially shown with the closed squares. [13] We performed fitting of the data by direct measurements in the incident energy region higher than 100keV using a cubic polynomial. In this energy region the screening enhancement is estimated to be 1% at utmost. The fitting parameters are shown in the first row of the table 2. The corresponding S-factor is shown with the thin solid curve in the figure 1. The curve supposed to give an extrapolation of low energy data, however it strays away from the trend of experimental data in the lower energy region. Instead of fitting higher-energy data, if we fit whole data by direct measurements in the form of Eq. 6 and $U_e^{(AD)} = 175\text{eV}$ [8], we obtain the fitting parameters in the second row of the table 2. The corresponding S-factor is shown with thick dashed curve in Fig. 1. Note that there is a big difference between these two curves. The extrapolation of the S-factor to the lower energy region should be done, therefore, with caution. The

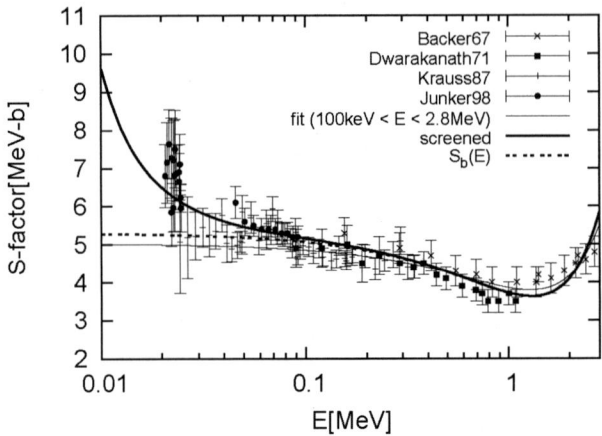

**FIGURE 2.** S-factor for the reaction $^3\text{He}(^3\text{He},2p)^4\text{He}$ as a function of the incident center-of-mass energy. The experimental points are from [27](Backer67), from [28](Dwarakanath71), from [29](Krauss87) and from [26](Junker98).

S-factor at zero-energy $S_b(0)=0.066 \pm 0.002$[MeVb] from our result is in agreement with the result from R-matrix fit in Ref. [15] $S_b(0)=0.067\pm0.004$[MeVb] but higher than $S_b(0)=0.055\pm0.003$[MeVb] by THM [13].

## $^3\text{He}(^3\text{He},2p)^4\text{He}$

The S-factor of the reaction $^3\text{He}(^3\text{He},2p)^4\text{He}$ from several measurements are shown with error bars in Fig. 2. At the minimum incident energy, which has been reached in an experiment by the LUNA collaboration [26], the screening enhancement is estimated to be more than 20%. In the case of the reaction $^3\text{He}(^3\text{He},2p)^4\text{He}$, $^3\text{He}$ projectiles are likely to be $^3\text{He}^+$ or charge neutral state in the target medium. For the $^3\text{He}$ neutral projectile the adiabatic screening potential is $U_e^{(AD)} = 246.8$ eV [26].

$$f_e^{(AD)}(^3He) = e^{\pi\eta(E)\frac{U_e^{(AD)}}{E}}. \tag{7}$$

For the $^3\text{He}^+$ projectile the adiabatic screening potential is calculated considering the charge symmetry of the system [30]. $U_e^{(AD)1} = 255.5$ eV and $U_e^{(AD)2} = 122.2$ eV in the cases where the system ends up with $^6\text{Be}^+(1s)^2(2s)$ state and $^6\text{Be}^+(1s)(2p)^2$ state respectively. The corresponding enhancement factor within the adiabatic limit is written as [4]

$$f_e^{(AD)}(^3He^+) = \frac{1}{2}\left(\exp\left[\pi\eta(E)\frac{U_e^{(AD)1}}{E}\right] + \exp\left[\pi\eta(E)\frac{U_e^{(AD)2}}{E}\right]\right). \tag{8}$$

**TABLE 3.** Fitting parameters of the reaction $^3$He($^3$He,2p)$^4$He.

|  | $S_b(0)$[MeVb] | $S_1$[b] | $S_2$[MeV$^{-1}$b] |
|---|---|---|---|
| $S_b(E)$ | 5.03 ± 0.07 | -1.9 ± 0.2 | 0.75 ± 0.08 |
| $S_b(E) \cdot f_e^{(AD)}$($^3$He) | 5.31 ± 0.05 | -2.6 ± 0.2 | 1.0 ± 0.1 |
| $S_b(E) \cdot f_e^{(AD)}$($^3$He$^+$) | 5.41 ± 0.06 | -2.8 ± 0.3 | 1.1 ± 0.1 |

**TABLE 4.** Fitting parameters of the reaction D(p,γ)$^3$He.

|  | $S_b(0)$[eVb] | $S_1$[b] | $S_2$[eV$^{-1}$b] |
|---|---|---|---|
| $S_b(E) \cdot f_e^{(AD)}$ | 0.22 ± 0.03 | 6.0 ± 0.3 | 3.0 ± 0.2 |

The results of fitting are shown in the table 3. The parameters in the second row are for $^3$He neutral projectile and ones in the bottom row are for $^3$He$^+$ projectile. The corresponding curve for $^3$He neutral projectile case are shown in Fig. 2 together with the experimental points. We, again, see the difference between the extrapolation from high energy data(the top row in Tab. 3, the thin solid curve in Fig. 3) and the bare S-factor obtained by fitting the whole data, including low energy region(the bottom row in Tab. 3, the thick dashed curve in Fig. 3). It is clear that if we derive the screening potential by comparing the dashed and the thin curves, as it is often done in the previous studies, its screening potential will be larger than that in the adiabatic limit.

## D(p,γ)$^3$He

The S-factor of the reaction D(p,γ)$^3$He from several measurements are shown with error bars in Fig. 3. At the minimum incident energy, which has been reached in an experiment by the LUNA collaboration [31], the screening enhancement is estimated to be 7% at utmost. The obtained fitting parameters fot the bare S-factor are shown in the table 4 and its is shown with thick dashed curve in Fig. 3. The S-factor at zero-energy $S_b(0)$=0.22 ± 0.03[eVb] from our result is in agreement with the result from R-matrix fit in Ref. [15] $S_b(0)$=0.223±0.010[eVb].

## $^3$He($\alpha$,γ)$^7$Be

The screening potential for the reaction $^3$He($\alpha$,γ)$^7$Be is estimated in the same way with the reaction $^3$He($^3$He,2p)$^4$He. The estimated enhancement at the minimum incident energy within the adiabatic limit is 2%. In Fig. 4 experimental data are shown with crosses, squares, triangles with error-bars. The thick curve is again the result of fitting by polynomial expression and its fitting parameters are shown in Tab. 5. The S-factor at zero-energy $S_b(0)$=0.49 ± 0.01[keVb] from our result is in agreement with the result from R-matrix fit in Ref. [15] $S_b(0)$=0.51±0.04[keVb] within the error-bars.

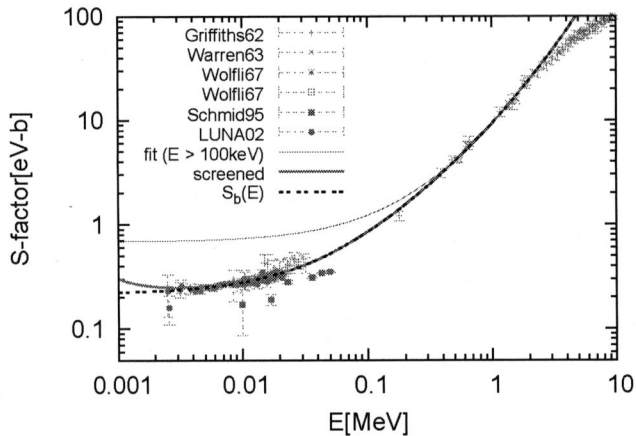

**FIGURE 3.** S-factor for the reaction D($p,\gamma$)$^3$He as a function of the incident center-of-mass energy. The experimental points are from [32](Griffiths62), from [33](Warren63), from [34](Berman64), from [35](Wolfli67), from [36](Schmid95) and from [31](Casella02).

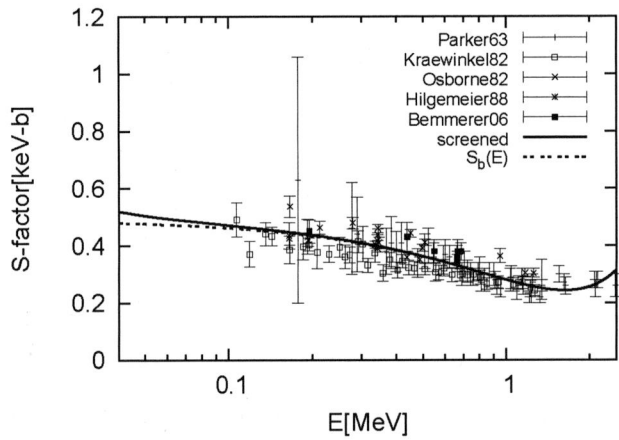

**FIGURE 4.** S-factor for the reaction $^3$He($\alpha,\gamma$)$^7$Be as a function of the incident center-of-mass energy. The experimental points are from [37](Parker63), from [38](Kraewinkel82), from [39](Osborne82), from [40](Hilgemeier88) and from [17](Bemmerer06).

**TABLE 5.** Fitting parameters of the reaction $^3$He($\alpha,\gamma$)$^7$Be.

|  | $S_b(0)$ [keVb] | $S_1$ [b] | $S_2$ [keV$^{-1}$b] |
| --- | --- | --- | --- |
| $S_b(E) \cdot f_e^{(AD)}(^4\text{He})$ | 0.49 ± 0.01 | -0.30 ± 0.02 | 0.09 ± 0.01 |

## $^7\text{Be}(p,\gamma)^8\text{B}$

The reaction $^7\text{Be}(p,\gamma)^8\text{B}$ is a key process to produce the high energy solar neutrino through the $\beta$-decay of $^8\text{B}$ and its S-factor is studied intensively by many groups by means of the direct capture reaction and the coulomb dissociation method. For this reaction we only give that the screening enhancement factor is of the order of 1% at the minimum incident energy.

## CONCLUSIONS

We discussed the bound electron screening corrections to the PP-chains reactions. We have assumed the screening enhancement within the adiabatic limit and have determined the bare S-factors for the reactions through the polynomial fitting of the experimental data. The obtained bare S-factors are significantly different from the simple extrapolation from high energy region, but $S_b$ at the zero incident energy is in agreement with the results from R-matrix fitting in Ref. [15]. We conclude that the polynomial expression of the bare S-factor in the low energy region is appropriate, if the S-factor data in this region are available.

## ACKNOWLEDGMENTS

The authors acknowledge Prof. S. Kubono for suggestion of the problem. One of us(S. K.) thanks Dr. H. Costantini and Dr. R. G. Pizzone for stimulating discussions and for providing us experimental data.

## REFERENCES

1. C. Angulo, M. Arnould, M. Rayet, P. Descouvemont, D. Baye, C. Leclercq-Willain, A. Coc, S. Barhoumi, P. Aguer, C. Rolfs, R. Kunz, J. Hammer, A. Mayer, T. Paradellis, S. Kossionides, C. Chronidou, K. Spyrou, S. Degl'Innocenti, G. Fiorentini, B. Ricci, S. Zavatarelli, C. Providencia, H. Wolters, J. Soares, C. Grama, J. Rahighi, A. Shotter, Rachti, and M. Lamehi, *Nucl. Phys. A* **656**, 3 (1999).
2. G. Fiorentini, C. Rolfs, F. L. Villante, and B. Ricci, *Phys. Rev. C* **67**, 014603 (2003).
3. T. D. Shoppa, S. E. Koonin, K. Langanke, and R. Seki, *Phys. Rev. C* **48**, 837 (1993).
4. S. Kimura, N. Takigawa, M. Abe, and D. Brink, *Phys. Rev. C* **67**, 022801(R) (2003).
5. M. Papa, T. Maruyama, and A. Bonasera, *Phys. Rev. C* **64**, 024612 (2001).
6. S. Kimura, and A. Bonasera, *Phys. Rev. A* **72**, 014703 (2005).
7. S. Kimura, and A. Bonasera, *Phys. Rev. Lett.* **93**, 262502 (2004).
8. S. Kimura, and A. Bonasera, *Nucl. Phys. A* **759**, 229 (2005).
9. A. Balantekin, C. Bertulani, and M. Hussein, *Nucl. Phys. A* **627**, 324 (1997).
10. K. Hagino, M. S. Hussein, and A. B. Balantekin, *Phys. Rev. C* **68**, 048801 (2003).
11. F. C. Barker, *Nucl. Phys. A* **707**, 277 (2002).
12. C. Spitaleri, S. Typel, R. G. Pizzone, M. Aliotta, S. Blagus, M. Bogavac, S. Cherubini, P. Figuera, M. Lattuada, M. Milin, D. Miljanic, A. Musumarra, M. G. Pellegriti, D. Rendic, C. Rolfs, S. Romano, N. Soic, A. Tumino, H. H. Wolter, and M. Zadro, *Phys. Rev. C* **63**, 055801 (2001).
13. M. Lattuada, R. G. Pizzone, S. Typel, P. Figuera, D. Miljani, A. Musumarra, M. G. Pellegriti, C. Rolfs, C. Spitaleri, and H. H. Wolter, *Astrophys. J.* **562**, 1076 (2001).

14. A. Musumarra, R. G. Pizzone, S. Blagus, M. Bogovac, P. Figuera, M. Lattuada, M. Milin, D. Miljanic, M. G. Pellegriti, D. Rendic, C. Rolfs, N. Soic, C. Spitaleri, S. Typel, H. H. Wolter, and M. Zadro, *Phys. Rev. C* **64**, 068801 (2001).
15. P. Descouvemont, A. Adahchour, C. Angulo, A. Coc, and E. Vangioni-Flam, *Atomic Data and Nuclear Data Tables* **88**, 203 (2004).
16. J. N. Bahcall, W. F. Huebner, S. H. Lubow, P. D. Parker, and R. K. Ulrich, *Rev. Mod. Phys.* **54**, 767 (1982).
17. D. Bemmerer, F. Confortola, H. Costantini, A. Formicola, G. Gyurky, R. Bonetti, C. Broggini, P. Corvisiero, Z. Elekes, Z. Fülöp, G. Gervino, A. Guglielmetti, C. Gustavino, G. Imbriani, M. Junker, M. Laubenstein, A. Lemut, B. Limata, V. Lozza, M. Marta, R. Menegazzo, P. Prati, V. Roca, C. Rolfs, C. R. Alvarez, E. Somorjai, O. Straniero, F. Strieder, F. Terrasi, and H. P. Trautvetter, *Phys. Rev. Lett.* **97**, 122502 (2006).
18. L. T. Baby, C. Bordeanu, G. Goldring, M. Hass, L. Weissman, V. N. Fedoseyev, U. Koster, Y. Nir-El, G. Haquin, H. W. Gaggeler, and R. Weinreich, *Phys. Rev. C* **67**, 065805 (2003).
19. A. R. Junghans, E. C. Mohrmann, K. A. Snover, T. D. Steiger, E. G. Adelberger, J. M. Casandjian, H. E. Swanson, L. Buchmann, S. H. Park, A. Zyuzin, and A. M. Laird, *Phys. Rev. C* **68**, 065803 (2003).
20. N. Iwasa, F. Boué, G. Surówka, K. Sümmerer, T. Baumann, B. Blank, S. Czajkowski, A. Förster, M. Gai, H. Geissel, E. Grosse, M. Hellström, P. Koczon, B. Kohlmeyer, R. Kulessa, F. Laue, C. Marchand, T. Motobayashi, H. Oeschler, A. Ozawa, M. S. Pravikoff, E. Schwab, W. Schwab, P. Senger, J. Speer, C. Sturm, and A. Surowiec, *Phys. Rev. Lett.* **83**, 2910–2913 (1999).
21. H. Esbensen, G. F. Bertsch, and K. A. Snover, *Phys. Rev. Lett.* **94**, 042502 (2005).
22. Y. Cassagnou, J. M. Jeronymo, G. S. Mani, A. Sadeghi, and P. D. Forsyth, *Nucl. Phys.* **33**, 449 (1962).
23. C. Rolfs, and R. W. Kavanagh, *Nucl. Phys. A* **455**, 179 (1986).
24. S. Engstler, G. Raimann, C. Angulo, U. Greife, C. Rolfs, U. Schröder, E. Somorjai, B. Kirch, and K. Langanke, *Z. Phys. A* **342**, 471 (1992).
25. S. Engstler, G. Raimann, C. Angulo, U. Greife, C. Rolfs, U. Schröder, E. Somorjai, B. Kirch, and K. Langanke, *Phys. Lett. B* **279**, 20 (1992).
26. M. Junker, A. D'Alessandro, S. Zavatarelli, C. Arpesella, E. Bellotti, C. Broggini, P. Corvisiero, G. Fiorentini, A. Fubini, G. Gervino, U. Greife, C. Gustavino, J. Lambert, P. Prati, W. Rodney, C. Rolfs, F. Strieder, H. Trautvetter, and D. Zahnow, *Phys.Rev. C* **57**, 2700 (1998).
27. A. D. Backer, and T. A. Tombrello, *Astrophysical Problems in "Nuclear research with Low-Energy Accelerators*, Academic Press, 1967.
28. M. R. Dwarakanath, and H. Winkler, *Phys. Rev. C* **4**, 1532 (1971).
29. H. P. T. A. Krauss, H. W. Becker, and C. Rolfs, *Nucl. Phys. A* **467**, 273 (1987).
30. W. Lichten, *Phys. Rev.* **131**, 229 (1963).
31. C. Casella, H. Costantini, A. Lemut, B. Limata, R. Bonetti, C. Broggini, L. Campajola, P. Corvisiero, J. Cruz, A. D'Onofrio, A. Formicola, Z. Fülöp, G. Gervino, L. Gialanella, A. Guglielmetti, C. Gustavino, G. Gyurky, G. Imbriani, A. Jesus, M. Junker, A. Ordine, J. Pinto, P. Prati, J.P.Ribeiro, V. Roca, D. Rogalla, C. Rolfs, M. Romano, C. Rossi-Alvarez, F. Schuemann, E. Somorjai, O. Straniero, F. Strieder, F. Terrasi, H. Trautvetter, and S. Zavatarelli, *Nucl. Phys. A* **706**, 203 (2002).
32. G. Griffiths, E. Larson, and L. Robertson, *Can. J. Phys.* **40**, 402 (1962).
33. J. Warren, K. Erdman, L. Robertson, D. Axen, and J. MacDonald, *Phys. Rev.* **132**, 1691 (1963).
34. B. Berman, L. K. Jr., and J. Smith, *Phys. Rev.* **133**, B117 (1964).
35. W. Wolfli, R. Bosch, J. Lang, R. Muller, and P. Marmier, *Helv. Phys. Acta* **40**, 946 (1967).
36. G. Schmid, R. Chastler, C. Laymon, and H. Weller, *Phys. Rev. C* **52** (1995).
37. P. Parker, and R. Kavanagh, *Phys. Rev.* **131**, 2578 (1963).
38. H. Kraewinkel, H. W. Becker, L. Buchmann, J. G. K. U. Kettner, W. . Kieser, R. Santo, P. Schmalbrock, H. P. Trauttvetter, A. Vielks, C. Rolfs, J. W. Hammer, R. E. Azuma, and W. S. Rodney, *Z. Phys. A* **304**, 307 (1982).
39. J. L. Osborne, C. A. Barnes, R. W. Kavanagh, R. M. Kremer, G. J. Mathews, J. L. Zyskind, P. D. Parker, and A. J. Howard, *Phys. Rev. Lett.* **48**, 1664 (1982).
40. M. Hilgemeier, H. W. Becker, C. Rolfs, H. P. Trautvetter, and J. W. Hammer, *Z. Phys. A* **329**, 243 (1988).

# The electrostatic screening of nuclear reactions in dense stellar plasma

Giora Shaviv

*Dept. of Physics
and Asher Space Research Institute
Israel Institute of Technology
Haifa Israel*

**Abstract.** We review the derivation of the electrostatic screening from first principles and show the basic properties of the screening process.

We start with a numerical experiment in which a simplified finite range potential interaction between two particles is assumed and show how the screening builds up as the range of the interaction increases.

We then turn to the screening of nuclear reactions in the Sun and show that under the conditions prevailing in the Sun in particular and in stars in general, the number of particles in the Debye sphere is of the order of unity. Consequently, fluctuations play a dominant role in the screening process. The fluctuations lead to an effective time dependent potential. The consequence is an energy exchange between the scattering particles and the surrounding plasma which depends on the energy of the particles. Particles with low kinetic energy lose on the average energy to the plasma and vice versa with high energy particles.

Next, we adopt a second approach to the screening problem. The approach is based on the Langevin equation for charged particles with the Rosenbluth potential. We show how the two completely independent methods yield the same physical results.

We then review the arguments for a static screening based on a static potential and show its basic assumptions and shortcomings. The particular assumptions leading to the Salpeter formula are discussed along with the approximations involved in its derivation. One of the fundamental assumptions in the Salpeter approximation is that the scattering is fully elastic. The inelastic nature of the collisions, which is dominant in the present calculations, is clarified.

The variation of the screening of the proton-proton reaction with the plasma parameter is given.

**Keywords:** Screening,nuclear reactions, plasma
**PACS:** 26.20.+f,52.65.Yy,96.60.Jw

## INSTODUCTION

The problem of screening by electrons and ions in nuclear reactions was recently revived due to the solar neutrino problem. In view of the apparent resolution of the solar neutrino problem the interest in the accurate value of the screening factor is due to the desire to obtain accurate reaction rate to help in establishing accurate value of various parameters.

The screening problem carried a controversy about the mode of calculation and in particular to what extent the basic assumptions are valid. The necessary condition for the Debye theory to be valid is:

$$nR_D^3 \gg 1, \qquad (1)$$

where $n$ is the number of particles per cc and $R_D$ is the Debye radius. In the core of the sun we have $\rho = 155 gm/cc$ and $T = 15 \times 10^6 K$ and hence, $N_D$, the number of particles in the Debye sphere is

$$N_D \sim 2-3. \qquad (2)$$

Debye theory is a mean field theory and hence one cannot expect it to be valid with such a small number of particles per Debye sphere. Moreover, fluctuations should dominate. Bahcall et al. (2001) ignored the mean field condition and criticized Shaviv & Shaviv (2000) for assuming dynamic effects to be of importance. As a matter of fact, already Salpeter 195x in the original paper about the screening warns that the condition for the Debye theory to be valid is the above condition.

Bahcall et all (2001) assume that the scattering particles always gain energy from the plasma (the screening energy) and this energy is exactly returned to the plasma when they separate. This implies a detailed balance, not a global one, in contrast with statistical mechanics.

Here we check the basic assumption of the Debye screening theory by means of Molecular Dynamics. Under the conditions of the sun it is safe to treat the electrons and the ions as classical particles.

## NUMERICAL EXPERIMENT

The invalidity of the mean field theory in the case of stellar cores is easily demonstrated in a simple numerical experiment. We assume a binary interaction potential of the shape:

$$V_b(r) = C_f \left( \frac{1}{r} + \frac{r}{R_n^2} - \frac{2}{R_n} \right) \quad for \ r \leq R_n; \quad V_b(r) = 0 \quad for \ r > R_n \qquad (3)$$

This is not a substitute for a Debye potential and the cut-off is only to simplify the calculation. We carried out extensive Molecular Dynamic calculation assuming the above potential interaction between the particles. In fig 1 we show how the screening energy $V_s$ varies with the relative kinetic energy of the particles. The parameter is $R_n$ which is given in units of the mean inter particle distance. Thus, $R_n = 0.4$ means $0.4^3$ particles per Debye sphere. It is clear that the statistical mechanics limit is obtained for large $R_n$.

The distribution of the relative potential energy is shown in fig. ?? for the case $R_n = 5 <r>$. The letters $f$ and $c$ refer to when the particles are far and close. We notice the difference between the two cases. This difference vindicates the assumption that at this value of $R_n$ we are still not yet at the statistical mechanics limit.

In fig 2 we show the comparison between the distributions of the potential energy of the particles (without sorting out those particles which are close and those which are far as was done before). We find that as $R_n$ increases the distribution becomes narrower so that the fluctuations decrease. However, for $R_n \sim 1$ the fluctuations are of the order of the potential energy itself.

The approach of the statistical limit is shown in fig. 3. $R_n$ has to be at least of the order of $10^0$ for the mean field to become a sensible approximation.

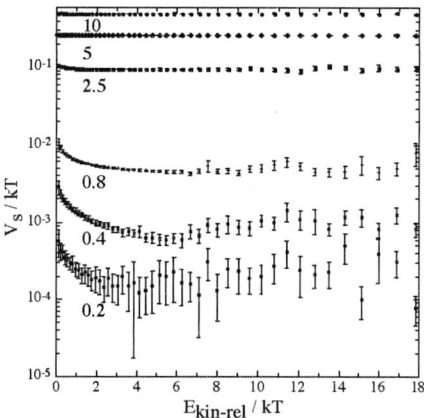

**FIGURE 1.** The screening energy (in units of kT) as a function of the relative kinetic energy for two identical particles scattering off each other according to the special potential. The numbers near the curves are the value of $R_n$ in units of the mean interparticle distance.

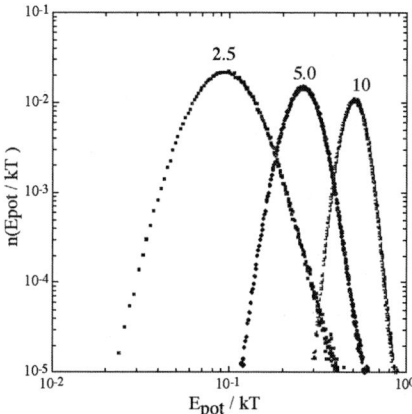

**FIGURE 2.** The distribution of the potential energy of the particles. The numbers near the curves are the radius $R_n$.

## A STOCHASTIC POTENTIAL

When the number of particles in the Debye sphere is small, the force acting on a given particle in the plasma fluctuates. Consequently, the potential should fluctuate as well. An example of the potential acting on a certain particle is shown in fig 4 over many time steps. The calculation time step is $10^{-3}$ of the dynamic time (the time for a particle with energy $kT$ to cross the distance $<r>$). As expected, the small $N_D$ causes temporal fluctuations in the potential felt by every particle. Note that the potential calculated is the potential felt by the particle, namely the Lagrangian potential and not the Eulerian

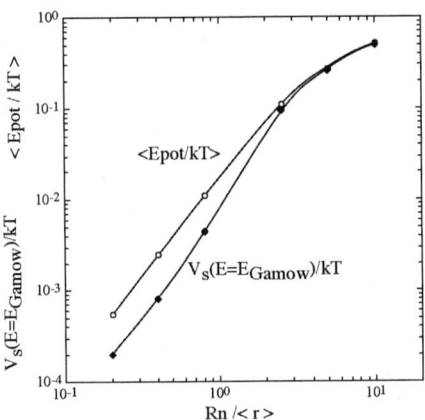

**FIGURE 3.** The approach to the statistical limit.

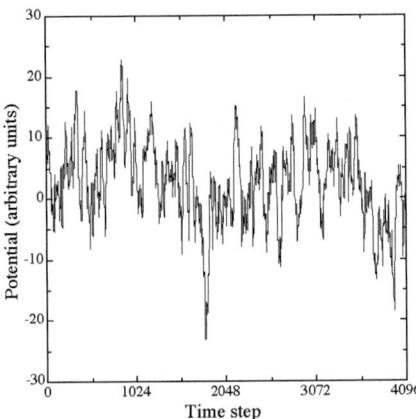

**FIGURE 4.** The time dependence of the potential of a single particle. The time steps are constant at $10^{-3}$ of the time units. The plasma frequency is beyond the range.

one (the potential at a given location in space).

The emerging physical picture from the MD calculations is that of protons interacting with a fluctuating cloud of positive and negative charges as well as with the target proton.

At the high densities that prevail in stellar cores, we have to see the interaction between the two protons as the effective interaction between two particles embedded in a stochastic fluctuating medium. Hence a static potential, which is the average over a long thermodynamic time, cannot provide the entire picture. The stochastic approach can be described with a single particle Hamiltonian of the following form:

$$H_{i,p} = \frac{p^2}{2m} + V_0(r) + V_1(r,t) = H_0 + V_1(r,t), \qquad (4)$$

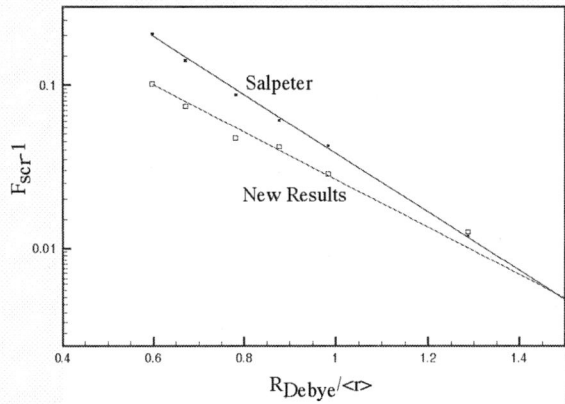

**FIGURE 5.** The comparison between the present full molecular dynamic calculation and the Salpeter Debye theory.

where $V_0$ is the static limit to the true potential. $H_0$ is the Hamiltonian which would lead to a mean field like a Debye potential for example. $V_1$ represents the time-dependent part of the fluctuating environment around the interacting protons. A possible way to obtain the time dependent potential is:

$$V_1(r,t) = e^2 \int dr \frac{\delta n(r,t) - n_D(r)}{|r-r'|}, \qquad (5)$$

where the fluctuation (positive charge minus negative one) beyond the local Debye potential is given by $\delta n(r,t)$. *Once the potential is a function of time, the distribution function is not separable any longer and the effect of the screening on the scattering particles depends on the momentum of the particle.* Moreover, the time dependence is required to describe the energy exchange between the variable environment and the scattered particle.

The particular assumption about the approximation required for the potential depends on the problem. Because of the mass difference between the electrons and the protons, the electrons move in a quasi static protons field while the ions move in the mean field of the electrons and the fluctuating field of the protons. The time dependence of the potential explains the results of the molecular dynamics nicely.

# SELF CONSISTENT LANGEVIN SIMULATION OF COULOMB COLLISIONS IN PLASMA

Since the background generates a fluctuating potential, the problem of the scattering of two protons inthe plasma can be attacked using the Langevin formalism. For details see: Shaviv & Shaviv (2001).

# THE ARGUMENTS JUSTIFYING THE SALPETER APPROXIMATION

Bahcall et al. (2001) explain why the Salpeter formula is the correct and the accurate one under the conditions prevailing in the core of the Sun. They write: *'If one of the fusing ions has charge $Z_1 e$, it creates an electrostatic potential $\varphi_1 = (Z_1 e/r) exp(-r/R_D)$, where r is the distance from the ion. For $r << R_D$, $\varphi_1 = Z_1 e/r - Z_1 e/R_D$ is the Coulomb potential minus a constant potential drop. This potential drop increases the concentration of ions $Z_2$ in the neighborhood of $Z_1$ by the Boltzmann factor $\propto exp(-Z_2 e\varphi_1/kT)$, where T is the plasma temperature.*

$$f_0 = exp\left(\frac{Z_1 Z_2 e^2}{kT R_D}\right). \tag{6}$$

*Eq. 6 is the Salpeter formula. According to Salpeter, the quantity $f_0$ is the ratio of the true reaction rate to the reaction rate calculated using the ideal gas formula.*

*Salpeter's derivations makes physically clear that electrostatic screening causes an enhancement in the density of fusing partners by lowering the potential in the vicinity of a fusing ion'*. (end of citation)

As a matter of fact, this is not the original Salpeter (1954) argument. And so writes Salpeter: *'in stellar interiors ... when two nuclei approach each other in a collision, each of them carries its screening charge cloud with it and this screening affects the interaction energy between the nuclei. We write the total interaction energy as:*

$$U_{tot}(r_{12}) = Z_1 Z_2 e^2/r_{12} + U(r_{12}) \tag{7}$$

*The main aim of this paper is to discuss the screening term $U(r_{12})$ and its effect on the Barrier penetration factor.'* (end of citation) In principle, our definition of the screening energy conforms with Salpeter's definition.

Salpeter continues and derives the interaction energy assuming the Boltzmann-Poisson equation. Next, Salpeter argues about the validity of the approximations he implemented in his derivation: *'(1) We have used a continuous (average) charge density $\bar{\rho}(r)$ and in its evaluation have used the statistical Boltzmann factor $exp[-U(r)/kT]$ for particles at the point r. For this procedure to be strictly valid many nuclei and electrons should be contained in a volume small enough so that $\bar{\rho}(r)$ and $U(r)$ do not vary appreciably over this volume.'* In other words, Salpeter (1954) discusses density fluctuations and states that he ignores them in his approximation. Salpeter (1954) approximation boils down to assuming negligible density spatial fluctuations. Here we extend the treatment to include time fluctuations due to the motions of the protons and spatial density fluctuations.

The grand canonical distribution function is given by:

$$f_N \approx exp\left(-\sum E_{kin,i}/kT\right) exp\left(-\sum \phi_{ij}/kT\right) \tag{8}$$

where $\phi_{ij}$ is the potential between two protons (we have omitted the part of the electrons). Because of the Grand Canonical distribution, when we take a snap shot of the state, we expect a distribution in potential energies exactly as manifested by the MD calculation. The width of the distribution is inversely proportional to the number of particles in the Debye sphere. If we adopt the Bahcall et al. (2001) picture, then the two scattering protons should be treated as having a distribution of screening energies or else how come that in spite of the distribution in potential energy the energy gain is constant? If the penetration through the Coulomb potential had been linear, it would not have mattered that the potential energy has a distribution. However, the very non-linearity can easily lead to a significant effect.

We turn now to discuss the plausibility of the dynamic physical argument. According to Bahcall et al. (2001), as the two scattering particles approach each other they *always* gain energy from the plasma. It does not matter which particle gains the energy because the extra energy (the screening energy) appears in the relative kinetic energy so we do not know which particle or may be both, gained energy from the plasma.

After the scattering, when the particles separate, they return the energy to the environment, or else the particles will heat indefinitely. How come the particles return to the plasma exactly the energy they gained during the approach? When one considers the Debye cloud with few particles the problem becomes even more severe: the scattering particles (the two particles composing the pair) gained energy from few protons and now the pair has to return to these particles (or others?) the same amount of energy?

In the present physical picture, the particles with low relative kinetic energy gain *on the average* energy from the plasma upon close scattering while if their relative kinetic energy is large, they lose *on the average* energy to the plasma. In the thermodynamic steady state the integral over the entire population of particles vanishes. This result is a direct consequence of equilibrium and the mechanism under discussion (dynamic friction) is an integral part of the equilibration mechanisms. This mechanism does not depend on the particular shape of the interaction.

Let $\langle \delta E_{in}(E_k^{-\infty}) \rangle$ be the average energy gain/loss by two approaching particles, the energy of which at large separation is $E_k^{-\infty}$. Note that $\delta E_{in}(E_k^{-\infty})$ may have a distribution and we here use the average over this distribution.

Next, we assume that the average energy gain/loss by the particles depends as well on the relative kinetic energy at large separation. Similarly, let $\langle \delta E_{out}(E_k^{\infty}) \rangle$ be the average energy gain/loss by two particle as they move apart from the distance of closest approach to large separation. Again, we assume dependence on the relative kinetic energy at large separation. Define next by $\Delta E_{coll}(E_k^{-\infty}, E_k^{\infty})$ the energy change in a collision by a pair of particles that have before the scattering a relative kinetic energy $E_k^{-\infty}$ and after the collision they have a relative kinetic energy of $E_k^{\infty}$. In the general case (when the energy change depends on the relative kinetic energy at large separation) the total change in the relative energy of the two particles is given by:

$$\Delta E_{coll}(E_k^{-\infty}) = \int \delta E_{in}(E_k^{-\infty} - E) \delta E_{out}(E_k^{\infty} + E) dE, \quad (9)$$

where the integration is carried over all energies $E$. Let now $f(E_k)$ be the distribution function of kinetic energies in equilibrium. Clearly, in equilibrium the following condition must hold

$$\langle \Delta E_{coll}(E_{k,-\infty}) \rangle = \int_0^\infty dE_{kin,-\infty} f(E_k^{-\infty}) \Delta E_{coll}(E_k^{-\infty}) = 0. \tag{10}$$

As the distribution is a positive definite function, it follows that either $\Delta E_{coll}(E_k^{-\infty})$ changes sign at least once at some relative kinetic energy or $\Delta E_{coll}(E_k^{-\infty}) \equiv 0$. The first possibility about the change of sign does not depend on the particular interaction but on the condition of equilibrium. This is exactly what the MD results and the Langevin formalism shows. The Salpeter approximation corresponds to the second case namely:

$$\Delta E_{coll}(E_k^{-\infty}) = 0 \tag{11}$$
$$\delta E_{in}(E_k^{-\infty}) = \delta_D(\delta E_{in} - U) \tag{12}$$
$$\delta E_{out}(E_k^{\infty}) = \delta_D(\delta E_{out} + U), \tag{13}$$

where $\delta_D$ is the Dirac $\delta$ function. (Note that the last two expressions are not average values). The first condition implies no change of energy in a collision irrespective of the relative energy at large separation. The second condition means that the energy gained by the two particles from the plasma upon approaching each other is constant and equal to the mean potential energy per particle. The third condition means that as the particles separate they lose always the same energy, namely the energy gained upon approach. Another way to put it is as follows: In the Salpeter approximation all proton-proton collisions are fully elastic, no energy is lost/gained to the plasma. The present results are the extension to non inelastic collisions.

## THE SCREENING ENERGY FOR THE PP REACTION

In fig 5 we compare our results for the screening with those of the Salpeter theory. The new screening is lower than the one predicted assuming the validity of the Debye theory. The results can be summarized as

$$F_{sc} = F_{sc-Sal}(X=1)(0.59 * R_D + 0.183)(1 \pm 0.05) \tag{14}$$

Calculation of the solar neutrino flux with the new screening factor is under way. Also, the calculations are being extended to include heavy ions in the plasma.

## ACKNOWLEDGMENTS

The author is happy to acknowledge a discussion with M. Fisher about fluctuations. Extensive discussions with N.J. Shaviv are acknowledged. This research was partly supported by the Foundation for the Promotion of Research at the Technion and by the Asher Foundation at the Asher Space Research Institute.

# REFERENCES

Bahcall, N.J., Brown, L.S., Gruzinov, A. & Sawyer, R.F. A & A, 383, p291, (2001)
Salpeter,E.E., Aust. J. Physics, 7, 373, (1954)
Shaviv, G. & Shaviv, N.J., ApJ. 529. p1054, (2000), ApJ. 558, p925, (2001)

# Microscopic description of fission properties

H. Goutte[*], J.-P. Delaroche[*], M. Girod[*] and J. Libert[†]

[*]CEA/DAM Ile de France, DPTA/Service de Physique Nucléaire,
BP 12, 91680 Bruyères-le-Châtel, France
[†]Institut de Physique Nucléaire, CNRS-IN2P3 and Université Paris XI,
15 rue Georges Clémenceau, 91406 Orsay, France

**Abstract.** Microscopic results on fission barriers, partial $\gamma$-back and fission lifetimes of shape isomers are presented. They have been obtained from mean-field and beyond mean-field calculations using the effective D1S Gogny force.

**Keywords:** fission process, Hartree-Fock-Bogoliubov approach, lifetimes, Gogny force, actinides
**PACS:** 21.60.Jz, 21.10.Tg

## 1. INTRODUCTION

The fission process is of fondamental importance in astrophysics, as:
- fission limits the production of Super Heavy Elements (SHE) from the r-process,
- the fission cycling introduces a structure in nuclide distribution (it depopulates the heavy region and enhances the production in the fission-fragment region),
- and fission is in competition with the beta-decay process towards stability.

As a matter of fact, many fission data intervene in the modeling of the aboundances of elements such as cross-sections, fragment yields and half-lives. Among the different approaches able to describe the fission process, microscopic ones have the advantage of making reliable predictions where few or no data exist. Even if full parameter-free evaluations are not possible nowadays, some noteworthy progress in both structure and reaction fields has been accomplished recently. Calculations have been performed using the effective D1S Gogny force and the Genererator Coordinate Method with the Gaussian Overlap Approximation based on Hartree-Fock-Bogoliubov results. We recall here some results on fission barriers, and partial $\gamma$-back and fission lifetimes.

## 2. FISSION BARRIERS

The present work relies on two major hypothesis; first we suppose that low-energy fission is an adiabatic process where collective and intrinsic degrees of freedom are decoupled, and second we assume that the collective motion of the system can be described in terms of a few collective variables. As a matter of fact, in our self-consistent study of fission barriers, the nuclear shapes are generated by only the axial quadrupole operator:

$$\hat{Q}_{20} = \sqrt{\frac{16\pi}{5}} \sum_{i=1}^{A} r_i^2 Y_{20}, \qquad (1)$$

where $Y_{20}$ is a spherical harmonics. The intrinsic deformed nuclear state is then solution of the constrained Hartree-Fock-Bogoliubov (HFB) equation:

$$\delta \langle \Phi(q_{20})| \hat{H} - \lambda_N \hat{N} - \lambda_Z \hat{Z} - \sum_i \lambda_{20} \hat{Q}_{20} |\Phi(q_{20})\rangle = 0 , \qquad (2)$$

the Lagrange parameters $\lambda_N$, $\lambda_Z$, and $\lambda_{20}$ being deduced from:

$$\langle \Phi(q_{20})|\hat{N}|\Phi(q_{20})\rangle = N,$$
$$\langle \Phi(q_{20})|\hat{Z}|\Phi(q_{20})\rangle = Z, \qquad (3)$$
$$\langle \Phi(q_{20})|\hat{Q}_{20}|\Phi(q_{20})\rangle = q_{20}. \qquad (4)$$

In such an approach, all the deformations that are not constrained take on values that minimize the total energy.

In ref. ([1]) we have calculated fission barriers for 55 even-even actinides from $^{226}$Th to $^{262}$No. Between the first and the second wells the triaxial degree of freedom has been left free so that triaxial inner barriers are obtained, whereas beyond the second well parity has been broken instead in the HFB calculations.

The main features of these fission barriers are :

- triaxial inner barriers are systematically lowered by up to 4 MeV when compared to the axial ones,
- inner barrier heights display maxima at N = 146 for the U, Pu, and Cm isotopic chains,
- the outer barrier is found to be asymmetric for systems with N < 152 and symmetric for more neutron-rich systems, and finally iv) super deformed minima appear to be washed out for N > 156.

The pairing energy $E_{pair} = \frac{1}{2} Tr(\Delta \kappa)$ has been calculated along the barrier, where $\Delta$ and $\kappa$ are the pairing field and pairing tensor, respectively. As expected, pairing has been found not constant as a function of the elongation: minima are found inside the wells and maxima at the top of barriers. These variations of the pairing energy are very important since inertia are known to be very sensitive to pairing correlations. These changes will affect among other things the kinetics of the fission process. This is illustrated for $^{238}$U in fig. (1) where symmetric (dashed lines) and asymmetric (solid lines) fission barriers, pairing energy and mass parameter $\mathcal{M}_{20}$ are plotted as functions of $q_{20}$ in panels a), b), and c), respectively.

In our case $\mathcal{M}_{20}(q_{20})$ is calculated using a formula based on the Inglis-Belyaev approximation [2]:

$$(\mathcal{M}_{20})(q_{20}) = \frac{\hbar^2}{2} \frac{M_{(-3)}}{(M_{(-1)})^2}. \qquad (5)$$

In Eq. (5) the moments of order -k are calculated as:

$$M_{(-k)} = \sum_{\mu\nu} \frac{|\langle \phi(q_{20}) | \eta_\mu \eta_\nu \hat{Q}_{20} | \phi(q_{20}) \rangle|^2}{(E_\mu + E_\nu)^k},\qquad(6)$$

where $\eta_\mu$, $\eta_\nu$ are destruction operators of quasi-particles with energies $E_\mu$ and $E_\nu$, and $\hat{Q}_{20}$ is the quadrupole deformation operator defined in Eq. (1).

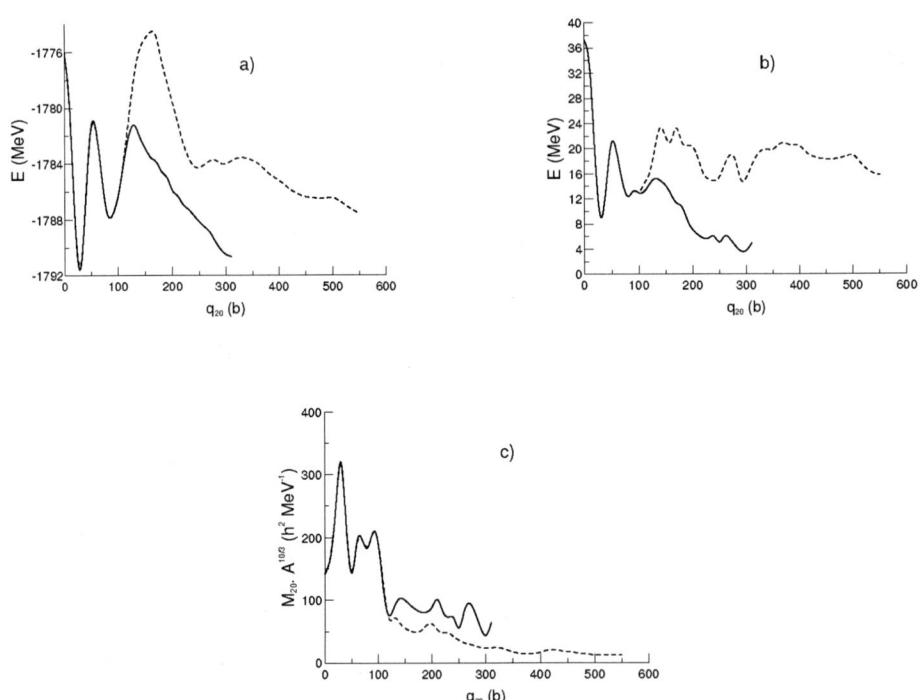

**FIGURE 1.** Potential energy (Panel a)), pairing energy (Panel b)) and mass parameter (Panel c)) as functions of $q_{20}$ along the asymmetric (solid line) and the symmetric (dashed line) fission paths in $^{238}$U

## 3. LIFETIMES

Partial $\gamma$-back decay ($T_\gamma$) and fission ($T_f$) half-lives of shape-isomers have been determined using the well-known one-dimensional WKB method [3]:

$$T_{(\gamma,f)} = \frac{2.87\,10^{-21}}{E_0}[1 + exp(2S_{(\gamma,f)})],\qquad(7)$$

where $E_0$ is the assault energy (in MeV) here calculated as the lowest energy sustained by a one dimension SD potential V(q), and $S_i$ is the action calculated separately for $\gamma$-

back and fission decay modes. In Eq. (7), the action is calculated along a given trajectory $L_i$ [4] as

$$S_i = \int_{L_i} \sqrt{2M_s(s)\left[V(q(s))-E\right]}ds, \qquad (8)$$

where E is the energy of the fissioning system, s is a curvilinear coordinate which specifies the position of a point on the trajectory $L_i$ ($i = \gamma$ or $f$), and $M_s(s)$ is the effective inertia along $L_i$:

$$M_s(s) = \sum_{ij} M_{ij}(q_i, q_j) \frac{dq_i}{ds}\frac{dq_j}{ds}. \qquad (9)$$

In eq. (9), $q_i$ stands for $q_{20}$, $q_{22}$, and $q_{30}$ the axial and triaxial quadrupole moment and the octupole one, respectively.

The trajectories $L_f$ associated to the fission decay are here taken as the lowest energy paths in the ($q_{20}$,$q_{30}$) plane. The calculated fission half-lives are found much longer than the experimental data in $^{236-238}$U [5].

For $\gamma$-back decay the optimum pathes $L_\gamma$ connecting second well and first well are taken as the least action paths (LAP) in the ($q_{20}$,$q_{22}$) plane. For most of the actinides under study these LAP's go through axial saddles, contrary to the least energy trajectories. This feature illustrates how important is the role played by collective masses in the determination of dynamical paths. Examples of least action and least energy paths between the first well and the second well are displayed in Fig. 2 for $^{242-246}$Cf nuclei in the cartesian ($\beta_0$,$\beta_2$) plane as dark and red lines, respectively.

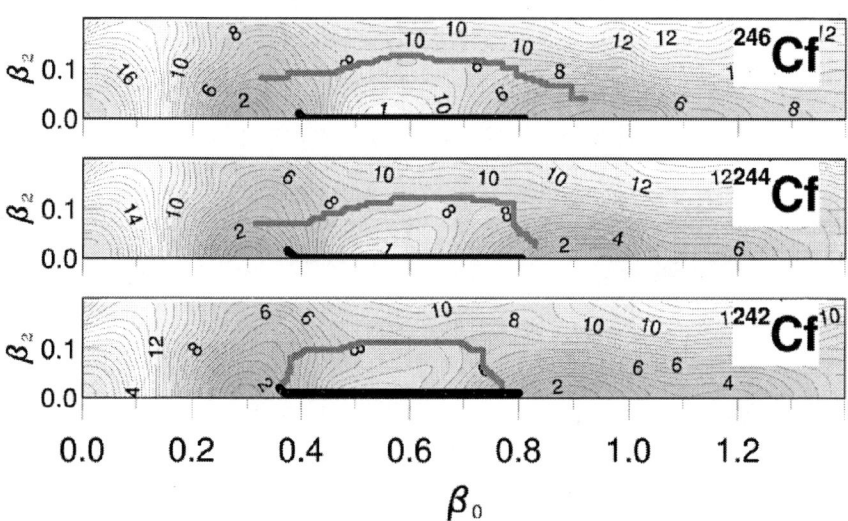

**FIGURE 2.** Potential energy surfaces as functions of $\beta_0$ and $\beta_2$ for $^{242-246}$Cf. Least action paths are displayed as thick dark lines and least energy pathes as grey lines.

As mentioned in ref. [3], taking dynamical path (LAP) in the calculation of lifetime leads to a decrease of the half-lives by a few orders of magnitude compared to static paths. The overall predictions for $T_\gamma$ are in qualitative agreement with experimental data (in $^{236-244}$Pu, $^{240-244}$Cm).

When comparing the two partial half-lives we have found that:

- shape isomers in Th and U decay by $\gamma$-emission,
- fission and $\gamma$-back decay are competing for Pu and Cm,
- and finally shape isomers in Cf, Cm and No decay through fission.

These overall predictions are in qualitative agreement with experimental evidences.

## 4. FISSION DYNAMIC

More sophisticated dynamical description of the fission process have been undertaken, where quadrupole and octupole collective deformation parameters have been explicitely taken into account. The time evolution in the fission channel is based on the time dependent generator coordinate method with the gaussian overlap approximation. The fission yields are then derived from the calculation of the flux for all scission points. Examples of scission configurations can be found in the proceeding of Noel Dubray [6]. In ref. [7] we focussed on several physical aspects of low-energy fission: dynamical effects on mass distributions, influence of initial state on symmetric fragmentation. We plan now to go deeper in this study in order to determine effective fission barriers which take into account transverse modes.

## 5. CONCLUSION

From a theoretical point of view, fission appears as one of the most complex process in nuclear physics as it involves, among others, large amplitude motion, coupling between different collective modes and intrinsic excitations. Results presented here represent a first step in the direction of a fully microscopic description of this process. The overall agreement of the predictions with experimental data for low-energy fission encourages us to continue further studies along these lines with some additional improvements.

## REFERENCES

1. J.-P. Delaroche, M. Girod, H. Goutte, J. Libert, Nucl. Phys. A 771 (2006) 103
2. J. Libert, M. Girod, J.-P. Delaroche, Phys. Rev. C 60 (1999) 054301, and references therein.
3. A. Baran, K. Pomorski, A. Lukasiak, A. Sobiczewski, Nucl. Phys. A 361 (1981) 83.
4. K. Yoshida, M. Matsuo, Y.R. Shimizu, Nucl. Phys. A 696 (2001) 85.
5. B. Singh, R. Zywina, R.B. Firestone, Table of Superdeformed Nuclear Bands and Fission Isomers, Nucl. Data Sheets 97 (2002) 241.
6. N. Dubray, H. Goutte and J.-F. Berger, proceeding of this conference.
7. H. Goutte, J.-F. Berger, P. Casoli, and D. Gogny, Phys. Rev. C71 (2006) 024316.

# Photoneutron Cross Sections of Astrophysical Significance

## H. Utsunomiya

*Department of Physics, Konan University, Higashinada, Kobe 658-8501, Japan*

**Abstract.** Presented in this paper are some of the latest measurements of photoneutron cross sections of direct relevance to the p-process nucleosynthesis in the context of the statistical model of compound nuclear reactions. We discuss the p-process origin of the rarest nuclide and the only naturally occurring isomer $^{180}$Ta$^m$, a serious underproduction problem of $^{138}$La, and the nuclear level density of $^{180}$Ta determined from the partial photoneutron cross section for $^{180}$Ta$^m$. As the laser-Compton scattering γ ray at AIST has enabled one to directly determine (γ,n) cross sections, the blackbody synchrotron radiation to be produced by a ten-Tesla superconducting wiggler at SPring-8 is expected to be a promising tool for exploring (γ,α) and (γ,p) reactions in the future.

**Keywords:** photonuclear reactions, p-process nucleosynthesis, statistical model
**PACS:** 25.20.-x, 26.30.+k, 23.20.-g, 21.10.Ma

## 1. P-PROCESS NUCLEOSYNTHESIS

Most of heavy nuclides are produced by neutron capture followed by β⁻ decay; either s-process or r-process [1]. However, 35 neutron-deficient nuclides (p-nuclei) from Se to Hg with rare solar abundances are bypassed by the s-process flow and shielded from the β⁻ decay after the r-process. Photodisintegration of pre-existing seed nuclei of s and/or r-process origin is a primary process of producing the p-nuclei [2]. An extensive reaction network is required to describe the p-process nucleosynthesis, consisting of (γ,n), (γ,α) (γ,p) reactions complemented by n, p, and α-captures, neutrino capture, and weak transformation (electron capture and β⁺ decay) [2]. The reaction network involving some 2,000 nuclei and 20,000 reactions effectively produces the p-nuclei in a delicate balance of temperature and time-scale. A narrow temperature window of $T = (1.5 - 3.5) \times 10^9$ K is required, on the one hand, to ensure efficient photodisintegrations and prevent the photo-erosion of all the heavy nuclei to ion-peak species. On the other hand, the freeze-out of photodisintegrations on a short time-scale of the order of 1 second is also necessary. These conditions are met in the deep Oxygen and Neon-rich layers of massive stars during their explosions as type II supernovae satisfactorily [3-7]. Other plausible sites include pre-supernova phases of massive stars [3,4,8] or type Ia supernovae (see [2] for a more complete review).

## 2. LABORATORY PHOTONEUTRON RATE

Experimentally, the quasi-monochromatic γ ray produced from laser inverse-Compton scattering (LCS γ rays) at AIST [9] has enabled one to directly determine (γ,n) cross sections [10]. Those cross sections determine the photoreaction rate for nuclei in the ground state, which is defined by

$$\lambda_{\gamma n} = \int_0^\infty c\, n_\gamma(E,T)\sigma_{\gamma n}(E)\, dE, \qquad (1)$$

where $c$ is the speed of light, $n_\gamma(E,T)$ is the number of photons per unit volume and energy in a stellar photon bath at a temperature $T$, and $\sigma_{\gamma n}(E)$ is the photoneutron cross section. In the interior of stars, $n_\gamma(E,T)$ is remarkably close to the Planck distribution

$$n_\gamma(E,T)\,dE = \frac{1}{\pi^2}\frac{1}{(\hbar c)^3}\frac{E^2}{\exp(E/kT)-1}\,dE. \qquad (2)$$

For the s-wave nature of $\sigma_{\gamma n}(E)$, i.e., $\sigma_{\gamma n}(E) \propto \sqrt{(E-S_n)/S_n}$ with the neutron separation energy $S_n$ [11], the integrand of Eq. (1) defines the most effective energy region for (γ,n) reactions immediately above the neutron threshold [12].

## 3. STELLAR PHOTONEUTRON RATE AND HAUSER-FESHBACH MODEL

In the stellar photon bath, nuclei are in thermal equilibrium in their ground state and excited states $\mu$. The occupation probability governed by the Boltzmann factor $\exp(-\varepsilon^\mu/kT)$ decreases with increasing the excitation energy $\varepsilon^\mu$. This decrease is however compensated by the Planck distribution governed by $\exp(-E_\gamma/kT)$ where the photon energy $E_\gamma$ is reduced by $\varepsilon^\mu$. Thus, photoreactions on nuclei in excited states remarkably contribute to the stellar rate. As a result, the stellar photoreaction rate can become 100 – 10,000 times larger than the laboratory rate. Because of this, one must rely on the Hauser-Feshbach model of compound nuclear reactions.

Let us consider that photoabsorption by a nucleus excited in a state $\mu$ forms a compound nucleus at an excitation energy $E$ and decays by emitting a neutron. The HF model gives cross sections for this process by

$$\sigma^\mu{}_{\gamma n}(E_\gamma) = \pi \lambdabar_\gamma^2 \frac{1}{2(2j^\mu+1)}\sum_{J^\pi}(2J+1)\frac{T_\gamma^\mu(E_\gamma,J^\pi)T_n(E,J^\pi)}{T_{tot}(E,J^\pi)}. \qquad (3)$$

Here $E_\gamma$ is the photon energy, $j^\mu$ is the spin of the state μ, $J^\pi$ represents the spin and parity of the compound states. $E$ is given by $E = \varepsilon^\mu + E_\gamma$ with the excitation energy of the excited state $\varepsilon^\mu$. $T_\gamma^\mu$ and $T_n$ are the photon and neutron transmission coefficients in the formation and the decay of the compound nucleus, respectively, while $T_{tot}$ is the total transmission coefficient summed over all decay channels. Note that the helicity of photons is 2. $T_n$ is estimated by summing the neutron transmission coefficient $t_n^\mu(E_n, J_n, \pi_n)$ for a final state $f$ over all experimentally known levels and by integrating over the level density $\rho(\varepsilon, J_f, \pi_f)$ at excitation energies $\varepsilon$ for which experimental data are not available.

$$T_n(E, J^\pi) = \sum_{\nu=0}^{\omega} t_n^\nu(E_n, J_n, \pi_n) + \int_{\varepsilon^\omega}^{\varepsilon^{max}} \sum_{J_f, \pi_f} t_n(E_n, J_n, \pi_n) \rho(\varepsilon, J_f, \pi_f) d\varepsilon \qquad (4)$$

The stellar rate is given

$$\lambda_{\gamma n}^* = \frac{\sum_\mu (2j^\mu + 1) \lambda_{\gamma n}^\mu(T) \exp(-\varepsilon^\mu / kT)}{\sum_\mu (2j^\mu + 1) \exp(-\varepsilon^\mu / kT)}, \qquad (5)$$

with the Boltzmann factor $\exp(-\varepsilon^\mu / kT)$ and the rate $\lambda^\mu$ for a state $\mu$ similar to Eq. (1). One can see that there are three important nuclear ingredients in the HF model; the photon $T_\gamma^\mu(E_\gamma, J^\pi)$ and neutron $t_n(E_n, J_n, \pi_n)$ transmission coefficients and the nuclear level density $\rho(\varepsilon, J_f, \pi_f)$.

As seen in Eq. (3), the photon transmission coefficient $T_\gamma^\mu(E_\gamma = E > S_n, J^\pi)$ above $S_n$ is involved in the ($\gamma$,n) reaction on nuclei in the ground state. In contrast, $T_\gamma^\mu(E_\gamma = E - \varepsilon^\mu < S_n, J^\pi)$ below $S_n$ is involved in the ($\gamma$,n) reaction on nuclei in excited states μ. The importance of measurements of photoneutron cross sections near $S_n$ lies in the fact that they determine the laboratory rate and more importantly provide strong constraints on the photon transmission coefficient below $S_n$.

## 4. PHOTOPRODUCTION OF $^{180}$TA$^m$

Photodisintegration of pre-existing $^{181}$Ta of the s- and r-process origin is a primary process of producing the rarest isotope and the only naturally occurring isomer $^{180}$Ta$^m$. The stellar photon bath at temperature of billions of Kelvin is hot enough to achieve thermalization of $^{180}$Ta between the ground state and the isomeric state through mediating states above 1 MeV [13]. Total photoneutron cross sections for $^{181}$Ta is necessary to determine photo-production of $^{180}$Ta$^m$, where the thermal condition determines the partition of the total cross sections to the isomeric state.

Fig. 1 shows total photoneutron cross sections measured for $^{181}$Ta in comparison with theoretical ones obtained by using three different prescriptions of calculating the E1 γ strength function. Both the Lorentzian [14] and the Hybrid [15] models considerably underestimate the cross sections near $S_n$, whereas the QRPA model [16] which is based on a microscopic description of the single-particle and collective excitations reproduces the cross section by providing some extra strength near $S_n$.

Fig. 2 shows the overproduction factors for the p-nuclei obtained for the model star with the solar metallicity and 25 solar mass. The model star evolves from the core-helium burning to the supernova explosion. During its explosion as a type-II supernova, we consider the 25 O/Ne-rich zones with explosion temperatures peaking in the T = (1.7 – 3.5) × $10^9$ K range responsible for the production of the p-nuclei. All reaction rates except that for $^{181}$Ta are taken from the HF code MOST [2]. For the stellar $^{181}$Ta($\gamma$,n)$^{180}$Ta, we adopt the QRPA prediction shown in Fig. 1.

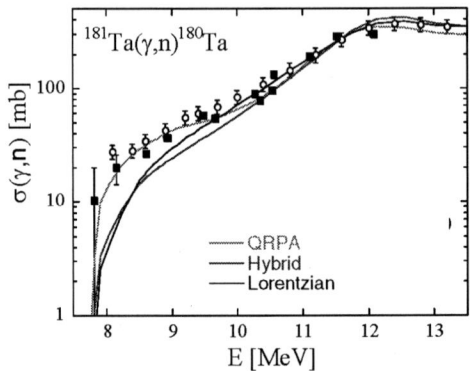

**FIGURE 1.** Present total photoneutron cross sections for $^{181}$Ta (solid squares) and the data reconstructed from the Livermore data [17] and the Saclay data [18] by IAEA (open circles) [19].

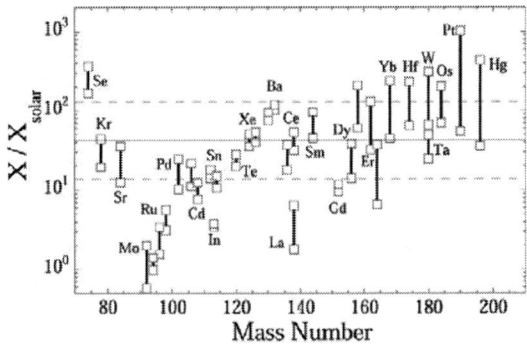

**FIGURE 2.** p-nuclides overproduction factors obtained for the model star with the solar metallicity and 25 solar mass. The MOST rates and the present $^{181}$Ta($\gamma$,n)$^{180}$Ta rate are used.

The overproduction factors resulted from a sensitivity test using 14 sets of reaction rates with different nuclear parameters such as the nuclear structure properties, nuclear level density, optical potential and γ strength. The error bars represent the sensitivity. Note that the error bar for $^{180}$Ta is significantly reduced because of the experimentally constrained $^{181}$Ta(γ,n)$^{180}$Ta, rate. The remaining error bar originates from the still uncertain $^{180}$Ta(γ,n)$^{179}$Ta photo-destruction rate.

Like most of other p-process elements, the overproduction factor for $^{180}$Ta is as large as 40, a typical value for a main product of massive stars $^{16}$O, and safely stays within a factor of 3, uncertainties associated with the nuclear ingredients and the stellar model. Thus, it is concluded that $^{180}$Ta is a natural thermonuclear product of the p-process.

It is to be noted that the final $^{180}$Ta production may be increased by the neutrino nucleosynthesis, where ν capture (charged current capture) on $^{180}$Hf is the major contributor rather than the neutral current on $^{181}$Ta. This extra production is still subject to all the uncertainties related to the neutrino physics, in particular, the neutrino luminosity, temperature, oscillation and interaction cross section.

## 5. THERMONUCLEAR ORIGIN OF $^{138}$LA

$^{138}$La is underproduced in all p-process calculations performed so far [2], including the present one shown in Fig. 2. In view of the low $^{138}$La abundance, it has been attempted to explain its production by non-thermonuclear processes involving either stellar energetic particles [20] or neutrino-induced transmutations [21,22]. The former mechanism is predicted not to be efficient enough, while the latter through the ν-capture on $^{138}$Ba was shown by [21] to be so far the most efficient production mechanism of the solar $^{138}$La. Despite its promising character, the neutrino scenario still suffers from astrophysics as well as nuclear/neutrino physics uncertainties. In addition, it cannot be excluded that a substantial fraction of the solar $^{138}$La be of a thermonuclear origin if agents other than type-II Supernovae could have contaminated the solar system with p-nuclides. In particular, sub-Chandraskhar White Dwarf explosions could be significant contributors [2,23].

The thermonuclear origin of $^{138}$La is known to depend in a sensitive way on the competition between the $^{138}$La production and destruction by photodissociation. It is shown in [21] that a suitable $^{138}$La production could be obtained by adequate changes in the corresponding nominal $^{139}$La(γ,n) and $^{138}$La(γ,n) rates. Because the level of required changes is relatively high, only experimental measurements can possibly confirm this assumption.

Recently photoproduction of $^{138}$La was investigated with the LCS γ rays at AIST [24]. Fig. 3 shows photoneutron cross sections measured for two N=82 nuclei, $^{138}$La and $^{141}$Pr. As in the photodissociation data for $^{181}$Ta, a similar comparison was made with the predictions obtained with the three different prescriptions of the E1 γ strength functions [15-17]. The three models considered here reproduce almost identically the experimental data and therefore give rise to similar estimate for the photodissociation rate on the target in its ground-state. We find at a typical p-process temperature of T = $2.5 \times 10^9$ K a rate $\lambda^0_{\gamma n}$ = 0.18 ± 0.02 s$^{-1}$ for $^{139}$La and $\lambda^0_{\gamma n}$ = 0.015 ± 0.004 s$^{-1}$ for $^{141}$Pr, where the quoted uncertainties are associated with the adopted model for the E1-

strength function, nuclear level densities and optical potential. The errors are seen not to exceed some 10 and 30%, respectively.

When considering the total stellar rate, the contribution of the thermally populated state becomes dominant at temperatures of interest for the p-process nucleosynthesis, so that the photodissociation rate becomes sensitive to the tail of the γ-ray strength below the neutron threshold. These new conditions relax somehow the constraints related to experimental cross section measurements above the neutron threshold. We found at T = 2.5 × $10^9$ K stellar rates $\lambda^*_{\gamma n}$ = 27 ± 15 $s^{-1}$ for $^{139}$La and 4.5 ± 3 $s^{-1}$ for $^{141}$Pr. These rates are about 100 times larger than for the targets in their ground state.

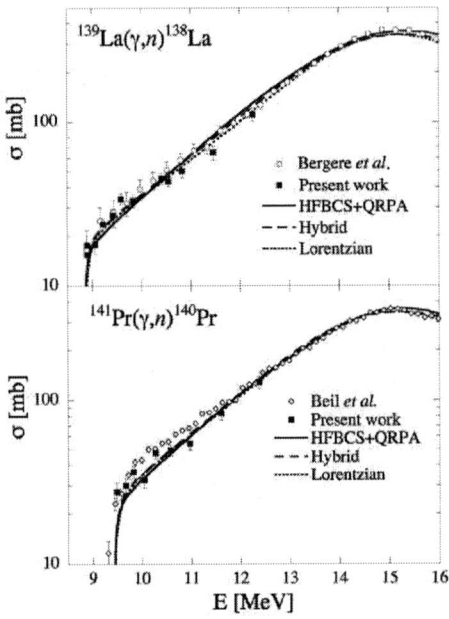

**FIGURE 3.** Photoneutron cross sections for $^{139}$La and $^{141}$Pr in comparison with the HF model cross sections obtained with the three different prescriptions of the E1 γ strength functions. For comparison, the previous data [18,25] are also shown.

The standard values predicted in [21] with the Hauser-Feshbach code MOST amount to $\lambda^*_{\gamma n}$ = 12 $s^{-1}$ for $^{139}$La which corresponds to the lower limit estimated in the present study and 1200 $s^{-1}$ for $^{138}$La at T = 2.5 × $10^9$ K. From the present analysis, an increase by a factor of 3.5 to reach the upper value of $\lambda^*_{\gamma n} \approx 42$ $s^{-1}$ found cannot be excluded when considering other nuclear inputs to the HF model, i.e more precisely, the Lorentzian-type E1 strength function [15].

The p-process $^{138}$La abundance also depends on the $^{138}$La($\gamma$,n)$^{137}$La destruction rate. In this case, no experimental data exists, neither to constrain the E1 strength function, nor the nuclear level densities. We estimate $\lambda^*_{\gamma n}(^{138}$La) to be in the range of 500 to 3000 s$^{-1}$. It should however be emphasized that by consistency the same model for the E1 strength function should be used to determine both the $^{138}$La production and destruction rates. The above-mentioned upper limit for the production rate implies a simultaneous increase in the destruction rate with respect to the above-quoted value of 1200 s$^{-1}$. It is therefore unlikely that an increase in the $^{138}$La production be accompanied by a simultaneous decrease in the $^{138}$La destruction. Hence, based on Fig. 2 of [21], the photoreaction mechanisms are not favored to explain the total $^{138}$La solar abundance. However, at this stage, no definite conclusion can be drawn due to the large uncertainties still affecting the unconstrained stellar photoneutron rate of $^{138}$La. Future experiments on the $^{138}$La nuclear properties will hopefully bring new insight.

## 6. NUCLEAR LEVEL DENSITY OF $^{180}$TA

The nuclear level density (NLD) is an important ingredient in the HF model calculation as seen in Eq. (4). The back-shifted Fermi gas model is widely used for NLD. Recently, partial photoneutron cross sections for $^{180}$Ta$^m$ measured in the photodisintegration of $^{181}$Ta shed new light on the spin- and parity-dependence of NLD [26]. The E1 photoexcitation of $^{181}$Ta in the ground state (7/2$^+$) populates states with 9/2$^-$, 7/2$^-$ and 5/2$^-$ spins in $^{181}$Ta. After an s-wave neutron emission, excited states in $^{180}$Ta are left with 5$^-$, 4$^-$, 3$^-$ and 2$^-$ spins and $\gamma$-decay to either the short-lived ground state (1$^+$: 8.152 h) or the long-lived isomeric state (9$^-$: > 1.2 × 10$^{15}$ y). It is interesting to note that the ground state transition occurs downward in terms of spin, while the isomeric transition occurs upward. In particular, the $\gamma$ transition to the isomer $^{180}$Ta$^m$ (9$^-$) proceeds in multisteps, for example, by E1 transitions, 5$^-$ → 6$^+$ → 7$^-$ → 8$^+$ → 9$^-$. Thus, partial photoneutron cross sections for $^{180}$Ta$^m$ can serve as a good probe of the spin- and parity-dependence of NLD of $^{180}$Ta.

Fig. 4 shows partial photoneutron cross sections measured with the LCS $\gamma$ rays [26]. The Talys code [27] for the statistical model calculation was used to predict the partial cross section. Two different models of the NLD were adopted. The first one used a statistical calculation that takes into account the discrete structure of the single-particle spectra associated with the HFBCS potential [28], while the second one is based on the combinatorial approach using the single-particle scheme and paring strength derived from an HFB calculation [29,30]. The E1 $\gamma$-strength function based on the HFBCS + QRPA model is adjusted to reproduce the total cross section which is insensitive to the adopted NLD model. In contrast, the partial cross section is sensitive to the NLD as expected. While the combinatorial NLD satisfactorily reproduces the partial cross section, the statistical NLD significantly underestimates. This difference originates from the large NLD obtained with the combinatorial approach for spins J >

5 at low energies, where the number of intermediate states feeding to the isomeric state increases.

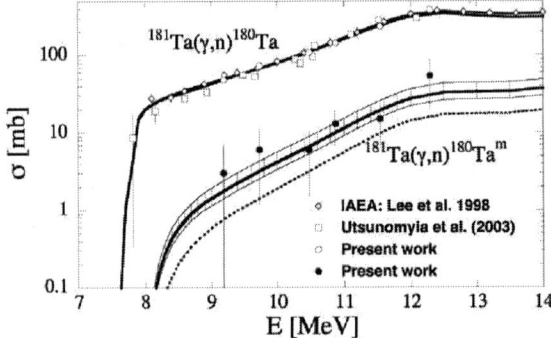

**FIGURE 4.** Partial photoneutron cross sections for $^{180}\text{Ta}^m$ (solid circles) with the statistical and systematic uncertainties linearly combined and total cross sections (open circles) in comparison with the theoretical cross sections. The data recommended for the total cross section by IAEA [19] is also shown.

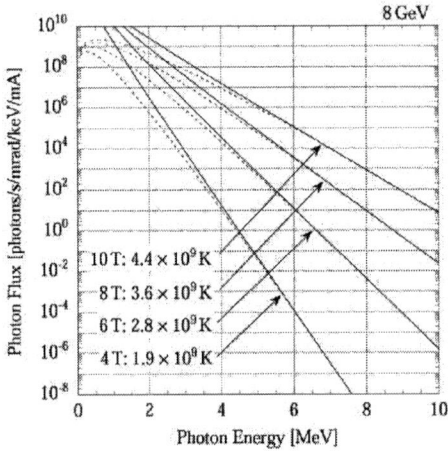

**FIGURE 5.** The energy distribution of the synchrotron radiation produced by a 10-T superconducting wiggler at SPring-8 (solid lines) as a function of the magnetic field strength in comparison with the equivalent Planck distribution (dotted lines).

## 7. BLACKBODY SYNCHROTRON RADIATION AT SPRING-8

A new concept of blackbody synchrotron radiation with equivalent temperatures of billions of Kelvin is discussed in reality that can be produced by 8-GeV electrons passing through a ten-Tesla superconducting wiggler (SCW) at SPring-8 [31]. Fig. 5 shows the synchrotron radiation produced by the SCW. The high-energy part of the synchrotron radiation remarkably mimics the Planck distribution, $n_\gamma(E,T)$ in Eq. (2), in a stellar photon bath at T = (1.9 – 4.4) × $10^9$ K depending on the magnetic field strength from 4 to 10 T. This temperature range of the synchrotron radiation overlaps with that expected in the p-process nucleosynthesis. The synchrotron radiation at an equivalent temperature T = 4.4 × $10^9$ K has the highest γ flux that is experimentally most favorable. The slight difference of the experimentally most favorable synchrotron radiation from the upper limit of the p-process temperature does not hamper the value of experimental data that constrain the laboratory cross section.

Because of the spectral equivalence between the SCW synchrotron radiation and the blackbody radiation, the laboratory reaction rate given by Eq. (1) can be directly obtained from the experimental yield. There are many (γ,n) reactions to be studied by photoactivation among which the most important one is the photo-destruction rate of $^{180}Ta^m(\gamma,n)^{179}Ta$ [31]. Very recently, it is shown that the SCW blackbody radiation can be extensively used to measure (γ,α) and (γ,p) cross sections by direct counting of α particles and protons (for details see [32]).

## 8. CONCLUSION

The quasi-monochromatic LCS γ rays produced from laser inverse-Compton scattering at AIST were used to directly determine photoneutron cross sections for $^{181}Ta$ and $^{139}La$. The thermonuclear origin of the two odd-odd p-nuclei, $^{180}Ta^m$ and $^{138}La$, was discussed. It is a robust conclusion that the only naturally occurring isomer $^{180}Ta^m$ with the rarest solar abundance is a natural thermonuclear product of the p-process, irrespective of uncertainties in photodisintegration rates responsible for the production and the destruction or the details of the type II supernova models in their spherically symmetric one-dimensional approximation. The nuclear level density of $^{180}Ta$ was also studied based on the partial photoneutron cross sections for $^{180}Ta^m$ measured in the photodisintegration of $^{181}Ta$. Photodestruction cross sections for $^{180}Ta^m(\gamma,n)^{179}Ta$ remain to be measured in the future.

In contrast, the underproduction problem of $^{138}La$ in its thermonuclear origin is far from a satisfactory solution. A thermonuclear solution requires an excess of about a factor of 25 in its net production in a balance of the production and the destruction processes. While the production rate was experimentally constrained, such large excess is unlikely from the viewpoint of the Hauser-Feshbach statistical model. However, it is premature to draw a definite conclusion before a further study of the nuclear properties of $^{138}La$, in particular, experimental constraints on the highly uncertain photodestruction rate for $^{138}La$.

The blackbody synchrotron radiation to be produced by a 10-Tesla superconducting wiggler at SPring-8 may enable one to proceed to a previously-unexplored field of (γ,α) and (γ,p) reactions.

# ACKNOWLEDGMENTS

This work was supported by the Japan Private School Promotion Foundation and the Japan Society of the Promotion of Science.

# REFERENCES

1. F.Käppeler, H. Beer and K. Wisshak, Rep. Prog. Phys. **52**, 945 (1989).
2. M. Arnould, and S. Goriely, Phys. Rep. **384**, 1 (2003).
3. M. Rayet et al., Astron. Astrophy **298**, 517 (1995).
4. T. Rauscher et al., Astrophys. J. **576**, 323 (2002).
5. S.E. Woosley and W.M. Howard, Astrophys. J. **36**, 285 (1978).
6. S.E. Woosley et al., Astroph. J. **356**, 272 (1990).
7. S. Goriely et al., Astron. Astrophys. **375**, L35 (2001).
8. M. Arnould, Astron. Astrophys. **46**, 117 (1976)
9. H. Ohgaki et al., IEEE Trans. Nucl. Sci. **38**, 386 (1991).
10. H. Utsunomiya, P. Mohr, A. Zilges, and M. Rayet, Nucl. Phys. **A777**, 459 (2006).
11. E.P. Wigner, Phys. Rev. **73**, 1002 (1948).
12. P. Mohr et al., Phys. Lett. **B488**, 127 (2000).
13. D. Belic et al., Phys. Rev. Lett. **83**, 5242 (1999).
14. C.M. McCullagh, M.L. Stelts, and R.E. Chrien, Phys. Rev. **C23**, 1394 (1981).
15. S. Goriely, Phys. Lett. **B436**, 10 (1998).
16. S. Goriely and E. Khan, Nucl. Phys. **A706**, 217 (2002).
17. R.L. Bramblett, J.T. Van Cladwell, G.F. Auchampaugh, and S.C. Fultz, Phys. Rev. **129**, 2723 (1963).
18. R. Bergère, H. Beil, and A. Veyssière, Nucl. Phys. **A121**, 463 (1968).
19. Photonuclear data for applications: Cross sections and Spectra, IAEA-Tecdoc-1178 (2000).
20. J. Audouze, Astron. Astrophys. **8**, 436 (1970).
21. S. Goriely, M. Arnould, I. Borzov and M. Rayet, Astron. Astrophys. **375**, L35 (2001).
22. S.E. Woosley, D.H. Hartmann, R.D. Hoffman, and W.C. Haxton, Astroph. J. **356**, 272 (1990).
23. S. Goriely, J. Josè, M. Hernanz, M. Rayet, and M. Arnould, Astron. Astrophys. **383**, L27 (2002).
24. H. Utsunomiya et al., Phys. Rev. **C74**, 025806 (2006).
25. H. Beil, R. Bergère, P. Carlos, A. Lepretre, and A. Veyssière, Nucl. Phys. **A172**, 426 (1971).
26. S. Goko et al., Phys. Rev. Lett. **96**, 192501 (2006).
27. A.J. Koning, S. Hilaire and M.C. Duijvestijn, Proc. Int. Conf. on Nuclear Data for Science and Technology (eds. C. Haight et al.) AIP Conference Vol. 769, p. 1154, 2005.
28. P. Demetriou, and S. Goriely, Nucl. Phys. **A695**, 95 (2001).
29. S. Hilaire and S. Goriely, Nucl. Phys. A (2006) submitted.
30. S. Hilaire et al., Eur. Phys. J. **A12**, 169 (2001).
31. H. Utsunomiya, S. Goko, K. Soutome, N. Kumagai, and H. Yonehara, Nucl. Instr. Meth. **A538**, 225 (2005).
32. P. Mohr, Zs. Fülop, and H. Utsunomiya, Euro. Phys. J. A (2006) submitted.

# Study of Collective Dipole Excitations below the Giant Dipole Resonance at HIγS

A.P. Tonchev[1], C. Angell[2], M. Boswell[2], A. Chyzh[3], C.R. Howell[1], H.J. Karwowski[2], J.H. Kelley[3], W. Tornow[1], N. Tsoneva[4], and Y.K. Wu[5]

[1]*Duke University and TUNL, Department of Physics, Box 90308, Durham, NC 27708-0308, USA*
[2]*University of North Carolina and TUNL, Department of Physics and Astronomy, Chapel Hill, NC 27599-3255, USA*
[3]*North Carolina State University and TUNL, Department of Physics, Box 8202, Raleigh, NC 27695-8202, USA*
[4]*Institute of Theoretical Physics, University of Giessen, Heinrich-Buff-Ring 16, D-35392, Giessen, Germany*
[5]*Duke University and Duke Free Electron Laser Laboratory, Durham, NC 27708-0319, United States*

**Abstract.** The High-Intensity Gamma-ray Source utilizing intra-cavity back-scattering of free electron laser photons from relativistic electrons allows one to produce a unique beam of high-flux gamma rays with 100% polarization and selectable energy and energy resolution which is ideal for low-energy γ-ray scattering experiments. Nuclear resonance fluorescence experiments have been performed on N=82 nuclei. High sensitivity studies of E1 and M1 excitations at energies close to the neutron emission threshold have been performed. The method allows the determination of excitation energies, spin, parities, and decay branching ratios of the pygmy dipole mode of excitation. The observations are compared with calculations using statistical and quasi-particle random-phase approximations.

**Keywords:** Nuclear reactions $^{138}$Ba(γ,γ′), E=8.5 MeV, monoenergetic and polarized gamma-ray beam, nuclear resonance fluorescence, dipole excitation, pygmy dipole resonance.
**PACS:** 25.20.-x, 21.10.Hw, 21.60.Ev

## INTRODUCTION

A wide variety of nuclear structure phenomena have been investigated with the nuclear resonance fluorescence (NRF) technique [1–4]. Experiments probing two-phonon excitations of even-even nuclei near closed shells revealed large magnetic dipole strengths in heavy deformed nuclei. The corresponding excitations have been associated with scissors-like oscillations of the deformed proton density distribution against the neutron distribution, and this excitation mode was, accordingly, called the "scissors mode" [5]. Large electric dipole transitions to the ground states have been observed in spherical nuclei near Z=50 and N=82. They are assumed to arise from the coupling of quadrupole and octupole vibrational modes of the nucleus [6,9]. In heavier nuclei a resonance like concentration of E1 strength has been observed below the neutron separation energy in semi-magic N=82 isotones [12].

Recent experimental activities at the High-Intensity Gamma-ray Source (HIγS) have focused on investigation of this collective mode, commonly referred to as a "pygmy" dipole resonance (PDR), which is observed as a clustering of states close to the neutron threshold at excitation energies E = 5.5 - 8 MeV. Some theoretical calculations indicate a correlation between the observed total B(E1) strength of the PDR and the neutron-to-proton ratio N/Z [10]. Although carrying only a small fraction of the full dipole strength, these states are of particular interest because they reflect the motion of the neutron skin against the isotropic symmetric core. The observation of this collective dipole mode near the neutron threshold might have important astrophysical implications. For example, the conditions governing thermal equilibrium of (γ,n) and (n,γ) reactions in explosive nucleosynthesis of certain neutron deficient heavy nuclei may be significantly modified [8]. Their low energy may well enhance their contribution to photo-dissociation processes in spite of their relatively low strength as compared to the Giant Dipole Resonance (GDR).

The purpose of this work is determine first, the character of this low-energy mode of excitation. Is it E1 or M1 mode of excitation? Second, what is the decay pattern of these collective states below the particle separation energy? Third, what are the strength, energy distribution, and nature of these collective phenomena? To answer these questions we focus our experimental activity on the $^{138}$Ba(γ,γ') reaction below the neutron separation energy.

## EXPERIMENTAL TECHNIQUE

The HIγS facility is used to produce high-intensity and nearly monoenergetic γ-ray beams by intracavity Compton backscattering [11]. In γ-ray production mode two electron bunches stored in the Duke storage ring are synchronized so that the lased photons from one electron bunch are reflected by the downstream mirror, and then collide head on with the second electron bunch. The backscattered photons are collimated with a 1.27 cm on-axis cylindrical lead collimator, located 60 m downstream of the collision point. In addition, the γ rays are highly polarized resulting from the Compton scattering process of ≈ 100% polarized photons. The gamma-ray spectrum measured with a 123% HPGe detector is shown in Fig.1. The detector was placed in the beam operated in a low flux mode. A Monte-Carlo simulated response function of the detector is presented in the insert. A deconvolution analysis provides the FWHM of the photoabsorption peak displayed by the hatching area. The analysis of this peak shows that the HIγS beam has an asymmetric Gaussian shape ($E_\gamma$= $8.5^{+0.064}_{-0.086}$ MeV) with total width of $\Delta E_\gamma$=140 keV at 8.5 MeV. The average γ-ray flux on the target position using a 1.27 cm diameter lead collimator exceeds $1\times10^6$ γ/s at 15 MeV [13]. The flux of γ rays can be increased in principle by increasing the current in the bunches in the storage ring, presently up to 30 mA, and/or by operating with more than two electron bunches.

The NRF method is used at HIγS to study low-multipolarity ground state transitions (*i.e.*, E1 and M1) with large partial widths. Due to the low detection limit the NRF technique represents an outstanding tool for measuring dipole transitions. The main advantage of this method is that both the excitation and the de-excitation processes proceed via the electromagnetic interaction, the best understood interaction in physics. Using this approach, the quality of the HIγS γ-ray beams provides tremendous advantages over bremsstrahlung beams. For example, the monoenergetic γ-ray beam allows excitation of only the desired levels of interest. In addition the parity of the excited state can be determined by measuring the angular distribution of the scattered photons with respect to the incoming unpolarized photon beam. For the case of even-even nuclei with ground state spin $J = 0$ it is sufficient to measure the scattered radiation at two different azimuthal angles. The NRF setup at HIγS consist of four 60% HPGe detectors, positioned at $90^0$ to the beam axis, two in the horizontal and two in the vertical plane.

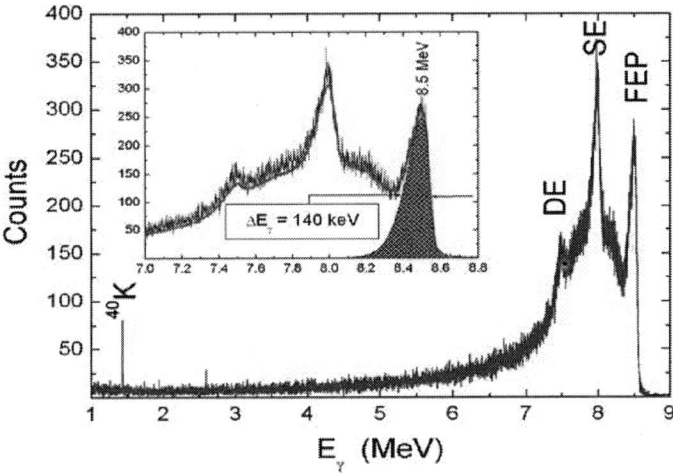

**FIGURE 1.** Gamma-ray spectrum measured with 123% HPGe detector at $0^0$ after 1.27 cm lead collimator. The full-energy peak (FEP), the single escape (SE), and double escape (DE) peaks are labeled. The insert shows the energy resolution of the FEP at $E_\gamma$=8.5 MeV obtained from MCNPX simulation.

The scattering target consisted of 8.24 g of $BaCO_3$ powder of natural isotopic abundance packed into a Lucite container 2.0 cm in diameter by 2.0 cm in height.

## EXPERIMENTAL RESULTS

From the recent photon scattering experiments at the TU Darmstadt bremsstrahlung facility, a large data set on E1 excitations in N=82 nuclei has been compiled [17]. For example, in $^{138}$Ba alone more than 70 states have been observed in the energy range from 4 to 8.5 MeV. For the vast majority of the observed transitions only one decay branch to the ground state has been measured. However, this data set is still incomplete since the polarization sensitivity is very low above 4 MeV, making parity assignment impossible. To obtain the parity information of the low-lying dipole states in the PDR region, NRF measurements have been performed on $^{138}$Ba using the linearly polarized and monoenergetic γ-ray beam at HIγS. Fig. 2 shows the experimental asymmetry of 13 dipole states in $^{138}$Ba in the energy range from 7.5 to

8.5 MeV. All the dipole states deexcite via E1 transitions ($J^\pi = 1^-$). Earlier NRF measurements at HIγS unambiguously showed that all of the dipole states in the energy region from 5.5 to 6.5 MeV are also $J^\pi = 1^-$ states [7]. Fig. 3 shows the low-energy part (top panel) and high energy part (bottom panel) of the γ-ray spectra in the vertical (left) and horizontal (right) detectors relative to the polarization plane at $E_\gamma$=8.5±0.075 MeV. There are only a few dipole states which decay to the ground state (bottom-left panel). These states are observed only in the vertical detectors located in the plane perpendicular to the polarization plane. According to the azimuthal distribution these states exhibit E1 character. However, strong E2 transitions from the lowest ($J^\pi = 2^+$) excited states in $^{138}$Ba at 1435.8, 2217.9, and 2639.5 keV to the ground state have been observed with the advantage of the pulsed γ-ray beam at the HIγS facility (top panel). This experimental information is usually difficult to obtain in bremsstrahlung experiments due to the increasing nonresonant scattering at low-energies and the DC beam from the electron linear accelerator [12,17]. Note the logarithmic scale on the top panel and the absence of non beam related lines in the low-energy region.

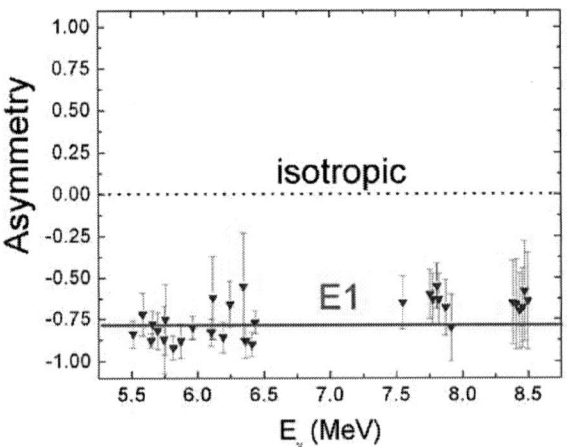

**FIGURE 2.** Asymmetry of the low-lying dipole states in $^{138}$Ba using the 100% linearly polarized γ-ray beam at HIγS.

The first question arises how these low-lying $2^+$ states were populated with the 100% polarized γ-beam at 8.5 MeV and total width of 140 keV? It has to be mentioned that the decay intensity from the first $2_1^+$ state, for example, is 4 times higher compared to the ground state decay of the 8.43 MeV level. There are three possible contributions to the population of the $2^+$ states. The first one is that the monoenergetic beam at the HIγS facility might have a low-energy tail. Such a tail could result from bremsstrahlung radiation of the relativistic electrons in the storage ring or from off-axis gamma rays entering into the 1.27 cm diameter Pb collimator. However, direct measurements of the γ beam with low flux show that the intensity at 8.5 MeV, for example, is 40 times higher than in the 2 MeV region (see Fig. 1). Hence, this potential low-energy tail contribution should be negligible in the present experiment. The second possible contribution could be from neutrons produced in the lead collimator or neutrons produced in the DFELL storage ring. However, this potential contribution is removed by the pulsed technique used in the present experiment. The third possibility is for the primary beam of $E_\gamma$=8.5 MeV to exhibit multiple Compton scattering in the barium sample itself. These multipole scattered photons can populate directly the $2^+$ states. This possibility was simulated by Monte-Carlo methods and a negligible ($10^{-6}$) contribution was obtained.

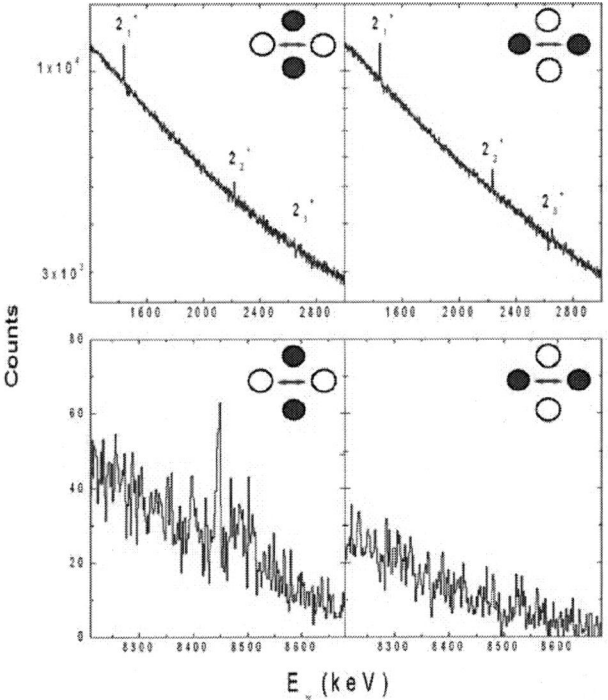

**FIGURE 3.** Low-energy part (top panel) and high-energy part (bottom panel) of the γ-ray spectra in the vertical (left) and horizontal (right) detectors at $E_\gamma$ = 8.5±0.075 MeV.

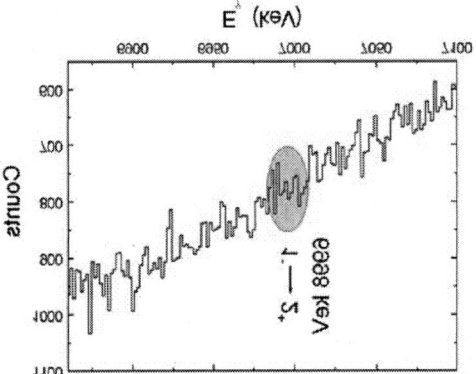

**FIGURE 4.** NRF spectrum of $^{138}$Ba between 6850 and 7100 keV. The 6988 keV line indicates the region where the primary γ–transition from the 8.433 MeV state to the $2^+_1$ states should appear.

There are other important consequences of these measurements. As can be seen from Fig. 3, the decays from the first three $J^\pi = 2^+$ states to the ground state is observed in both the horizontal and vertical detectors. The intensities of these transitions are almost identical for the horizontal and vertical detectors, indicating a loss of initial polarization. This observation is further supported by the lack of discrete inelastic transitions from the resonance energy to the $2^+$ states. In Fig. 4 the part of the spectrum taken with the vertical detectors is shown where a possible decay to the $2_1^+$ state should be visible. If indeed a branching were observed, such a line should be clearly visible above the hatched statistical background. However, no such line was observed. This measurement was followed with a measurement at the lower beam energy of $E_\gamma = 8.2 \pm 0.07$ MeV, where no resonance transitions were observed at the Darmstadt facility. No direct transitions were measured at the HIγS facility either to the ground state or the $J^\pi = 2^+$ states. However, the low-energy $J^\pi = 2^+$ state to ground state transitions still remain in both horizontal and vertical detectors. From this picture one can see that at $E_\gamma = 8.5 \pm 0.075$ MeV, there are only a couple of dipole states strongly connected with the ground state. According to the statistical model there should be 430 dipole states ($J^\pi = 1^+$ or $1^-$) in the 140 keV energy range. Therefore, there are a lot of unobserved dipole transitions and respectively γ-strength, which is hidden in the "background" of the previous nuclear resonance experiments. The advent of pulsed and monoenergetic HIγS beam allows for accounting of the hidden γ-strength in $^{138}$Ba at this energy. The ratio of inelastic versus direct transitions is $5.7 \pm 0.7$ at 8.5 MeV excitation energy. It should be noted that the resonance peak at 8496.0 keV is the highest dipole excited state in $^{138}$Ba that has been observed. Our experiments at excitation energies above the neutron threshold at $8.8 \pm 0.08$, $9.1 \pm 0.08$, $9.5 \pm 0.08$, and $10.0 \pm 0.085$ MeV, for example, did not identify any dipole transitions to the ground state or to any of the higher excited states in $^{138}$Ba.

## THEORETICAL CALCULATION

Microscopic calculations of the dipole strength distribution were performed for the $^{138}$Ba isotope within the framework of the quasiparticle-phonon model (QPM) [14]. The QPM analysis shows that the structure of the states contains a large neutron part (more than 90%). These states correspond mainly to oscillations of weekly bound neutrons from $s$-and $p$-shells. A microscopic study of the $1^-$ states close to the particle threshold revealed that their structure is mostly an admixture of complex configurations. As a result, direct transitions from these states to the ground state are strongly hindered. Only a few dipole states in this region decay by strong E1 transitions to the ground state due to the PDR or GDR component in the structure of the state vector. From a recent analysis of transition densities [10] these excitations have been related to the neutron PDR mode. The single phonon component in these dipole states, which is responsible for direct transitions to the ground state, is only a few percent.

As discussed in previous sections, a certain concentration of electric dipole strength was found around 6.5 MeV in N=82 nuclei. The E1 strength distribution, taken from Ref. 17, is shown in Fig. 5. The authors conclude the centroid energy of the higher-lying dipole excitations is shifted to lower energies for more proton-rich nuclei. In addition it was measured that the integrated total strength in $^{138}$Ba decreases by a factor of 4 in comparison to $^{144}$Sm. To understand this systematic behavior of the E1 strength distribution, statistical and QPM models were invoked. In Fig. 5 the dipole strength is observed to damp out as the excitation energy approaches either the neutron ($B_n$) or proton ($B_p$) binding energy. Q-value systematics for the N=82 nuclei show that $B_n$ increases with increasing proton number, while $B_p$ goes in the opposite direction. The different Q-values will provide $^{142}$Nd and $^{144}$Sm nuclei with less energy space in comparison to $^{138}$Ba, for example. Hence the total E1 strength will be in favour of high-Q value nuclei. The observed centroid shift is most probably due to the same Q-value systematics. The lowest-lying excitation in all N=82 isotones is the two-phonon excitation which determines the low-energy border of the PDR. The high-energy end is governed by the proton or neutron separation energy.

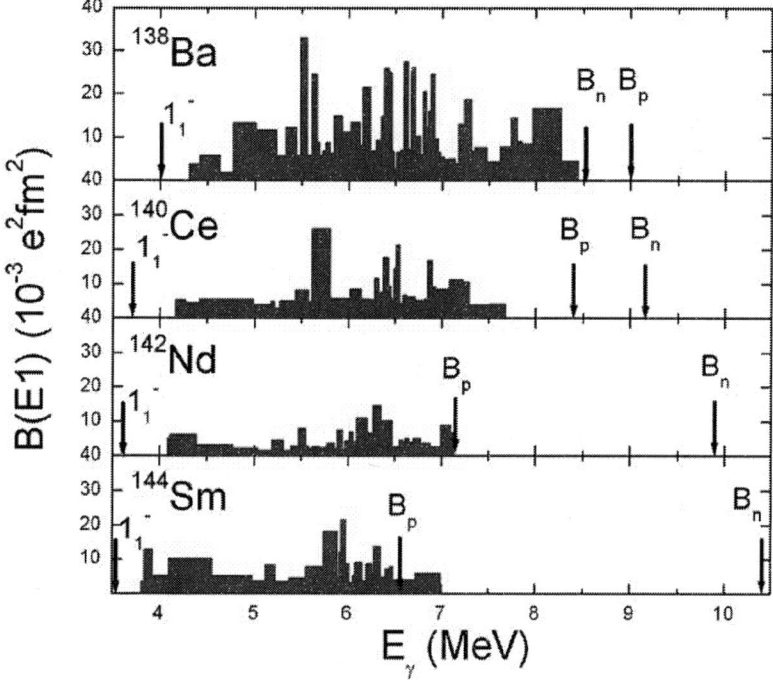

**Figure 5.** The B(E1) strength distribution in the N = 82 isotones. Data taken from [17].

Another important question that remains to be answered is how this low-energy collective PDR mode evolves to the GDR mode. Fig. 6 a) presents the photoneutron cross sections for all these N=82 nuclei in the GDR region. The total photoneutron cross sections is showing almost identical width, maximum, and strength of the giant resonance as one adds protons. Fig. 6 b) shows the total absorption cross section below the particle separation energy using the statistical model. All these calculations based on the statistical model in the PDR region have the same total absorption value.

Hence it will be very interesting to measure if the direct transitions in $^{142}$Nd or $^{144}$Sm will be accompanied with an even stronger statistical decay.

Figure 7 shows the constant temperature (CT) model fit [15] to the experimental cumulative number of dipole levels in $^{138}$Ba measured so far. This heavy mass nucleus has one of the most extensively studied level scheme up to 8.5 MeV excitation energy. Above 4.5 MeV the predicted level densities [15] increasingly surpass the number of measured dipole states. Apparently the NRF experiments start missing a lot of levels at E>4.5 MeV. Since the levels close to the neutron threshold produce many but weak branchings or cascades, a large portion of the strength of these states is likely to be missed experimentally. The intensity of these transitions is much smaller, ≈100, than the ground states transition and they do not produce identifiable photopeaks in the germanium detector. However, the fact that there are so many, at the end they will "rain" down by fast E1 or M1 transitions to the low-energy $2^+$ states. Statistical calculations shows that its takes 3-4 γ-ray transitions to release the initial excitation energy. In principle, also M1-transitions contribute to the dipole strength, but the average M1 strength is typically 1-2 orders of magnitude smaller compared to the E1 strength, and it is usually not taken into account.

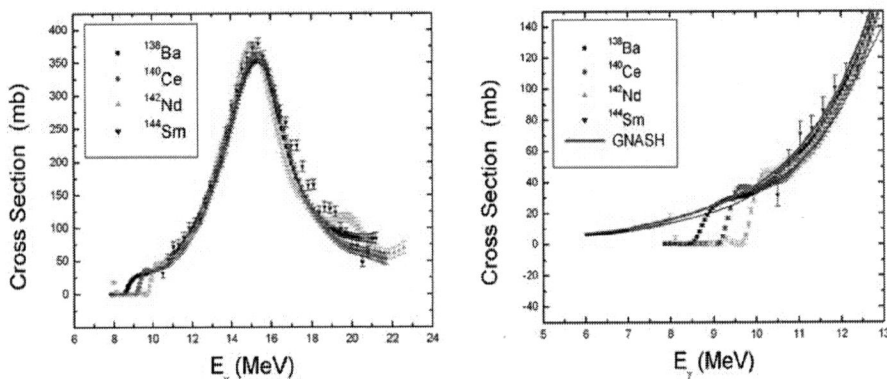

**Figure 6.** Photoneutron absorption cross section for N=82 nuclei in the GDR region (a). Low-energy tail of the GDR extended by statistical calculations below the particle separation energy (b).

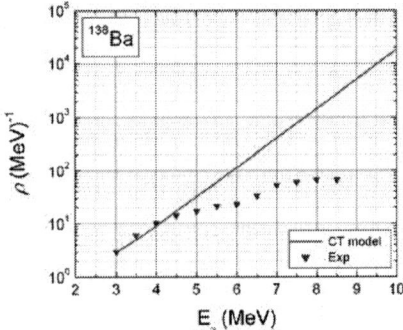

**Figure 7.** Calculated cumulative number of levels (continous curve) of $^{138}$Ba with spin $1^+$ or $1^-$ with the CT model formula [15]. Measured dipole levels are shown with triangles [7,12,16,17].

## CONCLUSION

On the basis of the present parity measurements in $^{138}$Ba it is fairly safe to assume that the observed dipole strength below the particle threshold is predominantly electric in nature. Our findings are in agreement with other theoretical predictions on the character of this dipole mode of excitation. The high level density at the particle separation energy leads to many weak transitions which are difficult to observe directly. In addition these weak transitions decay via cascades to many intermediate levels. Therefore much of the information on dipole strength is missing from observation of de-excitation only to the ground state. These transitions are so weak but numerous that their superposition is observed as a continuum of unresolved strength. Despite the fact that the PDR exhibits well pronounce collective structure below the neutron separation energy the statistical transitions will mostly dictate the reaction rate in this energy region.

The present measurements also show that the monoenergetic and pulsed beam from the HIγS facility opens up a new opportunity for direct measurements of the level density and the radiative γ-ray strength functions by measuring not only the direct gamma transitions but also the non-resonance part of the decay process.

## ACKNOWLEDGMENTS

We thank the technical staff at the DFELL for providing excellent γ-ray beams. We are also grateful to G.E. Mitchell, M. Krticka, and N. Pietralla for fruitful discussion. This work was partially supported by the Grant Nos. DE-FG02-97ER41033, DE-FG02-97ER41042, and DE-FG02-97ER41041.

## REFERENCES

1. F.R. Metzger, *Prog. in Nucl. Phys.* **7**, 54 (1959).
2. U.E.P. Berg and U. Kneissl, *Ann. Rev. Nucl. Part. Sci.* **37**, 33 (1987).
3. U. Kneissl et al., *Prog. Part. Nucl. Phys.* **37**, 349 (1996).
4. U. Kneissl et al., *J. Phys. G: Nucl. Part. Phys.* 32 R217–R252 (2006).
5. A. Richter, *Prog. Part. Nucl. Phys.* **13**, 1 (1985).
6. C. Fransen et al., *Phys. Rev.* C **70**, 044317 (2004).
7. N. Pietralla et al., *Phys. Rev. Lett.* **88**, 012502 (2002).
8. M. Arnould, S. Goriely., *Phys. Rep.* **384**, 1 (2003).
9. D. Savran et al., *Phys. Rev.* C **71**, 034304 (2005).
10. N. Tsoneva et al., *Phys. Lett.* B **586**, 213 (2004).
11. V. Litvinenko et al., *Phys. Rev. Lett.* **78**, 4569 (1997).
12. A. Zilges at al., *Phys. Lett.* B **542**, 43 (2002).
13. A.P. Tonchev et al., *Nucl. Instrum. Methods Phys. Res.* B **241**, 170 (2005).
14. V.G. Soloviev, *Theory of Atomic Nuclei: Quasiparticles and Photons*. Institute of Physics, Bristol, 1992.
15. T. von Egidy and D. Bucurescu, *Phys. Rev.* C **72**, 044311 (2005).
16. R.-D. Herzberg et al., *Nucl. Phys.* A **592**, 211 (1995).
17. S. Voltz et al., *Nucl. Phys.* A (2006) in press.

# Nuclear Level densities from drip line to drip line

S. Hilaire* and S. Goriely[†]

*CEA/DAM Ile de France, DPTA/Service de Physique Nucléaire,
BP 12, 91680 Bruyères-le-Châtel, France
[†]Institut d'Astronomie et d'Astrophysique, Université Libre de Bruxelles,
Campus de la plaine CP 226, B-1050 Brussels, Belgium

**Abstract.** New energy-, spin-, parity-dependent level densities based on the microscopic combinatorial model are presented and compared with available experimental data as well as with other nuclear level densities usually employed in nuclear reaction codes. These microscopic level densities are made available in a table format for nearly 8500 nuclei.

**Keywords:** Nuclear level densities, Hartree-Fock-Bogoliubov approach, Skyrme force, Combinatorial method
**PACS:** 21.10.Ma, 21.10.-k, 21.60.Jz

## 1. INTRODUCTION

The knowledge of nuclear level densities (LDs) has been a matter of interest and study for more than 50 years going back at least to 1936 with Bethe's pionnering work [1]. Since then, more or less sophisticated methods have been developed to reproduce the available experimental data [2]. For specific applications such as astrophysics or accelerator driven systems, a large number of data needs to be extrapolated far away from the experimentally known region. In such situations, extrapolations have to be as *reliable* and *accurate* as possible. Therefore, when no experimental informations exists to constrain analytical LD formulae, one should preferentially use microscopic or semi-microscopic global predictions, provided they compete with more phenomenological highly-parameterized formulae in the reproduction of experimental data. The microscopic character indeed ensures the required reliability, while the searched accuracy is garanted by the ability to reproduce experimental data as well as the best global phenomenological approaches.

Many global microscopic approaches have been developed in the past [2] but they are almost never used for practical applications, either because of their lack of accuracy in reproducing experimental data or because they do not offer the same flexibility as that of the analytical expressions. Recently, we have performed large scale calculation of LDs for nearly 8500 nuclei [2] modernizing the combinatorial method described in ref. [3] which had already proved its predictive power [3, 4]. The main advantage of our results is to provide not only the LDs tabulated as function of the excitation energy, but also the spin and parity distributions without any statistical assumption.

We first summarize the method we use to compute our new LDs. We then show that our new LDs reproduce the experimental data as well as other global approaches and discuss the parity and spin distributions obtained from our tables. We finally explain how our tabulated LDs can be adjusted - as the usually employed analytical expressions - to fit level density experimental data and/or cross sections.

## 2. THE MODERNIZED COMBINATORIAL METHOD

As precisely described in [2], our combinatorial method consists in calculating incoherent particle-hole state densities out of two single particle level schemes, one for the protons and the other for the neutrons of a nucleus with Z protons and N neutrons. All the possible particle-hole configuration having the same number of particles and holes are then summed to get the total state densities $\omega(U,K,\pi)$ as function of the excitation energy $U$, the spin projection $K$ and the parity $\pi$. Once these incoherent state densities have been obtained, we then have to account for collective effects - namely the vibrational effects and, when then nucleus under consideration is deformed, the rotational effects - to deduce the total LDs depending on the nucleus spin $J$ rather than on the spin projection $K$. On top of that we also deal with the disappearance of deformation effects at increasing energies.

If the nucleus under consideration displays spherical symmetry, the intrinsic and laboratory frames coincide, and the level density $\rho_{sph}(U,J,\pi)$ is trivially obtained through the relation
$$\rho_{sph}(U,J,\pi) = \omega(U,M{=}J,\pi) - \omega(U,M{=}J{+}1,\pi). \quad (1)$$
For deformed nuclei, rotational bands are explicitely constructed to get the deformed level density $\rho_{def}(U,J,\pi)$. In this case, any intrinsic state of specified spin projection $K$ and parity $\pi$ is the band head of a set of levels having the same parity and spins $J = K, K+1, K+2, \ldots$ if $K \neq 0$, and $J = 0, 2, 4, \ldots$ or $1, 3, 5, \ldots$ if $K^\pi = 0^+$ or $0^-$, respectively. These sequences of levels form rotational bands in which each member's energy can be deduced from the band-head energy provided the difference $E_{rot}^{J,K}$ between the energy of the level $J^\pi$ and that of the band-head state $K^\pi$ is known. In our approach, this rotational energy is obtained with the well-known expression [5],
$$E_{rot}^{J,K} = \frac{J(J+1) - K^2}{2\mathscr{J}_\perp}, \quad (2)$$
where $\mathscr{J}_\perp$ is the moment of inertia of a nucleus rotating around an axis perpendicular to the symmetry axis which reads in our case
$$\mathscr{J}_\perp = \frac{2}{5}mR^2\left(1 + \sqrt{\frac{5}{16\pi}}\beta_2\right),$$
for an ellipsoidal shape with axial quadrupole deformation parameter $\beta_2$.

Once this transition from state to level densities is done, vibrational effects are taken into account multiplying $\rho_{def}$ or $\rho_{sph}$ by the vibrational enhancement $K_{vib}$ approximated by the equation [6]

$$K_{vib} = \exp[\delta S - (\delta U/T)], \quad (3)$$

where $T$ is the nuclear temperature and

$$\delta S = \sum_i (2\lambda_i + 1)\left[(1+n_i)\ln(1+n_i) - n_i \ln n_i\right],$$
$$\delta U = \sum_i (2\lambda_i + 1)\omega_i n_i, \quad (4)$$

where $\omega_i$, $\lambda_i$ are the energies and multipolarities of the vibrational excitations and $n_i$ the occupation numbers that account for the disappearance of vibrational enhancement of the level density at high temperatures thanks to the equation

$$n_i = \frac{\exp(-\gamma_i/2\omega_i)}{\exp(\omega_i/T) - 1}, \quad (5)$$

where $\gamma_i$, the spreading widths of the vibrational excitations, read

$$\gamma_i = C(\omega_i^2 + 4\pi^2 T^2). \quad (6)$$

The originality of our treatment consists in accounting phenomenologically for shell effects in the analytical expressions of the vibrational energies $\omega_2$ and $\omega_3$ as function of the mass of the nucleus [2].

Finally, when the excitation energy increases, it is well known that deformed nuclei tend to become spherical. Such a shape transition significantly affects the LDs predictions. As suggested in Ref.[7], this effect is described by introducing a phenomenological damping function between the spherical and deformed expressions of the nuclear level densities.

## 3. RESULTS

This new microscopic approach has been used starting from the single particle level schemes and pairing properties generated from the Hartree-Fock-Bogoliubov (HFB) method based on the BSk13 interaction [8] which has proven its ability to reproduce nuclear ground state properties (and nuclear masses in particular) with a high degree of accuracy. The results obtained with the present method have been compared with the experimental s-wave neutron resonance spacings $D_0$ at the neutron separation energy $S_n$ of the compound nucleus $(Z, A+1)$ (resulting from the capture of a low-energy neutron by a target $(Z, A)$ with spin $J_0$ and parity $\pi_0$) corresponding to

$$D_0 = \frac{1}{\rho(S_n, J_0 + 1/2, \pi_0) + \rho(S_n, J_0 - 1/2, \pi_0)} \quad \text{for } J_0 > 0$$
$$= \frac{1}{\rho(S_n, 1/2, \pi_0)} \quad \text{for } J_0 = 0, \quad (7)$$

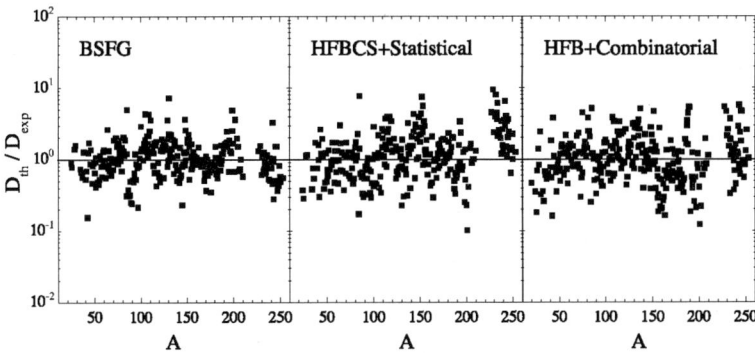

**FIGURE 1.** Ratio of theoretical ($D_{th}$) BSFG [9] (left), HFBCS plus statistical [7] (centre) or the present HFB plus combinatorial (right) to experimental ($D_{exp}$) s-wave neutron resonance spacings for the 295 nuclei compiled in [6].

and are shown in Fig. 1. The present approach using the BSk13 single particle levels and pairing properties clearly gives comparable predictions to the accurate global BSFG-type formula of [9] as well as the HFBCS plus statistical estimate of [7].

The HFB plus combinatorial model also gives satisfactory extrapolations to low energies. As an example, we compare in Fig. 2 the predicted cumulative number of levels $N(U)$ with the experimental data [6] for 9 nuclei, including light as well as heavy, spherical as well as deformed species. Globally, the present model is in rather good agreement with experimental data. The uncertainties affecting the exact determination of the ground-state energy as well as the pairing effect (typically around 1 MeV) lead to a corresponding error in the shift of the predicted excitation energy.

It is also of interest to discuss the spin and parity dependence of our LDs. Since no statistical assumption is used to derive these dependences, the spin distribution departs from the expected derivative of a gaussian law particularly at low energies and the parity equipartition is only obtained for high enough excitation energies.

Concerning parity distribution, the parity asymmetry $\mathscr{A}_\pi$ defined as the ratio

$$\mathscr{A}_\pi = \frac{\rho(U, \pi = +) - \rho(U, \pi = -)}{\rho(U, \pi = +) + \rho(U, \pi = -)} \quad (8)$$

is a practical measure of deviation from equipartition ($\mathscr{A}_\pi = 0$). As illustrated in Fig. 3, where the parity asymmetry is plotted for 4 nuclei, one observes that the excitation energy required to reach parity equipartition can be of the order of a few MeV for deformed nuclei ($^{66}$Zn and $^{240}$Pu) and even a few tens of MeV for spherical nuclei ($^{132}$Sn and $^{208}$Pb). The impact of such parity distributions on nuclear reaction can be illustrated for instance on the $^{56}$Fe radiative neutron capture, where the cross section obtained with the parity-dependent LDs is compared with that obtained assuming the

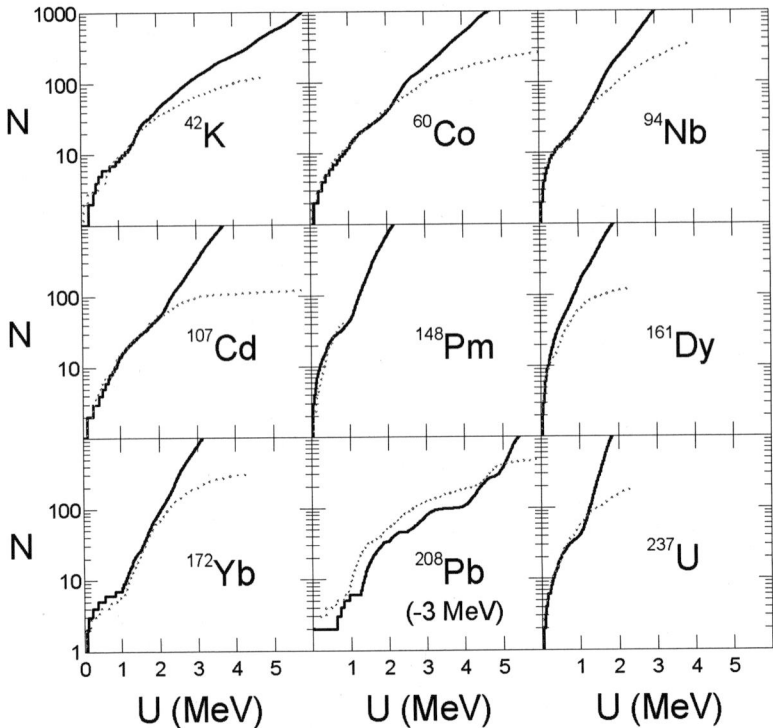

**FIGURE 2.** Comparison of the cumulative number of observed levels (dotted staircase) with the HFB plus combinatorial predictions (solid lines) as a function of the excitation energy $U$ for a sample of 9 nuclei. For $^{208}$Pb, *both* curves have been shifted by 3MeV, the energy range corresponding consequently to [0-6] MeV instead of [0-6] MeV.

LDs are given by the average value of both parities. As can be seen on Fig. 4, the fact parity equipartition for $^{57}$Fe is only obtained for excitation energies of the order of 10 MeV has a significant impact on the cross section predictions.

The impact of spin distribution at low excitation energies can also be of great importance for some reaction channel, like, for instance the isomeric to ground states production ratio in the $^{181}$Ta photoneutron reaction [10]. In this case, it has been found that our combinatorial approach predicts much more high-spin levels at low energy than what a statistical gaussian law would. As a consequence, the $^{181}$Ta photoreaction leads to a significantly larger production of the $9^-$ isomeric state of $^{180}$Ta, in agreement with experimental data.

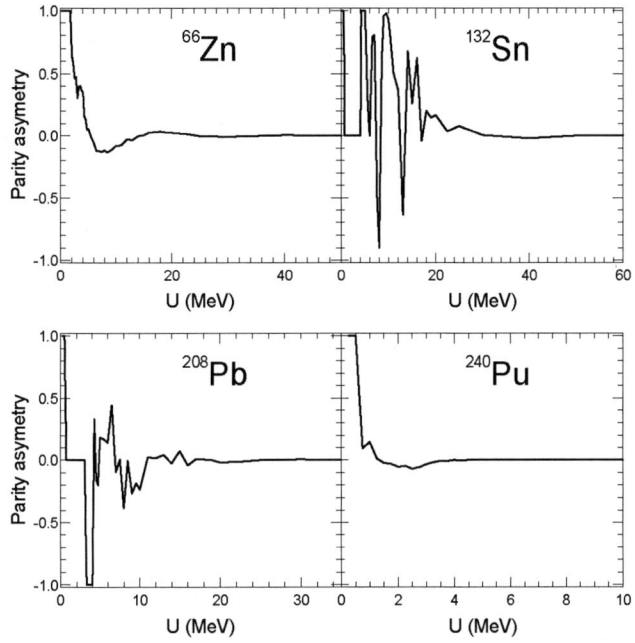

**FIGURE 3.** Parity asymmetry for several nuclei as function of the excitation energy $U$.

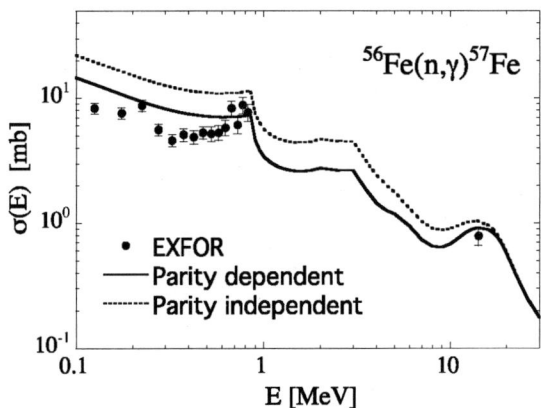

**FIGURE 4.** $^{56}$Fe(n,$\gamma$)$^{57}$Fe cross section with the parity-dependent combinatorial LD and the parity-independent LD obtained averaging both parities.

## 4. FITTING PROCEDURE

For many nuclear physics applications, a renormalization procedure of the nuclear LDs on experimental data (such as $D_0$ values or low-lying discrete levels) is required, in particular for nuclear data evaluation or for accurate and reliable estimate of cross sections. Also, when cross section have been measured for a few channel, one generally wants to fit these measured cross section and predict unmeasured ones with an improved confidence. In such situations, analytical LDs are generally prefered because of the freedom they provide to play with the parameters on which they depend. Even if our HFB combinatorial LDs are provided in a tabulated format, it is possible to renormalize them in a similar way to what is usually done with analytical formulae.

More precisely, as proposed in [2], it is possible to modify our level densities writing

$$\rho(U,J,\pi)_{renorm} = e^{\sqrt{\alpha\,(U-\delta)}} \times \rho(U-\delta,J,\pi). \qquad (9)$$

With such a modification, the two parameters $\alpha$ and $\delta$ can be adjusted so as to reproduce both the experimental low-lying states and the experimental $D_0$ values or used to fit cross sections as well as when analytical expression are employed.

## 5. CONCLUSION

Large scale Hartree-Fock-Bogoliubov plus combinatorial calculation of nuclear LDs have been performed up to 200 MeV and are available as tables at the website **http://www-astro.ulb.ac.be**. The resulting LDs do not rely on any statistical assumption and are energy-, spin- and parity-dependent. These LDs reproduce experimental data as accurately as other analytical or statistical approaches and can be renormalized in a similar way if required. Important efforts still have to be made to improve both the vibrational enhancement and the transition from deformed to spherical shapes of the nuclei with increasing excitation energy. Also the sensitivity to the nuclear structure input needs to be further studied.

## REFERENCES

1. H.A. Bethe, Phys. Rev 50 (1936) 332.
2. S. Hilaire and S. Goriely, in press in Nucl. Phys. A (2006) and references therein.
3. S. Hilaire, J.P. Delaroche and M. Girod, Eur. Phys. J. A12 (2001) 169
4. S. Goriely, In Proc. of the International Conference on Frontiers in Nuclear Structure, Astrophysics and Reactions (eds. S. Harissopulos, P. Demetriou, R. Julin, AIP Conference proceedings 831, New York) (2006) p. 8.
5. T. Døssing and A. S. Jensen, Nucl. Phys. A**222**, 493 (1974).
6. Reference Input Parameter Library, IAEA-Tecdoc (2005) in press (also available at http://www-nds.iaea.or.at/ripl2)
7. P. Demetriou and S. Goriely, Nucl. Phys. A695 (2001) 95.
8. S. Goriely, J. M. Pearson and M. Samyn, Nucl. Phys. A773 (2006) 279.
9. S. Goriely, J. Nucl. Science and Technology, Suppl. 2 (Ed. K.Shibata), p. 536 (2002)
10. S. Goko et al., Phys. Rev. Lett. 96 (2006) 192501.

# A status report of the data evaluation in the NACRE update and extension project

M. Katsuma

*Institut d'Astronomie et d'Astrophysique, Université Libre de Bruxelles,
Campus de la Plaine, 1050 Brussels, Belgium*

**Abstract.** Since its publication in 1999, the NACRE compilation of thermonuclear reaction rates of astrophysical interest has been widely used in studies of stellar evolution and concomitant nucleosynthesis processes. In consideration of ever-accumulating experimental data, the NACRE update and extension project is under way in the framework of a Konan-ULB convention. The goal is to "evaluate" the existing experimental data, and to provide reliable thermonuclear reaction rates at low-energies as astrophysical applications call for. The phenomenological approaches such as the potential model, R-matrix method, and DWBA are adopted in the evaluation. The astrophysical S-factors of the $^7$Be(p,$\gamma$)$^8$B, $^{20}$Ne(p,$\gamma$)$^{21}$Na, $^3$H(d,n)$^4$He, and $^3$He(d,p)$^4$He reactions are investigated with those three models, for instance. The experimental S-factor data can be reproduced well by our models.

**Keywords:** Thermonuclear reactions; Astrophysical S-factors
**PACS:** 95.30.-k; 29.87.+g

## INTRODUCTION

In the last decades, our basic understandings of the Big Bang nucleosynthesis, and of stellar evolution and concomitant nucleosynthesis have been deepened greatly. The achievement has been helped to a more or less large extent by the efforts to estimate thermonuclear reaction rates at very-low energies.

In this respect, the pioneering works of compilation by W.A. Fowler and his collaborators of the thermonuclear reaction rates (e.g. [1]) have contributed as a key element to the progress. More recently, the so-called "NACRE" [2] compilation was carried out, with its main features including the short documentations of the procedures of evaluation adopted in the individual cases, as well as the explicit references to the original experimental data. Since its publication in 1999, the NACRE compilation has been used in many nucleosynthesis calculations. Meanwhile, many precise and remarkable cross-section measurements of astrophysical interest have been performed. Also, some new compilations have been presented in relation to some cases of specific astrophysical interest (e.g. [3, 4, 5]). Under the circumstances, the NACRE update and extension project was launched by a convention between the Konan University and the Université Libre de Bruxelles (ULB) [6].

In this paper, we report the current status of the NACRE update and extension project by showing four examples of the data evaluation: Two radiative capture reactions, $^7$Be(p,$\gamma$)$^8$B and $^{20}$Ne(p,$\gamma$)$^{21}$Na, are analyzed with the potential model and the R-matrix method [7], and two transfer reactions, $^3$H(d,n)$^4$He and $^3$He(d,p)$^4$He, are with the distorted-waves Born approximation (DWBA) [8] and the R-matrix method. The as-

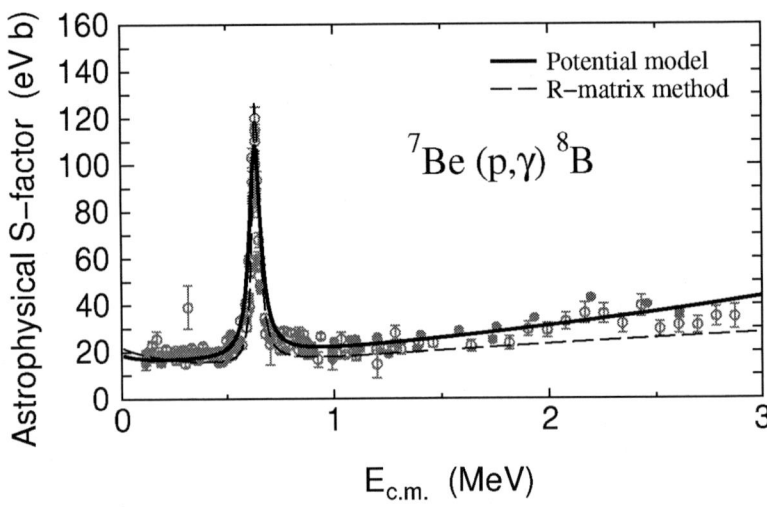

**FIGURE 1.** The astrophysical S-factor of the $^7$Be$(p,\gamma)^8$B reaction. The solid and dashed curves are the results of the potential model and R-matrix method, respectively. The experimental data are taken from [2, 9, 10, 11, 12, 13, 14].

trophysical S-factors obtained from the models are compared with the experimental data including those reported after the NACRE compilation. The low-energy S-factors are extrapolated from our models.

# EVALUATIONS FOR NUCLEAR REACTIONS RELEVANT TO NUCLEAR ASTROPHYSICS

## Astrophysical S-factors

Before moving on to the results of the data evaluation, let us recall the astrophysical S-factor, which is introduced to compensate for the fast energy dependence of the cross section below the Coulomb barrier, and is defined as follows:

$$S(E) = E\exp(2\pi\eta)\sigma(E), \quad (1)$$

where $\eta$ is the Sommerfeld parameter, $\eta = Z_p Z_t e^2/(\hbar v)$, and $\sigma(E)$ is the cross section. The astrophysical S-factor not only gives the smooth energy dependence of non-resonance at low-energies but also provides the better criterion for deciding the validity of the models than the thermonuclear reaction rates that are average quantities integrated over the energy with the Maxwell-Boltzmann distribution. In order to see whether our models are applicable to describe thermonuclear reactions relevant to nuclear astrophysics, we therefore compare the calculated S-factor with the data, but do not discuss the reaction rates explicitly here.

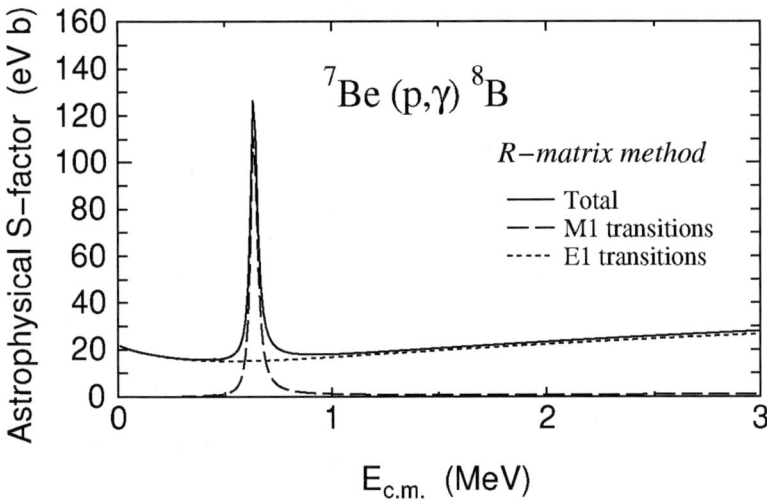

**FIGURE 2.** The decomposition of the transition in the $^7$Be(p,$\gamma$)$^8$B reaction. The dashed and dotted curves are the E1 and M1 components of the transition obtained from the R-matrix method.

## Radiative capture reactions: $^7$Be(p,$\gamma$)$^8$B and $^{20}$Ne(p,$\gamma$)$^{21}$Na

Figure 1 shows the astrophysical S-factor of the $^7$Be(p,$\gamma$)$^8$B reaction as a function of incident energy. The solid and dashed curves are the results of the potential model and R-matrix method, respectively. The resonance at $E_R \approx 630$ keV is reproduced by both models, and so are the direct capture (non-resonant) data. In the potential model, we adopt the Woods-Saxon form as the nuclear part of potentials. We use the radius and diffuseness parameters of the real central part of the global optical potential [15]. The strength parameter of the initial channel is adjusted to reproduce the experimental data. In the calculation with the R-matrix method, the resonance energy $E_r$ and width $\Gamma_i$ are taken from Ref. [16]. The $\gamma$ width is assumed to be $2.5 \times 10^{-3}$ eV.

In NACRE, the low-energy astrophysical S-factor is extrapolated by using a polynomial expansion leading to the recommended range of the zero-energy S-factor of $19 \leq S(0) \leq 23$ eV b. The S-factors obtained from our results are 18.9 eV b for the potential model and 22.1 eV b for the R-matrix method, which are approximately inside of the range given by NACRE.

The decomposition of the S-factor for the $^7$Be(p,$\gamma$)$^8$B reaction into the transition multipolarities is shown in Fig. 2. The dashed and dotted curves are the E1 and M1 components of the transition obtained from the R-matrix method. In this reaction, the direct capture results from the E1 component of the transition and the resonance from the M1 component of the transition. The decomposition of the transition obtained from the potential model is similar to that shown in Fig. 2.

The astrophysical S-factor of the $^{20}$Ne(p,$\gamma$)$^{21}$Na reaction is displayed in Fig. 3. The solid curve is the result obtained from the potential model. The non-resonant data can

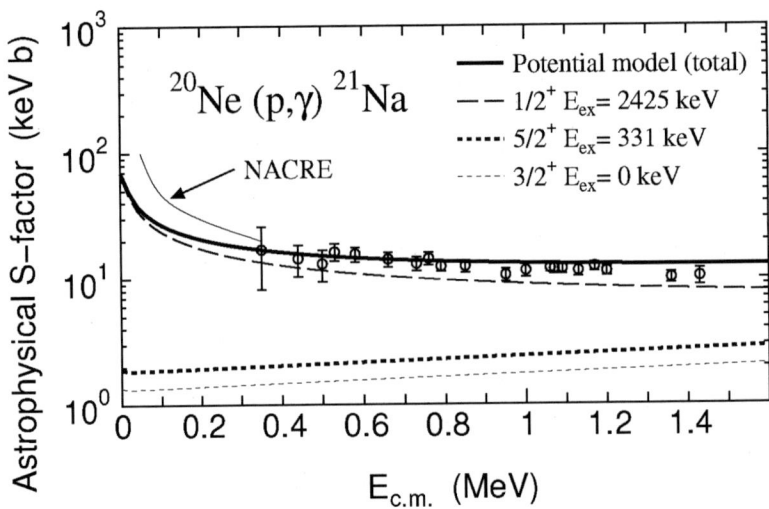

**FIGURE 3.** The astrophysical S-factor of the $^{20}$Ne(p,γ)$^{21}$Na reaction. The solid curve is the result obtained from the potential model. The experimental data have already been compiled in NACRE [2].

be reproduced by the potential model. We have not performed the analysis with the R-matrix method of the $^{20}$Ne(p,γ)$^{21}$Na reaction yet.

In the reproduction of the data, the ground state, 1st excited state (5/2$^+$, $E_{ex}$= 331 keV), and sub-threshold state (1/2$^+$, $E_{ex}$= 2425 keV) of $^{21}$Na are taken into account, i.e., the solid curve is the sum of the transitions to these three final states. The dashed curve is the component of the transition to the sub-threshold state. The dotted and thin dotted curves are the components of the transition to the 1st excited state and to the ground state, respectively. From Fig. 3, we find that the transition to the sub-threshold state is the dominant component in the $^{20}$Ne(p,γ)$^{21}$Na reaction, especially in the low-energy region.

The thin solid curve is the extrapolation of the data estimated in NACRE. Our result obtained from the potential model predicts smaller values of the astrophysical S-factor at low-energies.

## Transfer reactions: $^3$H(d,n)$^4$He and $^3$He(d,p)$^4$He

Figures 4 and 5 show the results of our analyses of the $^3$H(d,n)$^4$He and $^3$He(d,p)$^4$He reactions. The solid and dashed curves are the results obtained from the DWBA and the R-matrix method, respectively. The experimental S-factor data appear to be reproduced by both models. The evaluation shown in Fig. 5 is an example of the extension of NACRE as the $^3$He(d,p)$^4$He reaction was not included in NACRE. In the low-energy region of Fig. 5, one can see some difference between the experimental and theoretical S-factors. The difference may be interpreted in terms of the electron screening effect [17,

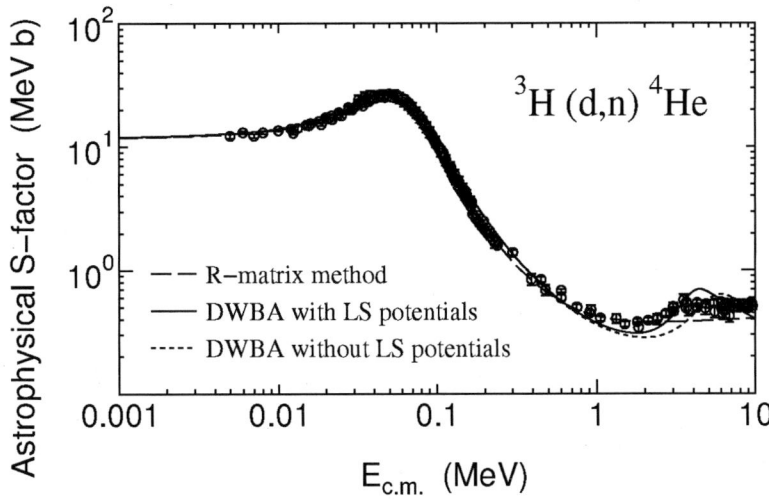

**FIGURE 4.** The astrophysical S-factor of the $^3$H(d,n)$^4$He reaction. The solid and dashed curves are the results obtained from the DWBA and the R-matrix method, respectively. The experimental data have already been compiled in NACRE [2].

25].

To generate the distorted-waves in the DWBA analyses, we adopt the global optical potentials as the nuclear part of potentials, which are defined in the mass and energy ranges of $6 < A < 16$ and $10 < E_p < 50$ MeV for nucleon [15], and of $27 < A < 238$ and $11.8 < E_d < 90$ MeV for deuteron [26]. Although the masses and energies of our current interest are out of those ranges, the parameters expected by extrapolation from the empirical expression of the global potential were adopted as the first trial. However, we could not obtain good reproductions of the experimental data by this choice. Thus, we adjust the strengths of the potentials so as to reproduce the experimental data. In the parameter optimization, we find that the reproductions of the resonance energies in the $^3$H(d,n)$^4$He and $^3$He(d,p)$^4$He reactions are sensitive to the variation of the potential strengths for the initial channel, but not to those for the exit channel.

In order to investigate the effect of the spin-orbit potentials in the distorted waves, we have performed the calculations using the DWBA without the spin-orbit potentials in distortion, the results of which are shown by the dotted curves in Figs. 4 and 5. The deviations between the solid and dotted curves indicate the effects of the spin-orbit potentials in the distorted waves for the initial and exit channels. From the small deviation between the two curves, we figure out that the spin-orbit potentials in distortion do not play a significant role in the present energy range. The small effect of the spin-orbit potentials can be understood since the $^3$H(d,n)$^4$He and $^3$He(d,p)$^4$He reactions in the low-energy region are expected to be dominated by the low partial waves.

In the literature, the spin and parity of the resonances at $E_{c.m.} \approx 64$ keV in $^3$H(d,n)$^4$He and at $E_{c.m.} \approx 250$ keV in $^3$He(d,p)$^4$He are apparently assigned to $J^\pi = 3/2^+$ by the use of the one-level nuclear dispersion formula and R-matrix method [22, 27, 28]. In

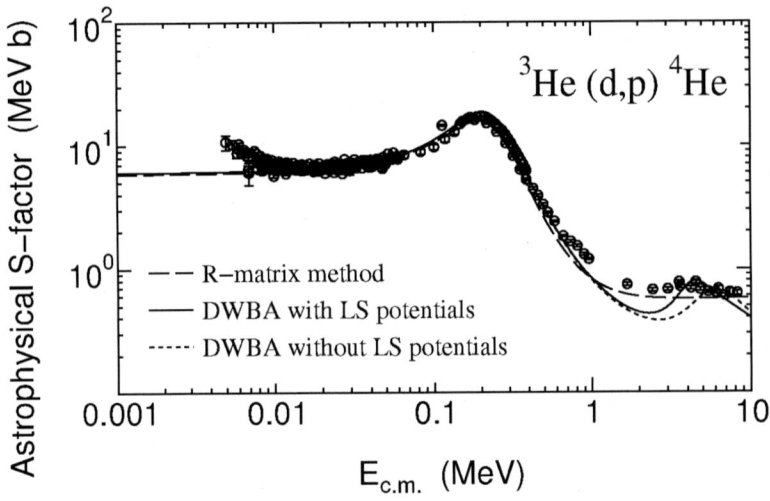

**FIGURE 5.** The astrophysical S-factor of the $^3$He(d,p)$^4$He reaction. The solid and dashed curves are the results obtained from the DWBA and the R-matrix method, respectively. The experimental data are taken from [17, 18, 19, 20, 21, 22, 23, 24].

contrast, our DWBA analyses have led to a different $J^\pi$ assignment, which warrants the necessity for a further scrutiny.

As for the extrapolation of the data to the low-energy in the $^3$H(d,n)$^4$He reaction, the solid and dashed curves of Fig. 4 give almost the same zero-energy S-factor around 11 MeV b, which is concordant with the value adopted by NACRE.

## The p-N and d-N potentials in the potential model and DWBA

To gauge the reliability of our calculations with the potential model and DWBA, we illustrate in Fig. 6 the volume integrals of the potentials used to derive the results shown in the previous two subsections. The open circles are the volume integrals per nucleon pair of the real central part of potentials in the scattering channel $J_R/(A_p A_t)$ for the proton-$^7$Be, proton-$^{20}$Ne, proton-$^4$He, deuteron-$^3$He, and deuteron-$^3$H systems. In comparison, the solid curve represents $J_R/(A_p A_t)$ for the proton-$^7$Be system obtained from the global optical potential. The dotted, dashed and dot-dashed curves are $J_R/(A_p A_t)$ for the proton-$^7$Be, proton-$^4$He, and proton-$^{20}$Ne systems calculated from the extrapolated parameter values of the global optical potential. Our results (open circles) seem to be consistent with the values expected from the higher-energy and neighboring reactions. The proton-$^7$Be system may have the smaller value of $J_R/(A_p A_t)$ than the one expected from the naive extrapolation.

We frequently encounter some ambiguities of the determination of the potential parameters in the low-energy reactions, e.g., small variations in the strength $V$ and radius

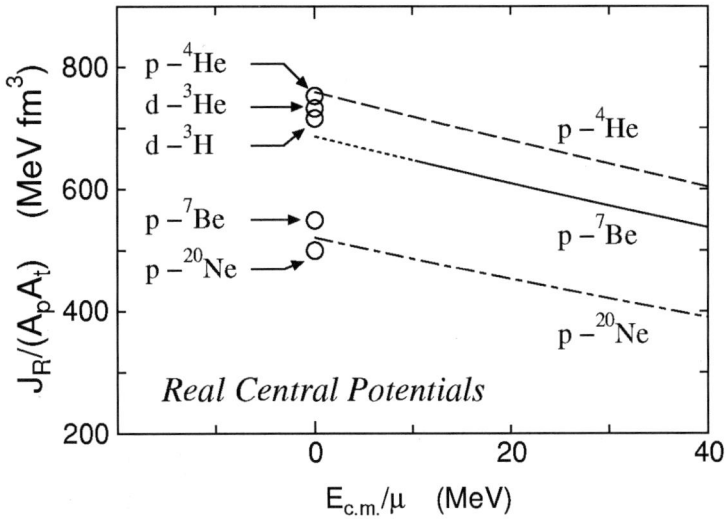

**FIGURE 6.** The volume integrals per nucleon pair of the real central part for the p-N and d-N potentials as a function of center-of-mass energy per reduced mass. The open circles are those for the potentials used in the present analyses. For comparison, the values inferred from the global optical potential are displayed (see text).

$R$ that preserve $VR^n =$ constant ($n \approx 2$) give similarly good reproductions of the data. The discrete large difference of potential depth can also be obtained from the optimized fits. In order to simplify the parameter optimization process, we use the fixed values of the radius and diffuseness parameters expected from the global optical potential in the present analyses. Namely, we adjust only the strengths to reproduce the data. In fact, other parameter sets could be found in the energy range of our current interest. Considering the consistency of the higher-energy and neighboring nuclear reactions, however, we think that the parameter values derived from the procedure mentioned above would be more acceptable.

Another feature of the resulting potentials in the DWBA is that the absorption is weak.

## SUMMARY

We have briefly reported the current status of the data evaluations for nuclear reactions relevant to nuclear astrophysics. As exemplifying cases, the astrophysical S-factors of the $^7$Be(p,$\gamma$)$^8$B, $^{20}$Ne(p,$\gamma$)$^{21}$Na, $^3$H(d,n)$^4$He, and $^3$He(d,p)$^4$He reactions have been analyzed with the potential model, R-matrix method, and DWBA. Through the present work, the astrophysical S-factor data of these reactions turn out to be reproduced well by the models.

The low-energy S-factors of the $^7$Be(p,$\gamma$)$^8$B and $^3$H(d,n)$^4$He reactions extrapolated from the models are approximately inside the range of the recommended values in

NACRE. In contrast, for the $^{20}$Ne(p,$\gamma$)$^{21}$Na reaction, our calculation predicts smaller values of the low-energy S-factor than those in NACRE.

This work has been carried out in the framework of the NACRE update and extension project that started two years ago. We are now in the stage of survey of experimental data and developing the evaluation procedure to make reliable thermonuclear reaction rates. Judging from the examples presented here, we expect that our models would be able to be applied extensively to the evaluations of various thermonuclear reaction data. In the near future, we plan to calculate more systematically the astrophysical S-factor for other reactions relevant to nuclear astrophysics.

## ACKNOWLEDGMENTS

This work has been supported in part by the Interuniversity Attraction Pole IAP 5/07 of the Belgian Federal Science Policy and by the Konan University – Université Libre de Bruxelles convention "Construction of an Extended Nuclear Database for Astrophysics". The author acknowledges the contributions from M. Arnould, K. Takahashi, H. Utsunomiya, K. Arai, and M. Aikawa.

## REFERENCES

1. G. R. Caughlan, and W. A. Fowler, *Atom. Data and Nucl. Data Tables* **40**, 283–334 (1988).
2. C. Angulo, M. Arnould, M. Rayet, P. Descouvemont, D. Baye, C. Leclercq-Willain, A. Coc, S. Barhoumi, P. Aguer, C. Rolfs, R. Kunz, J. W. Hammer, A. Mayer, T. Paradellis, S. Kossionides, C. Chronidou, K. Spyrou, S. Degl'Innocenti, G. Fiorentini, B. Ricci, S. Zavatarelli, C. Providencia, H. Wolters, J. Soares, C. Grama, J. Rahighi, A. Shotter, and M. L. Rachti, *Nucl. Phys. A* **656**, 3–183 (1999).
3. P. Descouvemont, A. Adahchour, C. Angulo, A. Coc, and E. Vangioni-Flam, *Atom. Data and Nucl. Data Tables* **88**, 203–236 (2004).
4. C. Iliadis, J. M. D'Auria, S. Starrfield, W. J. Thompson, and M. Wiescher, *Astrophys. J. Suppl.* **134**, 151–171 (2001).
5. S. O. Nelson, E. A. Wulf, J. H. Kelley, and H. R. Weller, *Nucl. Phys. A* **679**, 199–211 (2000).
6. M. Aikawa, K. Arai, M. Arnould, K. Takahashi, and H. Utsunomiya, *Proceedings of Int. Conf. on Frontiers in Nuclear Structure, Astrophysics and Reactions (FINUSTAR), AIP Conference Proceedings* **831**, 26–30 (2006); M. Aikawa, K. Arai, M. Katsuma, K. Takahashi, M. Arnould, and H. Utsunomiya, *Proceedings of Int. Symp. on Origin of Matter and Evolution of Galaxies 2005 (OMEG05), AIP Conference Proceedings* **847**, 359–361 (2006).
7. P. Descouvemont, *Theoretical Models for Nuclear Astrophysics*, Nova Science Publishers, 2003.
8. G. R. Satchler, *Direct Nuclear Reactions*, International series of monographs on physics, Oxford University Press, 1983.
9. M. Hass, C. Broude, V. Fedoseev, G. Goldring, G. Huber, J. Lettry, V. Mishin, H. J. Ravn, V. Sebastian, L. Weissman, and ISOLDE Collaboration, *Phys. Lett. B* **462**, 237–242 (1999).
10. L. Gialanella, F. Strieder, L. Campajola, A. D'Onofrio, U. Greife, G. Gyurky, G. Imbriani, G. Oliviero, A. Ordine, V. Roca, C. Rolfs, M. Romano, D. Rogalla, C. Sabbarese, E. Somorjai, F. Terrasi, and H. P. Trautvetter, *Eur. Phys. J. A* **7**, 303–305 (2000).
11. F. Hammache, G. Bogaert, P. Aguer, C. Angulo, S. Barhoumi, L. Brillard, J. F. Chemin, G. Claverie, A. Coc, M. Hussonnois, M. Jacotin, J. Kiener, A. Lefebvre, C. Le Naour, S. Ouichaoui, J. N. Scheurer, V. Tatischeff, J. P. Thibaud, and E. Virassamynaïken, *Phys. Rev. Lett.* **86**, 3985–3988 (2001).
12. F. Strieder, L. Gialanella, G. Gyürky, F. Schümann, R. Bonetti, C. Broggini, L. Campajola, P. Corvisiero, H. Costantini, A. D'Onofrio, A. Formicola, Z. Fülöp, G. Gervino, U. Greife,

A. Guglielmetti, C. Gustavino, G. Imbriani, M. Junker, P. G. P. Moroni, A. Ordine, P. Prati, V. Roca, D. Rogalla, C. Rolfs, M. Romano, E. Somorjai, O. Straniero, F. Terrasi, H. P. Trautvetter, and S. Zavatarelli, *Nucl. Phys. A* **696**, 219–230 (2001).
13. L. T. Baby, C. Bordeanu, G. Goldring, M. Hass, L. Weissman, V. N. Fedoseyev, U. Köster, Y. Nir-El, G. Haquin, H. W. Gäggeler, R. Weinreich, and the ISOLDE Collaboration, *Phys. Rev. C* **67**, 065805 (2003).
14. A. R. Junghans, E. C. Mohrmann, K. A. Snover, T. D. Steiger, E. G. Adelberger, J. M. Casandjian, H. E. Swanson, L. Buchmann, S. H. Park, A. Zyuzin, and A. M. Laird, *Phys. Rev. C* **68**, 065803 (2003).
15. B. A. Watson, P. P. Singh, and R. E. Segel, *Phys. Rev.* **182**, 977–989 (1969).
16. R. B. Firestone, V. S. Shirley, C. M. Baglin, S. Y. F. Chu, and J. Zipkin, *Table of Isotopes 8th Edition*, Lawrence Berkeley National Laboratory, University of California, John Wiley & Sons, Inc., New York, 1996.
17. M. Aliotta, F. Raiola, G. Gyürky, A. Formicola, R. Bonetti, C. Broggini, L. Campajola, P. Corvisiero, H. Costantini, A. D'Onofrio, Z. Fülöp, G. Gervino, L. Gialanella, A. Guglielmetti, C. Gustavino, G. Imbriani, M. Junker, P. G. Moroni, A. Ordine, P. Prati, V. Roca, D. Rogalla, C. Rolfs, M. Romano, F. Schümann, E. Somorjai, O. Straniero, F. Strieder, F. Terrasi, H. P. Trautvetter, and S. Zavatarelli, *Nucl. Phys. A* **690**, 790–800 (2001).
18. W. R. Arnold, J. A. Phillips, G. A. Sawyer, E. J. Stovall, Jr., and J. L. Tuck, *Phys. Rev.* **93**, 483–497 (1954).
19. T. W. Bonner, J. P. Conner, and A. B. Lillie, *Phys. Rev.* **88**, 473–476 (1952).
20. A. Krauss, H. W. Becker, H. P. Trautvetter, C. Rolfs, and K. Brand, *Nucl. Phys. A* **465**, 150–172 (1987).
21. L. Stewart, J. E. Brolley, Jr., and L. Rosen, *Phys. Rev.* **119**, 1649–1653 (1960).
22. W. H. Geist, C. R. Brune, H. J. Karwowski, E. J. Ludwig, K. D. Veal, and G. M. Hale, *Phys. Rev. C* **60**, 054003 (1999).
23. W. Grüebler, V. König, A. Ruh, P. A. Schmelzbach, R. E. White, and P. Marmier, *Nucl. Phys. A* **176**, 631–644 (1971).
24. S. Engstler, A. Krauss, K. Neldner, C. Rolfs, U. Schröder, and K. Langanke, *Phys. Lett. B* **202**, 179–184 (1988).
25. M. La Cognata, C. Spitaleri, A. Tumino, S. Typel, S. Cherubini, L. Lamia, A. Musumarra, R. G. Pizzone, A. Rinollo, C. Rolfs, S. Romano, D. Schürmann, and F. Strieder, *Phys. Rev. C* **72**, 065802 (2005).
26. W. W. Daehnick, J. D. Childs, and Z. Vrcelj, *Phys. Rev. C* **21**, 2253–2274 (1980).
27. H. V. Argo, R. F. Taschek, H. M. Agnew, A. Hemmendinger, and W. T. Leland, *Phys. Rev.* **87**, 612–618 (1952).
28. N. Jarmie, R. E. Brown, and R. A. Hardekopf, *Phys. Rev. C* **29**, 2031–2046 (1984).

# The Effects of Changes in Reaction Rates on Simulations of Nova Explosions

S. Starrfield[*], C. Iliadis[†], W. R. Hix[**], F. X. Timmes[‡] and W. M. Sparks[§]

[*]*School of Earth and Space Exploration, Arizona State University, Tempe, AZ 85287-1404:sumner.starrfield@asu.edu*
[†]*Department of Physics and Astronomy, University of North Carolina, Chapel Hill, NC27599-3255:iliadis@unc.edu*
[**]*Physics Division, Oak Ridge National Laboratory, Oak Ridge, TN 37831-6354:raph@ornl.gov*
[‡]*X-2, Los Alamos National Laboratory, Los Alamos, NM, 87545:fxt44@mac.com*
[§]*Science Applications International Corporation, San Diego CA, 92121 & X-4, Los Alamos National Laboratory, Los Alamos, NM, 87545:wms@lanl.gov*

**Abstract.** Classical novae participate in the cycle of Galactic chemical evolution in which grains and metal enriched gas in their ejecta, supplementing those of supernovae, AGB stars, and Wolf-Rayet stars, are a source of heavy elements for the ISM. Once in the diffuse gas, this material is mixed with the existing gases and then incorporated into young stars and planetary systems during star formation. Infrared observations have confirmed the presence of carbon, SiC, hydrocarbons, and oxygen-rich silicate grains in nova ejecta, suggesting that some fraction of the pre-solar grains identified in meteoritic material come from novae. The mean mass returned by a nova outburst to the ISM probably exceeds $\sim 2 \times 10^{-4}$ $M_\odot$. Using the observed nova rate of $35\pm11$ per year in our Galaxy, it follows that novae introduce more than $\sim 7 \times 10^{-3}$ $M_\odot$ yr$^{-1}$ of processed matter into the ISM. Novae are expected to be the major source of $^{15}$N and $^{17}$O in the Galaxy and to contribute to the abundances of other isotopes in this atomic mass range. Here, we report on how changes in the nuclear reaction rates affect the properties of the outburst and alter the predictions of the contributions of novae to Galactic chemical evolution.

**Keywords:** Nuclear Astrophysics; Nucleosynthesis; Nuclear Reaction Rates; Classical Novae; Cataclysmic Variables
**PACS:** 97.30.Qt;97.80.Gm;26.30.+k;26.50.+x

## 1. INTRODUCTION

The observable consequences of accretion onto white dwarfs (WDs) include the Classical (CN), Symbiotic, and Recurrent Nova (RN) outbursts, and the possible evolution of the Super Soft, Close Binary, X-ray Sources (SSS) to Type Ia Supernova (SN Ia) explosions. This diversity of phenomena occurs because of differences in the properties of the secondary star, the mass of the WD, and the stage of evolution of the binary system (the orbital period, the luminosity of the WD and the rate of mass accretion onto the WD). A CN explosion occurs in the accreted hydrogen-rich envelope on the low-luminosity WD component of a Cataclysmic Variable (CV) system. One dimensional (1D) hydrodynamic studies, which follow the evolution of the material falling onto the WD from a bare core to the explosion, show that the envelope grows in mass until it reaches a temperature and density at its base that is sufficiently high for ignition of the hydrogen-rich fuel to occur. Both observations of the chemical abundances in CN ejecta and theoretical studies of the consequences of the thermonuclear runaway (TNR) in the WD envelope strongly imply that mixing of the accreted matter with core matter occurs at some time during the evolution to the peak of the explosion. How and when the mixing occurs is not yet known (see, e. g., Starrfield, Iliadis, and Hix 2006:S06 for a discussion).

If the bottom of the accreted layer is sufficiently degenerate and well mixed with the core, then a TNR occurs and explosively ejects core plus accreted material in a fast CN outburst. The evolution of nuclear burning on the WD, and the total amount of mass that it accretes and ejects depends upon: the mass and luminosity of the underlying WD, the rate of mass accretion onto the WD, the chemical composition in the reacting layers (which includes the metallicity of the CV system), the mixing history of the envelope, and the outburst history of the system.

The high levels of enrichment of novae ejecta in elements ranging from carbon to sulfur confirm that there is significant dredge-up of matter from the core of the underlying WD and enable novae to contribute to the chemical enrichment of the interstellar medium (Gehrz et al. 1998: G98). Observations of the epoch of dust formation in the expanding shells of novae allow important constraints to be placed on the dust formation process and confirm that graphite, SiC, and $SiO_2$ grains are formed by the outburst (G98 and references therein). It is possible that grains from novae were injected into the pre-solar nebula and can be identified with some of the pre-solar grains or "stardust" found in meteorites (Zinner 1998, Amari et al. 2001, José et al. 2004). Finally, $\gamma$-ray observations during the first several years of their outburst, done with the next generation of satellite observatories, could confirm the presence of decays from $^7$Be and $^{22}$Na (Weiss and Truran 1990; Nofar et al. 1991; Jean et al. 2000, and references therein). In the next section we report on NOVA our one-dimensional (1D) hydrodynamic computer code that we have used for the new calculations done with the reaction rate library of Iliadis (2005, private communication). We follow that with a discussion of our evolutionary results and the implications of the new rates for the nova outburst. We end with a summary.

## 2. THE HYDRODYNAMIC COMPUTER CODE AND NUCLEAR REACTION RATE LIBRARIES

NOVA is a spherically symmetric, fully implicit, Lagrangian, hydrodynamic computer code that incorporates a large nuclear reaction rate network. It is described in detail in Starrfield et al. (1998: S98), Starrfield et al. (2000:S00; and references therein). As reported in those papers, we have found that improving the opacities, equations-of-state, and the nuclear reaction rates have had important effects on both the energetics and the nucleosynthesis. Similar results have been found in the calculations of the Barcelona group as reported elsewhere (Hernanz and Josè 2000, and references therein). Therefore, over the past few years we continued to improve the physics in NOVA and then determined the effects of the improved physics on simulations of the CN outburst (S06, and references therein). A major effort has been the effects of improving the reaction rates used in the calculations on the evolution of the CN outburst and the resulting nucleosynthesis. In this paper we compare our earlier studies to a recent reaction rate library of Iliadis (current as of August 2005). Since NOVA is always being updated and improved, for the work to be reported on in this paper we have made one major change and numerous minor changes.

The major change is that we no longer use the nuclear reaction network of Weiss and Truran (1990: WT90) but have switched to the modern nuclear reaction network of Hix and Thielemann (1999: HT99). While both networks utilize reaction rates in the common REACLIB format and perform their temporal integration using the Backward Euler method introduced by Arnett and Truran (1969: AT69), there are two important differences. First, WT90 implement a single iteration, semi-implicit backward Euler scheme, which has the advantage of a relatively small and predictable number of matrix solutions, but allows only heuristic checks that the chosen time step results in a stable or accurate solution. HT99 implement the iterative, fully implicit backward Euler scheme, repeating the Backward Euler step until convergence is achieved, providing a measure of the stability and accuracy. If convergence does not occur within a reasonable number of iterations, the time step is subdivided into smaller intervals until a converged solution is achieved. This allows the fully implicit backward Euler integration to respond to instability or inaccuracy in a way that is impossible with the semi-implicit backward Euler approach. As a result, the fully iterative approach can often safely employ larger time steps than the semi-implicit approach, obviating the speed advantage of the semi-implicit method's smaller number of matrix solutions per integration step.

Second, the HT99 network employs automated linking of reactions in the data set to the species being evolved. This is in contrast to the manual linking employed by WT90 and many older reaction networks. This automated linking helps to avoid implementation mistakes, as we discovered while performing tests of NOVA in order to understand the source of differences in the results of the simulations between two versions of the code which used the same reaction rate library but different nuclear reaction networks. In these tests, we discovered that while the REACLIB dataset used in our prior studies (S98, S00), included the *pep* reaction ($p + e^- + p \rightarrow d + v$), it was not linked to abundance changes in the WT90 network. While for Solar modeling energy generation from the *pep* reaction is unimportant (but not the neutrino losses), in the WD envelope the density can reach to values of $10^4$ gm cm$^{-3}$ which is about two orders of magnitude larger than in the core of the Sun. The increased density increases the rate of energy generation by about 40% over calculations with the *pep* reaction absent. The increased energy generation then has the effect of reducing the amount of accreted material since the temperature rises faster per gram of accreted material. (The effect of changes in the rate of energy generation on simulations of the CN outburst is discussed in detail in S98.) Given a smaller amount of accreted material at the time when the steep temperature rise begins in the TNR, the nuclear burning region is less degenerate and, therefore, the peak temperatures are lower for models evolved with the same nuclear reaction rate library used in our previous studies (see below).

Finally, we use the analytic fitting formulas of Itoh et al. (1996) for the neutrino energy loss rates from pair ($e^+ + e^- \rightarrow v_e + \bar{v}_e$), photo ($e^\pm + \gamma \rightarrow e^\pm + v + \bar{v}_e$), plasma ($\gamma_{plasmon} \rightarrow v_e + \bar{v}_e$), bremsstrahlung ($e^- + A^Z \rightarrow e^- + A^Z + v_e + \bar{v}_e$), and recombination ($e^-_{continuum} \rightarrow e^-_{bound} + v_e + \bar{v}_e$) processes. As stellar evolution codes generally require derivative information for the Jacobian matrix, our implementation of the Itoh et al. (1996) fitting formulas (available from http://www.cococubed.com) returns the neutrino loss rate and its first derivatives with respect to temperature, density, $\bar{A}$ (average atomic weight) and $\bar{Z}$ (average charge).

## 3. EVOLUTIONARY SEQUENCES USING FOUR NUCLEAR REACTION LIBRARIES

### 3.1. The Initial Models and Libraries

Our calculations were done with 95 zone, $1.35M_\odot$ complete WDs. As in S98 and S00, we assume that the material being accreted from the donor star is of Solar composition but that it has already mixed with the core material so that the actual accreting composition chosen for this study is 50% Solar and 50% ONeMg material. We assume a value of 2 for the mixing-length to scale height ratio ($l/H_p$). All other details of our calculations (opacities, equations of state, etc.) are described in S98 and S00.

We evolved four different sequences using a different reaction rate library for each sequence but the same nuclear reaction network (HT99). The reaction rate library used in Politano et al. (1995) included the rates from Caughlan and Fowler (1988) and Thielemann et al. (1987, 1988). They were compiled by Thielemann and made available to Truran and Starrfield and also used for the calculations reported in WT90 (P1995 in both plots and tables). S98 used an updated reaction rate library which contained new rates calculated, measured, and compiled by Thielemann and Wiescher (labeled S98 in both plots and tables). A discussion of the improvements is provided in S98. The third library is described in Iliadis et al. (2001) and was used for the simulations in Starrfield et al. (2001). It is labeled I2001. The last library used in "This Work" is the August 2005 compilation of Iliadis. This library is an updated version of the library described in Iliadis et al. (2001) and used in Starrfield et al. (2001). A detailed description of this library will appear in Starrfield et al. (2006, in prep.). There is one additional calculation in Table 1. A comparison calculation done with the Politano et al. (1995) reaction rate library and the WT90 nuclear reaction network. As noted above, this network does not include the *pep* reaction and we provide it here only for comparison. We have recalculated the simulation for this paper using the same equations of state and opacities as used for the other calculations.

### 3.2. The Evolutionary Results

The initial properties of the WD are provided in the table comments. We evolved five evolutionary sequences. In all cases, we assumed an initial WD luminosity of $\sim 4 \times 10^{-3} L_\odot$ and a mass accretion rate of $10^{16}$ gm s$^{-1}$ ($1.6 \times 10^{-10} M_\odot$ yr$^{-1}$). This mass accretion rate is 5 times lower than the lowest rate used in either S98 or S00 and was chosen to maximize the amount of accreted material given the increased energy generation from the *pep* reaction. Numerous studies of accretion onto WDs by many different authors demonstrate that the results of the evolution depend strongly on the initial WD luminosity and mass accretion rate (c.f., Yaron et al. 2005, and references

**TABLE 1.** Initial Parameters and Evolutionary Results

| Reaction Library:* | P1995[†] | P1995** | S1998[‡] | I2001[§] | This Work[¶] |
|---|---|---|---|---|---|
| $\tau_{acc}(10^5 \text{ yr})$ | 2.5 | 2.1 | 2.1 | 2.1 | 1.8 |
| $M_{acc}(10^{-5} M_\odot)$ | 3.9 | 3.3 | 3.3 | 3.3 | 2.8 |
| $T_{peak}(10^6 K)$ | 459 | 413 | 414 | 407 | 392 |
| $\varepsilon_{nuc-peak}(10^{17} \text{erg gm}^{-1} \text{s}^{-1})$ | 22.8 | 8.4 | 8.6 | 4.9 | 4.4 |
| $L_{peak}(10^5 L_\odot)$ | 8.0 | 9.6 | 8.0 | 7.3 | 5.9 |
| $T_{eff-peak}(10^5 K)$ | 20 | 13 | 13 | 8.8 | 8.8 |
| $M_{ej}(10^{-5} M_\odot)$ | 3.3 | 2.3 | 2.3 | 2.3 | 1.7 |
| $V_{max}(\text{km s}^{-1})$ | 6050 | 5239 | 4755 | 4787 | 4513 |

* The initial model for all 5 evolutionary sequences had $M_{WD}$=1.35$M_\odot$, $L_{WD}$=4.2 × $10^{-3} L_\odot$, $T_{eff}$=2.5 × $10^4$K, $R_{WD}$=2495 km, and a central temperature of 1.2 × $10^7$K
[†] Library used in Politano et al. (1995): pep reaction not included(WT90 network)
** Library used in Politano et al. (1995): pep reaction included (HT99 network)
[‡] Library used in Starrfield et al. (1998): pep reaction included (HT99 network)
[§] Library described in Iliadis et al. (2001): pep reaction included (HT99 network)
[¶] Iliadis (2005: this work) library: pep reaction included (HT99 network)

therein).

We use the same composition for the accreting material as in Politano et al. (1995; see also: S98; S00; and Starrfield et al. 2001: a mixture of half-solar and half-ONeMg by mass fraction). By using this composition, we assume that core material has mixed with accreted material from the beginning of the evolution. This composition also effects the amount of accreted mass at the peak of the TNR since it has a higher opacity than if no mixing were assumed. The results of our evolutionary calculations are given in Table 1 and 2. Table 1 gives the initial parameters and evolutionary results and Table 2 gives the abundances of the ejected material (by mass) for the 4 different simulations. We do not report the abundance results for the sequence done without the *pep* reaction since this calculation is not realistic. The evolutionary parameters are provided only to demonstrate the effects of including this reaction on the evolution.

The rows in Table 1 are the reaction rate library, the accretion time to the TNR($\tau_{acc}$), the accreted mass($M_{acc}$), peak temperature in the TNR ($T_{peak}$), peak rate of energy generation during the TNR, ($\varepsilon_{nuc-peak}$), peak luminosity ($L_{peak}$), peak effective temperature ($T_{eff-peak}$), ejected mass ($M_{ej}$), and the peak expansion velocity after the radii of the surface layers have reached ~ $10^{13}$cm ($V_{max}$). By this time the outer layers are optically thin, have far exceeded the escape velocity, and there is no doubt that they are escaping.

As noted above, we provide two different columns for Politano et al. (1995). The first, with the superscript †, is taken from a calculation done for this paper using the Politano et al. library and the WT90 network in NOVA. The second, with the superscript **, uses the same reaction rate library as Politano et al. but the energy generation and nucleosynthesis is obtained with the HT99 network. These two columns, therefore, show the effects of including the *pep* reaction on the TNR simulations.

Table 1 shows that the largest change in the results of the evolution occurs with the inclusion of the *pep* reaction. If we compare P1995[†] and P1995**, then the ~40% increase in energy production from just adding the *pep* reaction to the network results

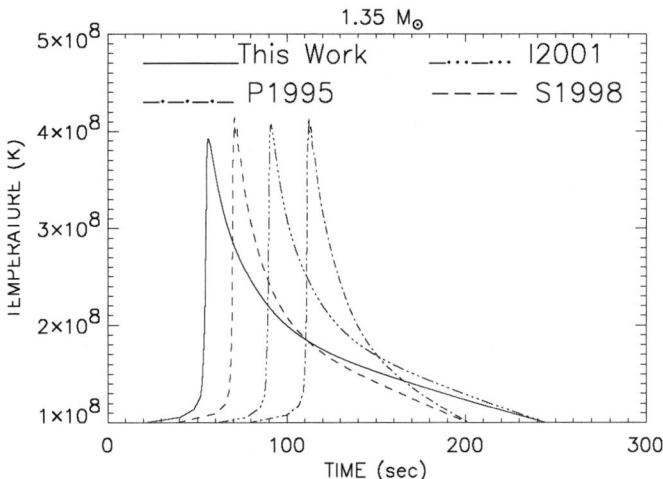

**FIGURE 1.** The variation with time of the temperature in the deepest hydrogen-rich zone around the time when peak temperature occurs. We have plotted the results for four different simulations on a $1.35 M_\odot$ WD. The identification with calculations done with a specific library is given on the plot. In this plot and all following plots, S1998 refers to Starrfield et al. (1998:S98), P1995 refers to Politano et al. (1995), I2001 refers to Iliadis et al. (2001), and This Work refers to this paper. The details of the associated reaction rate library are given in the text. The curve for each sequence has been shifted slightly in time to improve its visibility.

in a decrease of ∼19% in both accretion time and accreted mass. The large change must be caused by the addition of the *pep* reaction since the two reaction rate libraries are otherwise identical. Because there is less accreted mass on the WD at the time of the TNR (comparing the sequences done with the *pep* reaction included to the one without it), the peak temperatures do not reach to as high values as in the sequence with the *pep* reaction absent. If we compare the accretion time and accreted mass for the four sequences with the *pep* reaction included, we see that changes in the nuclear reaction rate library have a sizable effect. In addition, there have been some small changes to the reaction rates in the proton-proton chain and that is where the sequences spend most of their time during the accretion phase.

Figure 1 shows the variation of temperature with time for the deepest hydrogen-rich zone and we plot only the simulations done with the *pep* reaction included. The specific evolutionary sequence is identified on the plot. The reference to the reaction network used for that calculation is given in the caption for Figure 1. The time coordinate is arbitrary and chosen to clearly show each curve. This figure shows that there are important differences between the four simulations. Peak temperature drops from about 413 million degrees to 392 million degrees and peak nuclear energy generation drops by about a factor of 2.5 from the oldest library to the newest library ($8.4 \times 10^{17}$erg gm$^{-1}$s$^{-1}$ to $4.4 \times 10^{17}$erg gm$^{-1}$s$^{-1}$). The temperature declines more rapidly for the sequence computed with the oldest reaction library (Politano et al. 1995) because it exhibited a larger release of nuclear energy throughout the evolution which caused the overlying

**TABLE 2.** Comparison of the Ejecta Abundances for 1.35$M_\odot$ White Dwarfs (All abundances are mass fraction)

| Reaction Library: | P1995* | S1998† | I2001** | This Work‡ |
|---|---|---|---|---|
| H | 0.27 | 0.27 | 0.27 | 0.28 |
| $^4$He | 0.18 | 0.18 | 0.17 | 0.17 |
| $^{12}$C | $8.0 \times 10^{-3}$ | $1.2 \times 10^{-2}$ | $8.0 \times 10^{-3}$ | $6.2 \times 10^{-3}$ |
| $^{13}$C | $2.8 \times 10^{-3}$ | $4.0 \times 10^{-3}$ | $2.4 \times 10^{-3}$ | $2.4 \times 10^{-3}$ |
| $^{14}$N | $4.3 \times 10^{-3}$ | $4.8 \times 10^{-3}$ | $4.3 \times 10^{-3}$ | $8.4 \times 10^{-3}$ |
| $^{15}$N | 0.11 | 0.11 | $6.8 \times 10^{-2}$ | $6.0 \times 10^{-2}$ |
| $^{16}$O | $1.2 \times 10^{-3}$ | $1.1 \times 10^{-3}$ | $2.4 \times 10^{-3}$ | $2.4 \times 10^{-3}$ |
| $^{17}$O | $1.1 \times 10^{-3}$ | $1.0 \times 10^{-3}$ | $5.9 \times 10^{-2}$ | $6.7 \times 10^{-2}$ |
| $^{18}$O | $7.8 \times 10^{-3}$ | $6.7 \times 10^{-3}$ | $3.0 \times 10^{-3}$ | $1.5 \times 10^{-3}$ |
| $^{18}$F | $2.5 \times 10^{-3}$ | $2.3 \times 10^{-3}$ | $9.3 \times 10^{-4}$ | $5.9 \times 10^{-4}$ |
| $^{22}$Na | $3.5 \times 10^{-2}$ | $5.1 \times 10^{-2}$ | $3.0 \times 10^{-2}$ | $2.3 \times 10^{-2}$ |
| $^{24}$Mg | $2.8 \times 10^{-3}$ | $1.9 \times 10^{-3}$ | $2.1 \times 10^{-3}$ | $1.9 \times 10^{-3}$ |
| $^{26}$Mg | $1.6 \times 10^{-2}$ | $1.2 \times 10^{-2}$ | $2.2 \times 10^{-3}$ | $1.5 \times 10^{-3}$ |
| $^{26}$Al | $2.8 \times 10^{-3}$ | $2.1 \times 10^{-3}$ | $2.7 \times 10^{-3}$ | $3.0 \times 10^{-3}$ |
| $^{27}$Al | $2.8 \times 10^{-2}$ | $3.4 \times 10^{-2}$ | $1.4 \times 10^{-2}$ | $1.4 \times 10^{-2}$ |
| $^{28}$Si | $2.3 \times 10^{-2}$ | $3.5 \times 10^{-2}$ | $2.7 \times 10^{-2}$ | $2.9 \times 10^{-2}$ |
| $^{29}$Si | $6.3 \times 10^{-3}$ | $7.0 \times 10^{-3}$ | $1.9 \times 10^{-2}$ | $1.8 \times 10^{-2}$ |
| $^{30}$Si | $2.3 \times 10^{-2}$ | $2.4 \times 10^{-2}$ | $3.1 \times 10^{-2}$ | $3.8 \times 10^{-2}$ |
| $^{31}$P | $3.0 \times 10^{-2}$ | $3.0 \times 10^{-2}$ | $4.3 \times 10^{-2}$ | $3.7 \times 10^{-2}$ |
| $^{32}$S | $2.1 \times 10^{-2}$ | $2.8 \times 10^{-2}$ | $3.9 \times 10^{-2}$ | $4.0 \times 10^{-2}$ |
| $^{34}$S | $1.5 \times 10^{-3}$ | $1.3 \times 10^{-3}$ | $1.3 \times 10^{-3}$ | $7.2 \times 10^{-4}$ |
| $^{36}$Ar | $6.1 \times 10^{-4}$ | $2.2 \times 10^{-4}$ | $1.5 \times 10^{-4}$ | $6.8 \times 10^{-5}$ |
| $^{40}$Ca | $2.1 \times 10^{-5}$ | $2.4 \times 10^{-5}$ | $2.4 \times 10^{-5}$ | $1.8 \times 10^{-5}$ |

\* Library used in Politano et al. (1995): pep reaction included (HT99 network)
† Library used in Starrfield et al. (1998): pep reaction included (HT99 network)
\*\* Library described in Iliadis et al. (2001): pep reaction included (HT99 network)
‡ Iliadis (2005: this work) library: pep reaction included (HT99 network)

zones to expand more rapidly and the nuclear burning layers to cool more rapidly. In contrast, the newest library, with the smallest expansion velocities, cools slowly.

The differences in total nuclear energy generation ($L/L_\odot$) as a function of time for each mass is shown in Figure 2. The time coordinate is consistent with that used in Figure 1. Note that peak nuclear energy production for the latest library is definitely lower than seen in the earlier libraries. The changes in the libraries are more important for the more massive isotopes and become more important as higher temperatures are reached.

The abundance predictions, for the ejected material, from our four evolutionary sequences are given as mass fraction in Table 2. Here we discuss only the most important nuclei. The increase in the $^4$He abundance is small compared to the observed helium abundances in CN ejecta which can reach, if not exceed, 0.5 (G98). The large helium abundance, in combination with the large observed CNO abundances, is strong evidence for mixing of the accreted material with layers in the WD underlying the accreting material at some time during the outburst. The large helium abundances observed in RN such as U Sco or V394 CrA suggest that mixing has also occurred in these systems even

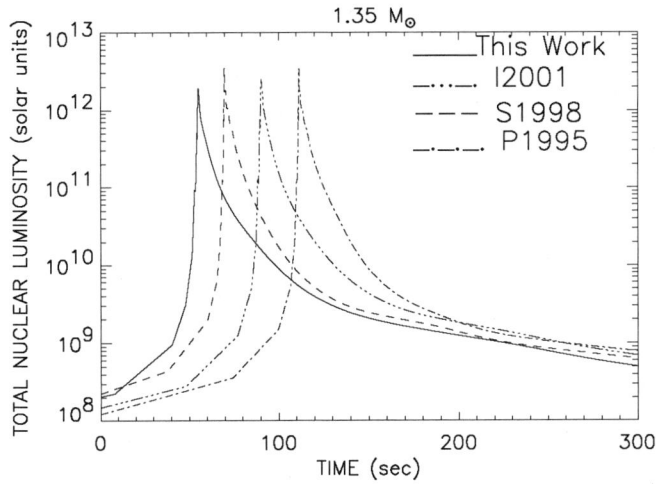

**FIGURE 2.** The variation with time of the total nuclear luminosity (erg s$^{-1}$) in solar units (L$_\odot$) around the time of peak temperature during the TNR on a 1.35M$_\odot$ WD. We integrated over all zones taking part in the explosion. The identification with each library is given on the plot. The time coordinate is chosen to improve visibility.

if their total CNO abundances are not dramatically enriched over solar. Examining the behavior of the individual abundances, we see that $^{12}$C and $^{13}$C are virtually unchanged by the updated reaction rates. In contrast, the abundance of $^{14}$N nearly doubles and that of $^{15}$N decreases by a factor of two going from the first to the latest reaction rate library. $^{16}$O also doubles in abundance while $^{17}$O grows by a factor of 60 and becomes the most abundant of the CNO nuclei in the ejecta. For this WD mass, the C/O ratio is 0.12.

The abundance of $^{22}$Na decreases with the library update and $^{24}$Mg is severely depleted by the TNR. $^{26}$Al is unchanged by the changes in the reaction rates while the abundance of $^{27}$Al drops by a factor of two. This result implies that TNRs on more massive WDs eject about the same fraction of $^{26}$Al as $^{27}$Al. We also find that contrary to Politano et al. (1995) that the amount of $^{26}$Al ejected is virtually independent of WD mass. All the Si isotopes ($^{28}$Si, $^{29}$Si, and $^{30}$Si) are enriched in the calculations done with the latest library. Finally, while the ejecta abundances of $^{36}$Ar, and $^{40}$Ca have declined as the reaction rate library has been improved, they are all produced in the nova TNR since their final abundances exceed the initial abundances.

## 4. SUMMARY

In this paper we examined the consequences of improving the nuclear reaction library on our simulations of TNRs on 1.35M$_\odot$ WDs. We have found that the changes in the rates have affected the nucleosynthesis predictions of our calculations but not, to any great extent, the gross features of the evolution. In addition, we have used a lower mass accretion rate than in our previous studies in order to accrete (and eject) more material. This has, as expected, caused the peak values of some important parameters to increase over our previous studies at the same WD mass. However, because some important reaction rates have declined in the new reaction rate library this has not increased the abundances for nuclei above aluminum and, in fact, they have declined while the abundances of both $^{26}$Al and $^{27}$Al have increased. In contrast, the abundance of $^{22}$Na has declined from the values predicted in our earlier work.

## ACKNOWLEDGMENTS

S. Starrfield acknowledges partial support from NSF and NASA grants to ASU. He also thanks J. Aufdenberg and ORNL for generous allotments of computer time. WRH is partly supported by the NSF under contracts PHY-0244783 and and by the DOE, through the Scientific Discovery through Advanced Computing Program. ORNL is managed by UT-Battelle, LLC, for the U.S. DOE under contract DE-AC05-00OR22725.

## REFERENCES

1. Amari, S., Gao, X. Nittler, L. R., Zinner, E., José, J. Hernanz, M. Lewis, R. 2001, ApJ, 551, 1065
2. Arnett, W. & Truran, J. 1969, ApJ, 157, 339
3. Caughlan, G. R., & Fowler, W.A., 1988, At. Data Nucl. Data Tables, 40, 283
4. Gehrz, R.D., Truran, J.W., Williams, R.E., & Starrfield, S. 1998, PASP, 110, 3
5. Hix, W. R. & Thielemann, F.-K. 1999, ApJ, 511, 862
6. Hernanz, M., & José, J. 2000, in Cosmic Explosions, Ed. S. S. Holt & W. W. Zhang, AIP Conference Proceedings # 522, p. 339
7. Iliadis, C., D'Auria, J.M., Starrfield, S., Thompson, W.J., & Wiescher, M. 2001, ApJS, 134, 151
8. Itoh, N. et al. 1996, ApJS, 102, 411
9. Jean, P., Hernanz, M., Gómez-Gomar, J., & José, J. 2000, MNRAS, 319, 350
10. José, J. Hernanz, M., Amari,S., Lodders, K., Zinner, E. 2004, ApJ, 612, 614
11. Nofar, I., Shaviv, G., & Starrfield, S. 1991, ApJ, 369, 440
12. Politano, M., Starrfield, S., Truran, J. W., Weiss, A., & Sparks, W. M. 1995, ApJ, 448, 807
13. Starrfield, S., Iliadis, C., & Hix, W. R. 2006, in Classical Nova II, ed. M. F. Bode & A. Evans, Cambridge, University Press, in press.
14. Starrfield, S., Iliadis, C., Truran, J. W., Wiescher, M., & Sparks, W. M. in Proceedings of the 6th International Conference on Nuclei in the Cosmos 2001, ed. J. Christensen-Dalsgaard & K. H. Langanke, (North-Holland, Elsevier, New York,) Nuclear Physics A, 688, 110c
15. Starrfield, S., Sparks, W. M., Truran, J.W., Wiescher, M.C. 2000, ApJS, 127, 485
16. Starrfield, S., Truran, J.W., Wiescher, M.C., & Sparks, W. M. 1998, MNRAS, 296, 502
17. Thielemann, F.-K., Arnould, M., & Truran, J. W. 1987, in advances in Nuclear Astrophysics, ed. E. Vangioni-Flam, Gif-sur-Yvette, Editions Frontieres, 525
18. Thielemann, F.-K., Arnould, M., & Truran, J. W. 1988, in Capture Gamma-Ray Spectroscopy, ed. K. Abrahams & P. Van assche, Bristol, IOP, 730
19. Weiss, A., & Truran, J. W. 1990, A&A, 238, 178
20. Yaron, O., Prialnik, D., Shara, M. M., Kovetz, A. 2005, ApJ, 623, 398
21. Zinner, E. 1998, Annual Review of Earth and Planetary Sciences, 26, 147

# Understanding nuclear "pasta": current status and future prospects

Gentaro Watanabe

*NORDITA, Blegdamsvej 17, DK-2100 Copenhagen Ø, Denmark*
*The Institute of Chemical and Physical Research (RIKEN), Saitama 351-0198, Japan*

**Abstract.** In cores of supernovae and crusts of neutron stars, nuclei can adopt interesting shapes, such as rods or slabs, etc., which are referred to as nuclear "pasta." In recent years, we have studied the pasta phases focusing on their dynamical aspects with a quantum molecular dynamic (QMD) approach. In this article, we review these works. We also focus on the treatment of the Coulomb interaction.

**Keywords:** pasta phases, neutron star crusts, supernova cores, quantum molecular dynamics
**PACS:** 26.60.+c, 26.50.+x, 97.60.Jd, 97.60.Bw

## I. INTRODUCTION

In ordinary matter on the earth, atomic nuclei are roughly spherical. This fact can be understood in the liquid drop picture of the nucleus as being a result of the surface tension of nuclear matter, which favors a spherical nucleus, being greater than the Coulomb repulsion between protons, which tends to make the nucleus deform. However, in supernova cores and neutron stars, the situation changes completely (see, e.g., Refs. [1, 2]): when the density of matter approaches that of atomic nuclei, i.e., the normal nuclear density $\rho_0$, nuclei are closely packed and the effect of the electrostatic energy becomes comparable to that of the surface energy. Consequently, around a density $\rho \simeq \rho_0/2$, the energetically favorable configuration could be rod-like or slab-like nuclei embedded in the gas phase, or rod-like or spherical bubbles of the gas phase embedded in nuclear matter [3, 4]. Such phases with exotic shapes of nuclei are referred to as nuclear "pasta" phases.

Properties of the pasta phases in equilibrium states have been investigated by many authors so far [5, 6, 7, 8, 9, 10]. These earlier works have confirmed that, for various nuclear models, the nuclear shape changes in the sequence sphere, cylinder, slab, cylindrical hole, spherical hole, uniform with increasing density. This conclusion holds for the case of non-zero temperatures with a constant entropy [6] as well as for the zero temperature case. However, in these works (except for Ref. [5]) they took account of only several specific nuclear shapes and determined the energetically favorable one by calculating the energy for these assumed structures within a liquid drop model or the Thomas-Fermi approximation. Thus the phase diagram at subnuclear densities and the existence of the pasta phases should be examined without assuming the nuclear shape. It is also noted that at typical temperatures of the collapsing cores, several MeV, effects of thermal fluctuations on the nucleon distribution are significant. However, these thermal fluctuations cannot be described properly by mean-field theories such as the Thomas-

Fermi and the Hartree-Fock approximations.

In the processes of supernova explosion and succeeding neutron star formation, the pasta phases can be formed in two stages. During the gravitational collapse, matter in the collapsing core is compressed and the central region starts to solidify into a bcc lattice because the Coulomb repulsion among nuclei gets stronger. In the later stage of the collapse, the central density reaches $\rho_0$ and the pasta phases could be formed from the bcc lattice due to the compression. After a bounce of the core takes place, the temperature of the core increases to $O(10)$ MeV and nuclei melt completely. As the cooling process of the protoneutron star proceeds later on, the pasta phases as well as normal spherical nuclei could be produced again from hot uniform nuclear matter at subnuclear densities.

The previous studies, which are based on a comparison of the energy with an assumption of the nuclear shape, cannnot answer whether or not the pasta phases can be formed dynamically within the time scale of the neutron star cooling nor whether or not the transitions between them, which is accompanied by drastic changes of nuclear shape, can be realized under nonequilibrium conditions in the collapsing cores. To solve these problems, molecular dynamic methods for nucleon many-body systems are suitable. They treat the motion of the nucleonic degrees of freedom and can describe thermal fluctuations and many-body correlations beyond the mean-field level.

Using one of the molecular dynamic methods, the quantum molecular dynamics (QMD) [11], we have arrived at the two following major conclusions [12, 13, 14]. (1) The pasta phases are formed from hot uniform nuclear matter by cooling it down within a time scale of $\sim O(10^3 - 10^4)$ fm/c. This supports the idea that the pasta phases exist in neutron star crusts. (2) The transition from rod-like nuclei to slab-like nuclei and that from slab-like nuclei to rod-like bubbles can be realized by compression of matter. This suggests that the pasta phases could be formed in collapsing cores.

Throughout the present article, we set the Boltzmann constant $k_B = 1$.

## II. METHOD: QUANTUM MOLECULAR DYNAMICS

Among various versions of the molecular dynamic models, QMD [11] is the most practical and suitable for studying the pasta phases. This method is so efficient that we can describe rod-like and slab-like nuclei in terms of nucleon degrees of freedom even though they consist of macroscopic number of nucleons. In addition, we focus on the pasta structure and the configuration of nuclei on a macroscopic scale, where the shell effects [15, 16, 17], which cannot be described by QMD, may be less important (in the case of supernova matter, shell effects are always small). Thus QMD is a good approximation for our present problem.

### 1. Model Hamiltonian

In our studies, we use a nuclear force given by a QMD model Hamiltonian with the medium-equation-of-state parameter set in Ref. [18]. This Hamiltonian contains the momentum-dependent "Pauli potential" to reproduce the effects of the Pauli principle phenomenologically. The Pauli potential generates repulsive forces between two identical particles close together in phase space. Parameters in the Pauli potential are determined by fitting the kinetic energy of the free Fermi gas at zero temperature.

This Hamiltonian reproduces the binding energy of symmetric nuclear matter, 16 MeV per nucleon, at the normal nuclear density $\rho_0 = 0.165$ fm$^{-3}$ and other saturation properties: the incompressibility is set to be 280 MeV and the symmetry energy is 34.6 MeV [18]. This Hamiltonian also well reproduces the properties of stable nuclei relevant for the situations we consider: the binding energy (except for light nuclei from $^{12}$C to $^{20}$Ne) [18], and the rms radius of the ground state of heavy nuclei with $A \gtrsim 100$ [19]. It is also confirmed that another QMD Hamiltonian close to this model provides a good description of nuclear reactions including the low energy region (several MeV per nucleon) [20], which is one of the essential elements when one studies the dynamical processes as in Sect. III-2.

## 2. Coulomb Interaction and its Screening Effect

The Coulomb interaction is one of the essential elements for the pasta phases. One must carefully treat the long-range nature of the Coulomb interaction since electron screening is negligibly small in this system [1, 21, 22]. A key quantity in determining the importance of electron screening is the ratio of the scale of the inhomogeneity, typically half the internuclear spacing $R$, to the Thomas-Fermi screening length $\lambda_{\rm TF} = [4\pi e^2(\partial n_e^{(0)}/\partial \mu_e^{(0)})_{n_e^{(0)}}]^{-1/2} = \sqrt{\pi/4\pi}(k_e^{(0)})^{-1}$, where $n_e^{(0)}$ and $\mu_e^{(0)}$ are the averaged electron number density and chemical potential, respectively, and $k_e^{(0)} = \sqrt{3\pi^2 n_e^{(0)}}$ is the electron Fermi wave vector. We note that, in the density region of the pasta phases, $R/\lambda_{\rm TF} < 1$ [23, 1], which means that it is a good approximation to assume that the electron density distribution is uniform and neglect the screening effect; but one must take account of the long-range nature of the Coulomb force by, e.g., the Ewald summation procedure [12].

It is fortunate that the screening effect is small; otherwise one has to solve equations of motion for the electronic degrees of freedom together with that for the nucleonic ones because the coupling between these two components cannot be neglected. Thus introducing a screening length shorter than the actual value for calculating the Coulomb force between protons keeping the density of electrons uniform is inconsistent. Let us restrict ourselves to the equilibrium state and discuss this point more quantitatively using a liquid drop model in the Wigner-Seitz approximation. The energy density of a relativistic degenerate electron gas with an inhomogeneous density distribution $n_e(\mathbf{r}) = n_e^{(0)} + \delta n_e(\mathbf{r})$ is given by $E_e[n_e(\mathbf{r})] \equiv E_e^{(0)}(n_e^{(0)}) + \delta E_e \simeq E_e^{(0)}(n_e^{(0)})[1 + (2/9V_c)\int_{\rm cell} d^3r\, (\delta n_e(\mathbf{r})/n_e^{(0)})^2]$, with $E_e^{(0)}(n_e^{(0)}) = (3/4)\hbar c k_e^{(0)} n_e^{(0)}$, where $V_c$ is the volume of the Wigner-Seitz cell. If the screening effect is significant, the Coulomb energy among protons decreases, but on the other hand, the $(\delta n_e(\mathbf{r})/n_e^{(0)})^2$ term cannot be neglected. Calculations assuming a small $\lambda_{\rm TF}$ and uniform $n_e$ artificially underestimate the Coulomb energy among protons but discard the significant contribution of $\delta E_e$. Now we estimate $\delta E_e$ for an unrealistically small value of $\lambda_{\rm TF}$. We take the phase with spherical nuclei in the density region where the pasta phases start to appear, i.e., the volume fraction of nuclei is $u = (r_N/r_c)^3 \simeq 1/8$ according to the condition of the fission instability of spherical nuclei [1]. Here $r_N$ and $r_c$ are the radii of nuclei and the Wigner-Seitz cell, respectively. Taking supernova matter for simplicity, in which the density of dripped neutrons is negligible, with the nucleon density in nuclei $n = 0.1$ fm$^{-3}$, proton fraction in nuclei $x_N = 0.3$, $r_c = 15$ fm, and $r_N = 15/2$ fm, and using expressions in Ref. [21], $\delta E_e$ per nucleon, $\delta E_e/n_N$, is estimated as $\delta E_e/n_N \simeq 0.23$ MeV [1] for $\lambda_{TF} = 10$ fm (this value is adopted in Ref. [24]). Note that this is much larger than the energy difference between different pasta phases in the same density region in supernova matter, $O(10)$ keV per nucleon.

## III. SIMULATIONS AND RESULTS

In the present section, let us review our previous works [12, 13, 14]. In our QMD simulations, we treated a system which consists of neutrons, protons, and electrons in a cubic box with periodic boundary conditions. The system is not magnetically polarized, i.e., it contains equal numbers of protons (and neutrons) with spin up and spin down. The relativistic degenerate electrons which ensure charge neutrality are regarded as a uniform background [1, 21, 22]. The Coulomb interaction is calculated by the Ewald method taking account of the Gaussian charge distribution of the proton wave packets.

### 1. Realization of the Pasta Phases and Equilibrium Phase Diagrams

In Refs. [12, 13], we showed that the pasta phases are produced from hot uniform nuclear matter by cooling it down and we studied phase diagrams at zero and non-zero temperatures. In these works, we first prepared a uniform hot nucleon gas at a temperature $T \sim 20$ MeV. We then cooled it down slowly until the temperature got $\sim 0.1$ MeV or less for $O(10^3 - 10^4)$ fm/$c$, keeping the nucleon number density constant. Note that this cooling time scale is much larger than the time scale $\tau_{\text{relax}}$ for relaxation of the system, which is estimated as $\tau_{\text{relax}} \sim$ (length scale of the inhomogeneity)/(sound velocity) $\sim 10$ fm/$0.1\ c = 100$ fm/$c$. In the cooling process, we mainly used the frictional relaxation method, which is given by the QMD equations of motion plus small friction terms: $\dot{\mathbf{R}}_i = \partial \mathcal{H}/\partial \mathbf{P}_i - \xi_R \partial \mathcal{H}/\partial \mathbf{R}_i$, $\dot{\mathbf{P}}_i = -\partial \mathcal{H}/\partial \mathbf{R}_i - \xi_P \partial \mathcal{H}/\partial \mathbf{P}_i$, where $\mathcal{H}$ is the QMD Hamiltonian, $\mathbf{R}_i$ and $\mathbf{P}_i$ are the position and momentum of nucleon $i$, respectively, and $\xi_R, \xi_P > 0$ are the frictional coefficients.

The resulting nucleon distributions of cold matter at $x = 0.5$ are shown in Fig. 1. We see from these figures that the phases with rod-like and slab-like nuclei, cylindrical and spherical bubbles, in addition to the phase with spherical nuclei are reproduced. The above simulations show that the pasta phases can be formed dynamically from hot uniform matter within a time scale $\sim O(10^3 - 10^4)$ fm/$c$.

---

[1] This calculation is based on the linearized Thomas-Fermi approximation and thus the true value of $\delta E_e/n_N$ would be larger than this value.

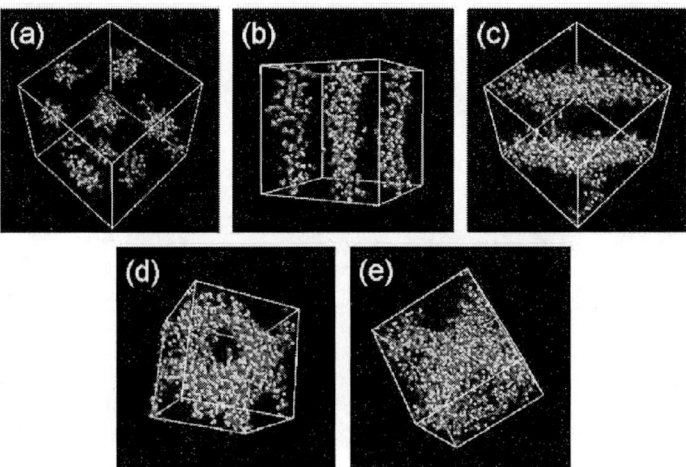

**FIGURE 1.** Nucleon distributions of the pasta phases in cold matter at $x = 0.5$; (a) sphere phase, $0.1\rho_0$; (b) cylinder phase, $0.225\rho_0$; (c) slab phase, $0.4\rho_0$; (d) cylindrical hole phase, $0.5\rho_0$; (e) spherical hole phase, $0.6\rho_0$. The darker particles represent protons and the brighter ones neutrons. Taken from Ref. [12].

In Fig. 2, we show snapshots of the nucleon distributions for $\rho = 0.225\rho_0$ at $T = 1, 2$ and 3 MeV. This density corresponds to the phase with rod-like nuclei at $T = 0$. We have observed the following qualitative features: at $T \simeq 1.5 - 2$ MeV the number of evaporated nucleons becomes significant; at $T \gtrsim 3$ MeV, nuclei almost melt and the nuclear surface is hard to identify.

The phase diagram for $x = 0.5$ is plotted in Fig. 3. The critical temperature of this model is $\gtrsim 6$ MeV. In the region below the dotted lines at $T \lesssim 3$ MeV, where we can identify the nuclear surface, we have obtained the pasta phases in the same sequence as in the earlier works: from lower densities, spherical nuclei [region (a)], rod-like nuclei [region (b)], slab-like nuclei [region (d)], cylindrical holes [region (f)], and spherical holes [region (g)]. In addition to these pasta phases, structures with multiply connected nuclear and bubble regions (i.e., sponge-like structure) such as branching rod-like nuclei, perforated slabs and branching bubbles, etc., have been obtained in the regions (c) and (e). Further study using a larger system is necessary to conclude the existence of these sponge-like phases (see also Refs. [1, 25, 26]).

## 2. Structural Transitions between the Pasta Phases

In Ref. [14], we demonstrated the second point mentioned at the beginning of this article. We performed QMD simulations of the compression of dense matter and observed the transitions from rod-like nuclei to slab-like ones and from slab-like nuclei to rod-like bubbles.

The initial conditions of the simulations are samples of the phase with rod-like nuclei

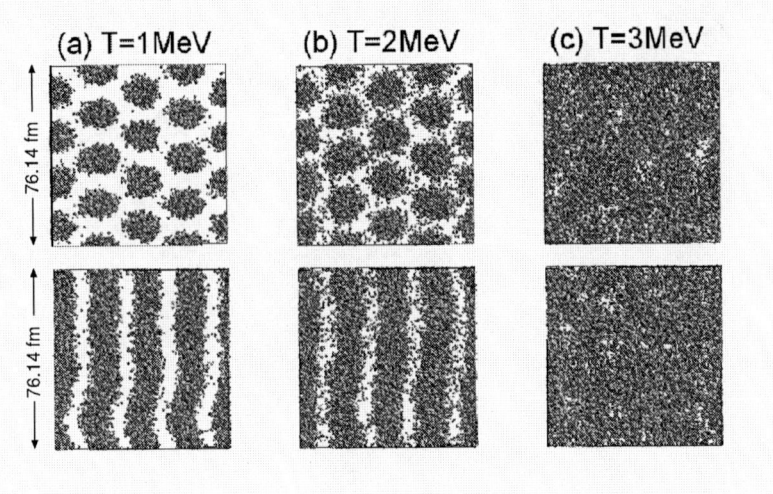

**FIGURE 2.** Nucleon distributions for $x = 0.5$, $\rho = 0.225\rho_0$ at temperatures of 1, 2 and 3 MeV. The total number of nucleons $N = 16384$ and the box size $L_{box} = 76.14$ fm. The upper panels show top views along the axis of the cylindrical nuclei at $T = 0$, and the lower ones show side views. Protons are represented by the darker particles, and neutrons by the brighter ones. Taken from Ref. [13].

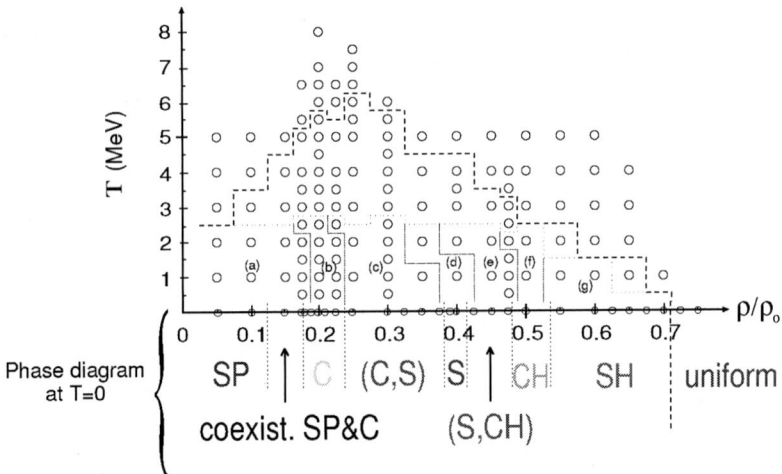

**FIGURE 3.** Phase diagram of matter at $x = 0.5$ plotted in the $\rho$ - $T$ plane. The dashed and the dotted lines on the diagram show the phase separation line and the limit below which the nuclear surface can be identified, respectively. The dash-dotted lines are the phase boundaries between the different nuclear shapes. The symbols SP, C, S, CH, SH, U stand for nuclear shapes, i.e., sphere, cylinder, slab, cylindrical hole, spherical hole and uniform, respectively. The parentheses (A,B) denote an intermediate phase between A and B-phases with a multiply connected structure. Simulations have been carried out at points denoted by circles. Adapted from Ref. [13].

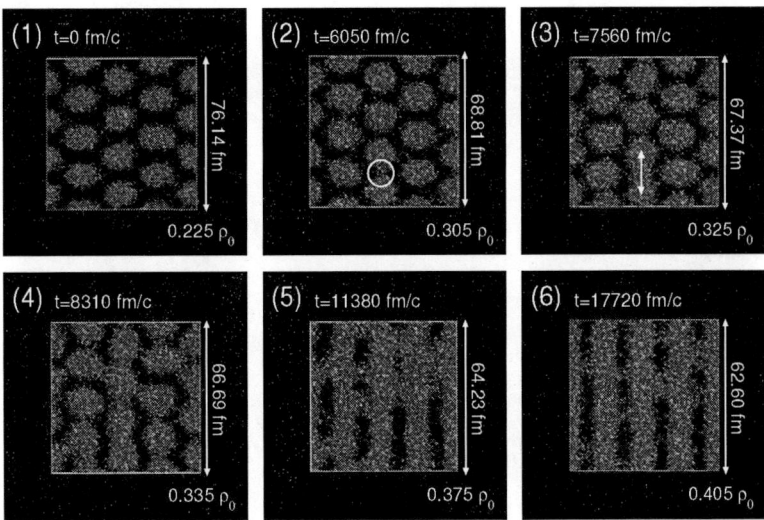

**FIGURE 4.** Snapshots of the transition process from the phase with rod-like nuclei to the phase with slab-like nuclei. The darker particles show protons and the brighter ones neutrons. The box size is rescaled to be equal in this figure. Adapted from Ref. [14].

($\rho = 0.225\rho_0$) and of the phase with slab-like nuclei ($\rho = 0.4\rho_0$) with 16384 nucleons at $x = 0.5$ and $T \simeq 1$ MeV. We adiabatically compressed these samples by increasing the density to the value corresponding to the next pasta phase taking $O(10^4)$ fm/$c$. This time scale is much larger than the typical time scale of the deformation and structural transition of nuclei (e.g., that of nuclear fission is $\sim 1000$ fm/$c$). Therefore, the compression in the simulations is adiabatic with respect to the change of the nuclear structure, so that the dynamics of the structural transition of nuclei observed in the simulations is physically meaningful, and is essentially independent of the compression rate, etc.

The resulting time evolution of the nucleon distribution is shown in Figs. 4 and 5. In Fig. 4, we see that the phase with slab-like nuclei is finally formed [Fig. 4-(6)] from the phase with rod-like nuclei [Fig. 4-(1)]. We note that the transition is triggered by thermal fluctuations, not by the fission instability: when the internuclear spacing becomes small enough and once some pair of neighboring rod-like nuclei touch due to the thermal fluctuations, they fuse [Figs. 4-(2) and 4-(3)]. Then such connected pairs of rod-like nuclei further touch and fuse with neighboring nuclei in the same lattice plane like a chain reaction [Fig. 4-(4)]; the time scale of the each fusion process is $O(10^2)$ fm/$c$.

The transition from the phase with slab-like nuclei to the phase with cylindrical holes is shown in Fig. 5. When the internuclear spacing decreases enough, neighboring slab-like nuclei touch due to the thermal fluctuation as in the above case. Once nuclei begin to touch [Fig. 5-(2)], bridges between the slabs are formed at many places on a time scale of $O(10^2)$ fm/$c$. Initially, the bridges cross the slabs almost orthogonally [Fig. 5-(3)]. Nucleons in the slabs continuously flow into the bridges and the bridges become wider

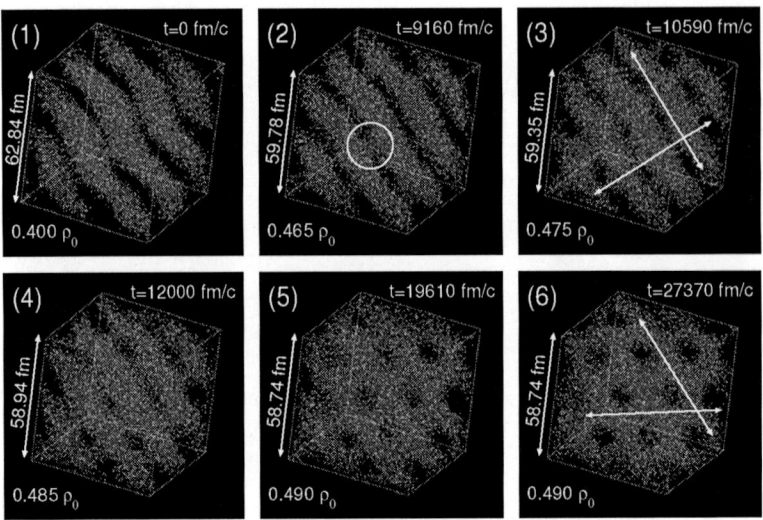

**FIGURE 5.** The same as Fig. 4 for the transition from the phase with slab-like nuclei to the phase with cylindrical holes (the box size is not rescaled in this figure). Adapted from Ref. [14].

and merge together to form cylindrical holes. Then the cylindrical holes finally relax into a triangular lattice [Fig. 5-(6)].

## IV. SUMMARY AND FUTURE PROSPECTS

Using QMD, we have shown the following two things: formation of the pasta phases in neutron star crusts by cooling and formation of rod-like bubbles from slab-like nuclei and that of slab-like nuclei from rod-like ones in supernova cores by compression of matter. In closing, we list important issues to be clarified in the future.

1. Formation of the pasta phases by compression [27]
   A remaining problem is a transition from a bcc lattice of spherical nuclei to a triangular lattice of rod-like nuclei induced by compression. If this process is confirmed, existence of the pasta phases in supernova cores will be almost established.
2. Effects of uncertainties of nuclear forces & properties of nuclei
   We employed a specific nuclear force. Although we have confirmed that this nuclear interaction yields reasonable values for the surface energy and the proton and neutron chemical potentials in neutron matter, etc. [12], in addition to reproducing saturation properties [18], further systematic survey is needed to understand effects of uncertainties in nuclear properties on the pasta phases [28]. Especially, uncertainties in the surface properties are the most important elements to be examined.
3. Detailed study of astrophysical consequences
   If formation of the pasta phases in actual astrophysical situations is established, quantitative discussion about effects of the pasta phases on astrophysical phenom-

ena will be increasingly important. As we have stressed previously [10, 12], effects of the pasta structure on coherent scattering of neutrinos in the stage of the neutrino trapping is an interesting problem [24, 13, 28, 29].

## ACKNOWLEDGMENTS

The author is grateful to Chris Pethick for helpful comments. The research reported in this article grew out of collaborations with H. Sonoda, K. Sato, T. Maruyama, K. Yasuoka, T. Ebisuzaki, and K. Iida. Further research currently in progress is performed using the RIKEN Super Combined Cluster System with MDGRAPE-2. This work was supported in part by a JSPS Postdoctoral Fellowship for Research Abroad, by the Nishina Memorial Foundation, by the Japan Society for the Promotion of Science, by the Ministry of Education, Culture, Sports, Science and Technology through Research Grant No. 14-7939, and by RIKEN through Research Grant No. J130026.

## REFERENCES

1. C. J. Pethick and D. G. Ravenhall, Annu. Rev. Nucl. Part. Sci. **45**, 429 (1995).
2. N. Chamel, contribution to this volume.
3. D. G. Ravenhall, C. J. Pethick, and J. R. Wilson, Phys. Rev. Lett. **50**, 2066 (1983).
4. M. Hashimoto, H. Seki, and M. Yamada, Prog. Theor. Phys. **71**, 320 (1984).
5. R. D. Williams and S. E. Koonin, Nucl. Phys. **A435**, 844 (1985).
6. M. Lassaut et al., Astron. Astrophys. **183**, L3 (1987).
7. C. P. Lorenz, D. G. Ravenhall, and C. J. Pethick, Phys. Rev. Lett. **70**, 379 (1993).
8. K. Oyamatsu, Nucl. Phys. **A561**, 431 (1993).
9. K. Sumiyoshi, K. Oyamatsu, and H. Toki, Nucl. Phys. **A595**, 327 (1995).
10. G. Watanabe, K. Iida, and K. Sato, Nucl. Phys. **A676**, 455 (2000); *ibid* **A687**, 512 (2001); Erratum, *ibid* **A726**, 357 (2003).
11. J. Aichelin and H. Stöcker, Phys. Lett. **B176**, 14 (1986); J. Aichelin, Phys. Rep. **202**, 233 (1991).
12. G. Watanabe et al., Phys. Rev. C **66**, 012801(R) (2002); *ibid.* **68**, 035806 (2003).
13. G. Watanabe et al., Phys. Rev. C **69**, 055805 (2004).
14. G. Watanabe et al., Phys. Rev. Lett. **94**, 031101 (2005).
15. K. Oyamatsu and M. Yamada, Nucl. Phys. **A578** (1994) 181.
16. P. Magierski and P. -H. Heenen, Phys. Rev. C **65** (2002) 045804.
17. N. Chamel, Nucl. Phys. **A747**, 109 (2005).
18. T. Maruyama et al., Phys. Rev. C **57**, 655 (1998).
19. T. Kido et al., Nucl. Phys. **A663 & 664**, 877c (2000).
20. K. Niita, in the Proceedings of the Third Simposium on *"Simulation of Hadronic Many-body System"*, A. Iwamoto et al., Eds., JAERI-conf. **96-009**, 22 (1996) (in Japanese).
21. G. Watanabe and K. Iida, Phys. Rev. C **68**, 045801 (2003).
22. T. Maruyama et al., Phys. Rev. C **72**, 015802 (2005).
23. G. Baym, H. A. Bethe, and C. J. Pethick, Nucl. Phys. **A175**, 225 (1971).
24. C. J. Horowitz et al., Phys. Rev. C **70**, 065806 (2004).
25. G. Watanabe and H. Sonoda, to appear in "Soft Condensed Matter: New Research", ed. K. I. Dillon (cond-mat/0502515).
26. M. Matsuzaki, Phys. Rev. C **73**, 028801 (2006).
27. G. Watanabe et al., in preparation.
28. H. Sonoda et al., in preparation.
29. H. Sonoda, contribution to this volume.

# The crust of neutron stars

N. Chamel

*Institut d'Astronomie et d'Astrophysique, Université Libre de Bruxelles, CP226, Boulevard du Triomphe, 1050 Brussels (Belgium)*

**Abstract.**
The structure of the crust of a neutron star is completely determined by the experimentally measured nuclear masses up to a density of the order of $10^{11}$ g.cm$^{-3}$. At higher densities, the composition of the crust still remains uncertain, mainly due to the presence of "free" superfluid neutrons which affect the properties of the nuclear "clusters". After briefly reviewing calculations of the equilibrium structure of the crust, we point out that the current approach based on the Wigner-Seitz approximation does not properly describe the unbound neutrons. We have recently abandoned this approximation by applying the band theory of solids. We have shown that the dynamical properties of the free neutrons are strongly affected by the clusters by performing 3D calculations with Bloch boundary conditions.

**Keywords:** neutron star, neutron superfluidity, pasta phase, neutron star crust, band theory
**PACS:** 26.60.+c, 71.20.-b, 71.18.+y, 97.60.Jd

## INTRODUCTION

At the end point of the stellar evolution, neutron stars are the compact remnants of core collapse supernova explosions. Born with temperatures as high as $10^{11} - 10^{12}$ K, the star rapidly cools down by emitting neutrinos and photons. A few hours after its birth, the temperature of the star falls below about $10^9$ K and the external layers crystallize into a solid crust (for a general review of neutron star crust, see [1, 2]). Within hundreds of years, the interior of the star becomes isothermal with temperatures typically less than $10^6$ K ($\sim 0.1$ keV).

From the nuclear physics point of view, a neutron star is a huge nucleus containing about $A \sim 10^{57}$ baryons and a proton fraction of the order of 10%. A rough estimate of the radius and the mass of the star from the liquid drop model yields $R = r_0 A^{1/3} \sim 10$ km and $M = A m_p \sim M_\odot$ respectively. The solid crust surrounding the star plays more or less the role of the neutron skin in heavy nuclei. Indeed correlations have been established between the neutron skin in lead $^{208}$Pb and the density at which the crust melts into a uniform liquid [3]. This gross picture however should not be taken too far since the conditions prevailing inside neutron stars are very different from those inside isolated nuclei.

The crust of a neutron star represents only a few percent of the mass of the star but plays a crucial role for its evolution (magnetic field, cooling, bursts, starquakes, spinning-down, glitches, free precession, non axial deformations giving rise to the emission of gravitational waves). We shall briefly review calculations of the equilibrium structure of the crust in the first section. The next section will be devoted to the neutron superfluidity in the crust. In the last section, we will show how the description of the inner crust can be improved by applying the band theory of solids and we will discuss some recent results.

## STRUCTURE AND COMPOSITION OF THE CRUST

In the following it will be assumed that the complete thermodynamical equilibrium at zero temperature with respect to all interactions has been reached and therefore the matter is in its lowest energy state (this excludes newly-born hot neutron stars and neutron stars accreting matter from a companion star which will not be discussed here). This assumption is usually known as the cold catalyzed matter hypothesis. The ground state of the crust is obtained by minimizing the total energy per nucleon under the assumption of beta equilibrium and electroneutroneutrality. The crust is further supposed to be formed of a perfect crystal with a single nuclear species at lattice sites (for a discussion of possible deviations from this idealized model, see for instance [4] and references therein).

### outer crust

At densities below $\sim 10^7$ g.cm$^{-3}$, the ground state of matter is a mixture of electrons and iron $^{56}$Fe (the atoms are fully ionized at densities above $\sim 10^4$ g.cm$^{-3}$). At higher densities, nuclei become increasingly neutron rich due to inverse beta decay. Following the classical paper of Baym, Pethick and Sutherland [5], the total energy density in a given layer can be written as

$$\varepsilon_{\text{tot}} = n_N E\{A,Z\} + \varepsilon_e + \varepsilon_L \quad (1)$$

where $n_N$ is the number density of nuclei, $E\{A,Z\}$ is the energy of a nucleus with $Z$ protons and $A-Z$ neutrons, $\varepsilon_e$ is the electron energy density and $\varepsilon_L$ is the lattice energy density. At densities $\rho \gg 10^6$ g.cm$^{-3}$ the electrons can be described as a relativistic Fermi gas. Assuming point like nuclei (the lattice spacing being very large compared to the size of the nuclei), the lattice energy density can be expressed as

$$\varepsilon_L = -c \left(\frac{4\pi}{3}\right)^{1/3} Z^2 e^2 n_e^{4/3}, \quad (2)$$

where $c$ is a coefficient which depends on the lattice structure. For cubic structures, this coefficient is respectively equal to 0.89593, 0.89588 and 0.88006 for body centered, face centered and simple cubic lattice which suggests that the crust crystallizes in a body centered cubic lattice.

The main physical input is therefore the energy of a nucleus. The structure of the outer crust is completely determined by the experimental nuclear masses up to a density of the order $\rho \sim 6 \times 10^{10}$ g.cm$^{-3}$ [6]. At higher densities the nuclei are so neutron rich that the energy $E\{A,Z\}$ must be extrapolated. The composition of the nuclei in these layers is thus model dependent. Nevertheless most models predict the existence of nuclei with the magic neutron numbers $N = 50, 82$, thus revealing the crucial role played by shell effects.

**TABLE 1.** Sequence of nuclei in the outer crust of non-accreting cold neutron stars calculated by Rüster et al. [6] (a theoretical nuclear mass table was used for the lower part). The last line corresponds to the neutron drip point.

| $\mu$ [MeV] | $\mu_e$ [MeV] | $\rho_{max}$ [g/cm$^3$] | $n_b$ [cm$^{-3}$] | Element | Z | N |
|---|---|---|---|---|---|---|
| 930.60 | 0.95  | $8.02 \times 10^6$    | $4.83 \times 10^{30}$ | $^{56}$Fe  | 26 | 30 |
| 931.32 | 2.61  | $2.71 \times 10^8$    | $1.63 \times 10^{32}$ | $^{62}$Ni  | 28 | 34 |
| 932.04 | 4.34  | $1.33 \times 10^9$    | $8.03 \times 10^{32}$ | $^{64}$Ni  | 28 | 36 |
| 932.09 | 4.46  | $1.50 \times 10^9$    | $9.04 \times 10^{32}$ | $^{66}$Ni  | 28 | 38 |
| 932.56 | 5.64  | $3.09 \times 10^9$    | $1.86 \times 10^{33}$ | $^{86}$Kr  | 36 | 50 |
| 933.62 | 8.38  | $1.06 \times 10^{10}$ | $6.37 \times 10^{33}$ | $^{84}$Se  | 34 | 50 |
| 934.75 | 11.43 | $2.79 \times 10^{10}$ | $1.68 \times 10^{34}$ | $^{82}$Ge  | 32 | 50 |
| 935.89 | 14.61 | $6.07 \times 10^{10}$ | $3.65 \times 10^{34}$ | $^{80}$Zn  | 30 | 50 |
| 936.44 | 16.17 | $8.46 \times 10^{10}$ | $5.08 \times 10^{34}$ | $^{82}$Zn  | 30 | 52 |
| 936.63 | 16.81 | $9.67 \times 10^{10}$ | $5.80 \times 10^{34}$ | $^{128}$Pd | 46 | 82 |
| 937.41 | 19.16 | $1.47 \times 10^{11}$ | $8.84 \times 10^{34}$ | $^{126}$Ru | 44 | 82 |
| 938.12 | 21.35 | $2.11 \times 10^{11}$ | $1.26 \times 10^{35}$ | $^{124}$Mo | 42 | 82 |
| 938.78 | 23.47 | $2.89 \times 10^{11}$ | $1.73 \times 10^{35}$ | $^{122}$Zr | 40 | 82 |
| 939.47 | 25.77 | $3.97 \times 10^{11}$ | $2.38 \times 10^{35}$ | $^{120}$Sr | 38 | 82 |
| 939.57 | 26.09 | $4.27 \times 10^{11}$ | $2.56 \times 10^{35}$ | $^{118}$Kr | 36 | 82 |

The nuclei present at the bottom of the outer crust may be experimentally studied in the near future by several facilities, such as for instance FAIR at GSI [1], ISAC at TRIUMPH [2], SPIRAL 2 at GANIL [3] and by the RIA project [4]. The structure of the outer crust is shown in table 1 for one particular representative recent model.

## inner crust

At the bottom of the outer crust, the nuclei become so neutron rich that some neutrons are no longer bound. The nuclear lattice then coexists with a neutron gas. The transition occurs when the electron Fermi energy becomes comparable to the binding energy of the protons in the nuclei. At the drip, the neutron excess $\delta = (1 - 2A/Z)$ can be estimated as [4]

$$\delta_{drip} = \sqrt{1 - E_0/S_0} - 1, \qquad (3)$$

where $E_0$ is the energy per nucleon of infinite symmetric nuclear matter and $S_0$ the symmetry energy. To lowest order in $\delta$, the density threshold for the onset of neutron drip is approximately given by

$$\rho_{drip} \simeq (S_0 \delta_{drip})^3 \times 10^9 \, \text{g.cm}^{-3}. \qquad (4)$$

---

[1] http://www.gsi.de/fair/index_e.html
[2] http://www.triumf.info/public/about/isac.php
[3] http://www.ganil.fr/research/developments/spiral2/index.html
[4] http://www.phy.anl.gov/ria/

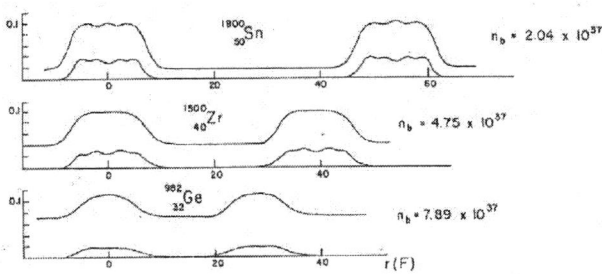

**FIGURE 1.** Nucleon number densities (in fm$^{-3}$) along the axis joining two adjacent Wigner-Seitz cells of the inner crust of neutron stars for a few baryon densities $n_b$ (in cm$^{-3}$) as calculated by Negele&Vautherin [9].

For $E_0 = -16$ MeV and $S_0 = 32$ MeV, we find $\delta_{\text{drip}} \simeq 0.225$ and $\rho_{\text{drip}} \simeq 4 \times 10^{11}$ g.cm$^{-3}$ which is in remarkable agreement with the value obtained from more realistic nuclear models. The simple estimates (3) and (4) illustrate the importance of the symmetry energy on the physics of neutron stars (see also [7] and references therein). It has been very recently shown using a liquid drop model that the composition of the inner crust is very sensitive to the density dependence of the symmetry energy [8].

The inner crust of a neutron star is a unique environment which is not accessible in the laboratory due to the presence of the "free" neutron gas. In the following we shall thus refer to the "nuclei" in the inner crust as "clusters" in order to emphasize these peculiarities. The structure of the inner crust has been studied using various approaches, mainly liquid drop and semi-classical models (for a recent review, see for instance [4]).

The most realistic calculations of the structure of the inner crust were pioneered by the work of Negele&Vautherin [9]. Until very recently[10], it was the only fully self-consistent quantum calculation. Expanding the density matrix in relative and c.m. coordinates, Negele&Vautherin derived a set of effective equations for the nucleons, which closely resemble those obtained in the Hartree-Fock approximation with Skyrme forces. They determined the structure of the inner crust by minimizing the total energy per nucleon in the Wigner-Seitz sphere, treating the electrons as a relativistic Fermi gas. As pertains the choice of boundary conditions, they imposed that wavefunctions with even parity (even $\ell$) and the radial derivatives of wavefunctions with odd parity (odd $\ell$) vanish on the sphere $r = R_{\text{cell}}$. This prescription yielded roughly uniform density outside the nuclear clusters. The remaining spurious fluctuations were removed at each iteration step by averaging the densities in the vicinity of the cell edge.

The composition of the crust is shown on table 2. These results are qualitatively similar to those obtained with liquid drop and semiclassical models. The remarkable distinctive feature is the existence of strong proton shell effects with a predominance of nuclear clusters with $Z = 40$ and $Z = 50$. Neutron shell effects are also important (while not obvious from the table) as can be inferred from the density fluctuations inside the clusters on figure 1. This figure also shows that shell effects disappear at high densities where the matter becomes nearly homogeneous.

**TABLE 2.** Sequence of nuclei in the inner crust of non-accreting cold neutron stars calculated by Negele & Vautherin [9]. $N$ is the total number of neutrons in the W-S sphere.

| $\mu_n$ [MeV] | $\mu_p$ [MeV] | $n_b$ [cm$^{-3}$] | Element | Z | N |
|---|---|---|---|---|---|
| 0.2 | -26.8 | $2.79 \times 10^{35}$ | $^{180}$Zr | 40 | 140 |
| 0.3 | -29.4 | $4 \times 10^{35}$ | $^{200}$Zr | 40 | 160 |
| 0.6 | -29.5 | $6 \times 10^{35}$ | $^{250}$Zr | 40 | 210 |
| 1.0 | -28.5 | $8.79 \times 10^{35}$ | $^{320}$Zr | 40 | 280 |
| 1.4 | -29.4 | $1.59 \times 10^{36}$ | $^{500}$Zr | 40 | 460 |
| 2.6 | -33.6 | $3.73 \times 10^{36}$ | $^{950}$Sn | 50 | 900 |
| 3.3 | -34.5 | $5.77 \times 10^{36}$ | $^{1100}$Sn | 50 | 1050 |
| 4.2 | -35.8 | $8.91 \times 10^{36}$ | $^{1350}$Sn | 50 | 1300 |
| 6.5 | -43.6 | $2.04 \times 10^{36}$ | $^{1800}$Sn | 50 | 1750 |
| 10.9 | -54.0 | $4.75 \times 10^{37}$ | $^{1500}$Zr | 40 | 1460 |
| 15 | -68.3 | $7.89 \times 10^{37}$ | $^{980}$Ge | 32 | 950 |

## "pasta" phases

The equilibrium composition of the clusters is the result of the competition between Coulomb and surface energies. At low densities, the lattice energy (2) is a small contribution to the total Coulomb energy and nuclei are spherical. However at the bottom of the crust, the size of the cluster is of the same order of magnitude as the lattice spacing and consequently the lattice energy represents a large reduction of the total Coulomb energy (this reduction is about 15 % at the neutron drip). This means that the nuclear clusters may be strongly deformed in the high density layers of the inner crust. Reasoning by analogy with percolating networks, Ogasawara&Sato[11] suggested that a transition to an "infinite network of linked nuclei" might occur at the bottom of the crust. The possibility of non spherical nuclear clusters, referred as "pastas", was considered by Ravenhall et al. [12] and Hashimoto &Yamada [13] who found from compressible liquid drop models, that as the density increases, the nuclei merge into cylinders ("spaghetti") followed by slabs ("lasagna"), cylindrical tubes and bubbles ("swiss cheese"). The pasta phases cover a small range of densities near the crust-core interface. Nevertheless they may represents up to half of the mass of the crust. The existence of these phases may have important astrophysical consequences for the gravitational wave emission and for pulsar glitches by changing the elastic properties of the crust[14], for the cooling of neutron stars by allowing direct URCA processes [15] and enhancing the heat capacity [16, 17] and for core-collapse supernovae [18]. These pasta phases have been studied by various nuclear models, from liquid drop calculations to quantum molecular dynamic simulations (for a recent review see for instance [19] and references therein). However a few models do not predict the existence of such phases [20, 21, 22, 23]. The energy differences between the various shapes are very small, typically less than $\sim$ keV.fm$^{-3}$, and as a result the structure of the crust is very sensitive to small differences between nuclear models.

# NEUTRON SUPERFLUIDITY IN THE CRUST

The possibility of superfluidity in neutron stars was suggested a long time ago by Migdal [24], only two years after the formulation of the BCS theory of electron superconductivity. Microscopic calculations performed in pure neutron matter indicate that at densities below saturation density, neutrons are bound in Cooper pairs which condense into a $^1S_0$ superfluid phase. However the density range for superfluidity and the magnitude of the pairing gap still remain uncertain due to different approximations of medium polarization and self-energy effects which tend to suppress the pairing as compared to mean field calculations with bare nucleon-nucleon forces (for a review see for instance [25, 26]).

The situation is even more uncertain in the crust owing to the small nuclear asymmetry and to the presence of inhomogeneities (for a recent review, see [26]). Superfluidity of the "free" neutrons induces superfluidity of the bound neutrons inside the clusters and *vice versa*. This proximity effect tends to smooth the spatial variations of the pairing field (see [27] and references therein). The effects of neutron superfluidity on the equilibrium structure of the crust have been considered by Baldo *et al.* [28] in the density functional theory generalized to account for nucleon pairing. The nuclear lattice was treated in the Wigner-Seitz approximation. They determined the ground state of the crust at the baryon density $\rho \simeq 1.9 \times 10^{13}$ g.cm$^{-3}$, for which the neutron pairing is expected to be the strongest. They found that the structure of the crust is significantly affected by the pairing. Indeed, the charge of the cluster $Z \simeq 52$ and the radius of the Wigner-Seitz sphere $R_{\text{cell}} \simeq 32$ fm are increased by about $\delta Z \simeq 8$ and $\delta R_{\text{cell}} \simeq 3.5$ fm respectively compared to calculations without including the pairing. However they emphasized that the results are very sensitive to the choice of the energy functional. These conclusions have been very recently confirmed by calculations at other densities [10]. This urge the need for a better understanding of pairing correlations in inhomogeneous neutron star crust matter.

# FROM NUCLEAR TO SOLID STATE PHYSICS

## Wigner-Seitz approximation

In the quantum calculations briefly reviewed in the previous sections, the nuclear lattice was treated in the Wigner-Seitz approximation. This approximation however does not properly takes into account the unbound neutrons which are artificially confined inside the Wigner-Seitz sphere. This approximation leads to spurious fluctuations of the neutron densities and of the neutron pairing field [29]. As a result calculations of the equilibrium structure of the crust and of crustal superfluidity are contaminated by unphysical shell effects which are very sensitive to the choice of boundary conditions that are imposed on the sphere [30]. These shell effects of the order of $\hbar^2/2mR_{\text{cell}}^2$ where $R_{\text{cell}}$ is the radius of the Wigner-Seitz sphere, may be very large in the deep layers of the crust where the nuclear clusters are very close to each other. A correct treatment of the "free" neutrons requires the application of the band theory originally proposed for describing electrons in solids and recently applied to neutron star crust [31, 32].

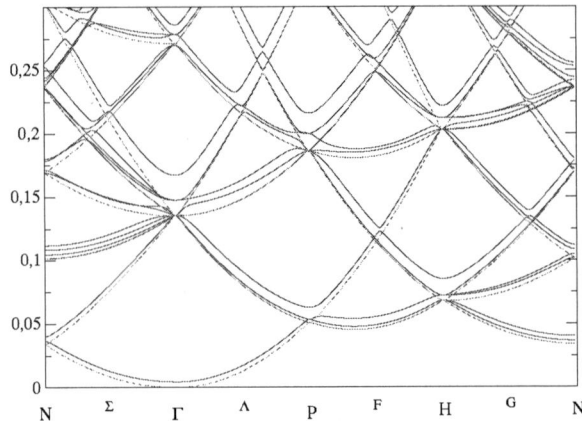

**FIGURE 2.** Energy spectrum (in MeV) of unbound neutrons in the outermost layers of the inner crust of neutron stars for different symmetry directions in **k**-space. The dashed line is the energy spectrum of the ideal Fermi gas. See reference [32] for details.

## Band theory

The nuclear lattice can be partitioned into identical Wigner-Seitz cells whose shape is determined by the geometry of the lattice. The boundary conditions on the cell are not arbitrary but are fixed by the Floquet-Bloch theorem. Each single particle quantum state is described by a discrete index $\alpha$ and by a wave vector **k**. The energy spectrum is thus formed of a series of sheets or "bands" in **k**-space as illustrated on figure 2. We have recently applied the band theory by performing 3D calculations with Bloch boundary conditions (see [32] and references therein), in order to investigate the effects of the nuclear clusters on the dynamical properties of the neutron gas. For this purpose, first ignoring the effects of pairing, we have studied the topology of the Fermi surface by computing its area and comparing it to the area of the corresponding sphere (remembering that the Fermi volume only depends on the density). We found that at low densities, meaning that the Fermi wave length of the unbound neutrons is much larger than the lattice spacing, the Fermi surface is nearly spherical, which is analogous to alkali metals. However at higher densities the Fermi surface is strongly deformed owing to Bragg diffraction as it is also observed for transition metals. This entails a renormalisation of the neutron mass into an effective mass defined by $m_\star = p_n/v_n$ and given by a Fermi surface integral [33]. This effective mass can take very large values, up to about $\sim 15 m_n$ at the baryon density $n_b = 0.03$ fm$^{-3}$ [31]. We have also shown that in nearly homogeneous neutron star crust matter for which the BCS approximation is valid, the effects of neutron pairing on the effective mass are small and vanish in the limit of a uniform system [34]. However as pointed out recently by Magierski [35], strong inhomogeneities may lead to neutron localisation, which might further enhance the effective neutron mass.

# CONCLUSION

The evolution of a neutron star is intimately related to the properties of its solid crust. Up to a density of about $10^7$ g.cm$^{-3}$, the crust is formed of a body centered cubic crystal of iron $^{56}$Fe, which are fully ionized above $\sim 10^4$ g.cm$^{-3}$. With increasing density, the nuclei immersed in a relativistic electron gas, become more and more neutron rich owing to electron capture. At densities around $10^{11}$ g.cm$^{-3}$ neutrons start to drip out of nuclei and may form Cooper pairs which condense into a superfluid phase. At the bottom of the crust at densities of the order of $10^{14}$ g.cm$^{-3}$, some calculations predict that nuclei may adopt exotic non spherical shapes, referred as "pastas".

Unlike the nuclei in the shallower layers, the nuclear "clusters" in the inner crust cannot be studied in the laboratory due to the presence of "free" superfluid neutrons. The structure of the crust has been studied using different approximations and nuclear models. The current state-of-the-art of self-consistent quantum calculations which were pioneered by Negele&Vautherin[9], is the Hartree-Fock-Bogoliubov approximation. It has been recently shown in this framework that the neutron superfluidity greatly affects the composition of the crust[28].

However much remains to be done. Indeed, in all these calculations the lattice is treated in the Wigner-Seitz approximation in which the dripped neutrons are artificially confined in spheres. A much more accurate description of the crust, which is essential in order to interpret observations of neutron stars, should rely on the band theory of solids. We have recently shown that the dynamical properties of the superfluid neutrons are strongly affected by the nuclear clusters by carrying out 3D calculations with Bloch boundary conditions [32]. This work is a first step towards realistic calculations of the properties of neutron star crust.

# ACKNOWLEDGMENTS

The author gratefully acknowledges financial support from a Marie Curie Intra-European Fellowship of the European Union (contract MEIF-CT-2005-024660).

# REFERENCES

1. C. J. Pethick, and D. G. Ravenhall, *Annual Revue of Nuclear Particle Science* **45**, 429–484 (1995).
2. P. Haensel, "Neutron Star Crusts," in *LNP Vol. 578: Physics of Neutron Star Interiors*, edited by D. Blaschke, N. K. Glendenning, and A. Sedrakian, ECT*, Springer, 2001, pp. 127–174.
3. C. J. Horowitz, and J. Piekarewicz, *Physical Review Letters* **86**, 5647–5650 (2001).
4. P. Haensel, A. Potekhin, and D. Yakovlev, *Neutron Stars 1: Equation of State And Structure*, Springer Verlag, 2006.
5. G. Baym, C. Pethick, and P. Sutherland, *Astrophysical Journal* **170**, 299–+ (1971).
6. S. B. Rüster, M. Hempel, and J. Schaffner-Bielich, *Physical Review C* **73**, 035804–+ (2006).
7. A. W. Steiner, M. Prakash, J. M. Lattimer, and P. J. Ellis, *Physics Reports* **411**, 325–375 (2005).
8. K. Oyamatsu, and K. Iida, *ArXiv Nuclear Theory e-prints* (2006), `nucl-th/0609040`.
9. J. W. Negele, and D. Vautherin, *Nuclear Physics A* **207**, 298–320 (1973).
10. M. Baldo, E. E. Saperstein, and S. V. Tolokonnikov, *ArXiv Nuclear Theory e-prints* (2006), `nucl-th/0609031`.
11. R. Ogasawara, and K. Sato, *Progress of theoretical physics* **68**, 222–235 (1982).

12. D. G. Ravenhall, C. J. Pethick, and J. R. Wilson, *Physical Review Letters* **50**, 2066–2069 (1983).
13. M. Hashimoto, H. Seki, and M. Yamada, *Progress of theoretical physics* **71**, 320–326 (1984).
14. C. J. Pethick, and A. Y. Potekhin, *Physics Letters B* **427**, 7–12 (1998).
15. M. E. Gusakov, D. G. Yakovlev, P. Haensel, and O. Y. Gnedin, *Astronomy&Astrophysics* **421**, 1143–1148 (2004).
16. F. V. de Blasio, and G. Lazzari, *Physical Review C* **52**, 418–420 (1995).
17. Ø. Elgarøy, L. Engvik, E. Osnes, F. V. de Blasio, M. Hjorth-Jensen, and G. Lazzari, *Physical Review D* **54**, 1848–1851 (1996).
18. C. J. Horowitz, M. A. Pérez-García, D. K. Berry, and J. Piekarewicz, *Physical Review C* **72**, 035801–+ (2005).
19. G. Watanabe, and H. Sonoda, "Recent progress on understanding "pasta" phases in dense stars," in *AIP Conf. Proc. 791: Reaction Mechanisms for Rare Isotope Beams*, edited by A. Brown, 2005, pp. 101–111.
20. C. P. Lorenz, D. G. Ravenhall, and C. J. Pethick, *Physical Review Letters* **70**, 379–382 (1993).
21. K. S. Cheng, C. C. Yao, and Z. G. Dai, *Physical Review C* **55**, 2092–2100 (1997).
22. F. Douchin, and P. Haensel, *Physics Letters B* **485**, 107–114 (2000).
23. T. Maruyama, T. Tatsumi, D. N. Voskresensky, T. Tanigawa, and S. Chiba, *Physical Review C* **72**, 015802–+ (2005).
24. A. B. Migdal, *Nuclear Physics* **13**, 655 (1959).
25. D. J. Dean, and M. Hjorth-Jensen, *Reviews of Modern Physics* **75**, 607–656 (2003).
26. M. Baldo, E. E. Saperstein, and S. V. Tolokonnikov, *Nuclear Physics A* **749**, 42–52 (2005).
27. N. Sandulescu, N. van Giai, and R. J. Liotta, *Physical Review C* **69**, 045802–+ (2004).
28. M. Baldo, U. Lombardo, E. E. Saperstein, and S. V. Tolokonnikov, *Nuclear Physics A* **750**, 409–424 (2005).
29. F. Montani, C. May, and H. Müther, *Physical Review C* **69**, 065801–+ (2004).
30. M. Baldo, E. E. Saperstein, and S. V. Tolokonnikov, *Nuclear Physics A* **775**, 235–244 (2006).
31. N. Chamel, *Nuclear Physics A* **747**, 109–128 (2005).
32. N. Chamel, *Nuclear Physics A* **773**, 263–278 (2006).
33. B. Carter, N. Chamel, and P. Haensel, *Nuclear Physics A* **748**, 675–697 (2005).
34. B. Carter, N. Chamel, and P. Haensel, *Nuclear Physics A* **759**, 441–464 (2005).
35. P. Magierski, *ArXiv Nuclear Theory e-prints* (2005), `nucl-th/0508023`.

# Burning Questions in Nuclear Astrophysics

Bradley S. Meyer

*Department of Physics and Astronomy, Clemson University, Clemson, SC 29634-0978, USA*

**Abstract.**
This paper presents a brief resumé of the round-table discussion at the conclusion of the Nuclear Astrophysics Symposium of the Tours 2006 meeting.

**Keywords:** Round, table, discussion
**PACS:** 26.20.+f,26.30.+k,26.35.+c,26.40.+r,26.60.+c,26.65.+t

The Sixth Tours Symposium on Nuclear Physics took place from September 5-8, 2006 in Tours, France. Sessions were held in the mornings and afternoons at the aptly named Hotel de l'Univers and were punctuated by wonderful lunches in a room set up in the hotel exclusively for us. In the evenings, we enjoyed dinner in the charming old part of Tours. Our conference outing was a visit to Château du Clos de Lucé, Leonardo da Vinci's last home, in Amboise. On the afternoon of the last day of the meeting, we held a round-table discussion on "Burning Questions in Nuclear Astrophysics". I had the privilege and pleasure to moderate the discussion, and I present a brief resumé here.

I readily agreed to be the moderator of the round table when asked by the organizers several months prior to the Symposium. I had never been to a Tours Symposium or to Tours, so I was eager to attend, and the role of moderator seemed like something I could do. Nevertheless, after I'd begun to consider the task ahead of me, I began to get a little nervous. Racine's wonderful line "Brulé de plus de feux que je n'en allumai ..." from his play *Andromache* began running through my mind. I wanted to remind myself of the full provenance of that line, so I picked up my collection of Racine and found the relevant scene. In it, Pyrrhus, the leader of the Greeks who sacked Troy, is declaring his violent passion for his Trojan captive Andromache. Andromache is the widow of Hector, the great Trojan hero, and mother of Hector's son. She is obviously reluctant to requite the love of a man who led the destruction of her city and the murder of her relatives. Pyrrhus pursues her love and declares his passionate anguish:

> Je souffre tous les maux que j'ai faits devant Troie:
> Vaincu, chargé de fers, de regrets consumé,
> Brulé de plus de feux que je n'en allumai ...

I translate this as

> I suffer all the evils I visited upon Troy:
> Conquered, clapped in irons, consumed with regrets,
> Burned by more fires than I lit ...

To equate the suffering from one's passion to the suffering of slaughtered thousands is outrageous, but such evocations of extreme emotion expressed in such perfect, disciplined Alexandrine couplets are what make Classical French drama so great for me. Perhaps I'm being a little outrageous too, but I did fear that, if I didn't do a good job in igniting discussions of "Burning Questions in Nuclear Astrophyiscs", I could well get burned by more fires than I lit.

I decided that the most important thing to do ensure a good discussion was to prepare. I did this by setting and circulating a tentative agenda prior to the meeting. This is available at

http://www.ces.clemson.edu/ mbradle/DOWNLOAD/TOURS2006/agenda.html

I believe this was helpful because it did get participants to think ahead about the discussion. When we actually held the round table on the afternoon of September 8, I began by following the agenda. I presented a few welcoming comments and then briefly discussed my views on when it is appropriate to suggest or make a new nuclear physics measurement for astrophysics. These followed the four "rules" we set out in 1997[1]. This touched off a lengthy discussion that essentially consumed the entire afternoon.

It became clear from the comments that there is a tension between theorists and experimentalists, which I characterize somewhat crudely as follows. Experimentalists complain that theorists study models and conclude that a particular reaction cross section needs to be measured to better accuracy. They urge their experimentalist colleagues to make that measurement. After many months or years in the lab, the experimentalist emerges with a hard-won measurement only to find to his or her great irritation that the models have changed and the theorists no longer need that reaction cross section to such great accuracy. On the other hand, theorists complain that experimentalists make a measurement and then claim that that measurement solves a long-standing problem in astrophysics when in fact the non-linear effects in the models means the new measurement has a much less significant effect than claimed.

I believe all participants recognized this tension as a problem. It suggests a lack of trust which ultimately can hinder proper choices of theoretical or experimental study. In addition, all participants in the discussion recognized that the solution to this problem was better dialog and collaboration; however, when we turned to the agenda item of how to improve that dialog, there was little new that we came up with. All agreed that web sites can be useful. I suggested mail lists, and, prior to the meeting, I set one up:

http://www.ces.clemson.edu/mailman/listinfo/nuclear-astro

with a view towards facilitating and archiving such dialog. Unfortunately, this has not really caught on–I think everyone is busy, and participants fear they don't have the time to devote to such fora. We will nevertheless continue to maintain the list, and any reader should feel free to join and post to it. The general consensus was that the old-fashioned approach of person-to-person emails and phone calls and discussions at conferences are probably still the best way to go, and that we just need to be respectful and considerate and willing to spend a little extra time in discussions with our colleagues.

Another area we spent a fair amount of time discussing was the application of new technology to old problems. There are two principal fears that people expressed. First, resources are limited, so when new tools come on line, old ones often have to be shut down. The fear is that the old tools may be disappear prematurely. Some discussion participants expressed concern, for example, that the current push for radioactive beams will cause the shutdown of currently functioning stable beam facilities before a number of important stable nucleus experiments have been carried out. This seems like valid concern, and the community needs to keep a proper balance of facilities in mind.

The second general fear concerns the new tools themselves. I got the sense from the discussions that some people feared that new techniques, either theoretical or experimental, can be oversold. On the other hand, those with a new tool seem to fear, no doubt in many cases with justification, that others will not understand or appreciate it. This, of course, can lead to lack of funding for an otherwise promising new technique. We need to be open to such new tools. For example, there were talks during the Symposium concerning the technique of producing $\gamma$ beams by Compton back-scattering of laser light from relativistic electrons. This looks like a promising scientific approach to studying nuclear properties. The underlying tools for such research were developed for other purposes, but they are now being applied to nuclear astrophysics, e.g., [2]. If we fail to appreciate the possibilities of such new tools, we risk stagnating. Here I am reminded of an amusing proof, I think by Spinoza, that hammers cannot exist. The wonderful, but absurd, argument runs that it takes a hammer to make a hammer. One then recurses back to the time when there were no hammers. The first hammer could then never have been made, and thus no hammers can exist. The proof is clearly facetious–the first hammer was obviously made with something other than a hammer–but it serves as a warning. We must keep in mind that the nuclear astrophysics tools of tomorrow are constructed from other tools of today.

I hope there will be another round table at the next Tours Symposium. It is useful to step back at the end of a meeting and let people present their views on topics discussed at the meeting itself or on the current state of the field. If indeed there is another discussion, I have a couple of suggestions. The first is to find some way to get all participants to stay for it. Many nuclear physicists and astrophysicists left prior to the round table, and we certainly missed their comments and insights. Perhaps the organizers can strongly encourage all Symposium participants to remain an extra afternoon. Also, the next moderator should find a way to focus on the scientific aspects. He or she might do this by noting significant scientific points during the Symposium talks, requesting slides from the speakers during the meeting that illustrate those points, and then presenting at the beginning of the round table a few comments on why those points were significant. This might stimulate the discussion better than a simple agenda.

In conclusion, I thank the organizers for the opportunity to attend the Symposium and to moderate the round-table discussion. I hope all the participants enjoyed themselves as much as I did!

## REFERENCES

1. L.-S. The, D. D. Clayton, L. Jin, and B. S. Meyer, *Astrophys. J.* **504**, 500–515 (1998).
2. S. Goko, H. Utsunomiya, S. Goriely, A. Makinaga, T. Kaihori, S. Hohara, H. Akimune, T. Yamagata, Y.-W. Lui, H. Toyokawa, A. J. Koning, and S. Hilaire, *Physical Review Letters* **96**, 192501 (2006).

# YOUNG SCIENTIST SESSION

# Neutron-Capture Nucleosynthesis in Extremely Metal-Poor Stars
— Application to the most iron-deficient stars HE0107-5240 and HE1327-2326 —

Takanori Nishimura*,†, Nobuyuki Iwamoto**, Masayuki Aikawa‡,
Takuma Suda§, Masayuki Y. Fujimoto* and Icko Iben Jr.¶

*Graduate School of Science, Hokkaido University
†Research Fellow of the Japan Society for the Promotion of Science
**Nuclear Data Center, Japan Atomic Energy Agency
‡Hokkaido University OpenCourseWare, Hokkaido University
§Research Center for the Early Universe, School of Science, the University of Tokyo
¶Departments of Astronomy and Physics, University of Illinois

**Abstract.** In extremely metal-poor stars ([Fe/H]≲ −2.5), hydrogen is mixed into the convection driven by helium flash and induces neutron-capture nucleosynthesis with the reactions, $^{12}C(p,\gamma)^{13}N(e^+v)^{13}C(\alpha,n)^{16}O$, as the neutron source. We investigate the progress of this nucleosynthesis with use of nuclear network for a wide range of model parameters such as the amount of mixed hydrogen and the strength of helium flash. We reveal the characteristic abundance pattern from the light elements through the s-process elements, produced by alpha- and neutron-capture reactions under the extremely metal-poor condition. On the basis, we explore the possible modifications of surface abundances in the metal-free, Population III stars and discuss their relevance to the two most iron-deficient stars, HE0107-5240 and HE1327-2326, known to date.

**Keywords:** stellar evolution, nucleosynthesis, Population III stars
**PACS:** 97.10.Cv;97.20.Wt

## Introduction

In a past decade, the number of known extremely metal-poor (EMP) stars, which have the iron abundance of [Fe/H]≲ −2.5, has increased owing to HK survey [1] and to Hamburg/ESO survey [2]. These EMP stars are expected to serve as a probe to inquiring into the formation of galaxies and/or the chemical evolution in early Universe. In order to draw the proper information, we have to understand the evolution and nucleosynthesis in these low-mass, metal-poor stars. In particular, two most iron-deficient stars of HE0107-5240 ([Fe/H] ∼ −5.3, [3]) and HE1327-2326 ([Fe/H] ∼ −5.4, [4]) share a peculiar abundance pattern that the light elements of carbon, nitrogen, oxygen, sodium and magnesium are enhanced relative to iron by large factors up to ten thousands as compared to the solar ratios. In addition, HE1327-2326 shows the surface enrichment of s-process elements, 1 dex enhancement of strontium and 1.4 dex upper limit for barium, relative to iron, which is smaller or absent in HE0107-5240. The origins of these peculiar abundance patterns have been attracting wide interest and possible mechanisms have been discussed. In this paper, we work on the binary scenario by Suda et al.[5], in which these patterns were produced in the primary star during the AGB phase and transferred onto the secondary star, now survived as an EMP star, through the wind accretion. It is known that in extremely metal-poor ($[\text{Fe/H}] \lesssim -2.5$) stars of mass $M \lesssim 3 M_\odot$, the convection driven by helium shell-flash extends outward through the bottom of hydrogen rich layer in early thermal-pulsating AGB phase [6]. Engulfed hydrogen is captured by $^{12}$C in the middle of helium convective zone to form $^{13}$C, and the latter is further carried inwards to burn and release neutron via the $^{13}$C$(\alpha,n)^{16}$O reaction. In the helium convective zone, the nucleosynthesis is promoted with the aid of these neutrons, and finally, the nuclear products are dredged up to the surface. We investigate the progress of nuclear reactions in the helium flash convective zone, induced by hydrogen mixing, to explore the new aspects of nucleosynthesis in extremely metal-poor stars. Based on the results, we reveal the current characteristics of such low-mass, EMP stars that can survive to date, and discuss their relevance to EMP stars observed in Galactic Halo.

## *Methods*

We compute the progress of nucleosynthesis in the helium flash-convection triggered by hydrogen mixing. We calculate the evolution of the temperature and density at the bottom of helium convective zone by using one-zone approximation (e.g., [7]), which is determined as a function of the core mass (and radius) and the initial pressure. We treat the mixing of $^{13}$C instead of the hydrogen and specify the amount of mixed $^{13}$C in the ratio to the abundance of $^{12}$C in the helium convective zone. The two nuclear networks are used for our investigations. The one covers 65 light elements from neutron and proton up to the $^{35}$S with the reactions of proton-, alpha- and neutron-captures and of beta-decays. The other covers elements up to bismuth (926 isotopes for $Z \gtrsim 16$) with the reactions of neutron captures and of beta-decays. The formation of the heavier elements is computed by the latter code from the $^{34}$S abundance and the neutron density, obtained by the former code.

## *Results and Discussion*

First, mixed $^{13}$C reacts with helium to produce neutron and $^{16}$O. Under the EMP condition, the released neutrons are mostly captured by $^{12}$C to form the neutron-recycling reactions of $^{12}$C$(n,\gamma)^{13}$C$(\alpha,n)^{16}$O. Thus, neutrons act as a catalyst and produce $^{16}$O in much exceeding the amount of mixed $^{13}$C. These neutron-recyling reactions once end when all the neutrons are captured by $^{16}$O and by the daughter of $^{17}$O and mixed $^{13}$C are mostly converted into $^{14}$C. Because of large ratio of $^{16}$O to $^{13}$C abundance, some of $^{17}$O survive to produce neutron by $(\alpha,n)$ reaction, though in much longer timescale as compared with carbon, and the liberated neutrons are likely to be absorbed again by $^{12}$C to start the doubly neutron-recycling reactions, $^{12}$C$(n,\gamma)^{13}$C$(\alpha,n)^{16}$O$(n,\gamma)^{17}$O$(\alpha,n)^{20}$Ne. $^{14}$C is also converted into $^{22}$Ne through the successive two $(\alpha,\gamma)$ reactions. Similaly, $^{22}$Ne may undergo the alpha captures to produce neutrons and produce magnesium isotopes. Because of the large Coulomb barrier, however, these alpha captures are dominated by the neutron captures except for the strong shell flashes of temper-

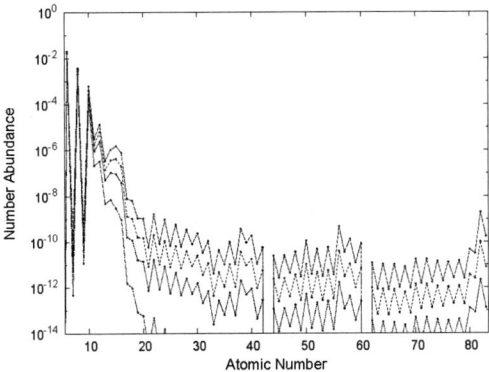

**FIGURE 1.** Abundance patterns from carbon to bithmus, resulting from neutron-capture nucleosynthesis in the helium flash convective zone triggered by hydrogen mixing for the different amount of mixed $^{13}$C of $^{13}$C/$^{12}$C $= 3 \times 10^{-3}$, 0.01, 0.02, 0.04, from the bottom to the top, respectively.

atures $T \gtrsim 3 \times 10^8$ K. Since $^{22}$Ne has the neutron capture cross section much smaller than nearby nuclides, the flow of neutron captures is bottlenecked and hence, sodium and magnesium isotopes can be produced in much smaller amount than the neon isotopes. For the temperatures, realized during helium shell flashes, the alpha captures are no longer effective beyond neon isotopes. Accordingly aluminium and heavier elements have to be formed solely only through successive neutron captures.

The Figure 1 shows the abundance patterns from carbon to bismuth, resulting from our calculations for initially metal-free stars with the different amounts of mixed $^{13}$C. The production of heavy elements increases steeply with the mixed $^{13}$C, and for the $^{13}$C mixing greater than $^{13}$C/$^{12}$C $\sim 0.01$, we can recognize three s-process peaks at strontium, barium and lead. They are synthesized with $^{22}$Ne and Mg isotopes as seed.

We compare our results with the abundances of the two most iron-deficient stars in Figure 2. The calculated abundances are normalized so as to give the observed carbon abundances of stars with taking into account the dredge-up by the surface convection and resultant dilution in the hydrogen-rich envelope. In the helium shell flash with hydrogen mixing, strong hydrogen injection causes the split of convective region into two, the upper one driven by CN-cycle and the lower one driven by helium burning. In this case, the surface convection deepen in mass and penetrates into the site of the upper convective region during the decay phase of the shell flash and carries the products of neutron-capture nucleosynthesis and also nitrogen produced by CN-cycle [5, 6].

The observed abundance pattern from HE1327-2326 can be reproduced by our results both for the light elements of carbon, nitrogen, oxygen, sodium, magnesium and aluminium, and for the heavy s-process element, strontium. When the CNO abundances in the envelope increase above [CNO/H] $\gtrsim -2.5$, the hydrogen mixing no longer occurs. And yet, stars of mass M $\gtrsim 1.5$ M$_\odot$ experience the third dredge-up during TP-AGB evolution and dredge up the nuclear products in the helium convective region. Without the hydrogen mixing, only carbon and oxygen are formed so that the products of neutron-capture nucleosynthesis are diluted relative to carbon and oxygen. The observed pattern

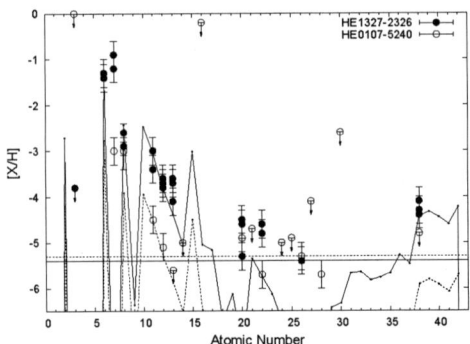

**FIGURE 2.** Comparison of the calculated abundance patterns with those observed from the two most iron-deficient stars. For HE1327-2326, the observed abundances of light elements, C through Al, and heavy element Sr (filled circles) are well reproduced by the model calculated with $^{13}C/^{12}C = 0.02$ (solid line), and for HE0107-5240, the observed abundance of light elements (open circles) are compatible with the same model, diluted by a factor of 30 relative to carbon due to the third dredge-up.

from HE0107-5240 can be well fitted with taking into account the effect of third dredge-up. In the binary scenario, the origin of iron group elements, e.g., Ca through Ni, is attributed to the accretion of interstellar gas, polluted by later supenovae during the long life of these low-mass survivors.

## Conclusions

Metal-free stars of low- and intermediate-masses can enrich the surface with not only the light elements of carbon, nitrogen, oxygen, sodium, magnesium, aluminium, but also with the s-process elements during the helium shell flashes attendant with hydrogen mixing. This is also true for stars of metallicity $[Fe/H] \lesssim -5$. The peculiar abundance patterns, observed from the two most iron-deficient stars, are explained in terms of the mass transfer of metal-free AGB stars in a close binary system and the accretion of the interstellar gas. Our results suggest these two stars to be survivors of the first generation stars; they have been low-mass companions in the close binary systems with the primary star of mass $M \lesssim 1.5 M_\odot$ for HE1327-2326 and of mass $M \gtrsim 1.5 M_\odot$ for HE0107-5240.

## REFERENCES

1. T.C.Beers, G.W.Preston, S.A.Shectman, ApJ. 103, 1987 (1992).
2. N.Christlieb, P.J.Green, L.Wisotzki, D.Reimers, A&A. 375, 366 (2001).
3. N.Christlieb et al., Nature. 419, 904 (2002).
4. A.Frebel et al., Nature. 434, 871 (2005).
5. T.Suda, M.Aikawa, M.N.Machida, M.Y.Fujimoto, I.Iben.Jr, ApJ. 611, 476 (2004).
6. M.Y.Fujimoto, Y.Ikeda, I.Iben.Jr, ApJ. 529, L25 (2000).
7. M.Y.Fujimoto, ApJ. 257, 752 (1982).
8. Z.Y.Bao, H.Beer, F.Kappeler, F.Voss, K.Wisshak, At. Data Nucl. Data Tables. 76, 70 (2000).

# Screening effects on neutrino-nucleus reactions

Futoshi Minato*, Kouichi Hagino*, Noboru Takigawa*, A. B. Balantekin[†]
and Ph. Chomaz**

*Department of Physics, Tohoku University, 980-8578 Sendai, Japan
[†]Department of Physics, University of Wisconsin, Madison, Wisconsin 53706, USA
**GANIL-CEA/DSM-CNRS/IN2P3, B.P. 55027, F-14076 Caen Cédex 5, France

**Abstract.** We discuss effects of the electron plasma on charged-current neutrino-nucleus reaction, $(\nu_e, e^-)$ in a core-collapse supernova environment. We first discuss the electron screening effect on the final state interaction between the outgoing electron and the daughter nucleus. To this end, we solve the Dirac equation for the outgoing electron with the screened Coulomb potential obtained with the Thomas-Fermi approximation. In addition to the screening effect, we also discuss the Pauli blocking effect due to the environmental electrons on the spectrum of the outgoing electron. We find that both effects hinder the cross section of the charged-current reaction, especially at low incident energies.

**Keywords:** neutrino-nucleus reaction, electron screening effect, Pauli blocking effect, r-process nucleosynthesis
**PACS:** 23.40.Bw,26.30.+k,26.50.+x,98.80.Ft

A large number of neutrinos are emitted from a core-collapse supernova. These neutrinos interact with nuclei through the weak interaction. Although their cross sections are small, it is agreed that their contribution to nucleosynthesis (that is, r-, $\nu$- and p-processes) is not negligible due to the large neutrino luminosity [1]. Neutrinos may even play a leading role in some cases. For instance, Yoshida et al. recently argued that the abundance ratio between $^7$Li and $^{11}$B is sensitive to the $\nu$-process and thus can be used to extract information on the neutrino mass hierarchy [2]. Also, the abundance ratio between U and Th elements, which has been used as a cosmochronometer, may be affected by the $\nu$-process. It is thus important to calculate with high accuracy the cross section of the neutrino-nucleus reactions in a dense star.

In the supernova nucleosynthesis, only the charged current reactions of the electron neutrinos, $\nu_e$, and the electron anti-neutrinos, $\bar{\nu}_e$, are relevant, since those of $\nu_\mu$ and $\nu_\tau$ (and their antineutrinos) are suppressed due to the threshold effects. The outgoing electron and positron produced by the weak interaction feel the Coulomb interaction from the daughter nucleus as they leave. This final state interaction affects the neutrino-nucleus reaction rate [3, 4]. In the supernova environment, the motion of the outgoing electron is further perturbed by environmental electrons. Such effects have been considered in Ref. [5] for electron capture rates in a dense star. Furthermore, in a high electron density, the charged-current reaction is suppressed because low energy electron states are Pauli blocked. It is crucial to take into account those two effects in order to accurately estimate the neutrino-nucleus reaction rate for nucleosynthesis. In this contribution, we present such calculations, taking into consideration both the electron screening and Pauli blocking effects [6].

Let us first discuss how we implement the electron screening and the Pauli blocking effects in our calculations. We assume that the electron charge distribution is homogeneous with density $\rho_e^0$ in the absence of the daughter nucleus. This charge distribution of the environmental electrons is modified to $\rho_e(\vec{r})$ due to the presence of the daughter nucleus. The Coulomb field $\phi(\vec{r})$ at $\vec{r}$ from the daughter nucleus reads

$$\phi(\vec{r}) = \int d\vec{r}' \frac{e\rho_N(\vec{r}') - e\delta\rho_e(\vec{r}')}{|\vec{r} - \vec{r}'|}, \tag{1}$$

where $\delta\rho_e(\vec{r}) \equiv \rho_e(\vec{r}) - \rho_e^0$ is the polarization charge. In order to evaluate this function, we assume a sharp-cut charge distribution for the nuclear charge density $\rho_N$, that is, $\rho_N(\vec{r}) = [3Z/(4\pi R^3)] \cdot \theta(R-r)$ with a nuclear radius of $R$. Here, $Z$ is the atomic number of the daughter nucleus. For the electron density $\rho_e$, we use the Thomas-Fermi theory. The polarization charge then reads $\delta\rho_e(\vec{r}) = (2m\varepsilon_F(\vec{r}))^{3/2}/(3\pi^2\hbar^3) - \rho_e^0$. Here, $m$ is the electron mass, and the local Fermi energy $\varepsilon_F(\vec{r})$ is given by $\varepsilon_F(\vec{r}) = \varepsilon_F^0 + e\phi(\vec{r})$ with $\varepsilon_F^0 = (3\pi^2\hbar^3\rho_e^0)^{2/3}/2m$. The boundary condition is imposed so that the Coulomb potential vanishes at the radius where the net negative charge inside is equal to the charge number of the daughter nucleus. Once the Coulomb field $\phi$ is obtained, we solve the Dirac equation for the outgoing election with the potential $V_C(r) = -e\phi(r)$. Using the solution of the Dirac equation, we estimate the cross sections of neutrino-nucleus reactions with the DWBA method [7]. In order to take into account the Pauli blocking effect for electrons, we multiply a factor $(1 - f_e(E_e, T_e, \mu_e))$ to the cross section, where $E_e$ is the energy of the outgoing electron and $f_e$ is the distribution function of the environmental electrons given by $f_e(E, T_e, \mu_e) = 1/(1 + \exp[(E - \mu_e)/T_e])$. For a given electron density $\rho_e^0$, we estimate the Fermi energy $\mu_e$ using the relativistic Fermi gas model,

$$\rho_e^0 = \frac{m^3 c^6}{\pi^2 \hbar^3} \int_0^\infty \frac{\sinh^2 x \cosh x}{\exp(\beta(mc^2\cosh x - \mu_e)) + 1} dx, \tag{2}$$

where $\cosh x = p/mc^2$, $p$ being the momentum of the electron.

We now evaluate numerically the electron screening and the Pauli blocking effects on the charged current $^{56}\text{Fe}(\nu_e, e^-)^{56}\text{Co}$ and $^{208}\text{Pb}(\nu_e, e^-)^{208}\text{Bi}$ reactions. We set the electron temperature to be $T_e = 0.5$ MeV. We consider the Fermi type transition to the $J^\pi = 0^+$ state at $E_x = 3.5$ MeV in $^{56}\text{Co}$ [8] and $E_x = 15.0$ MeV in $^{208}\text{Bi}$ [9]. For simplicity, we follow Ref. [7] and assume the transition density which is proportional to $\rho_{fi} \propto \delta(r - R)Y_{JM}(\theta, \phi)$. The differential cross sections $d\sigma/dE_e$ for the $^{56}\text{Fe}(\nu_e, e^-)^{56}\text{Co}$ and the $^{208}\text{Pb}(\nu_e, e^-)^{208}\text{Bi}$ reactions are shown in Figure 1. The solid line shows the results in the absence of the environmental electrons. The top, middle and bottom panels are for the electron density of $\rho_e^0 = 10^{32}, 10^{33}$, and $10^{34}$ cm$^{-3}$, respectively. The dotted line denotes the results with the electron screening effects, while the dashed line takes into account both the screening and the Pauli blocking effects. For the electron density smaller than $10^{31}$ cm$^{-3}$, we find that both the effects are marginal. The screening effect is larger in the $^{208}\text{Pb}(\nu_e, e^-)^{208}\text{Bi}$ than in the $^{56}\text{Fe}(\nu_e, e^-)^{56}\text{Co}$ reaction, as is expected. We confirm that Pauli blocking effects are important below the Fermi energies and that the screening and the Pauli blocking effects disappear in the high $E_e$ limit.

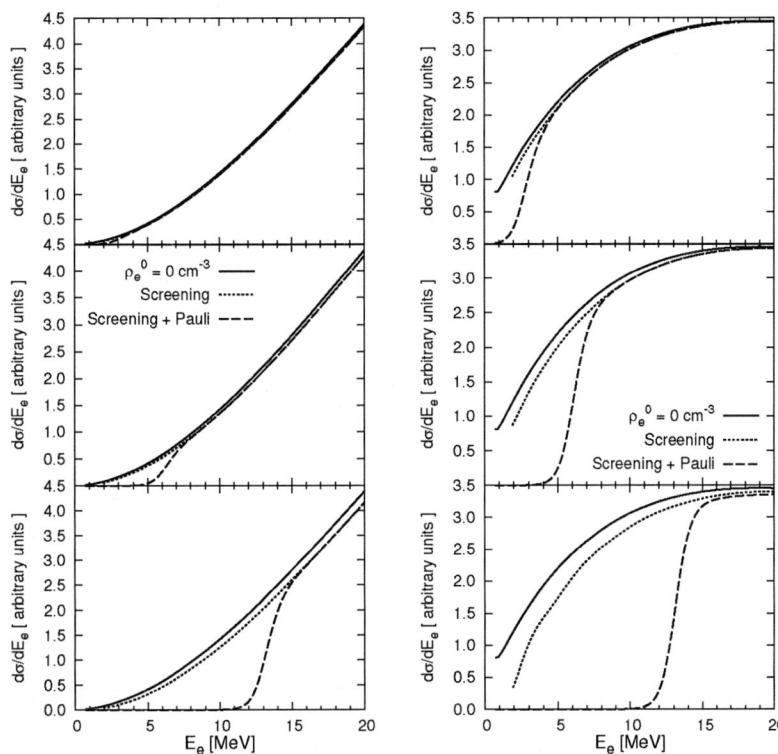

**FIGURE 1.** The cross sections for the charged current $\nu_e + {}^{56}\text{Fe} \to e^- + {}^{56}\text{Co}$ (left panel) and $\nu_e + {}^{208}\text{Pb} \to e^- + {}^{208}\text{Bi}$ (right panel) as a function of the energy of the outgoing electron. $\rho_e^0$. The top, middle, and bottom panels are for $\rho_e^0 = 10^{32}, 10^{33},$ and $10^{34}$ cm$^{-3}$, respectively.

We next discuss the total cross sections. In order to compute the total cross sections, we integrate the differential cross sections with a weight factor given by the energy distribution for the incident neutrino $n_\nu(E_\nu)$ [10]. Figure 2 shows the total cross sections for the $^{56}\text{Fe}(\nu_e, e^-)^{56}\text{Co}$ and the $^{208}\text{Pb}(\nu_e, e^-)^{208}\text{Bi}$ reactions as a function of the density of the environmental electrons, respectively. These are plotted as the ratio to the total cross sections in the absence of the environmental electrons, $\sigma_0$. The neutrino temperature $T_\nu$ is set to be 4 MeV. The dotted line takes into account only the screening effects, while the dashed line includes both the screening and the Pauli blocking effects. We see that the Pauli blocking effect influences the $^{208}\text{Pb}(\nu_e, e^-)^{208}\text{Bi}$ reaction much more significantly than the $^{56}\text{Fe}(\nu_e, e^-)^{56}\text{Co}$ reaction. This is due to the differences of the energy of the outgoing electron produced in each reactions.

In summary, we have discussed the electron screening as well as the Pauli blocking effects due to the environmental electrons on cross sections of the neutrino-nucleus reaction. For this purpose, we used the Thomas-Fermi theory for the screening potential,

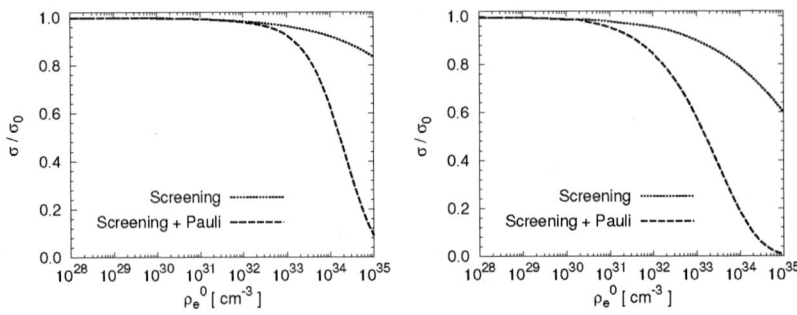

**FIGURE 2.** The total cross sections for the $v_e + {}^{56}\text{Fe} \to e^- + {}^{56}\text{Co}$ reaction (left panel) and $v_e + {}^{208}\text{Pb} \to e^- + {}^{208}\text{Bi}$ reaction (right panel) as a function of the density of the environmental electrons.

and the DWBA with the Pauli blocking factor for the cross sections. Our results for the $^{208}\text{Pb}(v_e, e^-)^{208}\text{Bi}$ and $^{56}\text{Fe}(v_e, e^-)^{56}\text{Co}$ reactions show that both the effects hinder the cross sections, especially at high electron densities. We have also shown that the Pauli blocking effect is more significant in the former reaction than in the latter reaction. The screening effect is also larger in the former reaction because of the larger atomic number.

We thank Y. Kato for useful discussions on electron screening. F. M. also acknowledges the 21st Century for Center of Excellence (COE) Program "Exploring New Science by Bridging Particle-Matter Hierarchy" and "Initiatives of Attractive Education in Graduate Schools" program at Tohoku University for financial support. This work was supported in part by the Grant-in-Aid for Scientific Research, Contract No. 16740139 from the Japanese Ministry of Education, Culture, Sports, Science, and Technology, in part by the U.S. National Science Foundation Grant No. PHY-0555231, and in part by the University of Wisconsin Research Committee with funds granted by the Wisconsin Alumni Research Foundation.

## REFERENCES

1. A. B. Balantekin and G. M. Fuller, J. Phys. G **29**, 2513 (2003).
2. T. Yoshida, T. Kajino, H. Yokomakura, et. al, Phys. Rev. Lett. **96**, 091101 (2006).
3. J. D. Walecka, *Theoretical Nuclear and Subnuclear Physics* (Oxford University Press, New York, 1995), Sec. 45.
4. M. Morita, *Beta decay and muon capture* (W.A. Benjamin Inc., Reading, 1973).
5. N. Itoh, N. Tomizawa, M. Tamamura and S. Wanajo, Astrophys. J. **579**, 380 (2002).
6. F. Minato, K. Hagino, N. Takigawa, A. B. Balantekin, Ph. Chomaz, arXiv:nucl-th/0608045.
7. J. Engel, Phys. Rev. C **57**, 2004 (1998).
8. T. Murakami, S. Nishihara and T. Nakagawa, Nucl. Phys. **A403**, 317 (1983).
9. A. Krasznahorkay, H. Akimune, M. Fujiwara et. al., Phys. Rev. C **64**, 067302 (2001).
10. Y.-Z. Qian and G. M. Fuller, Phys. Rev. D **51**, 1479 (1995).

# The importance and the sensitivity of the reaction $^{17}O(n\,\gamma)^{18}O$ in the s-process nucleosynthesis

K. Yamamoto, T. Wada, M. Ohta, T. Nishimura[1][†], M. Y. Fujimoto[†], K. Kato[†], T. Suda and M. Aikawa[‡]

*Department of Physics, Konan University, 8-9-1 Okamoto, Kobe 658-8501, Japan*
[†]*Division of Science, Hokkaido University, Sapporo 060-0810, Japan*
*Center for the Early Universe, University of Tokyo 113-0033, Japan*
[‡]*Hokkaido University OpenCourseWare, Sapporo 060-0811, Japan*

**Abstract.** We apply our Monte Carlo method to the s-process element synthesis in the helium shell flash model for extremely metal-poor stars. We emphasize the importance of the reaction $^{17}O(n\,\gamma)^{18}O$. It is presented that the uncertainty of this reaction rate affects the heaver elements abundance. We also discuss the variation of the O-Ne reaction paths with the uncertainty of the reaction rate of $^{17}O(n\,\gamma)^{18}O$.

**Keywords:** Neutron-capture reaction: Reaction rate: Nucleosynthesis:
**PACS:** 26.50.+f;26.30.+k

## INTRODUCTION

In the previous study, we demonstrated that we can determine the key reaction by the Monte Carlo method [1] in the nucleosynthesis in a certain phase of the stellar evolution. We applied our method to search the reaction paths and to determine the key reaction by combining it with the data from the network calculation. The combination of the Monte Carlo method and the network calculation gives us a better understanding of the mechanism of the stellar nucleosynthesis.

In this paper, we consider the s-process nucleosynthesis of the helium shell flash model in extremely metal-poor stars. We emphasize the importance of the reaction $^{17}O(n\,\gamma)^{18}O$. The neutron-capture reaction rate for $^{17}O$ affects the abundance of neon isotopes and hence it affects the heavy element abundance. Since there are no experimental data for this reaction, it is needed to give the theoretical prediction from the nuclear reaction theory in nuclear astrophysical energy region.

In order to check the sensitivity of the s-process nucleosynthesis in the helium shell to the reaction rate for $^{17}O(n\,\gamma)^{18}O$, we apply several values for this reaction rate in the Monte Carlo calculation of reaction network starting from $^{16}O$. We compare the results with different reaction rates and discuss the shift of the reaction paths.

---

[1] JSPS Research Fellow.

# FORMULATION AND APPLICATION

## Method for the reaction paths

In our Monte Carlo method, the element synthesis paths leading to specific elements under stellar nucleosynthesis can be elucidated. We consider the reaction $i(j\ l)k$, the weight function $R_j(t)$ of decay probability for the element $i$ is denoted as

$$R_j(t) = Y_j(t)\rho(t)N_A\ \sigma v_{\ i\to k} \tag{1}$$

where $Y_j(t)$[mol/g], $\rho(t)$[g/cm$^3$] and $N_A\ \sigma v_{\ i\to k}$[cm$^3$/sec/mol] mean the number abundance of an element $j$, the average density and the average reaction rate for the reaction $i(j\ l)k$, respectively. The weight function in Eq.1 has the unit of second inverse. Here, we use the set of $Y_j(t)$ and $\rho(t)$ given by the network calculation. The decay probability for the element $i$ is given by $1 - P_i$. We calculate the residual probability $P_i$ of the nucleus $i$ to survive the transmutation in the time interval $t - t_0$ as

$$P_i(t) = \exp\left[-\int_{t_0}^{t}\sum_j R_j(\tau)d\tau\right] \tag{2}$$

where $t_0$ is the initial time.

## Condition of the Monte Carlo calculation

We apply our Monte Carlo method to the case of the helium shell flashes in the extremely metal-poor stars experiencing $^{13}$C mixing [2]. The detailed description about the model is given in Refs[2, 3, 4]. The mixing amounts of $^{13}$C relative to $^{12}$C, $\Delta(^{13}\text{C}/^{12}\text{C})$, is set at $10^{-3}$ in this work to be consistent with the resultant abundance of HE0107-5240 [5]. The abundance in the helium convective zone before the mixing is taken from the result of the evolution of 2M$_\odot$ model.

We perform the Monte Carlo simulation from the initial time $t_0$ to the final time $t_f$ decided from the condition that the luminosity by helium burning decreases to $10^4 L_\odot$ when the nuclear products are sufficiently relaxed. The environment of the Monte Carlo calculation at time from $t_0$ to $t_f$ is indicated in Fig:1.

Especially for $^{17}$O, we used two kinds of the average reaction rate for the neutron-capture reaction. One is the reaction rate based on $^{18}$O(n $\gamma$)$^{19}$O from Bao's paper [6], because the reaction $^{17}$O(n $\gamma$)$^{18}$O has no experimental data. We can use this reaction rate as the lower limit data. The other is taken from the Brussels Nuclear Astrophysics Compilation of Reaction Rates (BRUSLIB) [7]. This reaction rate is taken from a theoretical estimation based on a microscopic model (HFB+QRPA). This data is used as the upper limit of the reaction rate $^{17}$O(n $\gamma$)$^{18}$O. The BRUSLIB data is about 1000 times larger than Bao's data.

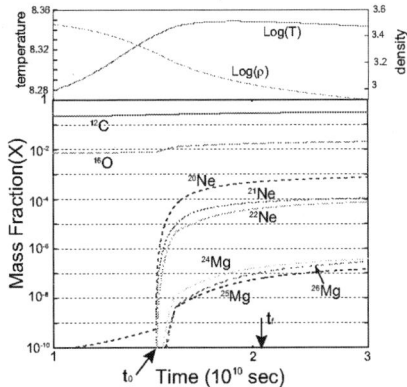

**FIGURE 1.** Time variation of the isotopic abundance calculated by network simulation. In the top panel, the solid and the dashed lines denote the temperature ($T$) and the average density ($\rho$), respectively. In the bottom panel, the abundances are drawn.

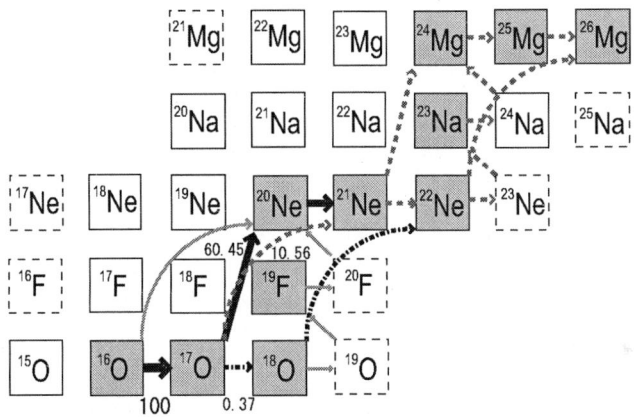

**FIGURE 2.** The element synthesis paths creating elements heaver than neon isotopes from $^{16}$O with our Monte Carlo calculation. The rate of reaction $^{17}$O(n $\gamma$)$^{18}$O is assumed by referring the data given by Bao *et al.* [7].

## RESULTS FOR THE REACTION PATHS

In Fig:2, we show the element synthesis paths from $^{16}$O to heaver than neon isotopes. We perform the Monte Carlo simulation from the initial time $t_0$ to the final time $t_f$ in Fig:1. In this case, we used the lower reaction rate $^{17}$O(n $\gamma$)$^{18}$O taken from Ref. [6]. As can be seen from the figure, the main path is $^{16}$O(n $\gamma$)$^{17}$O($\alpha$ n)$^{20}$Ne. This path works as a neutron source, the path was counted 60.45%. In the element synthesis path to heaver than neon isotopes, the path is $^{16}$O(n $\gamma$)$^{17}$O($\alpha$ $\gamma$)$^{21}$Ne by 10.56%. On the other hand, neutron-capture reaction by $^{17}$O was counted only 0.37%.

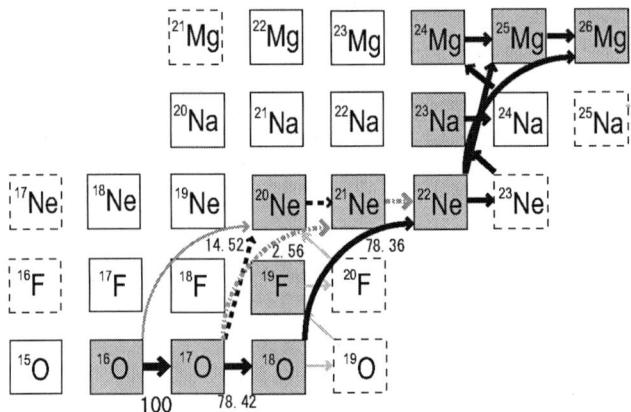

**FIGURE 3.** The synthesis paths creating elements from $^{16}$O with our calculation using BRUSLIB data for the reaction rate in $^{17}$O(n $\gamma$)$^{18}$O.

In Fig:3, the results with the higher reaction rate from BRUSLIB data are shown. In this case, the main path is shifted to $^{16}$O(n $\gamma$)$^{17}$O(n $\gamma$)$^{18}$O($\alpha$ $\gamma$)$^{22}$Ne, the path was counted 78.42%. The neutron-capture reaction of $^{17}$O becomes active.

## CONCLUSION

We discussed transition of the synthesis paths due to the variational reaction rates. There is a huge ambiguity in the reaction rate for $^{17}$O(n $\gamma$)$^{18}$O. The sensitivity of this reaction rate affects the neutron source reaction $^{17}$O($\alpha$ n)$^{20}$Ne and the heaver elements synthesis in the metal-poor star nucleosynthesis.

It is important to look for key reactions whose reaction rates are decisive for the shift of reaction paths in stellar nucleosynthesis.

## REFERENCES

1. K. Yamamoto, K. Hashizume, T. Wada, M. Ohta, T. Nishimura, M. Y. Fujimoto, K. Katō, T. Suda and M. Aikawa, Proc. Int. Symp. *"Origin of Matter and Evolution of Galaxies"*, ed. by Kobono *et al.*, AIP Conference proceedings **847** (2006) 497.
2. T. Suda, M. Aikawa, M. N. Machida and M. Y. Fujimoto, *Astrophys. J.* **611** (2004) 476.
3. M. Y. Fujimoto, M. Aikawa and K. Katō, *Astrophys. J.* **519** (1999) 733.
4. M. Y. Fujimoto, Y. Ikeda and I. Iben Jr., *Astrophys. J.* **529** (2000) L25.
5. T. Nishimura, N. Iwamoto, T. Suda, M. Aikawa, M. Y. Fujimoto and I. Iben Jr., Proc. the Int. Symp. *"Origin of Matter and Evolution of Galaxies"*, ed. by Kobono *et al.*, AIP Conference proceedings **847** (2006) 455.
6. Z. Y. Bao *et al.*, *At. Data Nucl. Data Tables* **76** (2000) 70.
7. S. Goriely and E. Khan, *Nucl. Phys.* **A706** (2002) 271.

# Microscopic Description of Scission Configurations

## N. Dubray, H. Goutte and J.F. Berger

*CEA/DAM Ile de France DPTA/Service de Physique Nucléaire, BP 12, F-91680 Bruyères-le-Châtel, France*

**Abstract.** Properties of $^{226}$Th, $^{256}$Fm, $^{258}$Fm and $^{260}$Fm nuclei in the scission region are described using a full-microscopic Hartree-Fock-Bogoliubov approach with the effective Gogny nucleon-nucleon interaction. In a first step, the Potential Energy Surfaces are computed in the ($q_{20}$, $q_{30}$) plane, the scission lines are found, fulfilling a given criterion on the density in the nuclear neck. Finally a few properties of the fragments along this line are presented.

**Keywords:** Hartree-Fock-Bogoliubov, fission, scission line, fragment properties
**PACS:** 21.10.Dr, 21.60.Jz, 21.10.Ft, 21.10.Gv

## INTRODUCTION

The description of the nuclear fission process can be considered as a good test for nuclear theories, especially for microscopic ones [1]. In this study we check the quality of reproduction of a few properties of fission fragments, for a given set of compound nuclei, using a Gogny-force-based Hartree-Fock-Bogoliubov self-consistent approach. Employing constraints on the elongation ($q_{20}$) and on the asymmetry ($q_{30}$) of the nuclear system, two-dimensional potential energy surfaces are obtained. A refinement process for the deformation mesh is used to get an accurate description of the scission line. Finally, fragment distance and total kinetic energy are computed along this line.

## THEORETICAL FRAMEWORK

We use the Hartree-Fock-Bogoliubov (HFB) method [2] with the Gogny effective nucleon-nucleon interaction [3]. This finite range and density dependent interaction allows the simultaneous treatment of the nuclear mean field and of pairing correlations. We have used the D1S set of parameters [4], which is known for its good reproduction of nuclear properties [5]. In order to obtain the total energy landscape of the nuclear system in the (elongation, asymmetry) coordinates, we introduce in the main HFB equation additional terms called *constraints*. This leads to a constrained HFB equation:

$$\delta \langle \varphi | \hat{H} - \lambda_N \hat{N} - \lambda_Z \hat{Z} - \lambda_{10} \hat{Q}_{10} - \lambda_{20} \hat{Q}_{20} - \lambda_{30} \hat{Q}_{30} | \varphi \rangle = 0. \quad (1)$$

The purpose of these constraints is to fix the mean values of the numbers of neutrons and protons of the nuclear system as well as those for the usual multipole operators $\hat{Q}_{10}$, $\hat{Q}_{20}$ and $\hat{Q}_{30}$ in the following way:

$$\langle \varphi | \hat{N} | \varphi \rangle = N, \quad (2)$$

$$\langle\varphi|\hat{Z}|\varphi\rangle = Z, \quad (3)$$
$$\langle\varphi|\hat{Q}_{10}|\varphi\rangle = 0, \quad (4)$$
$$\langle\varphi|\hat{Q}_{20}|\varphi\rangle = q_{20}, \quad (5)$$
$$\langle\varphi|\hat{Q}_{30}|\varphi\rangle = q_{30}, \quad (6)$$
$$\text{with} \quad \hat{Q}_{\lambda 0} = \sqrt{\frac{4\pi}{2\lambda+1}} \sum_{i=1}^{A} r_i^\lambda Y_{\lambda 0}(\theta_i, \phi_i). \quad (7)$$

The constraint on the dipole moment $\hat{Q}_{10}$ aims at imposing a fixed position to the center of mass of the system. By letting $q_{20}$ and $q_{30}$ take regularly spaced values, we are able to draw the total energy map of the system in the $(q_{20}, q_{30})$ plane.

Solving Eq. (1) is performed by expanding the quasi-particle operators onto axially-symmetric harmonic oscillator bases. As a consequence, the conservation of the $z$-axis symmetry of the system is enforced. For each $(q_{20}, q_{30})$ couple, the parameters describing the bases are optimized, i.e. they are chosen in order to minimize the total energy.

Finally, let us mention that a rotational zero-point energy correction is extracted from the potential energy of the system.

## DEFINITION OF THE SCISSION LINE

At large quadrupole moments, it becomes energetically more favorable for the nuclear system to appear as two separated fragments, rather than one elongated shape with a neck. In deformation space, this transition corresponds to the passage from the so-called fission valley to the so-called fusion valley. Several observables are affected by this process:

- The neck between the fragments vanishes. This can be observed by measuring the average number of nucleons in a small region in between the fragments (cf. the *slice operator* in [6]).
- The hexadecapole moment of the system decreases by a factor of 2.
- There is a drop of 10 to 15 MeV in the potential energy of the system.

If a point from the fission valley leads to a point in the fusion valley by a small increase in one deformation parameter, this point is called a *scission point*. Since we are working in a 2-dimensional deformation space $(q_{20}, q_{30})$, the ensemble of considered scission points forms a line, which we call the *scission line*.

## RESULTS

Computations have been performed for $^{226}$Th, $^{256}$Fm, $^{258}$Fm and $^{260}$Fm nuclei. Experimental data for the total kinetic energy of the fragments of $^{226}$Th are available [7]. This nucleus is known for presenting a three-humped structure in its fragment mass distribution. Concerning the Fm isotopes, their experimental fragment mass distributions show a transition from asymmetric fission ($^{256}$Fm) to symmetric fission ($^{260}$Fm).

As one can see in Fig. 1, there can exist several scission points with different elongations for a given value of the asymmetry. This can lead to multiple-valued curves when plotted against the number of particles in the fragments (cf. Fig. 2). To avoid such a feature, one could have used directly a constraint on the number of particles in one fragment, instead of the $q_{30}$ constraint.

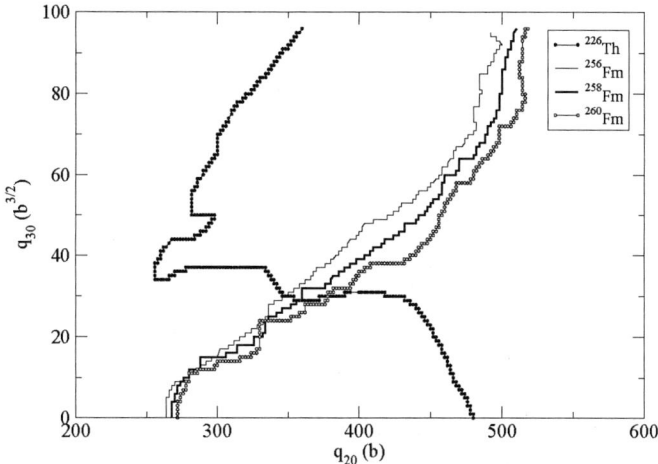

**FIGURE 1.** Scission lines in the ($q_{20}$, $q_{30}$) plane.

Plotting the potential energy along the scission lines (Fig. 2) seems to be a good indicator for the coexistence of two fission modes for $^{226}$Th, and for an increase in the ratio symmetric fission / asymmetric fission in Fm isotopes. Dynamical calculations will be performed to see if a transition from asymmetric to symmetric fission is obtained in this isotopic chain. The total kinetic energies of the fragments computed for $^{226}$Th (Fig. ??) are in rather good agreement with experimental data [7].

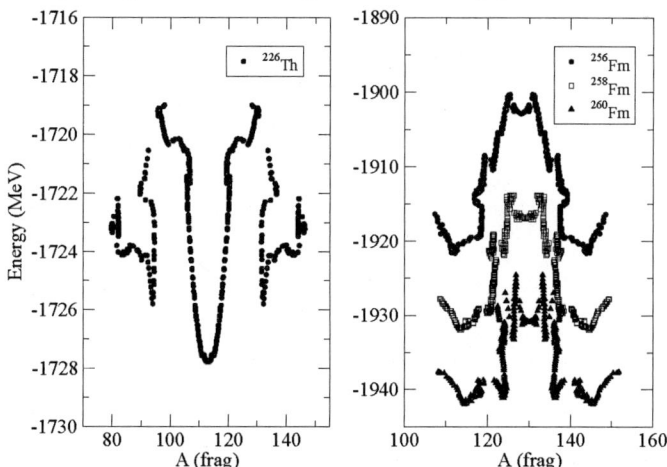

**FIGURE 2.** Potential energy along the scission lines.

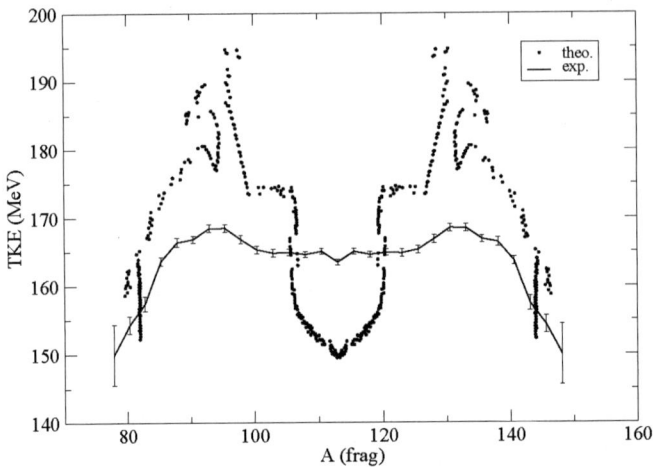

**FIGURE 3.** Total fragment kinetic energies in $^{226}$Th.

## CONCLUSION

The results obtained in the present study of Th and Fm nuclei clearly show that our fully microscopic approach is able to provide a quantitative account of scission properties of actinide nuclei. Additional observables, as fragment deformation energies and number of emitted neutrons are being computed, with first results in good agreement with data. The description of time-dependent fission dynamics and of fragment mass distributions with the method of Ref. [8] will be carried out in the near future. Finally, extensions of the present calculations to a three-dimensional mesh ($q_{20}$, $q_{30}$, $q_{40}$) and to non-axial nuclear shapes are envisaged.

## REFERENCES

1. C. Wagemans, *The nuclear fission process*, CRC Press (1991)
2. P. Ring and P. Schuck, *The Nuclear Many Body Problem* (1980), Springer-Verlag Edt. Berlin.
3. J. Dechargé and D. Gogny, Phys. Rev. **C21** (1980) 302
4. J.F. Berger, M. Girod and D. Gogny, Nucl. Phys. **A428** (1984) 23c
5. M. Warda, J.L. Egido and L.M. Robledo and K. Pomorski, Phys. Rev. **C66** (2002) 014310
6. L.M. Robledo and J.L. Egido, *Nuclear Fission and Fission-Product Spectroscopy*, AIP Conference Proceedings **798** (2005) 103
7. K.-H. Schmidt et al., Nucl. Phys. **A665** (2000) 221
8. H. Goutte, J.F. Berger, D. Gogny and W. Younes, *Low energy fission: dynamics and scission configurations*, AIP Conference Proceedings **798** (2005) 69

# Quantum diffusion approach to the formation of a heavy compound nucleus by heavy-ion fusion reactions

Kouhei Washiyama*, Noboru Takigawa* and Sakir Ayik[†]

*Department of Physics, Tohoku University, 980-8578 Sendai, Japan*
[†]*Physics Department, Tennessee Technological University, Cookeville, Tennessee 38505, USA*

**Abstract.** We discuss quantum effects in the diffusion process which is used to describe the shape evolution from the touching configuration of fusing two nuclei to a compound nucleus. Applying the theory with quantum effects to the case where the potential field, the mass and the friction parameters are adapted to realistic values of heavy-ion collisions, we show that the quantum effects play significant roles at low temperatures which are relevant to the synthesis of superheavy elements. We also discuss the mass distribution of the quasi-fission fragments calculated in the two-dimensional quantum Langevin approach.

**Keywords:** Quantum diffusion, Non-Markovian effect, Langevin equation, Superheavy elements
**PACS:** 25.70.Jj, 02.50.Ga, 05.40.-a, 05.60.Gg

It is now well accepted that it is not sufficient for the two nuclei in heavy-ion collisions to overcome the Coulomb barrier to form a compound nucleus such as superheavy elements. This is because the conditional saddle is located inside the Coulomb barrier for collisions between two heavy nuclei. This provides an origin of the so-called fusion hindrance phenomena [1]. We thus need to describe the shape evolution from the touching configuration of fusing two nuclei to a more compact spherical-like compound nucleus by overcoming a potential barrier near the conditional saddle point.

A diffusion model has been applied to describing this process, especially to describing the formation of superheavy elements [2, 3, 4]. In these studies, so far the standard classical fluctuation-dissipation relation which holds at high temperatures has been postulated to relate the diffusion coefficients to the friction coefficients. Although these studies provide some illuminating information and look to be successful to some extent in the data analysis, one needs to carefully examine the validity of the standard fluctuation-dissipation relation in order to apply to the diffusion process at low temperatures which are relevant to the synthesis of superheavy elements. Since superheavy elements are stabilized by shell correction energies, one has to synthesize them at reasonably low excitation energies, that is, at low temperatures as low as 1 MeV or below. On the other hand, the barrier curvature around the conditional saddle point is also of the order of 1 MeV. It is thus likely that quantum effects play an important role in the compound nucleus formation process, especially in the synthesis of superheavy elements.

One can find a diffusion theory with quantum effects in some literatures. However, most of them handle the quantum diffusion process in a potential well. To the contrary, our problem is the quantum diffusion process along a potential barrier. In order to adapt to the diffusion process along a potential barrier, especially at low temperatures, we have developed a quantum diffusion theory that takes the quantum fluctuation due to the finite curvature of a potential barrier into account [5, 6].

For the diffusion process in a potential well, it is known that the ratio of the diffusion to the friction coefficients is given by,

$$\frac{D}{\gamma} = \frac{1}{2}\hbar\Omega\coth(\frac{\hbar\Omega}{2T}) \quad (1)$$

if the quantum fluctuation is taken into account. In Eq. (1) $D$, $\gamma$, and $T$ are the diffusion and friction coefficients and the temperature, respectively. The $\Omega$ is defined by $\Omega = \sqrt{V''(R_b)/M}$, where $V''(R_b)$ and $M$ are the second derivative of the potential well at the bottom position of the potential $R_b$ and the mass parameter, respectively. The relevant formula for the diffusion process along a potential barrier can be obtained by analytic continuation of Eq. (1) with respect to the frequency parameter $\Omega$ [5, 7]. The result reads,

$$\frac{D}{\gamma} = \frac{1}{2}\hbar\Omega\cot(\frac{\hbar\Omega}{2T}), \quad (2)$$

with $\Omega = \sqrt{|V''(R_B)|/M}$, where $V''(R_B)$ is the second derivative of the potential at the barrier top position $R_B$. One can easily confirm that both formulas, Eqs. (1) and (2), reduce to the classical fluctuation-dissipation theorem, $D/\gamma = T$, in the high temperature limit where the thermal fluctuation far dominates the quantum fluctuation.

Our theory incorporates also the second quantum effect, i.e., the memory effect, which leads to a colored noise problem. The detail is given in [5, 6].

In applying our theory to the compound nucleus formation process we use the liquid drop model [8] to calculate the potential energy surface in the space of nuclear deformation, the hydrodynamical mass [9] for the mass parameter, and the one-body dissipation [10, 11] for the friction tensor. The colored noise random force is handled by the spectral method in Ref. [12]. We use the separation distance between two fragments to describe the dynamics from inside the Coulomb barrier to inside the conditional saddle and determine its time evolution by solving the Langevin equation for single macroscopic variable one hundred thousands times. The other macroscopic degrees of freedom in the two-center shell model parametrization [13] are frozen during the compound nucleus formation process as; the mass partition parameter $\alpha = (A_1 - A_2)/(A_1 + A_2) = 0$ with $A_1$ and $A_2$ being the mass number of each fragment, the deformation parameter $\delta_1 = \delta_2 = 0$, and the neck parameter $\varepsilon = 0.9$. We initiate each trajectory from the touching configuration of the two fusing nuclei with zero momentum. This corresponds to assuming that a strong energy dissipation from the macroscopic motion, i.e., the relative motion between the fusing two fragments, to nuclear intrinsic motions takes place inside the Coulomb barrier. For simplicity, we ignore the change of the temperature of nuclear intrinsic degrees of freedom during the time evolution of the system.

The left panel of Fig. 1 compares the compound nucleus formation probability as a function of nuclear temperature calculated by the quantum diffusion theory (the solid line) and by the classical diffusion theory (the dashed line) which postulates the standard fluctuation-dissipation relation for the $^{110}$Pd + $^{110}$Pd reactions. The figure shows that quantum effects become significant at low temperatures relevant to the experiments

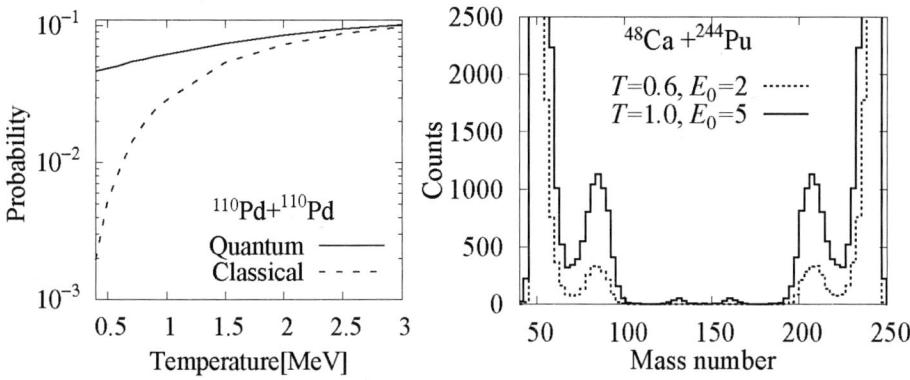

**FIGURE 1.** (The left panel) The compound nucleus formation probability as a function of nuclear temperature for the $^{110}$Pd $+^{110}$Pd. (The right panel) Mass distribution of quasi-fission fragments in the $^{48}$Ca$+^{244}$Pu reaction. Temperature ($T$) and initial kinetic energy ($E_0$) are in unit of MeV.

to synthesize superheavy elements. They increase the compound nucleus formation probability at low temperatures.

We now extend our theory to incorporate the mass partition degree of freedom in order to see the quantum effects on the mass distribution of quasi-fission fragments as well as on the compound nucleus formation probability. To that end, we solve two-dimensional Langevin equation to follow the time evolution of the mass asymmetry parameter $\alpha$ as well as the separation distance between two fragments. Since the shell correction energy plays an important role in determining the mass distribution, we calculate the potential energy as the sum of the energy in the liquid drop model and the shell correction energy. We calculate the potential energy by using the computer code of the two center shell model [14, 15]. We use the same models as those in the calculations with a fixed mass partition mentioned above for the mass and friction parameters.

The right panel of Fig. 1 shows the mass distribution of quasi-fission fragments in the $^{48}$Ca $+^{244}$Pu reaction which was used to synthesize the new element with the atomic number 114 in Ref. [16]. The dotted line is the result for the temperature ($T$) and the initial kinetic energy ($E_0$) being 0.6 MeV and 2 MeV, respectively, while the solid line for $T = 1.0$ MeV and $E_0 = 5$ MeV. The peak around the mass number 240 corresponds to deep inelastic collisions. In order to understand the peaks appearing around the mass number 210 ($\alpha \sim 0.42$) and 130 ($\alpha \sim 0.1$), we study the potential energy surface of the system with the total mass number 292 and the total atomic number 114 in the two-dimensional space, which is described by the separation distance between two potential centers and the mass partition parameter. The other degrees of freedom are frozen such that the deformation parameters are $\delta_1 = \delta_2 = 0$ and the neck parameter $\varepsilon = 1.0$. We find that the shell effects create valleys on the potential energy surface and the peaks in the mass distribution correspond to the mass flow along these valleys of the potential surface. As the temperature rises, the fluctuation of the trajectory becomes larger. Consequently, the mass distribution becomes broader. This tendency resembles that in the previous studies with the classical diffusion model [17].

The comparison of the results of the mass distribution of quasi-fission fragments and the compound nucleus formation probability with the quantum diffusion theory and those with the classical diffusion theory is now in progress. It is an interesting and important question to clarify whether one can probe the quantum effects in various observables in a consistent way.

In summary, we have discussed quantum effects in the formation process of a heavy compound nucleus described as a diffusion process along a potential barrier. We have shown that the quantum effects increase the compound nucleus formation probability at low excitation energies, which are relevant to the synthesis of superheavy elements. We also discussed the mass distribution of quasi-fission fragments in the mass region of superheavy elements using the two-dimensional quantum diffusion theory and discussed the relation between the structure of the potential energy surface and the peaks in the mass distribution. We are now working to explore the quantum effects in the mass distribution of quasi-fission fragments.

We wish to thank Prof. T. Wada and Dr. T. Asano for useful discussions. The work of K.W. was supported by the Japan Society for the Promotion of Science for Young Scientists.

## REFERENCES

1. W. J. Swiatecki, *Physica Scripta*, **24**, 113–122 (1981).
2. Y. Aritomo, T. Wada, M. Ohta, and Y. Abe, *Phys. Rev. C* **59**, 796–809 (1999).
3. C. Shen, G. Kosenko and Y. Abe, *Phys. Rev. C* **66**, 061602(R)1–5 (2002).
4. Y. Aritomo and M. Ohta, *Nucl. Phys.* **A744**, 3–14 (2004).
5. N. Takigawa, S. Ayik, K. Washiyama, and S. Kimura, *Phys. Rev. C* **69**, 054605-1–5 (2004).
6. S. Ayik, B. Yilmaz, A. Gokalp, O. Yilmaz, and N. Takigawa, *Phys. Rev. C* **71**, 054611-1–8 (2005).
7. H. Hofmann, *Phys. Rep.* **284**, 137–380 (1997).
8. H. J. Krappe, J. R. Nix, and A. J. Sierk, *Phys. Rev. C* **20**, 992–1013 (1979).
9. K. T. R. Davies, A. J. Sierk, and J. R. Nix, *Phys. Rev. C* **13**, 2385–2412 (1976).
10. J. Blocki, et. al., *Ann. Phys.* **113**, 330–386 (1978).
11. J. Randrup and W. J. Swiatecki, *Nucl. Phys.* **A429**, 105–115 (1984).
12. A. H. Romero and J. M. Sancho, *J. Comp. Phys.*, **156**, 1–11 (1999).
13. J. Maruhn and W. Greiner, *Z. Phys.* **251**, 431–457 (1972).
14. S. Suekane, A. Iwamoto, S. Yamaji, and K. Harada, *JAERI-memo*, 5818 (1974).
15. A. Iwamoto, S. Yamaji, S. Suekane, and K. Harada, *Prog. Theor. Phys.* **55**, 115–130 (1976).
16. Yu. Ts. Oganessian, et. al., *Phys. Rev. Lett.* **83**, 3154–3157 (1999).
17. Y. Aritomo and M. Ohta, *Nucl. Phys.* **A753**, 152–173 (2005).

# Variation of variance of fission fragment mass distribution: a probe to study the dynamics of fusion-fission reactions

Tilak Kumar Ghosh

*Variable Energy Cyclotron Centre*
*1/AF, Bidhannagar, Kolkata 700064, India*

**Abstract.** Fragment mass distributions in fusion-fission reactions of light projectiles ($^{12}$C, $^{16}$O and $^{19}$F) on thorium and bismuth targets in near and below Coulomb barrier energies are investigated. Precise and systematic measurements of mass distribution shows a sudden anomalous increase in variances of mass distributions ($\sigma_m^2$) near Coulomb barrier energies for all three projectiles with deformed thorium target, in contrast to a smooth variation of $\sigma_m^2$ with energy for spherical bismuth target. Microscopic effects due to change in entrance channel shape compactness for projectiles hitting the polar region of prolate thorium target is postulated to reach a almost symmetric saddle without complete fusion for events of anomalous fragment widths.

**Keywords:** Fusion-fission reactions, mass and angular distributions.
**PACS:** 25.70.Jj

The evolution of a super-heavy element in fusion of two nuclei is restricted by two key factors, one is the primary fusion process to produce the super-heavy element, and the other is the radioactive decay of the fused system to survive long enough. The fusion of the two nuclei from a touching configuration to a composite system equilibrated in all macroscopic degrees of freedom is governed by the path the system takes in a complicated multidimensional energy landscape [1]. The potential energy landscape in multi-dimensions depends critically on excitations and deformations of the fusing masses, the mass asymmetry, necking and the separation between two masses [2,3]. Depending upon the initial conditions the system can equilibrate to a compound nucleus. The fused system in fusion meadow may cool down with emission of a few particles and be referred as an evaporation residue (ER), or more frequently, with shape changes over a saddle ridge, slides down into a fission valley undergoing fission. However, it is also possible that the path in the energy landscape do not reach a fusion meadow, but re-separate into two fragments with altered mass asymmetry and excitation energies and deformations in a process which is known as a quasi-fission event.

The direct evidence of the system following a path to a fusion meadow is the observation of the ER's. In reactions in which fusion is followed by fission, the indirect evidence of the fusing system reaching a equilibrated compound nucleus is the angular distribution of the fission fragments following macroscopic statistical laws [4]. In many systems with a deformed target and/or projectile, it had been observed that at near and below Coulomb barrier energies, the angular anisotropy shows an anomalous increase over that predicted by the statistical models. The observed anomalous angular anisotropies are sought to be explained in terms of pre-equilibrium or quasi-fission model [5]. A critical test for the two possibilities (compound or non-compound) is a direct measurement of the ER's. Hinde et al reported hindrance in fusion of $^{220}$Th with different entrance channel target and projectiles of decreasing mass asymmetry [6]. Similar observations were reported in the fusion of $^{216}$Ra [7]. Recently, completely conflicting experimental evidence on the ER production in the deformed target of $^{238}$U with $^{16}$O has been reported [8]. It has been shown that even for projectiles hitting the polar region of the uranium nuclei, the fusion of the oxygen nuclei are unhindered when compared with those cases where the oxygen hits the equatorial region of deformed uranium nuclei, although angular anisotropy of fragments are reported to be anomalously large compared to macroscopic predictions [9]. So it is clear that more experimental probes, other than the angular anisotropy or the production of ER's are needed to understand the path of fusion of two nuclei in the energy landscape leading to different processes.

In the experimental studies, we have determined precisely the fission fragment mass distributions in reactions of spherical or very slightly deformed light projectiles of $^{19}$F, $^{16}$O and $^{12}$C on spherical $^{209}$Bi and deformed $^{232}$Th targets in near and below Coulomb barrier energies. Any departure of the smooth variation of the width of the mass distributions is looked for. The experiments were performed with pulsed heavy ions from the 15UD Pelletron at Nuclear Science Centre (NSC), New Delhi, India. The pulse width was about 0.8-1.5 ns with a pulse separation of 250 ns. The targets were either self-supporting $^{232}$Th of thickness 1.8 mg/cm$^2$ or a 500 µg/cm$^2$ thick self-supported $^{209}$Bi. Complimentary fission fragments were detected with two large area (24 cm x 10 cm) X-Y position sensitive multi-wire proportional counters (MWPCs) [10] developed at our laboratory. Folding angle technique [11] was used to differentiate between fusion-fission (FF) and transfer fission (TF) events. The masses of the fission fragments were determined event by event from precise measurements of flight paths and flight time differences of complimentary fission fragments. The estimated mass resolution for fission fragment was about 3 a.m.u. The details of experimental arrangements and data analysis and elimination of systematic errors were reported in ref [10, 12].

The measured mass distributions of $^{19}$F, $^{16}$O + $^{232}$Th and $^{16}$O + $^{209}$Bi [13, 14] and in case of $^{12}$C + $^{232}$Th and $^{19}$F + $^{209}$Bi at all energies are well fitted with single Gaussian distributions around the symmetric mass split for the target plus projectile systems. The variation of the square of the variance of the fission fragment mass distribution ($\sigma_m^2$) are shown in FIG 1 for $^{19}$F and $^{16}$O projectiles on the spherical bismuth nuclei. It has been observed that the mass variance ($\sigma_m^2$) shows a smooth variation with the excitation energy of the fused system across the Coulomb barrier. This is in qualitative agreement with the predictions of statistical theories. It is also noted that no significant departures are reported in the fragment angular anisotropy measurements for the spherical target and projectile systems [14 -16 ].

The variances of mass distribution ($\sigma_m^2$) for reactions of different projectiles on the deformed Thorium target are shown in FIG 2 for $^{19}$F, $^{16}$O, $^{12}$C projectiles. In all three cases, as the excitation energy is decreased, the $\sigma_m^2$ values decrease monotonically, but shows a sudden upward trend approximately around the Coulomb barrier energies. This is once again followed by a smooth decreasing trend as energy is further decreased. It is to be noted that anomalous increase in the angular anisotropy of fission fragments were observed in almost at the same beam energies at which anomalous increase in fragment angular anisotropies were observed in all these systems [17].

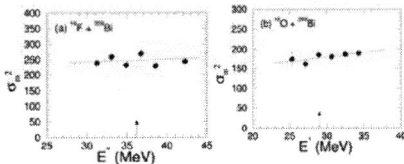

**FIGURE 1.** Mass variance $\sigma_m^2$ as a function of excitation energy (E*) for spherical bismuth target. The arrow points to excitation energy corresponding to Coulomb barrier. The solid lines show smooth variation of $\sigma_m^2$ with E*.

The observation of a sudden rise in $\sigma_m^2$ values as the excitation energy is lowered may signify a mixture of two fission modes, one following the normal statistical predictions of fusion-fission path and another following a different path in the energy landscape with zero or small mass asymmetry. The mixture of the two modes may give rise to wider mass distributions. Similar to the postulation of the orientation dependent quasi-fission [5], we postulate that for fusion-fission paths corresponding to the projectile orientations up to a critical angle ($\theta_c$) of impact on the polar region of prolate thorium, the width and energy slope of the symmetric mass distributions are different (shown by dot-dashed curves in FIG2 a, b & c) compared to those for the normal statistical fusion-fission paths (dotted curves). The mass widths weighted by the fission cross sections from earlier measurements [17] are mixed for the two fusion-fission modes and shown by different coloured continuous curves for different critical polar angles separating the two fission modes, for all three systems. The calculated $\sigma_m^2$ values quantitatively explain the observed increase in the widths of the mass distributions. It is interesting to note that the fusion-fission process is clearly dominated by the normal process at above Coulomb barriers and the "anomalous" fusion-fission process is dominant at lower energies. However, experimental evidence suggests that the variations of mass distributions with excitation energies are similar for the both processes, probably dominated by macroscopic forces, but differing quantitatively due to microscopic effects.

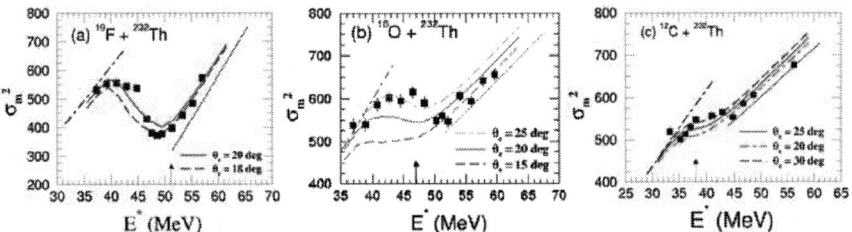

**FIGURE 2.** Variation of $\sigma_m^2$ with excitation energy for three system. The dotted and dot-dashed curves are postulated variation for normal and "anomalous" fission modes. Calculated $\sigma_m^2$ for two critical angles ($\theta c$) are indicated

Thus, we have established with the present set of measurements that widths of the mass distributions is a good tool to observe departure from the normal fusion-fission path in the fusion of heavy nuclei. The exact mechanisms for the departure from normal fusion-fission paths are not known accurately. However, macroscopic effects such as the direction of mass flow or the mass relaxation time being too prolonged may not be the cause. It has been established earlier from the experimental barrier distributions, the reaction cross sections in $^{19}$F, $^{16}$O, $^{12}$C + $^{232}$Th in near and below Coulomb barrier energies are mostly for impact of the projectiles on the polar regions of the thorium nuclei. Following the quantum mechanical effects favouring similar shapes in entrance and exit channels [1], we modify the simple postulation of the microscopic effects of the relative orientation of the projectile to the nuclear symmetry axes of the deformed target [5]. We assume that for the non-compact entrance channel shape, the impact of the projectile in the polar region of $^{232}$Th target drives the system to an almost mass symmetric saddle shape, rather than a compact equilibrated fused system. The observed fragment mass widths can be quantitatively explained under such assumptions. The above postulation is supported by the observation that for the spherical target $^{209}$Bi, where entrance channel compactness of shape is same for all relative target-projectile orientations, only normal fusion-fission paths, as characterized by the smooth variation of fragment mass widths with excitation energy, are observed. It is also worthwhile to note that effect of the anomalous mass widths increases with left-right mass symmetry in the entrance channel in case of $^{19}$F, $^{16}$O, $^{12}$C + $^{232}$Th system in consonance with our postulation. Our measurement indicates that higher entrance channel mass asymmetry and energies close to the Coulomb barrier may be preferable for synthesis of super heavy element.

## ACKNOWLEDGMENTS

I would like to thank the staff at NSC Pelletron for providing excellent beam. Help of Profs. P. Bhattacharya, S.Chattopadhyay, Drs. N. Majumdar, T. Sinha, A. Saxena, D. C. Biswas, Ms. K.S.Golda, Mr. P. K. Sahu, S. Pal, D. Das during the experiments and discussions are gratefully acknowledged.

## REFERENCES

1. P. Moller and A.J. Sierk, Nature 422, 485 (2003).
2. P. Moller, D.G. Madland, A.J. Sierk and A. Iwamoto, Nature 409, 785 (2001).
3. Peter Moller, Arnold J. Sierk and Akira Iwamoto, Phys. Rev. Lett. 92,072501 (2004).
4. I. Halpern and V. M. Strutinsky, in Proc. of 2nd International Conference on Peaceful Uses of Atomic Energy, (United Nations Publication, Geneva, 1958), Vol. 15, p 408.
5. D.J. Hinde et al., Phys. Rev. Lett. 74 ,1295 (1995).
6. D.J. Hinde. M Dasgupta and A. Mukherjee, Phys. Rev. Lett. 89, 282701 (2002).
7. A.C. Berriman et al., Nature 413, 144 (2001)
8. K. Nishio et al., Phys. Rev. Lett. 93, 162701 (2004).
9. D.J. Hinde et al., Phys. Rev. C 53, 1290 (1996).
10. T.K. Ghosh et al., Nucl. Instrum. Methods Phys. Res. A 540 , 285 (2005)
11. P. Bhattacharya et al., Nuovo Cimento Soc. Ital. Fis., A 108, 819 (1995).
12. T.K. Ghosh et al, Phys. Lett. B 627, 26 (2005).
13. T.K. Ghosh et al., Phys. Rev. C 69, 031603(R) (2004)
14. T.K. Ghosh et al., Phys. Rev. C 70, 011604(R) (2004).
15. A.M. Samant et al., Eur. Phys. J. A.7, 59 (2000).
16. S. Kailas , Phys. Rep. 284, 381 (1997).
17. N. Majumdar et al., Phys. Rev. Lett. 77, 5027 (1996).

# POSTER SESSIONS

# Fission fragment mass distribution for nuclei in the r-process region

S. Tatsuda, K. Hashizume, T. Wada, M. Ohta, K. Sumiyoshi[A,B],
K. Otsuki[E], T. Kajino[B,C,D], H. Koura[F], S. Chiba[F], Y. Aritomo[G]

*Department of Physics, Konan University, 8-9-1 Okamoto, Kobe 658-8501, Japan*
*Numazu College of Technology[A], NAO[B], GUSA[C], Univ. of Tokyo[D], Univ. of Chicago[E]*
*JAEA[F], FLNR(JINR)[G]*

**Abstract.** The fission fragment mass distribution is estimated theoretically on about 2000 nuclides which might have a critical role on the r-process nucleosynthesis through fission ($Z>85$). The mass distribution of fission fragment is derived by considering the location and the depth of valleys of potential energy surface near scission point of nuclei calculated by means of the liquid drop model with the shell energy correction by the Two-Center shell model. The guiding principle of determining the fission mass asymmetry is the behavior of the fission paths from the saddle to the scission point given by the Langevin calculation.

**Keywords:** r-process nucleosynthesis, mass distribution, fission
**PACS:** 25.85.-w, 26.30.+k

## 1. INTRODUCTION

The mass distribution of fission fragment is one of the most important elements on astrophysical condition of high temperature and high neutron density like supernova for the theoretical estimation of the abundance pattern of the r-process elements, together with β-decay rates and (n, γ) reaction rates. A number of observational data of the r-process elements in metal poor stars have been accumulated and they exhibit several abundance patterns beyond the universal abundance pattern which is applied to the elements $56 \leq Z \leq 75$ in the stars of extremely enhanced r-process elements. It is suggested theoretically that the fission process during and after the r-process may affect strongly the variation of these abundance patterns and even the universality. Especially, during the r-process the fission cycling may lead to very different r-process abundance pattern if the mass asymmetric fission data are taken into account. Even after the freeze-out of the r-process, the mass distribution of fission fragment will be useful for the estimation of products from various types of fission modes. Thus, our study of the fission fragment mass distribution is important to understand the role of fission in the r-process.

We calculated the potential energy surface (PES) by means of the liquid drop model with the shell energy correction by the Two-Center shell model[1][2][3] in 3-dimensional parameter space. The basic method to estimate the fission fragment mass distribution is to find the fission path by solving the Langevin equation using this PES. Since it is time consuming to solve the Langevin equation for about 2000 nuclides, we estimate the fission fragment mass distribution by considering the location and the depth of the valley of PES at the scission point in this study. The results of the Langevin calculation for several actinoid nuclei are used effectively to this estimation. [4][5]

The details of determining the fission asymmetry will be explained in Section 2, including the consistency of our estimation of mass asymmetry with the results of the Langevin calculation. In Section 3, we discuss the trend of the results.

## 2. Analysis of the fission asymmetry

We analyze the potential energy surface (PES) of nuclei calculated by using the liquid drop model with the microscopic correction of the Two-Center shell model in 3-dementional parameter space. The 3-dimentional parameter space is composed of the distance between the center of mass of the fragments $z$, the deformation of the fragments $\delta$ and the mass asymmetry parameter $\alpha$, which are defined by

$$z = \frac{z_0}{R}\frac{b}{a} \quad , \quad \delta = \frac{3(a-b)}{2a+b} \quad , \quad \alpha = \frac{A_1 - A_2}{A_1 + A_2} \tag{1}$$

where $R$ is the radius of the spherical compound nucleus, $A_1$ and $A_2$ are the mass number of fission fragments. $a$ and $b$ are shown in the Figure 1. There is only one valley at $\alpha=0$ and $\delta\sim0.4$ in the PES calculated by the liquid drop model without the shell correction in the 3-dimentional parameter space. On the other hand, there are three valleys in the PES including the shell correction by the two-center shell model. One of the valleys is at $\alpha\neq0$ and $\delta=0.2\sim0.3$ and corresponds to the asymmetric fission (we call it asymmetric valley). The others are at $\alpha=0$, $\delta\sim0.4$ and $\delta\sim0$ and correspond to the symmetric fission (we call them symmetric valleys).

We use two ways to estimate the mass asymmetry of fission fragment. One is the estimation of the mass asymmetry by solving directly the Langevin equation on the PES. We assume that the dynamics of the shape change of a nucleus is given by the exchange of energy between the internal energy and the deformation energy of a nucleus based on the fluctuation-dissipation theorem. This method gives the mass distribution which is consistent with the experimental results. However, the Langevin calculation is much time consuming.

The other is the indirect method where we estimate the mass distribution by the analysis of the valleys of the PES of nuclei at scission point. We have learned from the Langevin calculation that the location of the valley on PES corresponds to the fission modes and the depth of the valley corresponds to the fission ratio to another fission modes. (The depth of the valley is defined by the relative depth to the potential of the spherical mononucleus.) We therefore determine the mass asymmetry of the fission fragment by examining the valley depth and its location on PES. When the mass asymmetry $\alpha$ is determined, we get the mass number of the fission fragment $A_1$ and $A_2$ ($A_1>A_2$) from Eq. (1),

$$A_1 = A(1+\alpha)/2 \quad , \quad A_2 = A(1-\alpha)/2 \tag{2}$$

where $A$ is the mass number of the progenitor nucleus. If $\alpha=0$, $A_1=A_2$. We assume that the fission fragment mass distribution is given by the Gaussian distribution of its dispersion $\sigma^2=49$. If the symmetric fission is dominant, the mean value of the

Gaussian distribution is $A_1=A_2=A/2$. If the asymmetric fission is dominant, we estimate the fission fragment mass distribution by superposing two Gaussian distribution whose mean values are $A_1$ and $A_2$. When the symmetric fission and the asymmetric fission are comparable, we determine the fission fragment mass distribution by superposing the mass distributions which are multiplied by the weight factor according to the depth of the valleys. In this paper, we assume that the symmetric fission and the asymmetric fission occur at an equal mixture. The weight factors $\omega_{sym}$ and $\omega_{asym}$ of the symmetric and asymmetric mass distribution of fission fragment become $\omega_{sym}=\omega_{asym}=0.5$.

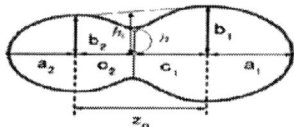

Figure 1. This figure represents the Two-Center parameterization of a nucleus.

## 3. RESULTS AND DISCUSSIONS

Figure 2 shows the analysis of the valleys of the PES of Fm-isotopes and Sg-isotopes. For the nuclei $^{258-269}$Fm, the symmetric valleys are as deep as the asymmetric valleys. For the nucleus $^{264}$Fm, the symmetric valley is the deepest. This situation can be clearly understood as follows. If the nucleus $^{264}$Fm decays into two in mass-symmetric fission, the neutron number and the proton number of the fission fragments become identical double-magic ($Z=50$, $N=82$). On the other hand, for the Sg-isotopes, the symmetric valleys are shallower than the asymmetric valleys and the differences between them are more than 5MeV. Therefore, the symmetric fission may hardly occur for the Sg-isotopes. Except for the nuclei in the neighborhood of $^{264}$Fm, the asymmetric valleys are usually deeper than the symmetric valley ($\delta\sim0$). In the special case that the nucleus may decay into identical double magic nuclei, the symmetric valley becomes the deepest by the shell structure, so the symmetric fission becomes dominant.

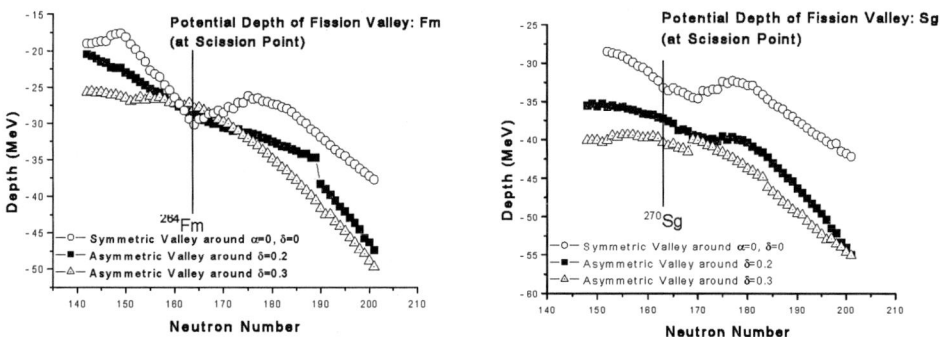

Figure 2. Depths of the valleys of the PES of Fm-isotopes and Sg-isotopes at scission point. The open circle indicates the symmetric valley around $\alpha=0$ and $\delta=0$. The closed square and the open triangle represent the asymmetric valley around $\delta=0.2$ and $\delta=0.3$, respectively.

Figure 3 shows the fission fragment mass distribution of Fm-isotopes. Since the symmetric fission may be in competition with the asymmetric fission for $^{258-269}$Fm from Figure 2, the fission fragment mass distribution for $^{258-269}$Fm is estimated by superposing the symmetric and the asymmetric mass distribution which are multiplied by the weight factor $\omega_{sym}=\omega_{asy}=0.5$.

We predict the fission fragment mass distribution for wide range of Fm-isotopes by analyzing the valleys of the PES at the scission point. This mass distribution is consistent with the results of the Langevin calculation for $^{256,258,264}$Fm. It is also consistent with the experimental results which show that the symmetric fission becomes remarkable. [6]

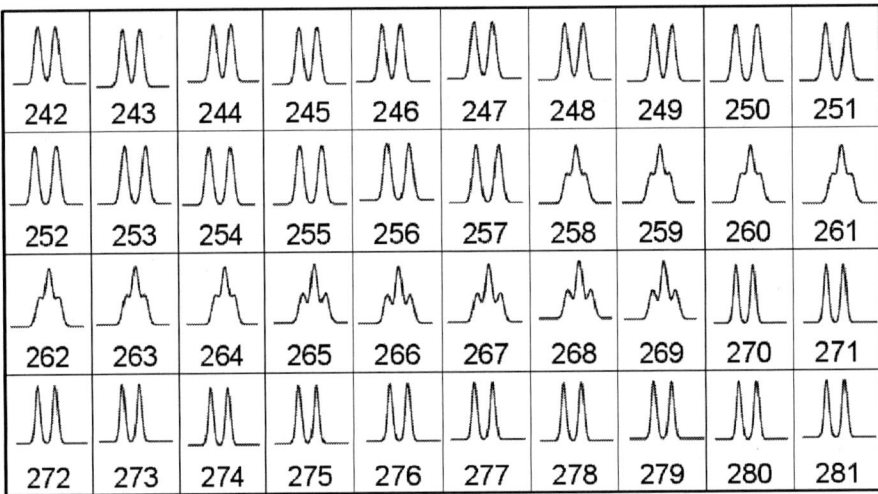

Figure 3. Fission fragment mass distribution of Fm-isotopes. The number indicates the mass number.

## ACKNOWLEDGMENTS

One of the authors (T. W.) is partially supported by a Grand-in-Aid for Scientific Research No. 17540270 provided by the Japan Society for the Promotion of Science.

## REFERENCES

1. J. Maruhn and W. Greiner, Z. Phys.251, 431(1972)
2. S. Suekane, A. Iwamoto, S. Yamaji and K. Harada, JAERI-memo 5918(1974)
3. A. Iwamoto, S. Yamaji, S. Suekane and K. Harada, Prog. Theor. Phys. 55, 115(1976)
4. T. Ichikawa, T. Asano, T. Wada and M. Ohta, J. of Nucl. and Radiochemical 3, 67(2002)
5. T. Asano, T. Wada, M. Ohta, T. Ichikawa, S. Yamaji and H. Nakahara, J. of Nucl. and Radiochemical5, 1(2004)
6. D. C. Hoffman and M. R. Lane, Radiochim. Acta 70/71, 135 (1995)

# Dipole Resonances in $^4$He

E. Matsumoto*, S. Nakayama*, R. Hayami*, K. Fushimi*, H. Kawasuso*,
K. Yasuda*, T. Yamagata$^\dagger$, H. Akimune$^\dagger$, H. Ikemizu$^\dagger$, M. Fujiwara**,
M. Yosoi**, K. Nakanishi**, K. Kawase**, H. Hashimoto**, T. Oota**,
K. Sagara$^\ddagger$, T. Kudoh$^\ddagger$, S. Asaji$^\ddagger$, T. Ishida$^\ddagger$, M. Tanaka$^\S$ and
M.B. Greenfield$^\P$

*Department of Physics, University of Tokushima, Tokushima 770-8502, Japan
$^\dagger$Department of Physics, Konan University, Kobe 658-8501, Japan
**Research Center for Nuclear Physics, Osaka University, Osaka 567-0047, Japan
$^\ddagger$Department of Physics, Kyushu University, Fukuoka 812-8581, Japan
$^\S$Kobe Tokiwa College, Kobe 654-0838, Japan
$^\P$Department of Physics, International Christian University, Tokyo 181-0015, Japan

**Abstract.** We investigated the analogs of the giant dipole resonance (GDR) and spin-dipole resonance (SDR) of $^4$He by using the $^4$He($^7$Li,$^7$Be) reaction at an incident energy of 455 MeV and at forward scattering angles. The $\Delta S=0$ and $\Delta S=1$ spectra for $^4$He were obtained by measuring the 0.43-MeV $^7$Be $\gamma$-ray in coincidence with the scattered $^7$Be. From the $\Delta S=0$ and $\Delta S=1$ spectra thus obtained, the strength distributions of the GDR and SDR in $^4$He can be derived and the results are compared with the previous data.

**Keywords:** ($^7$Li,$^7$Be) Reaction, GDR and SDR in $^4$He
**PACS:** 24.30.Cz, 25.55.Kr, 27.10.+h

Many studies were devoted to investigate experimentally as well as theoretically the cross section and shape of the giant dipole resonance (GDR) in $^4$He [1]. However, the existing data are in severe contradiction with each other, and not yet settled down to a reliable precision for test of charge symmetry. Recently Efros et al. [2] claimed that the previous data and calculation on the excitation function for the photodisintegration of $^4$He in the GDR energy region are not consistent with the result estimated from E1 sum rule, and predicted the peak energy for the total cross section of the photodisintegration of $^4$He to be about 30 MeV. It is much higher than the previously adopted value of 25 MeV. In addition Shima et al. [3] remeasured the photodisintegration of $^4$He and reported that the cross section monotonically increases with the excitation energy up to $E_\gamma=32$ MeV, and no pronounced peak of the GDR was observed at around $E_\gamma=25$ MeV. On the other side, there are scare data for the spin-dipole resonance (SDR) in $^4$He. The strength distribution of the SDR in $^4$He plays a critical role in neutrino-$^4$He scattering processes at the supernova explosion [4].

In the present experiment we investigated the analogs of the dipole resonances (GDR and SDR) of $^4$He by using the $^4$He($^7$Li,$^7$Be) reaction at an incident energy of 455 MeV and at forward scattering angles [5]. In the ($^7$Li,$^7$Be) reaction, the scattered $^7$Be populates either the ground state (3/2$^-$; $^7$Be$_0$) or the first excited state (1/2$^-$, 0.43 MeV; $^7$Be$_1$). The $^7$Be$_0$ state is produced when the reaction proceeds via a $\Delta S=0$ or $\Delta S=1$ transfer, and the $^7$Be$_1$ state via a $\Delta S=1$ transfer. The $^7$Be$_1$ decays to the $^7$Be$_0$ by emitting

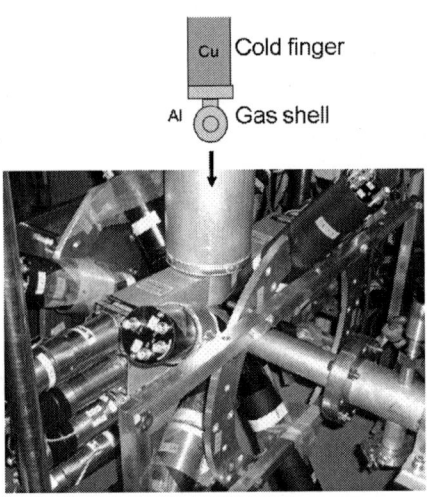

**FIGURE 1.** A scattering chamber surrounding with 18-GSO scintillators for detection of the 0.43-MeV γ-ray of $^7$Be.

the 0.43-MeV γ-ray. The $^7$Be$_1$ spectrum can be identified by tagging the 0.43-MeV γ-ray of $^7$Be. The $\Delta S$=0 and $\Delta S$=1 spectra can be obtained from the $^7$Be singles and $^7$Be$_1$ coincidence spectra. Details of the procedure for the separation of the $\Delta S$=0 and $\Delta S$=1 spectra are described in ref. [5]. From the $\Delta S$=0 and $\Delta S$=1 spectra presently obtained, the strength distributions of the GDR and SDR in $^4$He were derived and the results were compared with the previous data.

A 455-MeV $^7$Li$^{3+}$ beam was provided from the ring cyclotron at RCNP, Osaka University, and bombarded a $^4$He gas target. To increase the target thickness, we cooled the $^4$He gas down to ~10 K with a cryogenic refrigerator [6]. A thickness of $^4$He gas target was about 7 mg/cm$^2$ with a pressure of 1.5 atm. The gas shell has windows of Aramid (aromatic polyamid) foils with a thickness of 12 $\mu$m and with diameters of 10 and 12 mm at the beam entrance and exit. We measured the spectrum for an Aramid target with a thickness of 48 $\mu$m to evaluate backgrounds due to the Aramid windows. Reaction particles $^7$Be were analyzed by using the magnetic spectrometer "Grand RAIDEN" and detected with the focal plane detector system consisting of two multiwire drift chambers backed by a $\Delta E$-$E$ plastic scintillator telescope [7]. The aperture of the entrance slit of the Grand RAIDEN was ±20 mrad horizontally and ±10 mrad vertically. The 0.43-MeV γ-ray from $^7$Be was measured with 18 Gd$_2$SiO$_5$(Ce) scintillators as shown in Figure 1. The 0.43-MeV γ-ray was observed as a prominent peak in the coincident γ spectra. The coincident $^7$Be spectrum was obtained by gating on the photopeak of the 0.43-MeV γ-ray from the $^7$Be particles. The detection efficiency for the photopeak of the $^7$Be γ-ray was determined to be 12.5% by measuring the spin-flip transition of $0^+ \rightarrow 1^+$ in the $^{12}$C($^7$Li,$^7$Be)$^{12}$B reaction.

The ($^7$Li,$^7$Be) spectrum for $^4$He was obtained by subtracting the spectrum for the Aramid target from that for the $^4$He gas target and is shown in Figure 2. Here the two

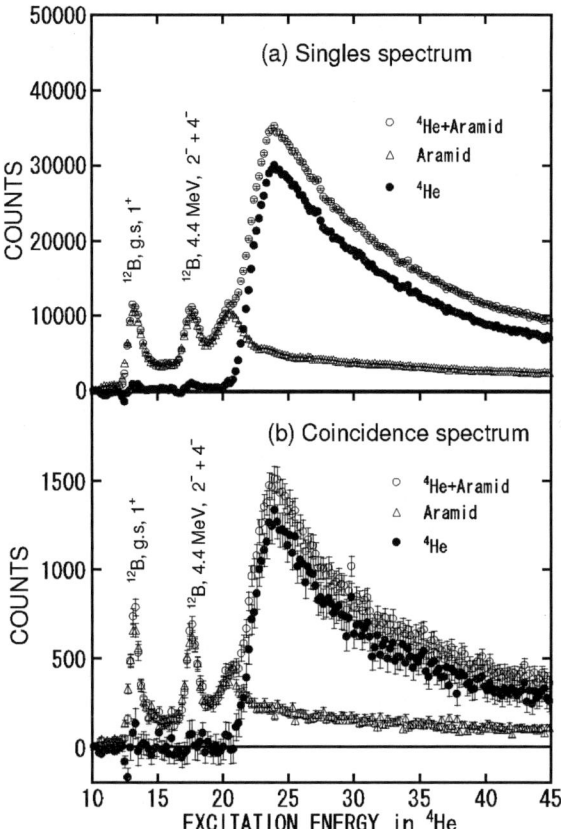

**FIGURE 2.** Singles (a) and coincidence (b) spectra for the $^4$He gas and Aramid targets in the ($^7$Li,$^7$Be) reaction at $E_L$=455 MeV and at $\theta_L = 0°$.

spectra were normalized by peak-yields due to $^{12}$C in the Aramid target. The $\Delta S=1$ spectrum was obtained from the coincidence spectrum. The $\Delta S=0$ spectrum was obtained by subtracting the $\Delta S=1$ spectrum from the $^7$Be singles one. Here the singles spectrum was renormalized to account for the detection efficiency of the $^7$Be $\gamma$-rays. The $\Delta S=0$ and $\Delta S=1$ spectra thus obtained are shown in Figure 3.

The $\Delta S=0$ spectrum reflects the strength distribution of the GDR in $^4$He. In Figure 3(a), we compare the $\Delta S=0$ spectrum with previous data of the total photodisintegration cross sections for $^4$He. Here the cross section presently observed was arbitrarily normalized to fit the previous data in the relevant energy region. The present results agree with previous data [1] and the peak of GDR was observed at around previously adopted excitation energy of $E_\gamma$=25 MeV, and are in deep contradiction with the work of Efros *et al.* [2] and the recent data obtained by Shima *et al.* [3].

The $\Delta S=1$ spectrum reflects the strength distribution of the SDR in $^4$He and is shown in Figure 3(b). We observed that the SDR is distributed as a compact resonance at

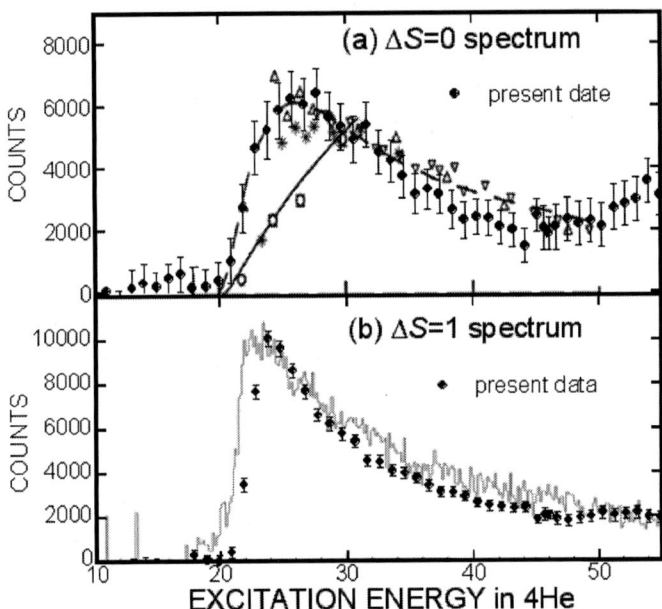

**FIGURE 3.** (a) The $\Delta S=0$ spectrum for $^4$He compared with available data ($\Delta$, $\nabla$, $*$ [1]; $\bigcirc$ [3]) of the $^4$He total photodisintegration cross sections. (b) The $\Delta S=1$ spectrum for $^4$He compared with the spectrum for the $^4$He$(p,p')$ reaction at $E_p=300$ MeV and at $10°$ (thin line) [6].

around $E_x=23$ MeV and with a width of about 8 MeV, and has a strength in a similar magnitude to that of the GDR. The $\Delta S=1$ spectrum was compared with the spectrum for the $^4$He$(p,p')$ reaction at 300 MeV which proceeds dominantly via a $\Delta S=1$ transfer in isovector excitations. The result follows well the $(p,p')$ spectrum.

This experiment was performed at RCNP, Osaka University under Program No.E263. The authors are grateful to the staff of the RCNP cyclotron for their support. This work was supported by a Grant-in-Aid for Scientific Research, No. 16540257, of the Japan Ministry of Education, Science, Sports and Culture.

## REFERENCES

1. J.T. Londergan and C.M. Shakin, Phys. Rev. Lett. **28**, 1729 (1972); B. Wachter *et al.*, Phys. Rev. **C38**, 1139 (1988); M. Unkelbach and H.M. Hofmann, Nucl. Phys. **A549**, 550 (1992).
2. V.D. Efros *et al.*, Phys. Rev. Lett. **78**, 4015 (1997).
3. T. Shima *et al.*, Phys. Rev. **C72**, 044004 (2005).
4. D. Gazit and N. Barnec, Phys. Rev. **C70**, 048801 (2004).
5. S. Nakayama *et al.*, Nucl. Instrum. Methods Phys. Res., Sect. A **404**, 34 (1998); Phys. Rev. Lett. **85**, 262 (2000).
6. T. Yamagata *et al.*, Phys. Pev. **C74**, 014309 (2006).
7. M. Fujiwara *et al.*, Nucl. Instrum. Methods Phys. Res., Sect. A **422**, 484 (1999).

# Photodisintegration of $^{80}$Se and its implications for s-process branching

A. Makinaga[1], H. Utsunomiya[1], S. Goko[1], T. Kaihori[1], S. Houhara[1],
S. Goriely[2], H. Toyokawa[3], H. Harano[3], T. Matsumoto[3], H. Harada[4],
F. Kitatani[4], K.Y. Hara[4], Y.-W. Lui[5]

[1] *Department of Physics, Konan University, Kobe 658-8501, Japan*
[2] *Institut d'Astronomie et d'Astrophysique, Université Libre de Bruxelle, Campus de la Place, CP-226, 1050 Brussels, Belgium*
[3] *National Institute of Advanced Industrial Science and Technology, 1-1-1 Umezono, Tsukuba, Ibaraki 305-8568, Japan*
[4] *Japan Atomic Energy Agency, 2-4 Tokai, Ibaraki 319-1195, Japan*
[5] *Cyclotron Institute, Texas A&M University, College Station, Texas 77843, USA*

**Abstract.** Photodisintegration cross sections were measured for $^{80}$Se with laser Compton scattering γ beams at $E_\gamma$=9.98-11.80 MeV near the neutron separation energy. The stellar neutron capture rate for $^{80}$Se was evaluated by using the photodisintegration data as constraints on the E1 γ strength function for $^{80}$Se within the framework of the Hauser-Feshbach statistical model. It was found that the rate at 30 keV is 0.7-1.1 times that of Bao and Käppeler. s-process temperatures deduced from the empirical solar abundances of the s-only nuclei $^{79}$Se and $^{80}$Kr with this neutron capture rate are 17-25 keV in the neutron density rage of $(0.8\text{-}1.9)\times 10^3/cm^3$.

**Keywords:** Photodisintegration, s-process
**PACS:** 25.20.-x, 23.20.-g, 28.20-v

## 1. INTRODUCTION

The physical condition for the s-process can be examined at s-process branching provided that a competition between $\beta^-$ decay and neutron capture is highly sensitive to the stellar temperature and neutron density. s-process branching at $^{79}$Se is known to be suited to a specification of the physical condition [1]. The half-life of $^{79}$Se in the ground state is less than 65,000 years [2] and greater than the assumed value, 1,700 years [3]. In contrast to the long half-life of the ground state, the first excited state of $^{79}$Se at 96 keV, which is thermally populated in the stellar condition, undergoes β-decay to $^{79}$Br with the half-life 7000 minutes [1]. As a result, the stellar β-decay rate is highly sensitive to temperature [1,4]. On the other hand, the stellar neutron capture rate for $^{79}$Se depends on both the neutron density of the relevant stellar site and neutron capture cross sections $\sigma_{n\gamma}$. Up until now, the $\sigma_{n\gamma}$ for $^{79}$Se has remained experimentally unknown though a direct measurement is planned at the CERN n-TOF facility in the future [5].

We have measured photoneutron cross sections ($\sigma_{\gamma n}$) for $^{80}$Se to evaluate $\sigma_{n\gamma}$ for $^{79}$Se within the framework of the Hauser-Feshbach model. In the model calculation, the experimental $\sigma_{\gamma n}$ is used as constraints on the low-energy E1 $\gamma$ strength function for $^{80}$Se. Uncertainties associated with the nuclear level density are taken into account. We remark that the present $\sigma_{\gamma n}$ constitutes the basic nuclear data for nuclear transmutation of a long-lived fission product $^{79}$Se.

## 2. EXPERIMENT

Beams of quasi-monochromatic $\gamma$-rays were produced in the energy range of 9.98-11.80 MeV from laser Compton scattering (LCS) in the electron storage ring TERAS at AIST. The LCS $\gamma$-ray beams were used to irradiate a sample of 1003.3mg $^{80}$Se enriched to 99.95% that is encapsulated in an aluminum container. A Nd:YVO$_4$ Q-switch laser was operated at 20 kHz in the second harmonics ($\lambda$=532 nm). The $\gamma$-ray beams had the same macroscopic time structure of 80 ms beam-on and 20 ms beam-off as that of the laser. The $^{80}$Se sample was mounted at the center of a 4$\pi$-type neutron detector consisting of 20 $^3$He counters embedded in a polyethylene moderator in a triple-ring configuration. Background neutrons were detected during the 20 ms beam-off. The neutron detection efficiency is more than 56% in the neutron energy range below 1 MeV. The so-called ring ratio technique [7] was used to determine the average neutron energies. The LCS $\gamma$ beam was measured with a 120% high-purity germanium detector (HPGe). The energy distribution of the LCS beam was determined by a least-squares analysis of the response function of the HPGe detector. The LCS beam was monitored with a large volume (8" in diameter times 12" in length) NaI(Tl) detector. Pile-up spectra were used to determine the number of the incident LCS $\gamma$ rays. Photoneutron cross sections were determined at the average $\gamma$-ray energies with the Taylor expansion method [8]. The systematic uncertainty for the cross section is 4.4% whose breakdown is 3.2% for the neutron detection efficiency and 3% for the number of incident $\gamma$ rays.

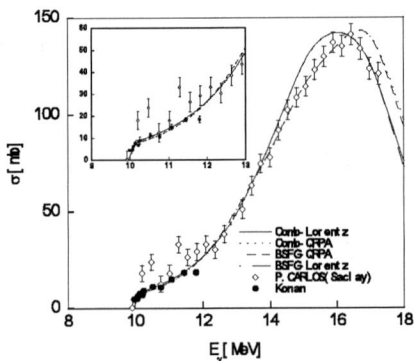

**FIGURE 1.** Photoneutron cross sections for $^{80}$Se in comparison with theoretical cross sections with different prescriptions of the E1 $\gamma$ strength function and the nuclear level density (see text for details).

## 3. PHOTONEUTRON CROSS SECTIONS FOR $^{80}$SE

Results of the present photoneutron cross section measurement for $^{80}$Se are shown in Fig.1. One can see that the present $\sigma_{\gamma n}$ are significantly smaller than those of the previous measurements [11] near the neutron threshold. Hauser-Feshbach model calculations were carried out based of different ingredients of model parameters: Lorentzian [9] and QRPA [10] models for the E1 $\gamma$ strength function and back-shifted Fermi gas (BSFG) and combinatorial models [12,13] for the nuclear level density. Two calculations with the Lorentzian-BSFG and the combinatorial–QRPA model parameters satisfactorily reproduce the present cross sections near the neutron threshold.

## 4. STELLAR NEUTRON CAPTURE RATES FOR $^{79}$SE

Neutron capture cross sections were evaluated for $^{79}$Se by using the same ingredients of the model parameters that were found for the present photoneutron cross section. Fig. 2 shows $<\sigma_{n\gamma}v>$ as a function of the stellar temperature in the unit of keV. The evaluated values at 30 keV are 0.7-1.1 times that of Bao and Käppeler [14]. The difference in the two model calculations stems from their different behaviors of the E1 $\gamma$ strength function below the neutron separation energy.

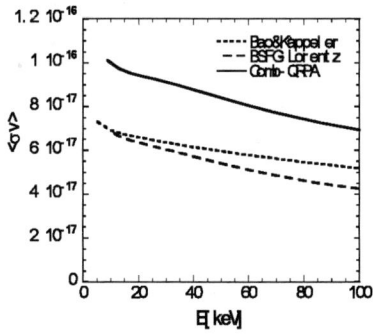

**FIGURE 2.** Neutron capture rate for 79Se in comparison with theoretical cross sections with different prescriptions of the E1 $\gamma$ strength function and the nuclear level density (see text).

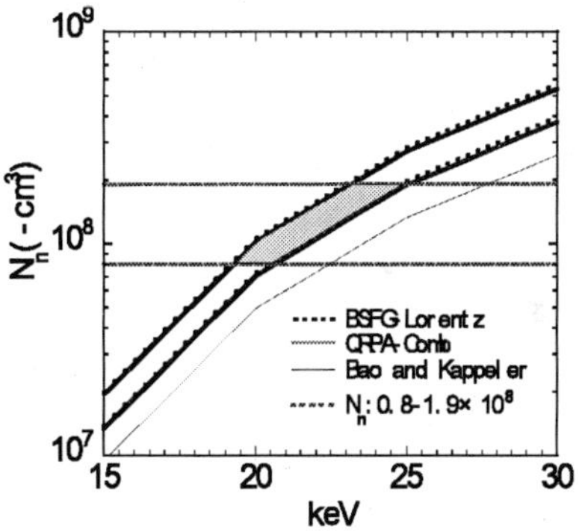

**FIGURE 3.** Range of temperatures and neutron densities allowed for the s-process branching at $^{79}$Se.

## 5. S-PROCESS BRANCHING AT $^{79}$SE

Let us consider the s-process flow starting from $^{76}$Se and ending to $^{80}$Kr. Both $^{76}$Se and $^{80}$Kr are s-only nuclei. In the s-process flow, a major branching takes place at $^{79}$Se and a minor one at $^{80}$Br. As a result, the solar abundance ratio between $^{76}$Se and $^{80}$Kr is expressed as

$$\frac{n(^{80}Kr)}{n(^{76}Se)} = \frac{\lambda_{n\gamma}(^{76}Se)}{\lambda_{n\gamma}(^{80}Kr)} f_\beta(^{79}Se) f_\beta(^{80}Br),$$

where $\lambda_{n\gamma}$ is the neutron capture rate as defined by $\lambda_{n\gamma}=N_n<\sigma_{n\gamma} v>$ with the neuron density $N_n$ and neutron velocity v, and $f_\beta$ is the branching ratio defined as $f_\beta=\lambda_\beta/(\lambda_\beta+\lambda_{n\gamma})$ with the β-decay rate $\lambda_\beta$. It is known that the $\lambda_\beta(^{79}Se)$ is highly sensitive to the stellar temperature [3]. At temperatures above 10 keV, the β decay rate is dominated by the first excited state at 96 keV with the population factor exp(-Ex/kT) and the fast decay character (7000 minutes), nearly independent of the long half life of the ground state. In contrast, $\lambda_\beta(^{80}Br)$ is 0.917, independent of the stellar temperature.

Fig 3 shows best temperature ranges for the present neutron capture rates obtained for $^{79}$Se. The empirical solar abundance ratio, n($^{80}$Kr)/n($^{76}$Se), is taken from [1]. A best temperature range is 17-25 keV in the neutron density range of (0.8-1.9) × $10^8$ cm$^{-3}$. Note that there are two different stellar contributors for the used solar abundances of $^{76}$Se and $^{80}$Kr: core-helium burning in massive stars and helium-shell burning in AGB stars corresponding to the weak and the main s-process components, respectively. The present temperature range is considered to represent a mixture of these two stellar sites.

## ACKNOWLEDGMENTS

This work was supported by the Japan Private School Promotion Foundation and the Japan Society of the Promotion of Science. We thank K.Takahashi for his comments of the s-process branching at $^{79}$Se.

## REFERENCES

[1] F.Käppeler, H. Beer and K. Wisshak, Rep. Prog. Phys. **52**, 945 (1989).
[2] C. M. Lederer and V. S. Shirley, Table of Isotopes, 7$^{th}$ ed. (Wiley, New York, 1978).
[3] N.Klay and F. Käppeler, Phys. Rev. **C38**, 295 (1988).
[4] K. Takahashi, Atomic Data and Nuclear Data Tables **36**, 375 (1987).
[5] H. Michael, private communications.
[7] B. L. Berman and S. C. Fultz, Rev. Mod. Phys. **47**, 713 (1975).
[8] H. Utsunomiya et al., Phys. Rev. **C74**, 025806 (2006).
[9] C. M. McCullagb, M. L. Stelts and R. E. Chrien, Phys. Rev. **C23**, 1394 (1981)
[10] S. Goriely and E. Khan, Nucl. Phys. **A706**, 217 (2002).
[11] P. Carlos et al., Nucl. Phys. **A258**, 365 (1976).
[12] S. Hilaire and S. Goriely (to be published).
[13] S. Hilaire et al., Eur. Phys. J. **A12**, 169 (2001).
[14] Z. Y. Bao, H. Beer, F. Käppeler, F. Voss, K. Wisshak, T. Rauscher, Atomic Data and Nuclear Data Tables **76**, 70 (2000).

# Langevin equation as a stochastic differential equation in nuclear physics

T. Asano[1], T. Wada[1], M. Ohta[1], and N. Takigawa[2]

1) Department of Physics, Konan University, 8-9-1 Okamoto, Kobe 658-8501, Japan
2) Department of Physics, Tohoku University, Sendai, 980-8578, Japan

**Abstract.** Two kinds of stochastic integrals, Ito integral and Stratonovich integral, are applied for solving Langevin equation. In the case of the simplified Langevin equation for over-damped motion, the fission rate obtained with Stratonovich integral is significantly larger than that with Ito integral. On the other hand, in the case where the random force acts on the momentum variables, the two integrals give essentially the same results. The condition for the difference with two integrals to appear is discussed. The proper treatment of the double stochastic integral is necessary to obtain a high numerical accuracy.

**Keywords:** Stochastic differential equation, Langevin equation, Nuclear fission, Fission rate.
**PACS:** 25.85.-w, 05.10.Gg

## 1. INTRODUCTION

Langevin equation (LE) is widely used in the study of physical and chemical reactions. Mathematically, LE is generalized down to a stochastic differential equation (SDE). SDE includes a random term which is absent in deterministic differential equation (DDE). The integration of the random term, namely the stochastic integral is very important for solving a SDE. The calculation of stochastic integral includes the following three problems.

The first problem concerns the definition of the stochastic integral. There are two typical calculi of the stochastic integral, Ito integral and Stratonovich integral. We must select the appropriate stochastic integral according to the problem in hand. The properties of Stratonovich integral are the followings;

1. The stochastic differential rule of Stratonovich type is formally the same as the deterministic differential rule.
2. Stratonovich integral is appropriate for the system which is continuous in time and space.

Since LE is reduced to Newton equation in the case of the no friction limit, the differential rule should formally the same as the deterministic differential rule. In addition, the physical system is continuous in time and space. Therefore, the use of

Stratonovich integral is more favorable than Ito integral. If a reader is interested in the details, you should refer the reference 3.

The second problem concerns that there are following two approximations of SDE;

1. Strong approximation which approximates the trajectories of SDE.
2. Weak approximation which approximates the statistical values, e.g., mean value and dispersion.

When we treat nuclear fission phenomena, Langevin trajectories bifurcate at saddle point. Therefore, the mean value of trajectories does not have a physical meaning. We should select the strong approximation. The third problem concerns numerical integration, which we will discuss in the next section.

The authors of references 4 and 5 did not consider the LE as a SDE explicitly and have not distinguished the difference between the two kinds of stochastic integrals. As a result, the numerical integral is not treated properly. In this study, we treat the LE as a SDE and discuss the proper formula to integrate the LE.

Section 2 gives a concise description of our framework. Results are shown in Section 3 concerning the fission of $^{264}$Fm nucleus. Summary is given in Section 4.

## 2. THEORETICAL FRAMEWORK

The one-dimensional LE has the following form;

$$\frac{dq}{dt} = \frac{p}{m} = f_q(t, q(t), p(t)),$$
$$\frac{dp}{dt} = -\frac{\partial V}{\partial q} - \frac{1}{2}\frac{\partial}{\partial q}\left(\frac{p}{m}\right)p^2 - \frac{\gamma}{m}p + gR(t) = f_p(t, q(t), p(t)) + g_p(t, q(t), p(t)), \quad (1)$$

where $q$ denotes a collective coordinate and $p$ is the conjugate momentum. $V(q)$ is the potential energy and $m(q)$ and $\gamma(q)$ are the shape-dependent collective inertia and dissipation, respectively. The normalized random force $R(t)$ is assumed to be a white noise, i.e., $<R(t)> = 0$, $<R(t_1)R(t_2)> = 2\delta(t_1 - t_2)$. The strength of the random force $g$ is given by Einstein relation $g = \sqrt{\gamma T}$.

As mentioned already, we have three important problems, selection of a stochastic integral and the approximation as well as a numerical accuracy. In the following, we discuss the third problem. In order to solve LE, many people have employed discretization using Taylor expansion. However, since LE is a SDE, we should not use Taylor expansion but should use the expansion according to Ito formula. We obtain the following expansions (called Milstein scheme) with Ito integral or with Stratonovich integral, respectively,

$$q(t_{n+1}) = q(t_n) + f_q(t_n, q(t_n), p(t_n))\Delta t, \quad (2)$$

$$p(t_{n+1}) = p(t_n) + f_p(t_n, q(t_n), p(t_n))\Delta t + g_p(t_n, q(t_n), p(t_n))\Delta W$$
$$+ \frac{1}{2}g_p(t_n, q(t_n), p(t_n))\frac{\partial g_p}{\partial p}(t_n, q(t_n), p(t_n))(\Delta W^2 - \Delta t) \quad \text{(Ito SDE)} \quad (3)$$

or

$$p(t_{n+1}) = p(t_n) + f_p(t_n, q(t_n), p(t_n))\Delta t + g_p(t_n, q(t_n), p(t_n))\Delta W$$
$$+ \frac{1}{2}g_p(t_n, q(t_n), p(t_n))\frac{\partial g_p}{\partial p}(t_n, q(t_n), p(t_n))\Delta W^2 \quad \text{(Stratonovich SDE)} \quad (4)$$

where $\Delta W = \sqrt{t}R$. We attain the order of strong convergence $\gamma=1.0$ with respect to $\Delta t$ from this discretization. In contrast, previous discretization using Taylor expansion attains the order of strong convergence $\gamma=0.5$ since it neglects the stochastic double integral.

## 3. RESULTS

In this section, we discuss the results with the two stochastic integrals by taking the nuclear fission as an example. In the present study, we use TWOCTR of two-center shell model code to calculate the potential energy, inertia and friction. First, we show the results of the over-damped motion for nuclear fission. In this case, LE is transformed to the following simplified form,

$$\gamma \frac{dq}{dt} = -\frac{\partial U}{\partial q} + gR(t). \quad (5)$$

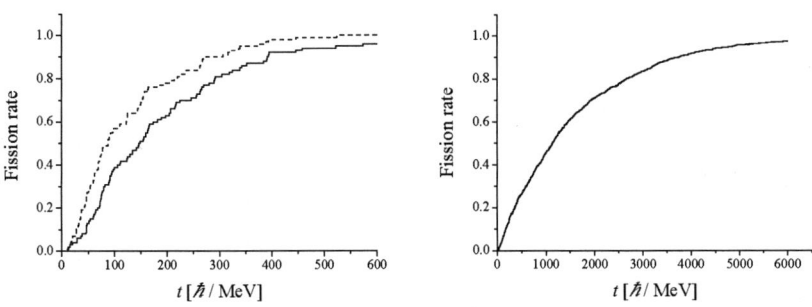

Figure 1. Time dependence of the fission rate for the symmetric fission of $^{264}$Fm. Solid lines are the fission rates of Ito type SDE and dotted lines are those of Stratonovich type SDE. Left panel shows the results of the simplified LE at $E$-$K$=100 MeV. Right panel shows the results of the LE at $E$=20 MeV.

As an example, we calculate the fission rate for the fission of $^{264}$Fm. Left side of figure 1 shows the fission rates obtained with the simplified LE as functions of time at $E$-$K$

$=U+aT^2=100$ MeV. The fission rate with Stratonovich integral is significantly larger than that with Ito integral. In order to analyze the origin of the difference, Stratonovich type SDE is transformed to Ito type SDE. Transformed Stratonovich type SDE has an additional term ($\frac{1}{2\gamma^2}\left(\frac{dT}{dq}\gamma - T\frac{d\gamma}{dq}\right)$) to Ito type SDE and the difference arises from this term. Next, we show the results of the normal LE (1). Right side of figure 2 shows the fission rates at $E=K+ U+aT^2=20$ MeV. It is found that both types of SDE give almost the same results. In this case, the transformed Stratonovich type SDE has an additional term in the form of ($-\frac{1}{4aT}\frac{\gamma}{m}p$). This term is found to be small because $aT$ is large, e.g., $aT$ is 26.4 when $T$ is 1 MeV. The difference can be significant at very low temperature. However, at low temperature, the classical treatment of LE has to be reexamined. We should consider the quantum treatment of LE.

## 4. SUMMARY

We have investigated the difference between two stochastic integrals, Stratonovich and Ito integral, in LE. By comparing the results with two kinds of the stochastic integral, we found that the effect of stochastic integral shows up for the simplified LE (5) while we did not find significant difference for normal LE (1). We plan to extend the study to include the microscopic energy and the microscopic inertia and dissipation. In the case which these inertia and dissipation depend on the temperature, we expect the difference to appear between two kinds of stochastic integral.

## ACKNOWLEDGEMENTS

The authors wish to express their gratitude to Dr. S. Yamaji, Prof. K. Hagino and Mr. K. Washiyama for their useful discussions. One of the authors (T. W.) is partially supported by a Grant-in-Aid for Scientific Research No. 17540270 provided by the Japan Society for the Promotion of Science.

## REFERENCES

1. K. Ito, Proc. Imp. Acad., Tokyo **20**, 519 (1944).
2. R.L. Stratonovich, "Topics in the Theory of Random Noise, vols. I and II", New York, Gordon and Breach, 1967.
3. E. Wong, and M. Zakai, Ann. Math. Stat. **36**, 1560–1564 (1965).
4. Y. Abe, S. Ayik, P. –G. Reinhard, and E. Suraud, Phys. Rep. **275**, 49 (1996).
5. P. Fröbrich and I. I. Gontchar, Phys. Rep. **292**, 131 (1998).
6. G. N. Milstein Theory Probab. Appl. **19**, 557-562 (1974).
7. S. Suekane, A. Iwamoto, S. Yamaji, and K. Harada, JAERI-memo 5918 (1974).
8. A. Iwamoto, S. Yamaji, S. Suekane, and K. Harada, Prog. Theor. Phys. **55**, 115-130 (1976).

# Fission barriers for neutron-rich nuclei by means of Skyrme-Hartree-Fock-Bogoliubov calculation

K.Hashizume, T.Wada, M.Ohta, M.Samyn* and S.Goriely*

*Department of Physic, Konan University, Okamoto, Kobe 658-8501 Japan*
*\*Institute d'Astronomie et d'Astrophysique, ULB-CP226, B-1050 Brussels, Belgium*

**Abstract.** The nuclear fission barrier height has been estimated by means of the constraint Skyrme Hartree-Fock-Bogoliubov method. The potential energy surfaces obtained by the method are analyzed with the flooding method to find several saddle points. The results for U, Np, Bk isotopes are compared with the barrier derived from the extended Thomas-Fermi plus Strutinsky integral method

**Keywords:** Fission barrier, Hartree-Fock-Bogoliubov
**PACS: 25.85.-w, 21.60.Jz, 27.90.tb**

## 1. INTRODUCTION

The information on the nuclear fission barrier height is one of the essential ingredients for the study of the r-process element abundance realized in the high neutron density environment such as in the supernova explosion. For the estimation of the reaction rate in the $\beta$ -delayed fission, the neutron induced fission, the $\beta$ -delayed fission after neutron emission and so on, the nuclear fission barrier height has an important role. The systematic analysis for the fission barrier height of nuclei relevant to the r-process nucleosynthesis has been reported by Mamdouh et al.[1]. Their calculation has been done by means of the extended Thomas-Fermi plus Strutinsky integral (ETFSI) method. The systematics is said to show an trend having a large fission barrier for the neutron number around 180 and larger than 200.

In order to confirm and improve the fission barrier height in the neutron rich region by means of ETFSI, the calculation has been started from the microscopic point of view by ULB group, i.e., by using the Skyrme Hartree-Fock-Bogoliubov method and the part of results has been published[2].

In the present paper, we present several devices for improving the constraint calculation. We report a part of the results and compare them with the ETFSI data.

## 2. METHOD

The details of the method of calculation are discussed in Ref.[2]. Here we make several comments on the practical calculation.

In the present calculation, use is made of the interaction BSk12 that is investigated recently by the ULB group. The nuclear shape are described by the (c, h, α) parametrization. The constraint for the HFB calculation are given by the multipole moment Q2, Q3 and Q4 corresponding to the specific parameter (c, h, α). Fundamentally, the calculation is performed in the (c, h) space and the asymmetric degree of freedom is introduced only for the area around the saddle point where the mass asymmetric division strongly affects to the potential value. This procedure is also effective to reduce the computational time and enable us to carry out the systematic analysys.

The convergence in the iterative calculation of the wave function is improved by taking account of the precise treatment of the center-of-mass of the system.

In general, it is a complicated problem to find the saddle point in the three dimensional deformation space. We adopted the flooding method to find automatically the lowest and other higher saddle points.

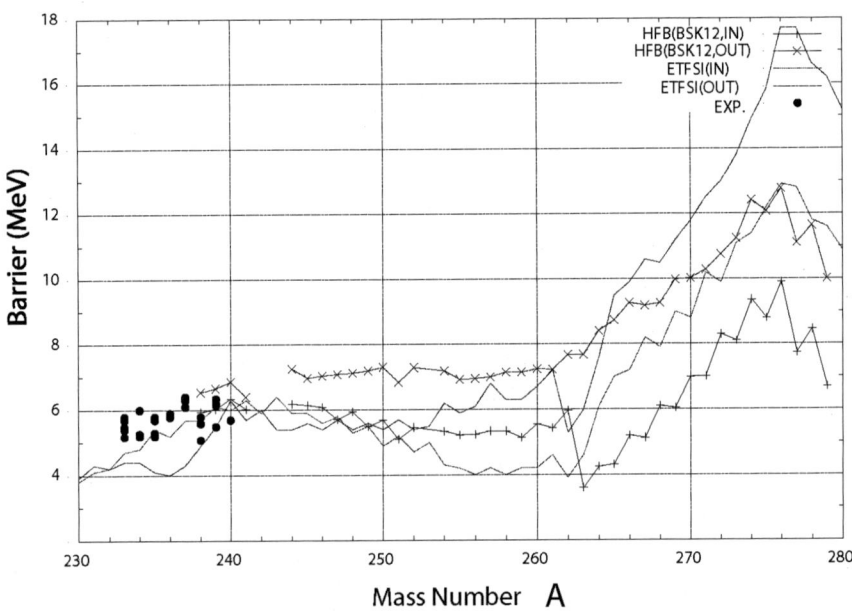

Fig.1. Fission barrier for U isotopes. The values of the inner and the outer barrier height given by the HFB calculation are plotted by + and × respectively. The corresponding barriers by the ETFSI method are drawn by the broken and the dotted lines which are indicated as ETFSI(IN) and ETFSI(OUT). The experimental values[3] are depicted by the solid circles. It can be seen that up to around A=265 the barrier

by the HFB method is comparable with or up to 1MeV higher than that of the ETFSI method. Beyond A~265, the situation is reverse.

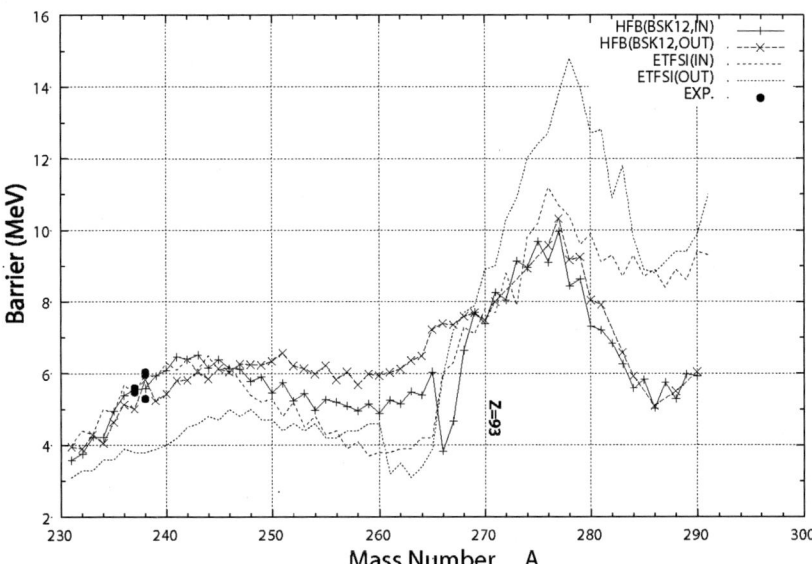

Fig.2. Fission barrier for Np isotopes. The symbols and lines used are the same as in Fig.1.

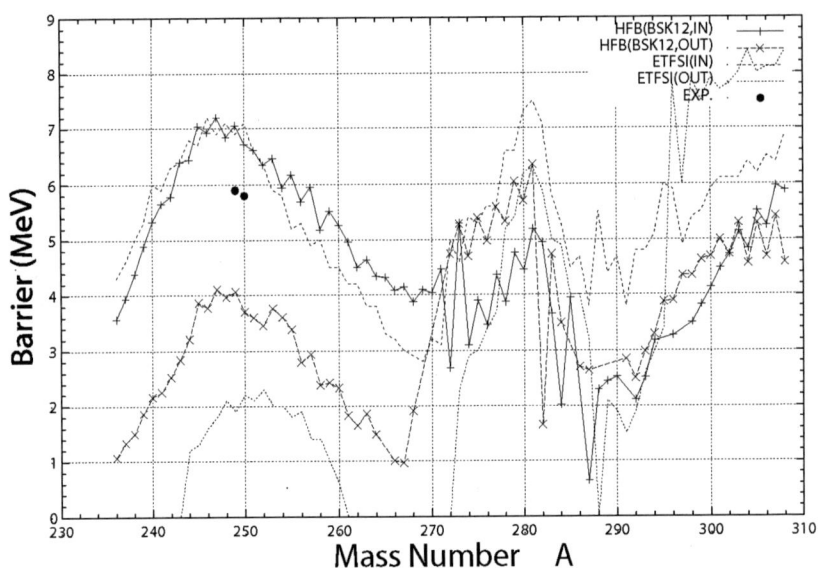

Fig. 3. Fission barrier for Bk isotopes. The symbols and lines used are the same as in Fig.1.

## 3. RESULTS and DISCUSSIONS

The barrier heights are presented for the inner and the outer barrier for U, Np and Bk isotopes in Figs. 1-3. The barrier height by ETFSI are also plotted in the same figures. In the case of U isotopes, both the inner and the outer barrier height given by ETFSI are comparable to the results by HFB up to the neutron number N about 170. For N>170, the results by the two methods deviate. The HFB calculation gives lower values than that of ETFSI in N>170. The difference is about 5MeV in maximum around N=184. As for the other nuclei, Np and Bk, the trends of the results are the same.

This means that the role of the fission process is more enhanced by using the fission barrier given by the Skyrme-HFB especially when the r-process path approaches to the neutron drip line.

The work of the systematic estimation for the fission barrier height is continuing.

## REFERENCES

1. A. Mamdouh, J.M.Pearson, M.Rayet and F.Tondeur, Nucl. Phys. A679 (2001) 337.
2. M.Samyn and S.Goriely, Phys. Rev. C72, (2005) 044316.
3. M.Dahlinger and D.Vermeulen, Nucl Phys. A376 (1982) 94.

# A reassessment of surface friction model for maximum cold fusion reactions in superheavy mass region

A. Fukushima, A. Nasirov*, Y. Aritomo*, T. Wada, and M. Ohta

*Department of Physics, Konan University, Okamoto, Kobe 658-8501, Japan*
*\*Flerov Laboratory of Nuclear Reactions, JINR, Dubna, Moscow region, 141980 Russia*

**Abstract.** We have made a study on the capture process of $^{40,48}$Ca+ $^{208}$Pb systems with a dynamical approach based on the surface friction model. The deformation of the nuclei due to the mutual excitation is taken into account. We have calculated the capture cross sections for several values of the friction coefficients. It was shown that, in the cold fusion reactions, the friction parameters of the surface friction model needs to be reexamined.

**Keywords:** friction, nuclear deformation, orientation, Langevin equation
PACS: 25.60.Pj, 25.70.Lm, 05.10.Gg

## 1. INTRODUCTION

In the research on synthesizing superheavy elements, the incident projectile energies or corresponding excitation energies of compound nuclei that give the maximum cross section are of great interest experimentally. There is a systematic research of excitation functions for superheavy elements production by means of cold fusion reaction reported by Hofmann [1] at GSI in Germany. As for the theoretical approach, there is an effective model to describe the fusion (capture) process, i.e. the surface friction model [2]. The multi-dimensional Langevin equations have been applied to calculate the capture cross-section [3], but deformations and orientations of the nuclei have not been included in the capture process so far. On the other hand, it has recently been noted that the nuclear deformation and orientation in capture process are very important for fusion reactions [4].

Following the theory of the compound nucleus reaction, we assume that the residue cross sections are given by the formula [3]

$$\sigma_{ER} = \frac{\pi \hbar^2}{2\mu E_{cm}} \sum_l (2l+1) P_{stick} P_{for} P_{sur} . \tag{1}$$

Here the factors $P_{stick}$, $P_{for}$, and $P_{sur}$ denote the sticking, formation and survival probabilities for the total angular momentum $l$, respectively. In this paper, we calculate the sticking probabilities with the Langevin equation involving the effects of quadrupole deformation and orientation, and compare the calculated capture cross-sections with the experiments [5]. We regard an event as capture when the relative distance $r$ becomes smaller than the geometrical contact of the surfaces of two nuclei with a negative relative velocity. We need to emphasize the importance of the combination of the nuclear deformations and orientations.

## 2. CALCULATION

We chose the systems $^{40,48}$Ca+ $^{208}$Pb, since we are interested in the effect of the induced deformation. We set the coordinate system [6] as is shown in FIGURE 1.

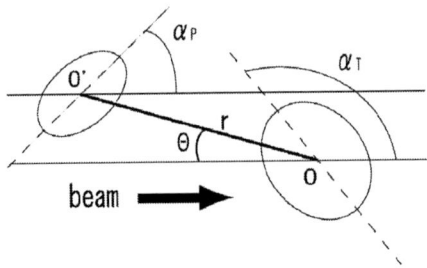

**FIGURE 1.** The coordinate system; $\alpha_P$, $\alpha_T$ are the orientations of projectile and target, respectively.

The multi-dimensional Langevin equations are given as

$$\dot{q}_i = \sum_j [m^{-1}(q)]_{ij} p_j$$

$$\dot{p}_i = -\frac{\partial V(q_i)}{\partial q_i} - \frac{1}{2} \sum_{j,k} \frac{\partial [m^{-1}(q)]_{jk}}{\partial q_i} p_j p_k - \sum_{j,k} \gamma_{ij}(q)[m^{-1}(q)]_{jk} p_k + \sum_j g_{ij}(q) \Gamma_j(t) \quad (2)$$

where $q_i$'s denotes the relative distance $r$, the polar angle to beam direction $\Theta$, the deformation variables ($\beta_P$, $\beta_T$), and the orientation angle variables ($\alpha_P$, $\alpha_T$). $p_i$ is the conjugate momentum for $q_i$. $V$ is the Gross-Kalinovski potential [2], which we take into account the nuclear deformation and orientation. $m_{ij}$ and $\gamma_{ij}$ are the collective inertia and friction tensor, respectively. We use the reduced mass for the inertia of the relative motion, since we stop our calculation at the contact point. The friction tensor $\gamma_{ij}$ can be written as

$$\gamma_{ij} = \begin{pmatrix} K_r & 0 & K_{r\beta_P} & K_{r\beta_T} \\ 0 & K_\Theta & 0 & 0 \\ K_{r\beta_P} & 0 & K_{\beta_P\beta_P} & K_{\beta_P\beta_T} \\ K_{r\beta_T} & 0 & K_{\beta_T\beta_P} & K_{\beta_T\beta_T} \end{pmatrix}$$

$$(i,j = r, \Theta, \beta_P, \beta_T). \quad (3)$$

Here the expressions for $K_{ij}$ are obtained in Ref.[2] with taking account of the orientation. We assume no frictions for the orientation variables.

The normalized random force $\Gamma_i(t)$ is assumed to be a white noise, i.e., $<\Gamma_i(t)> = 0$ and $<\Gamma_i(t_1)\Gamma_j(t_2)> = 2\delta_{ij}\delta(t_1 - t_2)$. The strength of random force $g_{ij}$ is given by the Einstein relation $\gamma_{ij}T = \Sigma_k g_{ik}g_{jk}$, where $T$ is temperature of the system which is brought out from the intrinsic energy as $E_{int} = aT^2$, with $a$ representing the level density parameter [7].

## 3. RESULTS

The sticking probability for orbital angular momentum $l$ gives the partial capture cross-section. By integrating the partial cross-section with respect to $l$, we obtain the capture cross-section.

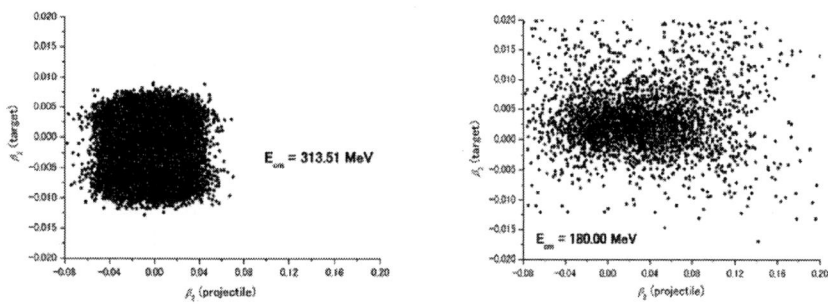

**FIGURE 2.** Distributions of deformation variables calculated for the $^{40}$Ca+$^{208}$Pb reaction at the contact point with the incident energies $E_{cm}$ = 313.51 MeV (left panel) and $E_{cm}$ = 180 MeV (right panel).

Distributions of deformations at contact point for the $^{40}$Ca+$^{208}$Pb reaction are shown in FIGURE2. At the high incident energy ($E_{cm}$ = 313 MeV), the deformation is small since the nuclei have no time to deform. On the other hand, the relative motion is slow near the Bass barrier energy and the distribution of deformation variables spreads more. Thus, the effect of deformation is stronger near the Bass barrier ($E_{cm}$ = 180 MeV). The effective barrier height is lowered by the nuclear deformation. The deformations of calcium are larger than those of lead. This is due to the Coulomb deformation of calcium by the large electric charge of lead.

In the left panel of FIGURE3, we show the capture cross sections for $^{40}$Ca+$^{208}$Pb plotted as a function of $E_{cm}$. Solid squares denote the experimental data [5], triangles denote the calculated results with the strength of the radial friction [2] $K_r$ = 10.5 [fm/c/MeV], while open circles denote those with $K_r$ = 2.1 [fm/c/MeV]. In the right panel of FIGURE3, the same is plotted as in the left panel but for the $^{48}$Ca+$^{208}$Pb reaction and open circles denote the calculated results with $K_r$ = 3.0 [fm/c/MeV]. These results suggest that we need a weak radial friction coefficient for the calculation of fusion cross sections. A large difference still exists in the sub-barrier region.

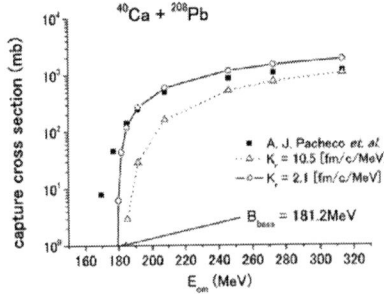
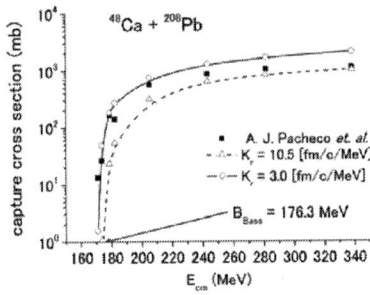

**FIGURE 3.** Capture cross sections for the $^{40}$Ca+$^{208}$Pb (left panel) and $^{48}$Ca+$^{208}$Pb (right panel) reactions as functions of $E_{cm}$.

## 4. SUMMARY

We have calculated the capture cross sections for $^{40,48}$Ca+ $^{208}$Pb systems with the multi-dimensional Langevin approach. The effect of the induced nuclear deformation and orientation is examined. We conclude that we need a weak radial friction coefficient for the calculation of capture cross sections. Our results are in good agreement with the experimental data near and above the Bass barrier. There still remains a large discrepancy between the experiments and our calculations for the incident energy well below the Bass barrier. We will have to discuss the friction of deformation because the sticking probability is very sensitive to nuclear deformation. In addition, effects of the zero point oscillation should be taken into account.

## ACKNOWLEDGMENTS

The authors would like to thank Dr. Hamada, Dr. Asano and Mr. Yamamoto for numerous enlightening discussions. One of the authors (T. W.) is partially supported by a Grand-in-Aid for Scientific Research No. 17540270 provided by the Japan Society for the Promotion of Science. A. F. and A. N. are grateful to the Advanced Science Research Center of JAEA for helpful discussions and supports.

## REFERENCES

1. S. Hofmann et al., Rev. Mod. Phys. 72, 733 (2000).
2. P. Fröbrich and I.I. Gontchar, Phys. Rep. 292 (1998) 131-237.
3. C. Shen, G. Kosenko, and Y. Abe, Phys. Rev. C 66, 061602(R) (2002).
4. K. Nisio et al., Phys. Rev. Lett. 93, 162701 (2004).
5. A. J. Pacheco et al. Phys. Rev. C 45 (1992) 2861.
6. A. Nasirov et al., Nucl. Phys. A759, 342 (2005).
7. J. Töke and W. J. Swiatecki, Nucl. Phys. A372, 141(1981).
8. P. Fröbrich, B. Strack, and M. Durand, Nucl. Phys. A406, 557 (1983).
9. P. Møller et al., At. Data Nucl. Data Tables 59, 185 (1995).
10. Y. Aritomo et al., Nucl. Phys. A759, 309 (2005).

# Coherent scattering of neutrinos by "nuclear pasta" in dense matter

Hidetaka Sonoda

*Department of Physics, University of Tokyo, Tokyo 113-0033, Japan*
*The Institute of Chemical and Physical Research (RIKEN), Saitama 351-0198, Japan*

**Abstract.** We examine coherent scattering cross section of neutrino and nucleon systems via weak-neutral current at subnuclear densities, which will be important in supernova cores. Below melting density and temparature of nuclei, nuclear shape becomes rodlike and slablike; this is called nuclear "pasta". Transition of structure will greatly influence coherent effects which can not easily be predicted. We calculate static structure factor of nuclear matter using data of several nuclear models, and discuss the effects of existence of nuclear pasta on neutrino opacity in hot dense matter.

**Keywords:** coherent scattering, neutrino, nuclear pasta

## INTRODUCTION

In the core collapse supernova, a large amount of neutrinos, which would play a critical role for success of supernova explosions, are mainly produced by electron capture by heavy nuclei. Above the density where neutrino trapping occurs interactions between neutrinos and supernova matter are dominated by coherent scattering of neutrinos and heavy nuclei [1, 2]. At subnuclear densities just below nuclei melt, nuclear shape would change from spheres to rods, slabs, rodlike bubbles and spherical bubbles [3, 4, 5, 6, 7, 8]. The existence of these phases called nuclear "pasta" would affect the cross section of neutrinos and matter, which is the point we focus on in this article. We calculate neutrino opacities of pasta phases with liquid drop model and quantum molecular dynamics (QMD), respectively. Especially, this is the first successful example to obtain neutrino opacity of pasta phases including rodlike and slablike nuclei within the framework of QMD [9], because Horowitz *et al.* attempted to obtain it in the same way but they can not have reproduced rodlike or slablike nuclei but only nonspherical ones [10].

## FORMULATION

When we consider neutrino and pasta scattering, its cross section per nucleon is written as follows,

$$\frac{1}{N}\frac{d\sigma}{d\Omega}(\mathbf{q}) = \frac{G_F^2 E_\nu^2}{4\pi^2}(1+\cos\theta)c_\nu^{(n)2}\overline{S}(\mathbf{q}), \tag{1}$$

where $N$ is the number of nucleons in the system, $G_F$ is the Fermi coupling constant, $E_\nu$ is the energy of neutrinos, $\theta$ is the scattering angle, $c_\nu^{(n)}$ ($c_\nu^{(p)}$) is the vector coupling constant of neutrons (protons) to the weak neutral current ($c_\nu^{(n)} = -1/2, c_\nu^{(p)} = 1/2 -$

$2\sin^2\theta_W$, $\sin^2\theta_W = 0.23$), $\mathbf{q}$ is the momentum transfer, and $\overline{S}(\mathbf{q})$ is the static structure factor averaged over those for proton-proton, proton-neutron, and neutron-neutron. The factor $\overline{S}(\mathbf{q})$ corresponds to the amplification because the neutrino cross section of an isolated neutron is expressed as the right hand side of Eq. (1) except for this factor. Since the static structure factor is related to the two-point correlation of nucleons, change of the lattice structure by transition to the pasta phases leads to that of the cross section of neutrino and matter. In the field of supernova simulation, transport cross section defined as

$$\frac{d\sigma_t}{d\Omega}(\mathbf{q}) \equiv \frac{d\sigma}{d\Omega}(1-\cos\theta), \qquad (2)$$

is often used. In the following analyses, we show the amplification factor of angle averaged transport cross section $\langle \overline{S}(E_v) \rangle$.

## RESULTS

We examine cross section using several nuclear models including a compressible liquid drop model based on an equation of state (EOS) by Baym, Bethe, and Pethick (BBP) [11, 12], that by Lattimer and Swesty (LSEOS) [13], and quantum molecular dynamics (QMD) [14]. Here we briefly summarize the tendencies we have found [9]. The common characteristics of $\langle \overline{S}(E_v) \rangle$ for all models is that the coherent effect cannot be seen below 20 MeV and above 100 MeV and that the opacity peak is located around 30–40 MeV. Peak energy shifts if we take account of the pasta phases, i.e., peak energy of the phase with rodlike nuclei is higher than that with spherical ones at a density where rodlike nuclei is energetically favored, for example. This is due to a smaller intespacing of pasta nuclei than that of spherical ones.

In Fig. 1 we demonstrate changes of $\langle \overline{S}(E_v) \rangle$ with density considering existence of pasta phases at zero temperature using BBP model at lepton fraction $Y_L = 0.3$. We naively expect that the cross section shows discontinuous change at transition points because of discontinuous change of lattice constant, lattice structure and nuclear size. However, contrary to the simple expectation, neutrino opacity changes almost continuously even at transition points as shown in Fig. 1. Although this phenomena can be predicted by the most simple nuclear model, imcompressible liquid drop model, we can not point out physical reasons for this phenomena.

We have studied $\langle \overline{S}(E_v) \rangle$ at finite temperatures using LSEOS and QMD simulations at proton fraction $x = 0.3$. Increasing temperatures basically decreases the cross section because of evaporation of nucleons and deviations of positions of nuclei from lattice points. However, QMD results shows that in some cases transitions between the pasta phases can increase $\langle \overline{S}(E_v) \rangle$. We can not conclude this is due to the advantage of QMD, which enables us to treat nuclear matter at finite temperature without assumptions on the nuclear shape, or due to the boundary effects of the simulations. We will investigate this point in the near future.

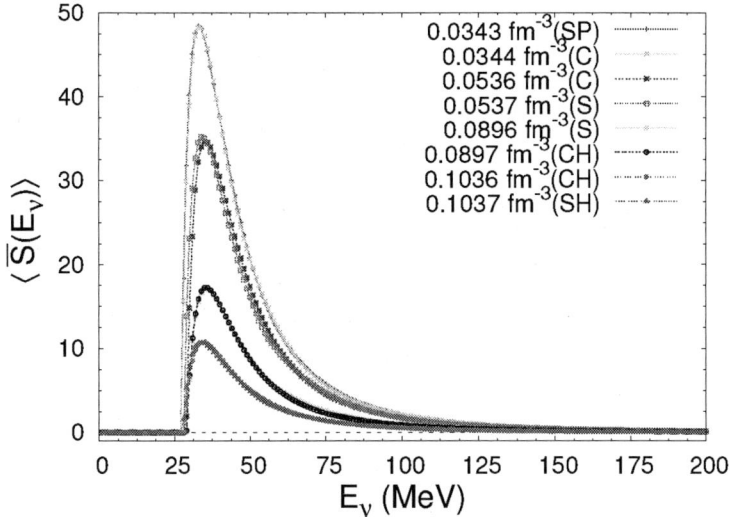

**FIGURE 1.** $\langle \overline{S}(E_\nu) \rangle$ at $Y_L = 0.3$ calculated for a model based on BBP with typical values of parameters given in Ref. [12]. Abbreviations SP, C, S, CH and SH mean phases with spherical nuclei, cylindrical nuclei, slablike nuclei, cylindrical holes and spherical holes, respectively.

## SUMMARY

In this article we have examined the cross section of the pasta phases in supernova environments. We have found out existence of pasta phases surely affects neutrino opacity, which can be seen from liquid drop approach. In addition, our QMD results show phase transitions between pasta phases induced by increasing temperature or density could discontinuously change or even increase neutrino opacity; these phenomena can not be predicted at all when we consider only spherical nuclei. We have to notice that our QMD calculations have some points to be improved; the most important one is boundary effects. In the near future, we will perform simulations with a larger number of particles to clarify the problems we cannot conclude here.

## ACKNOWLEDGMENTS

The study described in this article is supported by collaboration with G. Watanabe, K. Sato, K. Yasuoka and T. Ebisuzaki. The author also thanks RIKEN for providing computational resources as RSCC (RIKEN Super Combined Cluster).

## REFERENCES

1. K. Sato, Prog. Theor. Phys. **54** (1975) 1352.
2. D. Z. Freedman, Phys. Rev. D **9** (1974) 1389.

3. D. G. Ravenhall, C. J. Pethick and J. R. Wilson, Phys. Rev. Lett. **50** (1983) 2066.
4. M. Hashimoto, H. Seki and M. Yamada, Prog. Theor. Phys. **71** (1984) 320.
5. R. D. Williams and S. E. Koonin, Nucl. Phys. **A435** (1985) 844.
6. M. Lassaut, H. Flocard, P. Bonche, P. H. Heenen and E. Suraud, Astron. Astrophys. **183** (1987) L3.
7. G. Watanabe, K. Sato, K. Yasuoka and T. Ebisuzaki, Phys. Rev. C **69** (2004) 055805.
8. G. Watanabe, T. Maruyama, K. Sato, K. Yasuoka and T. Ebisuzaki, Phys. Rev. Lett. **94** (2005) 031101.
9. H. Sonoda, G. Watanabe, K. Sato, K. Yasuoka and T. Ebisuzaki, in preparetion.
10. C. J. Horowitz, M. A. Perez-Garcia, J. Carriere, D. K. Berry and J. Piekarewicz, Phys. Rev. C **70** (2004) 065806.
11. G. Baym, H. A. Bethe and C. J. Pethick, Nucl. Phys. **A175** (1971) 225.
12. G. Watanabe, K. Iida and K. Sato, Nucl. Phys. **A687** (2001) 512.
13. J. M. Lattimer and F. D. Swesty, Nucl. Phys. **A535** (1992) 331.
14. T. Maruyama, K. Niita, K. Oyamatsu, T. Maruyama, S. Chiba and A. Iwamoto, Phys. Rev. C **57** (1998) 655.

# LIST OF PARTICIPANTS

Akimune, Hidetoshi
Department of Physics, Konan Univ.
8-9-1 Okamoto Higashinada-ku
Kobe    658-8501
Japan

akimune@konan-u.ac.jp

Aoyama, Shigeyoshi
NIIGATA UNIVERSITY
Integrated Information Processing Center
Niigata University, Igarashi-2-8050
950-2181    Japan

aoyama@cc.niigata-u.ac.jp

Bansyo, Yasuyuki
Department of Physics, Konan Univ.
8-9-1 Okamoto Higashinada-ku
Kobe    658-8501
Japan

reaction_rate_1018@ybb.ne.jp

Binns, Walter
Washington University in Saint Louis
Dept. of Physics, CB1105 1 Brookings Drive
Saint Louis, MO    63130
USA

wrb@wuphys.wustl.edu

Boutin, David
Giessen University
Heinrich-Buff-Ring 14 Giessen
35392
Germany

d.boutin@gsi.de

Coc, Alain
CSNSM
Bat. 104 Orsay Campus
F-91405
France

coc@csnsm.in2p3.fr

Ando, Shin'ichiro
Theoretical Astrophysics, California Institute of Technology
Mail Code 130-33 1200 E. California Blvd.
Pasandena, CA    91125    USA

ando@tapir.caltech.edu

Arnould, Marcel
Institut d'Astronomie et d'Astrophysique
Universite Libre de Bruxelles - CP 226,
Boulevard du Triomphe
B-1050 Brussels
Belgium

marnould@astro.ulb.ac.be

Beaumel, Didier
IPN Orsay
15 rue G. Clemenceau
F-91406
Orsay

beaumel@ipno.in2p3.fr

Blumenfeld, Yorick
IPN Orsay
15 rue G. Clemenceau
F-91406
Orsay

yorick@ipno.in2p3.fr

Chamel, Nicolas
Institut d'Astronomie et d'Astrophysique
Universite Libre de Bruxelles CP226
Boulevard du Triomphe
B-1050Brussels    Belgium

nchamel@ulb.ac.be

Costantini, Heide
Univeristy of Notre Dame
Department of Physics 125 Nieuwland
Science Hall Notre Dame, Indiana
46656-5670    USA

hcostant@nd.edu

Düllmann, Christoph
Gesellschaft für Schwerionenforschung mbH
Planckstrasse 1
D-64291 Darmstadt
Germany

c.e.duellmann@gsi.de


Demetriou, Paraskevi
Institute of Nuclear Physics, NCSR
"Demokritos"
Agia Paraskevi Attikis Athens    15310
Greece

vivian@inp.demokritos.gr


Dubray, Noel
CEA
CEA-Bruyeres le Chatel B.P.12
Bruyeres-le-Chatel
F-91680   France

noel.dubray@cea.fr


Emling, Hans
GSI
Planckstr. 1
D-64291 Darmstadt
Germany

h.emling@gsi.de


Fukuda, Mitsunori
Department of Physics, Osaka University
1-1 Machikaneyama, Toyonaka, Osaka
560-0043
Japan

mfukuda@phys.sci.osaka-u.ac.jp


Ghosh, Tilak Kumar
Physics Group
Variable Energy Cyclotron Centre
1/AF, Bidhan Nagar, Kolkata - 700 064
INDIA

tilak@veccal.ernet.in


de Oliveira Santos, François
GANIL
Bd Henri Becquerel, BP 55027,
F-14076 CAEN Cedex 05
France

oliveira@ganil.fr


Doornenbal, Pieter
GSI
Planckstrasse 1
D-64291 Darmstadt
Germany

p.doornenbal@gsi.de


Elekes, Zoltan
Institute of Nuclear Research (ATOMKI)
H-4026 Debrecen Bem ter 18/c.

Hungary

elekes@atomki.hu


Fujimoto, Shin-ichiro
Kumamoto National College of Technology
2659-2 Suya, Goshi, Kumamoto 861-1102,
861-1102    Japan

fujishin@sci.kumamoto-u.ac.jp


Gales, Sidney
GANIL
Bd Henri Becquerel, BP 55027,
F-14076 CAEN Cedex 05
France

gales@ganil.fr


Goko, Shinji
Department of Physics, Konan Univ.
8-9-1 Okamoto Higashinada-ku
Kobe    658-8501
Japan

gokou@konan-u.ac.jp

Goriely, Stephane
Institut d'Astronomie et d'Astrophysique
Universite Libre de Bruxelles - CP 226,
Boulevard du Triomphe
B-1050 Brussels
Belgium

sgoriely@astro.ulb.ac.be

Goutte, Heloise
CEA Bruyeres-le-Chatel
CEA /DIF/DPTA/SPN BP 12
F-91680 Bruyeres-le-Chatel
FRANCE

heloise.goutte@cea.fr

Haba, Hiromitsu
Nishina Center for Accelerator Based
Science, RIKEN
Hirosawa 2-1, Wako, Saitama
351-0198   JAPAN

haba@riken.jp

Hagino, Kouichi
Department of Physics
Tohoku University
Sendai980-8578
Japan

hagino@nucl.phys.tohoku.ac.jp

Hanappe, Francis
Universite Libre de Bruxelles
PNTPM, CP229
Boulevard du Triomphe
B-1050 Bruxelles
Belgium

fhanappe@ulb.ac.be

Harissopulos, Sotirios
Institute of Nuclear Physics, NCSR
"Demokritos"
153.10 Aghia Paraskevi, Athens,
Greece

sharisop@inp.demokritos.gr

Heßberger, Fritz
Gesellschaft für Schwerionenforschung mbH
Abt. KP II Planckstraße 1
D-64291 Darmstadt
Germany

f.p.hessberger@gsi.de

Hilaire, Stéphane
DPTA/Service de Physique Nucléaire
CEA/DAM Ile de France BP12,
F-91680 Bruyères-le-Châtel
FRANCE

stephane.hilaire@cea.fr

Katsuma, Masahiko
Institut d'Astronomie et d'Astrophysique
Universite Libre de Bruxelles - CP 226,
Boulevard du Triomphe
B-1050 Brussels
Belgium

mkatsuma@ulb.ac.be

Kiener, Juergen
CSNSM
Bat. 104 Orsay Campus
F-91405
France

kiener@csnsm.in2p3.fr

Kimura, Sachie
LNS-Istituto Nazionale di Fisica Nucleare
Via S.Sofia 62,
I-5125 Catania
Italy

kimura@lns.infn.it

Kiselev, Oleg
Institute of Nuclear Chemistry,
University of Mainz
Fritz-Strassmann-Weg 2
D-55128 Mainz, Germany

O.Kiselev@gsi.de

Knoebel, Ronja
GSI, Darmstadt
Planckstr. 1
D-64291 Darmstadt
Germany

R.Knoebel@gsi.de

Lewitowicz, Marek
GANIL
Bd Henri Becquerel, BP 55027,
F-14076 CAEN Cedex 05
France

lewitowicz@ganil.fr

Makinaga, Ayano
Department of Physics, Konan Univ.
8-9-1 Okamoto Higashinada-ku
Kobe 658-8501
Japan

dn521001@center.konan-u.ac.jp

Matsumoto, Erika
University of Tokushima
1-8-5-3A,sumiyoshi,tokushima-shi
770-0861
Japan

c100632013@stud.tokushima-u.ac.jp

Michel, Rolf
University Hannover
Zentrum für Strahlenschutz und
Radioökologie Universität Hannover Am
Kleinen Felde 30 Hannover
D-30167    Germany

michel@zsr.uni-hannover.de

Morita, Kosuke
Nishina Center for Accelerator Based
Science, RIKEN
Hirosawa 2-1, Wako-shi, Saitama
351-0198   Japan

morita@rarfaxp.riken.jp

Kobayashi, Toshio
Department of Physics, Faculty of Science,
Tohoku University
Aramaki, Aoba-ku, Sendai,
Miagi 980-8578   Japan

kobayash@lambda.phys.tohoku.ac.jp

Münzenberg, Gottfried
Univ. Manipal
Bergfeldstr. 26 Diekholzen
D-31199
Germany

g.muenzenberg@gsi.de

Martín, Ana
GSI
Planckstr. 1
D-64291 Darmstadt
Germany

A.Martin@gsi.de

Meyer, Bradley
Department of Physics and Astronomy
Clemson University Clemson, South Carolina
29634-0978
USA

mbradle@clemson.edu

Minato, Futoshi
Nuclear Theory Group, Tohoku Univ.
6-3,Aoba, Aramaki, Aobaku, SENDAI,
MIYAGI   980-8578
JAPAN

minato@nucl.phys.tohoku.ac.jp

Nagai, Yasuki
Research Center for Nuclear Physics,
Osaka University
Mihogaoka 10-1, Ibaraki, Osaka
567-0047   Japan

nagai@rcnp.osaka-u.ac.jp

Nakayama, Shintaro
University of Tokushima
1-1 Minami-Josanjima-Cho Tokushima
770-8502
Japan

nakayama@ias.tokushima-u.ac.jp

Nishio, Katsuhisa
Japan Atomic Energy Agency
Advanced Science Research Center,
Tokai, Ibaraki
319-1195   Japan

nishio.katsuhisa@jaea.go.jp

Ohta, Masahisa
Department of Physics, Konan Univ.
8-9-1 Okamoto Higashinada-ku
Kobe   658-8501
Japan

masaota@konan-u.ac.jp

Scheid, Werner
Institut fuer Theoretische Physik
Justus-Liebig-Universitaet
Heinrich-Buff-Ring 16
D-35392 Giessen, Germany

Werner.Scheid@theo.physik.uni-giessen.de

Shaviv, Giora
Israel Institute of Technology
Dept. of Physics Technion, Israel Institute of Technology Technion City, Haifa,
32,000   Israel

gioras@physics.technion.ac.il

Starrfield, Sumner
School of Earth and Space Exploration,
Arizona State University
P.O. Box 871504
Tempe, Arizona   85287-1504   USA

starrfield@asu.edu

Nishimura, Takanori
Astrophysics Laboratory,
Hokkaido University
Kita 10 Nishi 8, Kita-ku, Sapporo 060-0810,
Japan

nishimura@astro1.sci.hokudai.ac.jp

Nomura, Toru
Nishina Accelerator Based Science Center,
RIKEN
2-15-1-618 Shirako, Wako-shi, Saitama-ken
351-0101   JAPAN

t-nomura@r5.dion.ne.jp

Saito, Akito
Department of Physics, University of Tokyo
RIKEN Campus, 2-1 Hirosawa, Wako,
Saitama
351-0198   Japan

akito@nucl.phys.s.u-tokyo.ac.jp

Schuck, Peter
Institut de Physique Nucleaire, Orsay
15 rue G.Clemenceau
F-91406 Orsay cedex
France

schuck@ipno.in2p3.fr

Sonoda, Hidetaka
Depertment of Physics, University of Tokyo
Hongo 7-3-1, Bunkyo-ku, Tokyo
113-0033   Japan

sonoda@utap.phys.s.u-tokyo.ac.jp

Stodel, Christelle
GANIL
Bd Henri Becquerel, BP 55027,
F-14076 CAEN Cedex 05
France

stodel@ganil.fr

Takahashi, Kohji
Institut d'Astronomie et d'Astrophysique
Universite Libre de Bruxelles · CP 226,
Boulevard du Triomphe
B-1050 Brussels
Belgium

ktakahas@ulb.ac.be


Tanaka, Masayoshi
Kobe Tokiwa College
Ohtani-cho 2-6-2, Nagata-ku,
Kobe 653-0838    Japan

mlg11946@nifty.com


Tonchev, Anton
Triangle Universities Nuclear Laboratory
Duke University BOX 90308 Durham NC,
27708-0308 USA

tonchev@tunl.duke.edu


Utsunomiya, Hiroaki
Department of Physics, Konan Univ.
8-9-1 Okamoto Higashinada-ku
Kobe    658-8501
Japan

hiro@konan-u.ac.jp


Villari, Antonio
GANIL
Bd Henri Becquerel, BP 55027,
F-14076 CAEN Cedex 05
France

villari@ganil.fr


Wagner, Andreas
Forschungszentrum Rossendorf
Bautzner Landstr. 128
D-01328 Dresden
Germany

a.wagner@fz-rossendorf.de


Takechi, Maya
Research Center for Nuclear Physics,
Osaka University
10-1 Mihogaoka, Ibaraki, Osaka
567-0047    Japan

takechi@rcnp.osaka-u.ac.jp


Tatsuda, Sayuki
Department of Physics, Konan Univ.
8-9-1 Okamoto Higashinada-ku
Kobe    658-8501
Japan

mn621010@center.konan-u.ac.jp


Ueno, Hideki
RIKEN
2-1 Hirosawa, Wako,
Saitama 351-0198    Japan

ueno@riken.jp


Uusitalo, Juha
University of Jyväskylä
Department of Physics
Survontie 9
40500 Jyväskylä    Finland

juha.uusitalo@phys.jyu.fi


Wada, Takahiro
Department of Physics, Konan Univ.
8-9-1 Okamoto Higashinada-ku
Kobe    658-8501
Japan

wada@konan-u.ac.jp


Washiyama, Kouhei
Nuclear Theory Group,
Department of Physics, Tohoku University,
6-3, Aoba, Aramaki, Aobaku,
Sendai    980-8578, Japan

washi@nucl.phys.tohoku.ac.jp

Watanabe, Gentaro
NORDITA
Blegdamsvej 17,
DK-2100 Copenhagen o.e.
Denmark

gentaro@nordita.dk


Wolski, Roman
Flerov Laboratory of Nuclear Reactions
Joint Institute for Nuclear Research,
141980 Dubna, Moscow region,
Russia

wolski@nrmail.jinr.ru


Yamamoto, Kazuyuki
Department of Physics, Konan Univ.
8-9-1 Okamoto Higashinada-ku
Kobe   658-8501
Japan

dn621003@center.konan-u.ac.jp


Winkler, Martin
Flerov Laboratory of Nuclear Reactions
Joint Institute for Nuclear Research,
141980 Dubna, Moscow region,
Russia

m.winkler@gsi.de


Yamagata, Tamio
Department of Physics, Konan Univ.
8-9-1 Okamoto Higashinada-ku
Kobe   658-8501
Japan

yamagata@center.konan-u.ac.jp


Yoshioka, Kazumasa
Department of Physics, Konan Univ.
8-9-1 Okamoto Higashinada-ku
Kobe   658-8501
Japan

yfa38584@yahoo.co.jp

## Author Index

### A

Ackermann, D., 19, 71
Adamian, G. G., 27
Aikawa, M., 397, 405
Akimune, H., 427
Akiyama, T., 3, 45
Amar, N., 55
Ando, S., 263
Angélique, J. C., 155
Angélique, M., 155
Angell, C., 339
Anne, R., 55
Antalic, S., 71
Antonenko, N. V., 27
Aoi, N., 122, 192, 205
Aoyama, S., 222
Arai, K., 222
Arai, T., 113
Aritomo, Y., 423, 443
Asahi, K., 113
Asaji, S., 427
Asano, T., 435
Auger, G., 55
Axiotis, M., 281
Ayik, S., 413

### B

Baba, H., 192, 205
Balantekin, A. B., 401
Beaumel, D., 192
Beckert, K., 199
Belhout, A., 254
Beller, P., 199
Berg, G. P., 131
Berger, J. F., 409
Berthoumieux, E., 155
Bhattacharyya, S., 60
Block, M., 19
Blumenfeld, Y., 147, 192
Boilley, D., 60
Bonasera, A., 306
Borcea, R., 155
Bosch, F., 199
Boswell, M., 339

Bouriquet, B., 55
Boutin, D., 199
Brandau, C., 199
Buta, A., 155

### C

Casandjian, J.-M., 55
Cee, R., 55
Chamel, N., 382
Chatillon, A., 55
Chaudhuri, A., 19
Chen, L., 199
Chiba, S., 423
Chomaz, Ph., 401
Chyzh, A., 339
Clément, E., 55
Coc, A., 155
Comas, V. F., 71
Costantini, H., 297
Cullen, I. J., 199

### D

Dalouzy, J. C., 155
Daugas, J. M., 155
Davinson, T., 155
Dayras, R., 55, 60
de France, G., 55, 60
de Grancey, F., 155
Delaroche, J.-P., 324
Demetriou, P., 281
Demichi, K., 122
de Oliveira Santos, F., 55, 155
Di, Z., 19
Dimopoulou, C., 199
Düllmann, C. E., 36
Dolinskii, A., 199
Dombrádi, Zs., 122
Dorvaux, O., 55
Drouart, A., 55, 60
Dubray, N., 409
Dumitru, G., 155

## E

Elekes, Z., 122, 192
Eliseev, S., 19

## F

Fabian, B., 199
Fadil, M., 155
Fülöp, Zs., 122
Fomichev, A. S., 226
Fortier, S., 192, 226
Frascaria, N., 192
Fujimoto, M. Y., 397, 405
Fujimoto, S., 272
Fujiwara, M., 427
Fukuda, M., 181, 187
Fukuda, N., 192
Fukushima, A., 443
Funaki, Y., 164
Fushimi, K., 427

## G

Gagyi-Palffy, Z., 27
Gan, Z., 71
Gaudefroy, L., 60
Geissel, H., 199
Ghosh, T. K., 417
Gibelin, J., 122, 192
Giot, L., 60
Girod, M., 324
Goko, S., 431
Golabek, C., 60
Golovkov, M. S., 226
Gomi, T., 122, 192, 205
Goriely, S., 246, 348, 431, 439
Gorshkow, A. V., 226
Goto, S., 3
Goutte, H., 324, 409
Greenfield, M. B., 427
Grévy, S., 55, 155
Grigorenko, L. V., 226
Gros, M., 254

## H

Haba, H., 3, 45, 122
Habs, D., 19
Hagino, K., 80, 401

Hannachi, F., 55
Hannappe, F., 55
Hara, K. Y., 431
Harada, H., 431
Harano, H., 431
Hasegawa, H., 122
Hashimoto, H., 427
Hashimoto, M., 272
Hashizume, K., 423, 439
Hauschild, K., 55
Hausmann, M., 199
Hayami, R., 427
Heinz, S., 60, 71
Heredia, J. A., 71
Herfurth, F., 19
Heßberger, F. P., 10, 19, 55, 71
Higurashi, H., 205
Hilaire, S., 348
Hix, W. R., 364
Hofmann, S., 19, 55, 71
Horiuchi, H., 164
Houhara, S., 431
Howell, C. R., 339

## I

Iben, Jr., I., 397
Ichihara, T., 131
Ideguchi, E., 3, 131
Ieki, K., 205
Igashira, M., 289
Ikemizu, H., 427
Ikezoe, H., 71
Iliadis, C., 364
Imai, N., 122, 205
Inoue, T., 113
Ishida, T., 427
Ishihara, M., 122
Ishikawa, K., 192
Ishimoto, S., 108
Ivanova, S. P., 27
Iwamoto, N., 397
Iwasa, N., 205
Iwasaki, H., 122, 205
Izumikawa, T., 181, 187

## K

Kaihori, T., 431
Kaji, D., 3, 45
Kajino, T., 423
Kameda, D., 113
Kanazawa, M., 181, 187
Kanno, S., 122, 205
Kanumgo, R., 3
Karwowski, H. J., 339
Kat, K., 405
Katori, K., 3
Katsuma, M., 355
Kawabata, T., 131
Kawai, S., 122
Kawamura, H., 113
Kawase, K., 427
Kawasuso, H., 427
Kelley, J. H., 339
Khuyagbaatar, J., 71
Kiener, J., 155, 254
Kikunaga, H., 3, 45
Kimura, S., 306
Kindler, B., 71
Kiselev, O. A., 172
Kishida, T., 122
Kitagawa, A., 181, 187
Kitamoto, Y., 45
Kitatani, F., 431
Kluge, H.-J., 19
Knöbel, R., 199
Kobayashi, K., 181, 187
Kobayashi, T., 108
Kojouharov, I., 71
Kondo, Y., 192
Korichi, A., 55
Korsheninnikov, A. A., 226
Kotake, K., 272
Koura, H., 3, 423
Kozhuharov, C., 199
Krupko, S. A., 226
Kubo, T., 122, 122, 131, 192
Kubono, S., 131, 205
Kudo, H., 3
Kudoh, T., 427
Kunibu, M., 205
Kurcewicz, J., 199
Kuribayashi, T., 45
Kurita, K., 122, 122
Kuusiniemi, P., 71

Kuzmin, E. A., 226

## L

Lefebvre-Schuhl, A., 155
Lenhardt, M., 155
Lewitowicz, M., 91, 155
Libert, J., 324
Lichtenhäler, R., 55
Lima, V., 192
Litvinov, S. A., 199
Litvinov, Yu. A., 199
Liu, Z., 199
Lojek, K., 55
Lommel, B., 71
Lopez-Martens, A., 55
Lui, Y.-W., 431

## M

Makinaga, A., 431
Marchix, A., 60
Martín, A., 19
Marx, G., 19
Maslov, V., 60
Matsuda, Y., 108
Matsumiya, R., 181, 187
Matsumoto, E., 427
Matsumoto, T., 431
Matsuo, K., 45
Matsuta, K., 181, 187
Matsuyama, Y., 122, 122, 205
Mazzocco, M., 19, 71, 199
Meyer, B. S., 391
Michel, R., 237
Michimasa, S., 122, 122, 205
Mihara, M., 181, 187
Miki, T., 108
Minamisono, T., 181, 187
Minato, F., 401
Minemura, T., 122, 122, 205
Mitsuoka, S., 71
Mittig, W., 60, 226
Münzenberg, G., 199
Momota, S., 181, 187
Montes, F., 199
Morimoto, K., 3, 45
Morita, K., 3, 45

Morjean, M., 60
Motobayashi, T., 122, 192, 205
Mukherjee, G., 60
Mukherjee, M., 19
Murata, J., 113
Musumarra, A., 199

## N

Nagae, D., 113
Nagai, Y., 289
Nagame, Y., 71
Nagatomo, T., 113
Nakajima, S., 181, 187, 199
Nakamura, T., 192, 205
Nakanishi, K., 131, 427
Nakayama, S., 427
Naoi, Y., 108
Narita, K., 113
Nasirov, A., 443
Navin, A., 60, 155
Negoita, F., 155
Neumayr, J. B., 19
Nikolskii, E. Yu., 226
Nishimura, T., 397, 405
Nishio, K., 71
Nociforo, C., 199
Nolden, F., 199
Notani, M., 122
Novatskii, B. G., 226

## O

Ohnishi, T., 3, 122
Ohsaki, T., 289
Ohta, M., 405, 423, 435, 439, 443
Ohtsubo, T., 181, 187, 199
Ohtsuki, T., 71
Ong, H. J., 122
Ooe, K., 45
Oota, T., 427
Ota, S., 122
Otsu, H., 108
Otsuki, K., 423
Ozawa, A., 3, 122, 199
Ozeki, K., 108

## P

Pantelica, D., 155
Patyk, Z., 199
Péghaire, A., 55
Pellegriti, M. G., 155
Penionzkevich, Yu., 60
Perrot, L., 155
Péter, J., 55
Plaß, W. R., 199
Plass, W., 19
Popeko, A. G., 71

## R

Rahaman, S., 19
Rauth, C., 19
Ray, I., 155
Rejmund, F., 60
Rejmund, M., 60
Rodin, A. M., 226
Rodríguez, D., 19
Roig, O., 155
Roussel, P., 226
Roussel-Chomaz, P., 60
Röpke, G., 164
Ryuto, H., 205

## S

Sagara, K., 427
Saika, D., 45
Saint Laurent, M. G., 55, 155
Saito, A., 122, 131, 192, 205
Sakai, H., 131
Sakai, H. K., 122
Sakurai, H., 122, 205
Samyn, M., 439
Saro, S., 71
Sasamoto, Y., 131
Sato, N., 3, 45
Sato, S., 181, 187
Satou, Y., 192
Scheid, W., 27
Scheidenberger, C., 19, 199
Schött, H. J., 71
Schuck, P., 164
Schweikhard, L., 19

Segawa, M., 289
Seki, Y., 108
Serata, M., 205
Shaviv, G., 315
Shima, T., 289
Shimada, K., 113
Shimakura, N., 138
Shimoda, T., 138
Shimoura, S., 122, 131, 205
Shindo, M., 199
Shinohara, A., 45
Shinohara, T., 108
Sidorchuk, S. I., 226
Slepnev, R. S., 226
Sonoda, H., 447
Sorlin, O., 155
Sosin, Z., 55
Sparks, W. M., 364
Stanoiu, M., 155
Starrfield, S., 364
Steck, M., 199
Stefan, I., 155
Stepanov, D. N., 226
Stepantsov, S. V., 226
Stodel, C., 55, 60, 155
Stuttge, L., 55
Suda, T., 3, 181, 187, 397, 405
Sueki, K., 3
Sugimoto, T., 113
Sulignano, B., 71
Sumiyoshi, K., 423
Sun, B., 199
Suzuki, S., 108
Suzuki, T., 181, 187, 199
Svirikhin, A., 71

## T

Takabe, T., 45
Takada, E., 108
Takahashi, Y., 108, 138
Takahisa, K., 138
Takase, K., 113
Takechi, M., 181, 187
Takemura, M., 113
Takeshita, E., 122, 192, 205
Takeuchi, S., 122, 192, 205
Takigawa, N., 401, 413, 435
Tamaki, M., 122

Tanaka, K., 181
Tanaka, M., 138, 427
Tashiro, Y., 45
Tatischeff, V., 155, 254
Tatsuda, S., 423
Temma, Y., 289
Ter-Akopian, G. M., 226
Teranishi, T., 192, 205
Theisen, Ch., 55
Thirolf, P. G., 19
Thomas, J. C., 155
Timmes, F. X., 364
Togano, Y., 122, 192
Tohsaki, A., 164
Tokanai, F., 3
Tomyo, A., 289
Tonchev, A. P., 339
Tornow, W., 339
Tourreil, R. de, 55
Toyokawa, H., 431
Toyoshima, A., 45
Tsoneva, N., 339
Tsukada, K., 71
Tsuruta, K., 71

## U

Uchida, M., 113
Ue, K., 205
Ueno, H., 113
Uesaka, T., 131
Utsunomiya, H., 329, 431

## V

Villari, A. C. C., 55, 60
Vinodkumar, A. M., 192
Vorobjev, G., 19

## W

Wada, T., 405, 423, 435, 439, 443
Walker, P. M., 199
Washiyama, K., 413
Watanabe, G., 373
Watanabe, K., 108
Weber, C., 19

Weick, H., 199
Weidenspointner, G., 254
Wieleczko, J.-P., 55
Wieloch, A., 55
Winckler, N., 199
Winkler, M., 60, 199
Wolski, R., 226
Wu, Y. K., 339

## X

Xu, H., 3

## Y

Yamada, K., 122, 205
Yamada, S., 272
Yamada, T., 164
Yamagata, T., 427
Yamaguchi, H., 131
Yamaguchi, T., 3, 181, 187, 199
Yamamoto, K., 405
Yanagisawa, Y., 122, 192, 205
Yasuda, K., 427
Yasui, S., 138
Yeremin, A. V., 71
Yoneda, A., 3, 45
Yoneda, K., 122
Yoshida, A., 3
Yoshida, K., 192
Yoshimi, A., 113
Yoshimura, T., 45
Yosoi, M., 138, 427

## Z

Zhao, Y.-L., 3
Zubov, A. S., 27